The Pheasants of the World

Also by Paul A. Johnsgard

Earth, Water, and Sky: Stories and Sketches by a Naturalist (1999)

Baby Bird Portraits by George Miksch Sutton: Watercolors in the Field Museum (1998)

The Avian Brood Parasites: Deception at the Nest (1997)

The Hummingbirds of North America (2nd ed. 1997; 1st ed. 1983)

Ruddy Ducks and Other Stifftails: Their Behavior and Biology (with M. Carbonell) (1996)

This Fragile Land: A Natural History of the Nebraska Sandhills (1995)

Arena Birds: Sexual Selection and Behavior (1994)

Cormorants, Darters, and Pelicans of the World (1993)

Ducks in the Wild: Conserving Waterfowl and Their Habitats (1992)

Bustards, Hemipodes, and Sandgrouse: Birds of Dry Places (1992)

Crane Music: The North American Cranes (1991)

Hawks, Eagles, and Falcons of North America: Biology and Natural History (1990)

Waterfowl of North America: The Complete Ducks, Geese, and Swans (with Robin Hill and S. D. Ripley) (1989)

North American Owls: Biology and Natural History (1988)

The Quails, Partridges, and Francolins of the World (1988)

Diving Birds of North America (1987)

Birds of the Rocky Mountains (1986)

The Pheasants of the World (1st ed. 1986)

Prairie Children, Mountain Dreams (1985)

The Platte: Channels in Time (1984)

The Cranes of the World (1983)

The Grouse of the World (1983)

Dragons and Unicorns: A Natural History (with Karin Johnsgard) (1982)

Teton Wildlife: Observations by a Naturalist (1982)

Those of the Gray Wind: The Sandhill Cranes (1981)

The Plovers, Sandpipers, and Snipes of the World (1981)

A Guide to North American Waterfowl (1979)

Birds of the Great Plains: Breeding Species and Their Distribution (1979)

Ducks, Geese, and Swans of the World (1978)

The Bird Decoy: An American Art Form (editor) (1976)

Waterfowl of North America (1975)

American Game Birds of Upland and Shoreline (1975)

Song of the North Wind: A Story of the Snow Goose (1974)

Grouse and Quails of North America (1973)

Animal Behavior (2nd ed. 1972; 1st ed. 1967)

Waterfowl: Their Biology and Natural History (1968)

Handbook of Waterfowl Behavior (1965)

THE
PHEASANTS
OF THE WORLD
SECOND EDITION BIOLOGY AND
NATURAL HISTORY

PAUL A. JOHNSGARD

SMITHSONIAN INSTITUTION PRESS • Washington, D.C.

Copy editor and typesetter: Princeton Editorial Associates
Production editor: Deborah L. Sanders
Designer: Janice Wheeler

Unnumbered illustrations appearing in the book are identified as follows: *on page i,* common pheasant, male; *on
page iii,* common pheasant, male; *on page v,* koklass, male;
on page vi, great argus, a male and a female; *on page 1,*
Malayan peacock pheasant, male; *on page 65,* wattled
pheasant, male.

Library of Congress Cataloging-in-Publication Data

Johnsgard, Paul A.
  The pheasants of the world : biology and natural history
/ Paul A. Johnsgard.—2nd ed.
      p.   cm.
  Includes bibliographical references and index.
  ISBN 1-56098-839-8 (cloth : alk. paper)
  1. Pheasants.   I. Title.
QL696.G27J64   1999
598.6′25—dc21   98-53497

British Library Cataloguing-in-Publication Data available

Manufactured in the United States of America

Color plates printed in Hong Kong by the South China
Printing Company, not at government expense

06 05 04 03 02 01 00 99   5 4 3 2 1

IN MEMORY OF JEAN DELACOUR (1890–1985)

Contents

Preface to the Second Edition

It has now been over 15 years since the publication of the first edition of this book. Since that time much has been learned of the biology and status of the pheasants, and unfortunately much has also been learned of their increasingly perilous status. For example, in the first edition of *Endangered Birds of the World* (King, 1981), 6 species of pheasants were listed as endangered and 8 more were considered vulnerable, which represented 28 percent of the entire pheasant group. More recently, McGowan and Garson (1995) listed 12 species as endangered or critically endangered (plus 3 more that are considered as subspecies in this volume) and 16 as vulnerable (plus 1 more taxon here regarded as a subspecies). These total 57 percent of the entire pheasant assemblage at the species level or twice the number listed in the earlier summary by King. McGowan and Garson also identified 2 additional species as lacking adequate information relative to their current status and 10 species as having one or more threatened subspecies within otherwise apparently secure species. Similarly, Collar et al. (1994) listed 20 species of pheasants as vulnerable, 9 as near-threatened, 2 as endangered, and 2 as critically endangered, excluding 3 forms considered as subspecies in this volume. Thus, 67 percent of the entire pheasant assemblage at the species level were considered by Collar et al. to be in some danger.

At the time I decided to write the first edition of this book, a major stimulus was the opportunity to use some of the wonderful and previously unpublished watercolors done by Major Henry Jones at the beginning of this century, which were housed in the Zoological Society of London. Although superb in their day, these paintings were of necessity done from museum skins and sometimes were not wholly accurate in terms of posture or soft-part coloration. As a result, I decided that any second edition would contain photographs of living birds rather than paintings of museum specimens. I have thus drawn on my own collection of color transparencies of pheasants and have been assisted when necessary by Kenneth

Fink, who surely has the finest selection of pheasant photographs in the world. Additionally, photographs and videotapes of tragopan display were sent to me for review and illustrative use by Kamal Islam. David Rimlinger likewise provided several videos and slides and was of tremendous assistance in providing translations and other obscure literature. Han Assink, Phan Viet Lam, and Dan Gia Tung of the Hanoi Zoological Gardens offered me photographs of the Vietnamese pheasant and crested argus, and Dr. Viet Lam provided avicultural information on this species. Jean Howman, Christopher Savage, and others also offered me photographs of rare pheasants.

I also wished to provide some paintings to supplement the color photographs and decided to use some that were done by Joseph Wolf for D. J. Elliot's great 1872 monograph on the pheasants of the world. These plates were originally published as hand-colored stone lithographs in elephant-folio format. Although they are often considered to be the finest pheasant illustrations ever made, Wolf's spectacular plates have been reprinted only as scattered illustrations or as fine-art reproductions for use by interior decorators. Those that I have selected are ones I consider to be among the best of the series. Many of them were photographed from an original set in the Field Museum of Natural History, Chicago, Illinois.

The first edition of this book was published in England, and so a minor part of the revisionary process involved converting most English spellings to their American counterparts. However, in bird names I have retained the use of "grey" rather than "gray," mostly to conform with usage by Sibley and Monroe's (1990) world checklist of birds. Also in common with that reference, I have retained traditional spellings of Chinese states (alternative spellings of these political entities are provided in figure 8), but have updated a few familiar locality spellings, such as Beijing for Peking. Likewise, I have retained a few other names reflecting re-

cent changes in long-standing and readily recognized geopolitical entities (such as Ceylon, Tibet, Cambodia, and Burma). English vernacular names of pheasants that were used in the earlier edition have been retained; these mostly, but not entirely, conform with those of Sibley and Monroe (1990). Further, several forms considered allospecies by these authors were retained as subspecies by me. I have also tentatively treated a recently described but rare and still inadequately studied taxon (*hatinhensis*) of *Lophura* from Vietnam as a subspecies of *edwardsi*, although Sibley and Monroe accepted it as a new species ("Vietnamese fireback"). Most of my revising efforts have been centered on bringing the status and distribution of the rarer species up to date (and correspondingly modifying nearly all of the range maps), adding to the species accounts, and attempting to correct any weaknesses or errors that were apparent in the earlier edition. I have added nearly 200 new references, which raises the total to more than 500. A substantial number of these new references reflect the very welcome recent surge in field research being done in China, which is becoming increasingly available in English-language reports or as translations.

Mainly because I wrote a monograph on the world's quails, partridges, and francolins (Johnsgard, 1988) subsequent to the first edition of this book and because several world checklists are now available, I decided that the inclusion of an annotated checklist of these additional members of the subfamily Phasianinae was no longer needed; I have thus eliminated it from this edition. After considerable reflection I also have eliminated the plumage descriptions for all species. They occupied much space in the first edition, and few readers have ever admitted to me that they used these descriptions, especially when diagnostic keys are provided. Interested persons can refer to descriptions in the first edition or in other monographs such as Delacour (1977). However, I have incorporated into this edition over 100 new drawings in 23 figures, modified several of the original figures, and added some new decorative drawings. As with my earlier books, all unsigned illustrations are my own.

Many valuable publications dealing with pheasant biology and distribution have appeared since the mid-1980s, among the most important of which is a summary of China's galliforms by Li (1996). My revision was also greatly facilitated by several other recent comprehensive publications, especially the world Galliformes survey by del Hoyo et al. (1994), a valuable status analysis of the rarer pheasant taxa by McGowan and Garson (1995), and the numerous publications of the World Pheasant Association (WPA). Among these are WPA annual reports, newsletters, and the proceedings or abstracts of several "Pheasants in Asia" symposia (Savage, 1981; Ridley, 1986; Savage and Ridley, 1987; Hill et al., 1990; Jenkins, 1993; Anonymous, 1997). Keith Howman (1993) has produced a very useful summary of pheasant aviculture and management in captivity that contains many splendid color photographs by Kenneth Fink and others. Finally, the classic, but rare, pheasant monograph by Beebe (1918–1922) has been recently (1990) reprinted in a slightly condensed form, and thus has become much more widely available.

As with the first edition, I have received the very valuable help and advice of many people, especially including my above mentioned friends Kenneth Fink, David Rimlinger, and Kamal Islam. Keith and Jean Howman were also a major source of advice and assistance in early planning. Christopher Marler provided me with a number of WPA publications, and Kathryn Gabig retyped several parts of the original text. Unpublished information and a master's thesis on the mikado pheasant were provided by Cara Lin Bridgman, some Chinese translations were done by Na Xu, and Christopher Eames sent me several very helpful reprints on Vietnamese birds. Portions or all of the revised manuscript were read by Peter Garson, David Rimlinger, Donald Bruning, and Christine Sheppard. Peter Garson kindly provided me with recent issues of the WPA/Species Survival Commission/BirdLife Pheasant Specialist Group's newsletter *Tragopan* (through no. 8), and the staff of the Wilson Ornithological Society's library at the University of Michigan loaned me several useful references.

From the first time I mentioned the possibility of revising and republishing this book, Peter Cannell of the Smithsonian Institution Press was an enthusiastic supporter. Indeed, David Hancock had suggested that I do a revision some years ago and had offered to publish it. At the time I was engaged in writing other books, but my delay in taking up the project has allowed me to take advantage of some of the most recent biological publications, such as those mentioned above. To all these people and organizations I offer my sincere thanks.

Preface to the First Edition

Because of their beauty, economic importance, and value as sporting birds, the pheasants have received more than their share of attention from writers. Their first major monographer, D. G. Elliot, authored what has often been described as the most beautiful monograph on birds ever produced when he published his *Monograph of the Phasianidae, or Family of the Pheasants,* between 1870 and 1872, with its 82 superb hand-colored lithographic plates by Joseph Wolf. Only four decades later (1918–1922) C. W. Beebe similarly produced a four-volume work titled *A Monograph of the Pheasants,* which was among the last of the great bird monographs of that era, and which employed the artistic talents of six of the finest bird artists of the period. Both of these monographs were produced in highly limited numbers, and are essentially unobtainable today.

However, Jean Delacour remedied this situation with a modern treatment of the pheasants in his book *The Pheasants of the World,* which was published initially in 1951, and republished in a slightly revised version in 1977. This volume had the stamp of Delacour's taxonomic authority as well as reflecting his vast background in both avicultural techniques and field experience with pheasants in Southeast Asia. Delacour's book is especially heavily oriented toward avicultural information rather than data on naturalistic ecology and breeding biology, although the limited amounts of information available even today in these areas makes such an orientation understandable. Nonetheless, it has appeared to me for some years that an alternative treatment, which emphasizes these aspects of pheasant biology, and devotes the least possible attention to descriptions of subspecies and avicultural histories, might be warranted. This idea was reinforced when I chanced on a very large series of unpublished pheasant watercolors made by Major Henry Jones at the beginning of the 20th century, while doing research in the library of the Zoological Society of London. My immediate hope was that a book might be built around these paintings, which are certainly the equal of many of the most famous earlier pheasant plates, such as those done by Joseph Wolf for Elliot's famous monograph. After extended discussions with the Zoological Society of London and Oxford University Press this hope finally materialized, and I was able to make a selection of more than 50 of the plates for illustrating my book. These paintings include 47 of the 49 forms of pheasants that I have regarded as full species. Two species that were not illustrated by Jones (because they remained undiscovered until a few decades later) have been commissioned by me, and I have been very fortunate in being able to obtain the help of a fine English artist, Timothy J. Greenwood, in achieving my goal of illustrating all the known species of pheasants with paintings. Not since Elliot's monograph have all the pheasant species been illustrated with individual color plates that show both scxcs in nearly all cases. Further, Delacour's (1977) monograph has only a few literature citations more recent than 1950, and none more recent than 1976. I have thus concentrated on summarizing the most recent available literature, especially of citations not to be found in Delacour, and on trying to summarize the current conservation and distributional situation as well as available information allows for each species.

One cannot produce a book unaided, and in addition to the artistic help of Timothy J. Greenwood I was particularly aided by various members of the World Pheasant Association. Dr. Timothy Lovel and Mr. and Mrs. Keith Howman were particularly influential in many ways, and the Howmans were unflaggingly helpful in providing advice and allowing me to photograph their marvelous pheasant collection. Similarly, Mr. John Bayliss loaned me a number of very useful photographs, Major Iain and Didy Grahame provided me with useful advice, Raymond Sawyer and Charles Sivelle provided me with access to their collections, and other WPA members helped in other ways. Additionally, Mr. Vern Denton let me observe

his collection of extremely rare pheasants, and David Rimlinger of the San Diego Zoo helped me in innumerable ways, by loaning photographs, providing otherwise unavailable data on pheasants, and reading an early draft of the manuscript. Mr. Kenneth Fink was always most generous in loaning photographs, providing advice, and in hosting me during a stay in San Diego. Mr. George Allen Jr. and Lincoln Allen also provided me with several useful photographs.

The use of several excellent libraries was invaluable to me during the preparation of this book, and I must in particular mention the assistance I received from the Edward Grey Institute of Oxford University, and the libraries of the Peabody Museum of Natural History, Yale University, the Zoological Society of London, the American Museum of Natural History, and the Van Tyne Memorial Library of the University of Michigan. I also acknowledge with gratitude the use of specimens in the collections of the British Museum (Natural History), the American Museum of Natural History, and the Peabody Museum of Natural History.

Work on this book was done between 1982 and 1984 at the University of Nebraska–Lincoln, and was greatly facilitated by a Faculty Development Leave provided me during the spring semester of 1983. I also appreciate the typing assistance given me by the secretaries of the School of Life Sciences, and by Janet Kumke.

Several persons at the Zoological Society of London were important in my work, particularly Mr. Reginald Fish, Librarian, who cheerfully complied with my every request for help or information. Dr. Marcia Edwards assisted me in the earlier stage of finding a publisher acceptable to all parties, and Dr. Peter Olney also provided me with various sorts of assistance. Finally, the foreword was very kindly written by Lord Zuckerman.

Introduction

The pheasants (tribe Phasianini as used here) comprise a group of 49 species of generally nonmigratory and terrestrially adapted birds of moderately large size. A hind toe is always present, and all the toes have short, fairly blunt claws that are suitable for scratching. The wings are relatively short and rounded, and in general extended flight is unusual or impossible. Thus, most species are relatively sedentary, although limited migrations do occur. Pheasants primarily consume vegetable foods plus variable amounts of arthropods. All species have well-developed crops for temporary storage of food, as well as muscular gizzards associated with grinding of hard food materials with the aid of grit. The feathers usually have well-developed aftershafts, but down is scanty or lacking. Their nests are typically unlined shallow scrapes on the ground (sometimes in low trees or bushes). The eggs are usually fairly numerous (rarely up to ten or more) and either unspotted or only slightly spotted. The young are precocial and often are able to fly short distances within a week or so after hatching. The species are variably monogamous or nonmonogamous, with the former condition much rarer and seemingly the more generalized form; polygyny (harem-formation) and promiscuity have evolved under appropriate ecological conditions.

Pheasants differ from other major subfamilial groups of the Phasianidae (the grouse and New World quails) in having tarsi and nostrils that are unfeathered, toes that are neither feathered nor have lateral pectinations, a lower mandible that is always smooth and unserrated, and in lacking the inflatable esophageal "air sacs" typical of male grouse. Complete distinction from the very closely related Old World partridges (tribe Perdicini) is difficult. However, in general, pheasants tend to be larger and exhibit greater sexual dimorphism in both size and plumage. Highly iridescent plumage coloration and sharply spurred tarsi are common in male pheasants, and males often have tails that are highly graduated, variably vaulted, or both.

All pheasant species are native to the Old World. The majority are found between the equator and the tropic of Cancer, in tropical to temperate climatic regions, and in variably forested habitats between sea level and 2,745 m (9,000 ft) elevation.

Although the pheasants thus occur in what are still some of the least-studied terrestrial environments in the world, they have by no means escaped the effects of man. Indeed, of all major bird groups pheasants are among the most seriously impacted by human exploitation and habitat destruction. As noted earlier, nearly two-thirds of the entire group of pheasant species were listed as critically endangered, endangered, vulnerable, or near-threatened in the most recent edition of *Birds to Watch* (Collar et al., 1994). Yet the pheasants have been of enormous benefit to humans and have provided the most numerous and widely domesticated species of bird, the domestic fowl. Similarly, the common pheasant is the single most widely distributed and abundant species of upland gamebird; estimated annual harvests in North America alone have at times approached 20 million birds. Several other species have been introduced in North America and elsewhere for sporting purposes (Long, 1981). Additionally, pheasants are among the most popular of all bird groups for aviculturists and exhibition in zoos. Based on an international survey by the WPA in 1979, no less than 46 of the 49 species of pheasants recognized here were known to be represented in captivity, and the total captive population was probably then in excess of 25,000 individuals. Since then several additional species or races have been brought into captivity, and as of a 1991 WPA census these included over 33,000 individuals. However, this captive population is small by comparison with the number of domestic fowl present in the world today, which certainly number several billion. Even the ring-necked pheasant has a captive population numbering the tens of millions; in the United Kingdom alone it is estimated that about 7 million of these birds are released every year (Savage, 1981).

PART ONE

COMPARATIVE BIOLOGY

CHAPTER 1

Relationships and Classification

Most current taxonomic classifications of the pheasantlike birds are at least partly dependent upon early studies by Beebe (1914a), who spent several months studying specimens in all the world's major museums in preparation for writing his large monograph on the group. After considering "several scores" of anatomical characters of possible taxonomic significance, Beebe settled on using the sequence of molting in the tail feathers (rectrices) as his primary criterion of generic groupings. He discovered that, at least in all the forms that he studied, molt in the Old World partridgelike species invariably begins with the central rectrices and proceeds regularly outward. With the exception of the genera *Ithaginis* and *Tragopan*, none of the pheasants that Beebe studied appeared to have such a molting pattern. Thus, he distinguished the Old World subfamily Perdicinae (but also including the two pheasantlike genera just mentioned) on the basis of their centrifugal (from the middle) pattern of rectrix molting. Furthermore, Beebe discovered that in most of the typical pheasants the molting pattern of the rectrices was exactly the reverse, from the outermost rectrices inward (centripetally). Beebe considered all the genera of this group to comprise the Phasianinae. A major exception that he found involved the typical peafowl (*Afropavo* was still undiscovered at the time), which he observed to molt from a locus beginning with the second rectrices from the outermost, with the outermost pair molting just prior to the inner ones. Beebe called this group the Pavoninae. Lastly, he observed that in the peacock pheasants and argus pheasants (*Polyplectron*, "*Chalcurus*," *Argusianus*, and *Rheinartia*) the tail molt begins with the third from central pair, and proceeds both outwardly and inwardly simultaneously, with the central rectrices being replaced just prior to the outermost pair. This group of genera was called the Argusianinae. Stresemann (1965) later supported Beebe's observations, although studies of the Reeves' pheasant indicate that juveniles have a centrifugal tail molt whereas adults molt their tails in a centripetal pattern (Mueller and Seibert, 1966).

Peters (1934) did not follow this taxonomic convention, but simply included all of the Old World partridges and their relatives as well as the typical pheasants in the subfamily Phasianinae, as have several more recent classifications. Delacour (1977) went even further in "lumping" groups. He included the New World quails in the subfamily Phasianinae as well, and did not directly address the question of possible criteria for distinguishing pheasants from these other quail-like or partridgelike groups. Delacour noted only that *Ithaginis* and *Tragopan* are "slightly related" to the partridgelike forms. He nevertheless considered them sufficiently pheasantlike to be included in his monograph, but excluded other similarly transitional forms such as *Galloperdix*.

The early observations of Beebe on pheasant molting patterns have largely gone unchallenged, although Marien (1951) has made some comments about their taxonomic utility. He observed that in at least two genera of partridges (*Perdix perdix* and *Ammoperdix griseogularis*) the postjuvenile tail molt begins with the third or third and fourth pair of rectrices and proceeds both laterally and medially—the pattern that Beebe described as typical of the subfamily Argusianinae. Marien also remarked that in one specimen of snow partridge (*Lerwa lerwa*) the molt pattern was imperfectly centripetal, casting further doubt of the universality of a centrifugal molting pattern in the Perdicinae. Marien also noted that the wing molt might offer only limited value in separating Phasianinae groups. For example, it has been suggested that the Old World partridges and quails and the New World quails agree in retaining the one or two (rarely three) outermost juvenile primaries during their postjuvenile molt, but that the true pheasants replace all of their juvenile primaries at this time. This situation does seem to apply to all of the New World quails so far studied (Johnsgard, 1973) and has been observed in several of the Old World partridge genera (*Perdix*, *Alectoris*, *Tetraogallus*, *Tetraophasis*, *Ammoperdix*). Yet

this molt pattern evidently is not invariable because *Francolinus* is alleged to have a complete postjuvenile wing molt (Marien, 1951). Although the Phasianini are supposed to have complete wing molts, Mueller and Seibert (1966) reported that the Reeves' pheasant retains its tenth juvenile primary through the first year, as do eared pheasants (Felix, 1964).

Verheyen (1956) reviewed these and various structural traits that have been suggested as having taxonomic importance in the Galliformes and proposed a new classification based largely on his proportional measurements of the skeleton. Within his family Phasianidae, Verheyen recognized the following subdivisions:

1. Numidinae (five genera)
2. Afropavoninae (monotypic, *Afropavo congensis*)
3. Meleagrininae (two genera)
4. Tetraoninae
5. Perdicinae
   Tribe Corturnicini (including *Coturnix*, *Synoicus*, *Excalfactoria*; probably also *Perdicula*, *Cryptoplectron*, and *Ophrysia*)
   Tribe Perdicini (incompletely studied, but including the following species groups)
   Tetraogallus–Tetraophasis–Lerwa
   Alectoris
   Perdix–Arborophila–Tropicoperdix–Caloperdix
   Ptilopachus
   Acentrortyx–Margaroperdix–Ammoperdix–Melanoperdix
   Rollulus
   Francolinus–Pternistis
   Bambusicola
   Anurophasis–Rhizothera
   Haematortyx
6. Phasianinae (including the following species groups)
   Galloperdix–Ithaginis
   Tragopan

Pucrasia–Houppifer (=*Lophura ery-*
thropthalma and *L. inornatus*)
Lophophorus
Lobiophasis (=*Lophura bulweri*)–
Lophura–Hierophasis
Lophura (including "*Lobiophasis*,"
"*Hierophasis*," "*Diardigallus*," and
"*Gennaeus*")–*Syrmaticus–Phasianus–*
Chrysolophus
Gallus
Crossoptilon–Catreus–Polyplectron
(including "*Chalcurus*")
Rheinartia–Argusianus
7. Pavoninae (including *Pavo* only)

Verheyen's strong separation of *Afropavo* from
the typical pheasants seems questionable, particularly because some detailed osteological
studies by Lowe (1939) suggest strongly that it
is simply a primitive or unspecialized type of
peacock that is fairly closely related to both
the Pavoninae and the Argusianinae. His inclusion of *Galloperdix* and *Ithaginis* in the
Phasianinae is also noteworthy. In an earlier
morphologic analysis of the Galliformes, Boetticher (1939) had divided the Phasianidae into
nine subfamilies: Odontophorinae; Tetraoninae; Lerwinae; Perdicinae (including *Ithaginis*);
Tragopaninae (*Tragopan* only); Phasianinae
(genera other than those listed separately);
Pavoninae (including *Afropavo, Chalcurus,*
Polyplectron, Rheinartia, Argusianus, and
Pavo); Numidinae; and Meleagrinae.

On the basis of immunological evidence,
Mainardi (1963) suggested that among the
pheasants *Gallus* has probably evolved from
early *Phasianus*-like stock and that both *Pavo*
and *Afropavo* are closely related to primitive
Phasianidae stock not very different from
present-day guineafowl. He also suggested that
the phasianids, cracids, and megapodes are all
fairly closely related and were derived from
this common ancestral stock.

In a major view of avian taxonomy, Sibley
and Ahlquist (1972) provided egg-white evi-

dence derived from electrophoretic studies and
concluded that variations in protein mobility
made taxonomic interpretation difficult within
the family Phasianidae. However, they noted
that a very similar electrophoretic pattern exists between *Coturnix* and *Phasianus* as well as
other pheasant genera, thus conflicting with
earlier biochemical and anatomical evidence
suggesting rather distant relationships between
them. However, *Perdix* exhibited a noticeably
compressed electrophoretic pattern. Sibley and
Ahlquist classified the pheasant group by distinguishing the typical pheasants (11 genera
and 20 species were studied) from the Old
World partridges as separate subfamilies
(Phasianinae and Perdicinae). However, they
suggested that additional studies would be
needed to establish relationships among the
genera of these groups.

Beyond the still-unresolved question of how
best taxonomically to recognize the pheasants
as distinct from the Old World partridges and
their relatives, there are a number of unsolved
problems concerning larger taxonomic groupings within the Galliformes. I reviewed this
general question earlier (Johnsgard, 1973,
1983*b*) and it has been very thoroughly discussed by Sibley and Ahlquist (1972), thus
there is little purpose served in repeating these
arguments. However, a comparison of some
representative classifications of the Galliformes
are presented in table 1 to indicate the kinds of
variations in proposals made within the past
half-century. As may be seen, there is general
agreement that the megapodes (Megapodidae)
and cracids (Cracidae) each deserve familial
recognition. Both are seemingly derived from
early, generalized galliform stock, although
Clark (1964) believes that some of the megapodes' "primitive" features are secondarily
evolved. The grouse, although often given familial status in earlier classifications, are now
generally thought to deserve no more than subfamilial separation from the pheasants (Johnsgard, 1983*b*). The turkeys (Meleagrididae of

Table 1

Some suggested classifications of Recent Galliformes (exclusive of *Opisthocomus*)

Peters (1934)	Mayr and Amadon (1951)	Wetmore (1960)	Johnsgard (1973)	Sibley and Monroe (1990)
Megapodiidae	Megapodiidae	Superfamily Cracoidea	Superfamily Cracoidea	Order Craciformes
Cracidae	Cracidae	Megapodiidae	Megapodiidae (10 spp.)	Cracidae (50 spp.)
Tetraonidae	Phasianidae	Cracidae	Cracidae (38 spp.)	Megapodidae
Phasianidae	Phasianinae	Superfamily Phasi-	Superfamily Phasi-	(19 spp.)
Odontophorinae	Numidinae	anoidea	anoidea	Order Galliformes
Phasianinae	Tetraoninae	Tetraonidae	Phasianidae	Parvorder Phasianida
Numididae	Meleagrididae	Phasianidae	Meleagrininae (2 spp.)	Phasianidae (177 spp.)
Meleagrididae		Numididae	Tetraoninae (16 spp.)	Numididae (6 spp.)
		Meleagrididae	Odontophorinae	Parvorder Odonto-
			(30 spp.)	phorida
			Phasianinae	Odontophoridae
			Perdicini (103 spp.)	(31 spp.)
			Phasianini (49 spp.)	
			Numididae (7 spp.)	
6 families	5 families	6 families	3 families	5 families
93 genera	—	—	—	73 genera
267 spp.	240 spp.	—	245 spp.	284 spp.

Peters [1934]) and New World quails (Odon-tophorinae of Peters [1934]) have been treated variably, but they too are now frequently recognized only as distinct subfamilies or, in the case of the New World quails, sometimes given no more than tribal distinction. Similarly, the guineafowl (Numididae of Peters [1934]) have usually been considered a full family or at least a distinct subfamily in most recent classifications. Their relationships to the more typical pheasants seem to be uncertain at present, the group perhaps having evolved from an early francolin-like precursor (Crowe, 1978).

Cracraft (1981) suggested that the guineafowl are the sister-group of the pheasants, grouse, turkeys, and New World quails, and considered them a separate family ("subfamily" in Cracraft) to the Phasianidae, which included all these other groups. However, Wolters (1975–1982) considered the guineafowl to be

but one of 15 subfamilies of the Phasianidae, and the typical pheasants were divided into eight separate subfamilies.

The most recent and significant change in higher-level galliform classification is that of Sibley and Monroe (1990), based in large part on DNA-DNA hybridization data and associated cladistic analyses. These authors grouped the pheasants, grouse, turkeys, Old World quails, partridges, and francolins within a single family (Phasianidae), without tribal or subfamilial subdivisions. They recognized the guineafowl group as constituting a separate family within this same parvorder, but the New World (odontophorine) quails were further distinguished as comprising a separate parvorder (see table 1). Sibley and Monroe also elevated four allopatric pheasant populations to the rank of allospecies that here are considered races (*Lophura inornata hoogerwerfi, L. edwardsi hatinensis,*

Crossoptilon c. harmani, and *Argusianus argus bipunctatus),* but conversely regarded the green pheasant as a subspecies of the common pheasant. Therefore, a total of 52 species of pheasants were accepted by Sibley and Monroe, rather than the 49 species-level pheasant taxa recognized in this book.

The New World or odontophorine quails were considered by Sibley and Alquist (1990) to be only very distantly related to (and thus warranting parvorder distinction from) the collective phasianid assemblage of Old World quails, partridges, francolins, pheasants, grouse, and turkeys. By contrast, using data from allozyme electrophoresis, Randi et al. (1991) judged that the Phasianidae consists of two phyletic lineages, with one containing the turkeys, grouse, pheasants, and the Old World partridge genus *Perdix,* and the other associating certain Old World partridges (at least *Alectoris*) with the Old World and New World quails. Recently Crowe (1996) (*World Pheasant Association News* 46:43, 1996) has surprisingly suggested on the basis of DNA-sequence data that the turkeys, peafowl, pheasants, and grouse may be more closely related to one another than they collectively are to junglefowl (*Gallus*) and that the nearest living relatives of the New World quails may be the guineafowl. More recently, Crowe and Bloomer (1997) have offered additional conclusions, including the association of the New World quails with some Old World partridges (*Rollulus* and *Arborophila*) and the typical francolins (*Francolinus*), but the African francolin genus *Pternistes* with the Eurasian *Alectoris* partridges. Clearly some of these conclusions based on biochemical data are seemingly incompatible with one another and with other evidence and much more comparative work needs to be done.

Generic and Species Limits

Like most groups of birds, the classification of the pheasants has tended to become modified in recent decades as ideas on the conceptual basis of the generic and species categories have undergone change and as new information on range limits and geographic variation has become available. Using Elliot's (1870–1872) monograph as a starting point, 17 genera and 77 species were recognized. Similarly, Ogilvie-Grant's (1893) catalogue of the then-known pheasant species included 17 genera and 77 species. Beebe (1918–1922) recognized a total of 19 genera (excluding the then-undiscovered *Afropavo*) and 61 species. Similarly, Peters (1934) recognized 21 pheasant genera (still exclusive of *Afropavo*), but only 49 species. Delacour (1977) likewise recognized 49 species (although not exactly the same as those of Peters), but accepted only 16 genera. Thus, over the past 125 years, the number of recognized pheasant genera has ranged from 16 to 21 and the species from 49 to 77. However, as the number of recognized species has tended to decline with time, the number of described subspecies has increased, reaching a maximum of 124 in Delacour's most recent (1977) summary. Like Delacour, I have accepted 49 species. However, he regards the perplexing and fragmental specimen (a single feather) of *Argusianus bipunctatus* as a full species, but I believe that, given the available information, it should be considered no more than a subspecies. On the other hand, in the first edition of this book I regarded the distinctive Bornean peacock pheasant (*schleiermacheri*) as a full species, as did Beebe, whereas Delacour considered it only a subspecies of *malacense*. I have generally continued to follow Delacour's sequence of genera, except for his placement of the genus *Crossoptilon,* and my sequence of species within genera also varies little from that used by Delacour. An evolutionary dendrogram (figure 1) is provided that indicates my interpretation of phyletic relationships of genera and species within the pheasant group, whereas figure 2 is intended to show more general relationships of the pheasants within the

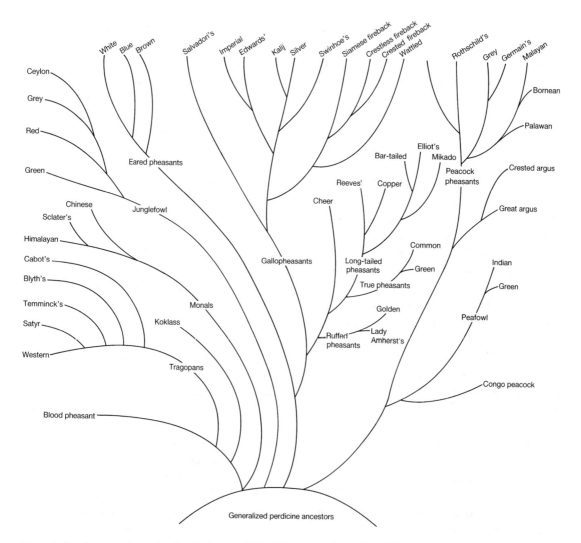

Figure 1. Dendrogram of postulated evolutionary relationships among the species of pheasants.

Galliformes, based on my own interpretation of the available information.

Table 2 provides a comparative summary of major anatomical and physiological or behavioral traits that seem to provide useful criteria for separating the pheasants from their seemingly nearest relatives. Most of these criteria are self-explanatory, but details of the behavioral traits are discussed in chapter 3. It seems clear that there are so many overlapping character traits between the pheasants and the Old World partridges that they should be regarded as no more than tribally separated. However, as will be made clear in the hybridization section (chapter 2), surprisingly few incidences of intergroup hybridization have been documented between the pheasants and Old World partridges.

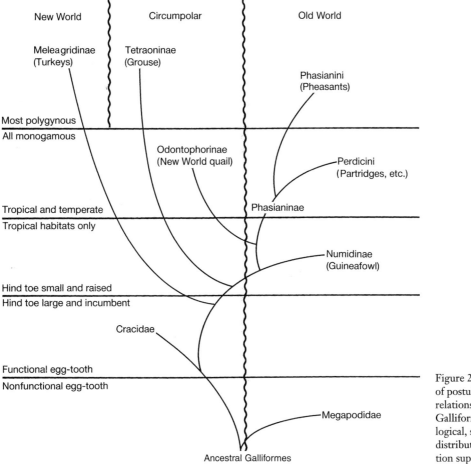

New World   Circumpolar   Old World

Meleagridinae
(Turkeys)

Tetraoninae
(Grouse)

Phasianini
(Pheasants)

Most polygynous

All monogamous

Odontophorinae
(New World quail)

Perdicini
(Partridges, etc.)

Phasianinae

Tropical and temperate

Tropical habitats only

Numidinae
(Guineafowl)

Hind toe small and raised

Hind toe large and incumbent

Cracidae

Functional egg-tooth

Nonfunctional egg-tooth

Megapodidae

Ancestral Galliformes

Figure 2. Dendrogram of postulated general relationships among the Galliformes, with ecological, structural, and distributional information superimposed.

Table 2
Comparison of major subgroups of the Phasianidae

| | Tetraoninae | Odontophorinae | Phasianinae | |
			Perdicini	Phasianini
Distribution	Holarctic	New World	Old World	Old World
Rectrices	16–22	10–14	8–22	14–32
Usual tail molt	Centripetal	Centrifugal	Centrifugal	Centripetal
Lower mandible	Smooth	Toothed	Smooth	Smooth
Nostrils	Feathered	Bare	Bare	Bare
Feathered tarsus	Yes	None	Rare (1 sp.)	None
Pectinated toes	Yes	None	None	None
Tarsal spurs	None	None	Few	Most
Usual pair bond	Polygynous	Monogamous	Monogamous	Polygynous
Sexual dimorphism	Variable	Slight	Slight	Variable
Postjuvenile primary molt	Incomplete	Incomplete	Incomplete	Variable
Sexual maturity	1 year	1 year	1 year	1–3 years
Average clutch size	5–12 eggs	10–15 eggs	4–16 eggs	1–12 eggs
Incubation period	21–27 days	22–30 days	16–25 days	18–29 days
Tidbitting display	No	Yes	Yes	Yes
Waltzing display	No	No	Yes	Yes

CHAPTER 2

Hybridization and Zoogeographic Patterns

Hybridization among the pheasants of the world has occurred under a variety of circumstances. In some instances it has resulted under natural conditions, for example, as between such locally sympatric species pairs as the kalij (*Lophura leucomelana*) and the silver pheasant (*L. nycthemera*). In most cases it has occurred "accidentally" among captive birds, especially when conspecific mates have been unavailable. Only in rare instances has it been specifically planned as part of an experimental program of hybridization for scientific purposes, such as for obtaining genetic, morphological, or biochemical data associated with hybridization. Some of these latter studies have not provided information of general interest to most ornithologists, but instead have involved such biochemical information as hybrid hemoglobins (Brush, 1967), transferrins (Crozier, 1967), or immunoelectrophoretic comparisons of blood sera (Sato et al., 1967).

The total literature concerning pheasant hybridization is thus very great, and has been admirably summarized by Gray (1958). Gray's review has provided the primary foundation for much of my own following summary, but has been supplemented by more recent information such as that provided by Delacour (1977) and Rutgers and Norris (1970). Some of the alleged hybrids mentioned by these authors have never been fully documented, and in a few cases such as a supposed hybrid between domestic fowl (*Gallus gallus*) and the lyrebird (*Menura novaehollandiae*) are sufficiently unlikely as to warrant discounting. Even by limiting the reported hybrids to those involving only members of the Galliformes, there are some cases that strain credulity. Thus, the primary emphasis in this review will be on intratribal hybrids. However, a brief survey of reputed examples of hybridization between pheasants and species representing other tribes and subfamilies of the Galliformes seems warranted.

Extratribal Hybridization

Included in this category are all hybrids between pheasants and other major taxonomic groups of the Galliformes. In all cases except those specifically noted, citations for their occurrence can be located in the reference by Gray (1958).

Phasianini × Perdicini

The pheasants and Old World partridges and their relatives are generally considered to be fairly closely related and in most classifications are included as members of the same subfamily (Phasianinae). Thus, one would expect a substantial number of hybrid records to have accrued between these two groups. However, such is not the case, and Gray lists only three such combinations. These include crosses of *Gallus gallus* (*G. "domesticus"* according to Gray [1958]) with *Alectoris graeca* and *Perdix perdix*, and one between *Phasianus colchicus* and *Perdix perdix*. These were all presumed hybrids; none was produced under controlled conditions. Likewise, none was proven to be fertile, although one of the presumed *Gallus × Perdix* hybrid males exhibited the sexual behavior of a "normal" domestic fowl.

Phasianini × Tetraoninae

By far the largest number of extratribal hybrid records involving pheasants have implicated various species of grouse. Except for an alleged, but seemingly unlikely, hybrid reported between a black grouse (*Tetrao tetrix*) and a silver pheasant, all have involved the domestic fowl or the common pheasant. Thus, domestic fowl have reportedly been hybridized with the ruffed grouse (*Bonasa umbellus*), hazel grouse (*Bonasa bonasia*), and willow ptarmigan (*Lagopus lagopus*, including *L. l. scoticus*), whereas pheasants have allegedly hybridized with ruffed grouse, pinnated grouse (*Tymphanuchus cupido*),

capercaillie (*Tetrao urogallus*), black grouse, rock ptarmigan (*Lagopus mutus*), red grouse (*Lagopus l. scoticus*), and blue grouse (*Dendragapus obscurus*). Unlikely as some of these combinations might seem, there is little doubt that at least some of them have occurred repeatedly. Thus, Boback and Müller-Schwarze (1968) provided a photograph of a hybrid pheasant × black grouse, and stated that at least 15 such specimens were reported between 1833 and 1854. Likewise, Jewett (1932) and Hudson (1955) described five apparently natural hybrids between pheasants and blue grouse, and this combination was originally described late in the 19th century (Anthony, 1899). There is no indication that any grouse × pheasant hybrids have proven fertile, nor have any even shown any signs of sexual activity. Probably the relatively promiscuous mating systems of most grouse as well as of pheasants and domestic fowl have facilitated this high incidence of intertribal hybridization. This promiscuous tendency might thus facilitate the production of such seemingly unlikely hybrids as occurring between the domestic fowl and the capercaillie (Skervold and Mjelstad, 1992).

Phasianini × Numidinae

Crosses between pheasants and guineafowl, although unlikely, have been unquestionably obtained. According to Gray (1958), domestic fowl have reputedly been hybridized with both the vulturine guineafowl (*Acryllium vulturinum*) and the domestic guineafowl (*Numida meleagris*). The latter cross has also been studied biochemically by Crozier (1967), as well as by Sato et al. (1967). Similarly, presumed hybrids between common pheasants and domestic guineafowl have been reported, and hybrids have been produced repeatedly between the Indian peafowl and domestic guineafowl. Hanebrinck (1973) has recently described the morphology and behavior of this combination. A fifth hybrid combination between pheasants and guineafowl is a reported cross between the Cabot's tragopan (*Tragopan caboti*) and the mitred guineafowl (*Numida mitrata*), which, like the other pheasant × guineafowl hybrids, appears to have been completely sterile. A "natural" hybrid between domestic fowl and a wild helmeted guineafowl (*N. meleagris*) has also been reported from Africa (Poda, 1985).

Phasianini × *Meleagridinae*

Pheasant × turkey hybrids have also undoubtedly been produced on various occasions in captivity. There are several apparent hybrids known involving domestic fowl and domestic turkey (*Meleagris gallopavo*). Four hybrids were reportedly reared (of a hatch of five) involving a domestic turkey and a peahen. Reciprocal crosses have also been obtained by artificial insemination between common pheasants and domestic turkeys (Asmundson and Lorenz, 1975). Birds obtained by this method have been found to be completely sterile. Presumed "natural" hybrids of this combination have also been reported occasionally.

Phasianini × *Cracidae*

Some rather dubious crosses between domestic fowl and various cracids have also been reported (Gray, 1958). There is an alleged early case of apparent hybridization between a male curassow (*Crax* sp.) and a female domestic fowl, another similar case of a male *Crax alberti* hybridizing with a female domestic fowl, and a third presumed case of hybridization between the domestic fowl and a guan (*Penelope* sp.). None of these cases was sufficiently well documented as to be accepted without additional proof.

Phasianini × *Megapodidae*

The only case of this highly unlikely cross was a reported example of hybridization between a male scrub turkey (*Alectura lathami*) and a domestic hen (G. A. Keartland cited in Gray, 1958). Three "alleged" hybrids were reported, including a female that laid eggs that were "not very large."

Summary of Extratribal Hybridization

It may be seen that a rather surprising array of intertribal and even a few interfamilial hybrids have been reported, although all of the interfamilial combinations are sufficiently vague and unsupported as to probably be discounted. What is surprising is the absence of any reported hybrids between the pheasants and the New World quails (Odontophorinae). Even more surprising, there are also no reported crosses between the New World quails and the Old World partridges (Perdicini), in spite of many species of both groups having been bred regularly in captivity. This might support the idea that the New World quails deserve at least subfamilial separation from the rest of the Phasianidae and may be no more closely related to the pheasants than are, for example, the grouse. As mentioned in chapter 1, this relatively distant relationship of the New World quails to the other galliform groups is supported by biochemical evidence (Sibley and Ahlquist, 1990).

Intratribal Hybridization

Hybridization within the pheasant tribe Phasianini is far more frequent than is intertribal hybridization and offers a much greater amount of information of significance from a taxonomic and ecological perspective. The summary provided (figure 3) lists all of the species of pheasants that have been implicated in interspecific hybridization in the summaries of Gray (1958), Rutgers and Norris (1970), and Delacour (1977). The domestic fowl (*Gallus* "*domesticus*") is here considered conspecific with the red junglefowl (*G. gallus*).

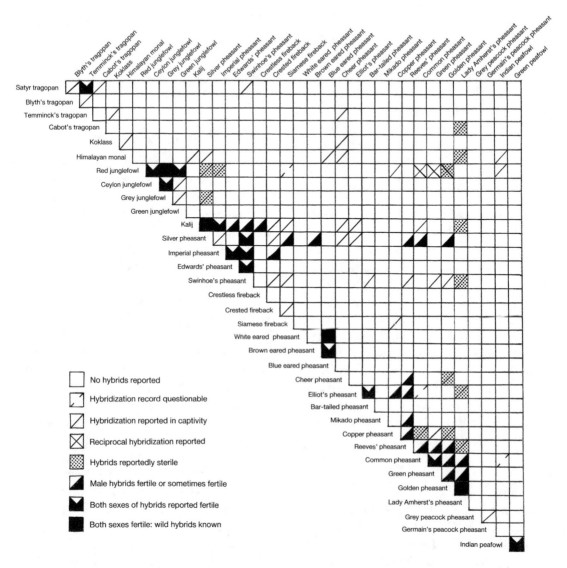

Figure 3. Records of interspecific hybridization reported among the pheasants. See text of Bornean peacock pheasant (in *Polyplectron* chapter) for two additional records.

Several interesting conclusions can be drawn from a study of this summary. The first is that fertility among intergeneric hybrids is relatively low and is seemingly limited only to males when it does occur. Male fertility has thus been reported for intergeneric hybrids between *Lophura* and *Crossoptilon*, *Lophura* and *Syrmaticus*, *Lophura* and *Phasianus*, *Lophura* and *Chrysolophus*, *Catreus* and *Syrmaticus*, *Syrmaticus* and *Phasianus*, and *Phasianus* and *Chrysolophus*.

Fertility involving both sexes is apparently limited to intrageneric hybrids, such as those between species of *Tragopan*, *Gallus*, *Lophura*, *Crossoptilon*, *Syrmaticus*, *Phasianus*, *Chrysolophus*, and *Pavo*. There are evidently only three possible instances of natural hybridization under wild conditions so far known among pheasants. These involve the red and grey junglefowl, the kalij and silver pheasants, and possibly the white and blue eared pheasants. Although white and blue eared pheasants have reportedly hybridized in the wild (Gray, 1958), secondary contact is still uncertain. The golden and Lady Amherst's pheasants are not yet known to exhibit secondary contact in the wild, but at least in captivity they have no apparent barriers to hybridization and little evidence exists of reduced hybrid fertility in either sex (Phillips, 1921; Danforth and Sandness, 1939; Danforth, 1950).

Another interesting point to be drawn from figure 3 is that *Gallus* seems to exhibit no intergeneric hybrid fertility whatsoever, suggesting that it occupies a somewhat isolated position in the pheasant tribe. Furthermore, this observation casts further doubt on the authenticity of the alleged hybrid mentioned earlier between a domestic fowl and a scrub turkey that reportedly was a "good layer."

However, the genus *Lophura* would seem to occupy a relatively central position in the pheasant assemblage, with hybrid combinations extending on the one extreme to the genus *Tragopan* and on the other to *Chrysolophus* and the other "long-tailed" pheasant genera. The peafowl and peacock pheasants seem to be relatively isolated, however, with sterile hybrids being reported between *Pavo* and the genera *Gallus* and *Phasianus* (Gray, 1958), as well as with *Lophophorus* (Delacour, 1977). So far, hybridization involving the genus *Polyplectron* seems to be limited to crosses between the grey and Germain's peacock pheasants, the Rothschild's and grey peacock pheasants, and the Malayan and Bornean peacock pheasants (Vernon Denton, pers. comm.; not included in diagram). Genera that have not been reported (by 1997) involved in hybridization include *Ithaginis, Pucrasia, Rheinartia, Argusianus,* and *Afropavo*. Of these, all but *Pucrasia* are only rarely maintained and bred under captive conditions.

Of the calculated 1,166 mathematically possible interspecific crosses that are possible within the 49 species of Phasianini, a total of 93 (8 percent) have actually been reported. This compares with 15 of 120 total possible combinations (12.5 percent) among the 16 species of grouse (Tetraoninae) as reported by Johnsgard (1982). Further, a total of 36 of the 49 pheasant species (73 percent of the total tribe) have been implicated in hybridization, whereas in the grouse subfamily 12 of 16 species (75 percent) have been so implicated. Of the pheasant hybrids, 48 percent have been intrageneric on the basis of current taxonomy and 52 percent intergeneric, whereas 43 percent (40 of 93) have been reported as being at least occasionally fertile. By comparison, 10 of the 15 known grouse combinations (67 percent) are intergeneric by current taxonomic standards and only 33 percent intrageneric. Most of these latter hybrids were wild birds, and thus their fertility is not generally known.

Distributional Patterns

It is quite apparent that the entire subfamily Phasianinae (Perdicini and Phasianini as recognized here) is centered in the Indo-Malaysian zoogeographic region. Except for the single anomalous case of *Afropavo* in Africa, all of the pheasants are limited to Southeast Asia, roughly between the vicinity of the Black Sea on the west and Japan on the east and extending northward as far as Mongolia and southward to the Lesser Sundas. If the collective native ranges of all the pheasants are plotted on a map (which is made somewhat difficult because of uncertainties as to the original range limits of *Phasianus colchicus* and *Gallus gallus*), this geographic relationship becomes very clear (figure 4). For example, 45 species of Phasianinae (18 Phasianini and 27 Perdicini) of an approximate world total of 174, or more than 25 percent, are native to the Indian subcontinent (Ali and Ripley, 1978). By comparison, sub-Saharan Africa has only a single species and genus of pheasant, but supports 40 additional species of Perdicini, nearly all of which are francolins (Snow, 1968). The former USSR supports six genera and 11 species of partridges, but has only a single species of pheasant (Dementiev and Gladkov, 1967). Thus the

Himalayan mountains and their associated deserts have evidently served as an effective barrier to northward expansion of the pheasants. The Himalayas themselves (as represented by Nepal) support 14 species of partridges and eight species of pheasants (Fleming, 1976). Likewise, Tibet supports 12 species of par-

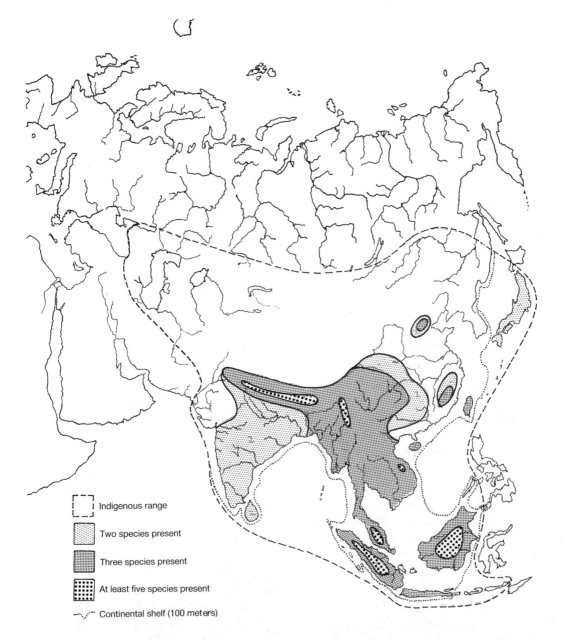

Figure 4. Species-density map of pheasants in Asia. Broken line indicates overall indigenous Asian range of pheasants; dotted line shows limits of continental shelf (Sunda Platform) in this region.

tridges and ten pheasants (Vaurie, 1972), and China has 22 species of partridges and 26 pheasants (Cheng et al., 1978). Southeast Asia (Burma, Indochina, and the Malay Peninsula) holds a total of eight genera and 16 species of partridges as well as 13 genera and 23 species of pheasants in an area somewhat smaller than that of the Indian subcontinent (King and Dickinson, 1975). Southeast Asia is thus relatively richer in pheasants than partridges, whereas Africa and southwestern Asia are considerably richer in partridges than pheasants.

Beyond these overall range aspects, it is of interest to note areas that are high in species diversity of pheasants, based on available information on individual species' ranges (figure 4). It is apparent that several such areas support five or more pheasant species. The first and most extensive of these are the Himalayas, where eight pheasant species occur. These include all of the most alpine-adapted and partridgelike of the pheasants, including the genera *Ithaginis, Tragopan,* and *Pucrasia* (table 3).

Table 3

Ecological distribution of pheasants in selected areas of high species density in Asia

	High montane forests	Mid-montane forests	Lowland forests
Central Himalayas	Blood pheasant Himalayan monal	Koklass Cheer pheasant Satyr tragopan Kalij	Indian peafowl Red junglefowl
Upper Burma/Yunnan	Kalij	Red junglefowl Bar-tailed pheasant Blyth's tragopan	Grey peacock pheasant Green peafowl
Annam (Vietnam)		Silver pheasant Imperial pheasant	Edwards' pheasant Red junglefowl Siamese fireback Grey peacock pheasant Green peafowl Crested argus
Malay Peninsula		Rothschild's peacock pheasant	Malayan peacock pheasant Red junglefowl Great argus Crested argus Green peafowl Crested fireback Crestless fireback
Sumatra		Bronze-tailed peacock pheasant Salvadori's pheasant	Great argus Crested argus Crestless fireback Red junglefowl Malayan peacock pheasant
Borneo			Great argus Crested fireback Crestless fireback Bornean peacock pheasant Wattled pheasant

A second center of pheasant diversity occurs in the general vicinity of northern Burma and adjacent Yunnan, in the upper reaches of the Yangtze, Mekong, Salween, and Irrawaddi Rivers. Six pheasant species occur in these temperate-zone mountain valleys and such essentially tropic-adapted genera as *Polyplectron* and *Pavo* exist in fairly close proximity to more montane-adapted types such as *Tragopan*.

A third area of high species diversity and endemicity occurs in Annam (now central Vietnam), which supports eight pheasant species, including two (*Lophura imperialis* and *L. edwardsi*) whose ranges apparently are the most limited of any mainland pheasant species. Delacour (1977) considered their closest living

relative to be the Swinhoe's pheasant, but zoogeographically it would seem more probable that they are offshoots of a generalized mainland kalijlike ancestor.

The Malay Peninsula, from southern Burma (Tenasserim) southward, represents an area of pheasant diversity that matches that of the central Himalayas. It supports eight native pheasant species, including one endemic (*Polyplectron inopinatum*) and one species shared only with Sumatra (*Polyplectron malacense*). This area would seem to be the center of evolutionary diversity of the highly specialized peacocklike pheasants (*Pavo, Argusianus, Rheinartia,* and *Polyplectron*), in the same way that the Himalayas obviously have served as the ancestral home of the more partridgelike genera.

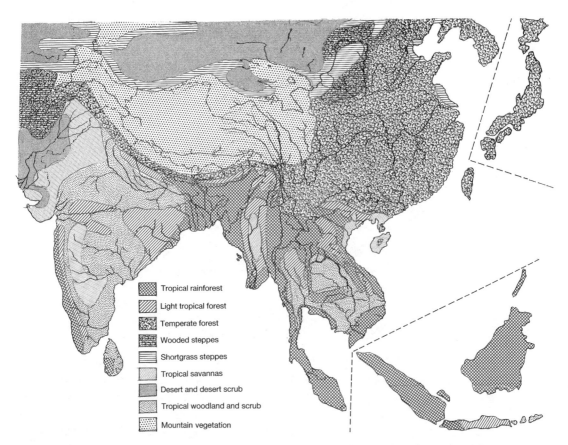

Tropical rainforest

Light tropical forest

Temperate forest

Wooded steppes

Shortgrass steppes

Tropical savannas

Desert and desert scrub

Tropical woodland and scrub

Mountain vegetation

Figure 5. Distribution of native vegetation types in Southeast Asia. Adapted from a map in the *Hammond Contemporary World Atlas.*

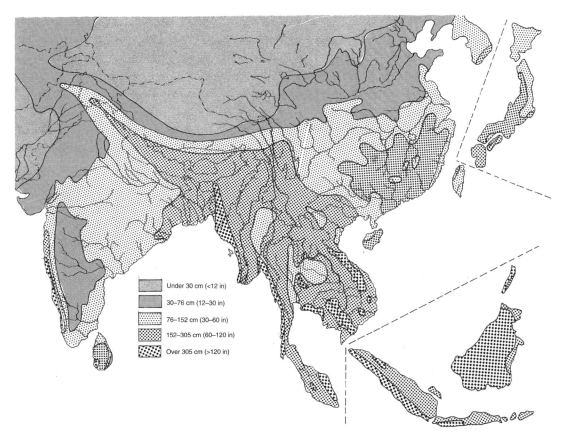

Figure 6. Distribution of annual precipitation patterns in Southeast Asia. Adapted from a map in the *Times Atlas of the World.*

The presence of an archipelago (Greater and Lesser Sundas plus Borneo) has probably facilitated speciation in this area. Both Borneo and Sumatra thus qualify as major centers of species diversity in pheasants, supporting seven and five species, respectively. Sumatra's pheasant fauna includes two endemics (*Polyplectron chalcurum* and *Lophura salvadori*), and Borneo likewise supports two endemics (*Lophura bulweri* and *Polyplectron schleiermacheri*, the latter considered by Delacour [1977] as only subspecifically differentiated). This general region of Indonesia from Malaya to Borneo also supports several endemic and distinctive genera of Perdicini (*Haematortyx, Caloperdix, Rhizothera, Melanoperdix*), further attesting to its importance as a center of phasianine evolutionary diversity. This entire region lies within the continental shelf of Asia (the Sunda Platform), and in general is separated from the mainland of Asia by current water depths of considerably less than 300 m (985 ft). Assuming that ocean levels during periods of maximum glaciation were at least 200 m (655 ft) lower than current ones, virtually all of the current collective Asian pheasant range would fall within the area then part of the Asian mainland, including the islands of Ceylon, Taiwan, the Greater Sundas (excepting the Celebes), and Palawan Island. Interestingly, Wallace's line (between Borneo and the Celebes) rather effectively separates the distribution patterns of the pheasants and the megapodes, which reach the coastal islands of Borneo and extend from there throughout much of the Australian region (Olson, 1980).

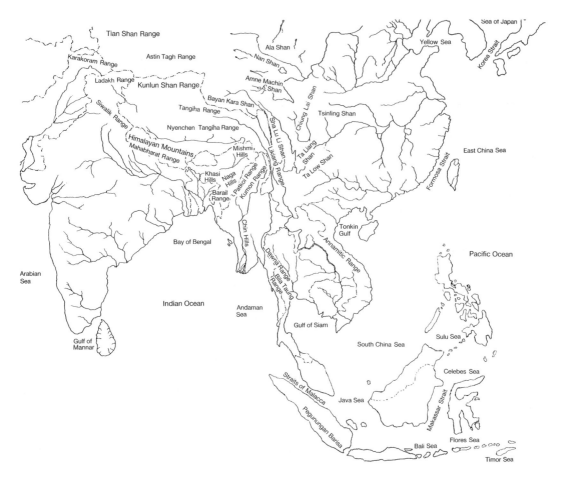

Figure 7. Distribution of river drainages, mountain ranges, and various oceanic features in Southeast Asia.

Finally, comparison of figure 5 with similar maps showing natural vegetation patterns and annual precipitation patterns further indicate that the richest areas of current pheasant species densities are those having at least 153 cm (60 in) of rainfall and regions that are at least partially wooded with temperate montane to tropical forests (figures 6–8). It seems probable that the singular distributional anomaly of an African peacock in the Congo Basin of Africa can be explained on the basis of isolation of a pre-*Afropavo* form in southwestern Europe during Miocene times, after which it was driven south into western Africa during the onset of colder periods in Europe (Lowe, 1939).

Figure 8. Map of Southeast Asia, showing various localities and elevations, with light shading indicating land areas above 1,830 m (6,000 ft) and heavier shading those above 3,660 m (12,000 ft).

CHAPTER 3

Growth and Behavioral Development

The ontogenetic development of pheasants, like that of other birds, follows a rather consistent pattern of changes in integumentary specializations (plumages and their intervening molts) and in behavioral capabilities. The former changes consist of a progressive series of feather generations that are highly adapted for needs associated with growth and maturity, such as flight, insulation, and advertisement or concealment, whereas the latter changes reflect a gradual refinement of motor abilities related to locomotion, foraging, and social interactions.

Plumage Development during Ontogeny

The galliform pattern of molt and plumage development has been reviewed in detail with particular reference to grouse and quails (Johnsgard, 1973, 1983*b*). Inasmuch as pheasant plumage development closely follows the same sequence, there is little purpose served in repeating all this information. However, the consistent pattern of growth and molting of the primary feathers has been a major basis for judging ages of pheasants by wildlife biologists, so it is appropriate that at least this aspect of plumage ontogeny be reviewed. Early work by Juhn (1938) on the domestic fowl provided the basic understanding of juvenile molting patterns in the Phasianidae. However, Buss (1946) was the first to develop a method of aging young ring-necked pheasants on the basis of growth and molting patterns of their primary feathers, and his methods were later amplified and modified by a variety of workers (e.g., Trautman, 1950; Westerskov, 1957; Etter et al., 1970). More recently, similar molt studies were performed using Reeves' pheasants (Mueller and Seibert, 1966), and work on the ring-necked pheasant has been expanded to include the molting patterns of all of the remiges and their coverts (Sutter, 1971). Felix (1964) has provided some data on molting patterns in the eared pheasants, and Durrer (1965) has analyzed the growth and microscopic structure of the "eye"

Table 4
Juvenile flight feather development in the ring-necked pheasant[a]

	Age at emergence (days)[b]	Period of growth (days)		Final feather length (mm, in)	
		Males	Females	Males	Females
Primaries					
no. 10	17.6	41.5	37.9	131, 5.1	119, 4.6
no. 9	11.5	40.0	38.1	137, 5.3	129, 5.0
no. 8	5.8	38.7	36.2	131, 5.1	124, 4.8
no. 7	0	36.7	35.2	123, 4.8	118, 4.6
no. 6	0	31.9	30.5	117, 4.6	112, 4.4
no. 5	0	28.1	26.8	109, 4.3	107, 4.2
no. 4	0	24.1	23.0	98, 3.8	94, 3.7
no. 3	0	19.9	10.1	85, 3.3	82, 3.2
no. 2	0	16.1	15.6	72, 2.8	70, 2.7
no. 1	0	13.7	12.8	61, 2.4	58, 2.3
Secondaries					
no. 1	13.3	25.7	25.4	97, 3.8	89, 3.5
no. 2	9.0	28.7	27.4	111, 4.3	104, 4.1
no. 3	0	18.2	18.2	74, 2.9	71, 2.8
no. 4	0	20.5	21.1	81, 3.2	79, 3.1
no. 5	0	—	—	—	—
no. 6	0	29.9	23.8	91, 3.5	89, 3.5
no. 7	0	—	—	—	—
no. 8	0	29.9	28.8	100, 3.9	98, 3.8
no. 9	0	—	—	—	—
no. 10	0	35.2	33.5	110, 4.3	105, 4.1
no. 11	0	—	—	—	—
no. 12	2.2	41.1	37.5	121, 4.7	110, 4.3
no. 13	6.3	—	—	—	—
no. 14	11.7	42.8	40.1	113, 4.4	106, 4.1
no. 15	17.3	—	—	—	—
no. 16	21.0	39.7	34.8	90, 3.5	79, 3.1
no. 17	23.4	37.0	32.3	76, 3.0	68, 2.7

[a] Data are given as averages and are from Sutter (1971). Dashes indicate that data were not available.

[b] Average of both sexes. A zero value indicates that the feather is present at hatching.

Juvenile Plumage

Although pheasants are covered in a downy (natal) plumage at the time of hatching, many species at that stage already exhibit several feathers associated with the subsequent juvenile plumage, including the first seven primaries and the third through 11th secondaries (Sutter, 1971). Similarly, in the Reeves' pheasant the first seven primaries are present at hatching. The remaining three juvenile primaries appear over the next 3–4 weeks. Growth in all these primaries occurs in an orderly, sequential fashion from the wrist outward, and in the secondaries from the third secondary both inwardly and outwardly (table 4). The

feathers of the male peafowl. Bura (1967) has performed a similar study of the juvenile and first-winter plumages of the common pheasant.

Table 5

Development rates of primary feathers in the Reeves' pheasant[a]

	Age when fully grown (days)		Final feather length (mm, in)	
	Males	Females	Males	Females
Juvenile primaries[b]				
no. 1	29	27	76, 3.0	68, 2.7
no. 2	29	29	83, 3.2	75, 2.9
no. 3	33	35	99, 3.9	90, 3.5
no. 4	33	39	106, 4.1	102, 4.0
no. 5	40	36	114, 4.4	109, 4.3
no. 6	40	38	114, 4.4	109, 4.3
no. 7	40	44	115, 4.5	113, 4.4
no. 8	44	47	115, 4.5	114, 4.4
no. 9	72	73	138, 5.4	130, 5.1
no. 10	92	91	152, 5.9	131, 5.1
First-winter primaries[c]				
no. 1	69	58	135, 5.3	121, 4.7
no. 2	74	66	153, 6.0	141, 5.5
no. 3	84	75	173, 6.7	154, 6.0
no. 4	96	87	180, 7.0	160, 6.2
no. 5	110	102	190, 7.4	169, 6.6
no. 6	123	118	194, 7.6	171, 6.7
no. 7	128	125	184, 7.2	172, 6.7
no. 8	149	139	184, 7.2	161, 6.3
no. 9	168	162	168, 6.6	152, 5.9

[a] Data from Mueller and Siebert (1966).

[b] Ages at which the juvenile primaries appear (average, both sexes): primary no. 1 through primary no. 7, day 1; no. 8, day 8; no. 9, day 16; and no. 10, day 29.

[c] No data are listed for first-winter primary no. 10, because juvenile primary no. 10 is retained throughout the first year in this species.

alular feathers also develop from the shorter ones to the longest, and both the major upper wing-coverts and under wing-coverts follow a sequence very similar to their corresponding remiges. Molt patterns of the lesser primary coverts are more complex and individually variable than are those of the major coverts (Sutter, 1971). Juvenile primary growth and molt in the Reeves' pheasant follows a very similar pattern (table 5).

Judging from limited observations on the Reeves' pheasant, growth and molt of the rectrices follow a somewhat different pattern. In that species none of the rectrices is present at hatching, but in one study they appeared during the 12th day in both sexes. All of the juvenile rectrices completed their growth between 40 and 60 days after hatching and were molted almost immediately thereafter. The second (from the middle) rectrix molted first and molt proceeded in a generally centrifugal pattern, with the outermost (ninth) rectrix molting on the 70th day after hatching (Mueller and Seibert, 1966).

Table 6
First-winter flight feather development in the ring-necked pheasant[a]

	Age at emergence (days)[b]	Period of growth (days)		Final feather length (mm, in)	
		Males	Females	Males	Females
Primaries					
no. 10	110.7	37.1	33.7	160, 6.2	140, 5.5
no. 9	96.7	36.9	32.5	181, 7.1	156, 6.1
no. 8	79.6	35.5	32.0	187, 7.3	163, 6.4
no. 7	69.2	34.6	31.4	189, 7.4	166, 6.5
no. 6	60.7	34.1	30.5	190, 7.4	167, 6.5
no. 5	53.3	32.5	29.1	197, 7.7	167, 6.5
no. 4	45.7	31.1	28.1	181, 7.1	164, 6.4
no. 3	38.8	28.8	26.6	167, 6.5	152, 5.9
no. 2	32.3	27.6	25.0	153, 6.0	139, 5.4
no. 1	26.9	26.5	24.6	140, 5.5	128, 5.0
Secondaries					
no. 1	102.9	21.1	20.9	100, 3.9	91, 3.5
no. 2	78.8	24.7	23.0	139, 5.4	124, 4.8
no. 3	36.6	25.3	23.8	137, 5.3	126, 4.9
no. 4	40.9	25.7	24.0	141, 5.5	129, 5.0
no. 5	46.1	—	—	—	—
no. 6	50.6	26.1	24.2	145, 5.7	131, 5.1
no. 7	55.1	—	—	—	—
no. 8	59.1	27.3	24.7	149, 5.8	133, 5.2
no. 9	63.4	—	—	—	—
no. 10	67.4	28.7	26.6	156, 6.1	138, 5.4
no. 11	73.1	—	—	—	—
no. 12	79.6	30.7	28.7	159, 6.2	139, 5.4
no. 13	90.5	—	—	—	—
no. 14	104.2	31.7	29.4	128, 5.0	110, 4.3
no. 15	117.3	—	—	—	—
no. 16	122.8	30.2	29.8	96, 3.7	87, 3.4
no. 17	117.3	28.3	28.0	81, 3.2	74, 2.9

[a] Data are given as averages and are from Sutter (1971). Dashes indicate that data were not available.

[b] Average of both sexes.

First-Winter Plumage

The juvenile wing plumage is lost during the postjuvenile molt, which in the ring-necked pheasant begins at the age of 23 days, with the appearance of the second (the first is rudimentary) alula quill, and proceeds next with the first primary (27 days) and greater secondary coverts (29 days). It terminates with the replacement of the lesser primary coverts and the innermost of the greater under secondary coverts at 4–5 months of age. The sequence of growth of the primaries and secondaries themselves follows a very similar pattern to those of the juvenile plumage (table 6), starting with the first primary and the third secondary and ending with the outermost primary and innermost (15th to 17th) group of secondaries.

Feather growth is virtually completed by the age of 150 days in the primaries and secondaries and by 170 days in the greater under secondary coverts. The remiges of the first-winter plumage grow considerably more rapidly than do those of the juvenile plumage and at a more constant maximum rate, of up to about 7 mm (0.3 in) daily in both primaries and secondaries. Sexual dimorphism is also less marked in the first feather generation than in later ones (Sutter, 1971).

The postjuvenile molt of the rectrices is apparently considerably different from the postnatal molt, at least in the Reeves' pheasant. All the rectrices of the first-winter plumage appear between the 50th and 70th days in both sexes; in this species the central pair of rectrices are by far the longest, whereas the ninth pair are the shortest. Although the exact order of molt has not been determined, the general pattern of the molt is clearly centripetal, beginning with the outer rectrices and proceeding centrally (Mueller and Seibert, 1966).

In the blue eared pheasant the postjuvenile tail molt apparently proceeds in a generally centripetal direction, although it begins at about 30 days with the pair on either side of the uropygial gland (Felix, 1964). The last to emerge is the tenth (third from innermost), and the innermost three pairs of feathers are not fully grown until the bird is about 140 days old. (The brown and white eared pheasants, with fewer rectrices, exhibit slightly different tail molting patterns that are neither strictly centripetal nor centrifugal.) At about the same time the innermost (13th) secondary has completed its growth, but the two outermost secondaries and the ninth primary do not complete their growth until the bird is about 180 days old. The juvenile tenth primary is retained in eared pheasants throughout the first winter, according to Felix (1964).

On the basis of these and similar studies, it is possible to estimate easily the age of young ring-necked pheasants from their first to at least their 24th week of life (table 7). Such information is extremely useful in estimating hatching dates in wild populations. Differences in the shape and color of the outer greater secondary coverts are also useful in aging and sexing ring-necked pheasants (figure 9). Thus, in males the third to fifth of these coverts are less adultlike than are the first and second of these coverts, which are molted later. In adult birds all five of these coverts are similar in shape and pattern (the third usually as long or longer than the second, as well as being broader and more strongly patterned). In females there is not such a striking contrast in pattern between the second and third coverts, although the third covert in adults is typically distinctly larger and broader than the second (Sutter, 1971).

Second-Winter Plumage

The first-winter plumage is carried not only through the winter but also through part or all of the following breeding season. It is not obviously different from definitive adult plumages. In the Reeves' pheasant the first postnuptial wing molt begins rather early in males, with the majority of them losing their first primaries by late April (in Ohio; Mueller and Seibert, 1966). Primary molt again follows a similar sequence, proceeding gradually to the outermost primaries, which are dropped by late August to early September. Females begin molting their innermost primaries approximately 2 months later than males in mid-June, and proceed correspondingly outwardly but at a more rapid rate, so that all of their primaries have been dropped by late September, or only about 2 weeks later than males. The first male primaries of the second-winter plumage appear (on average) in late April, and the first female primaries in early June. The outermost male primary appears in late August and completes growth in only 15 days, whereas the average female date for the outermost primary is

Table 7
Wing and tail criteria for aging young ring-necked pheasants[a]

Age (weeks)	Primary no.	Primary length range (mm, in)		Average tail length (mm, in)[b]
		Males	Females	
1	7 (juvenile)	1–28, 0.04–1.1	1–28, 0.04–1.1	0, 0.0
2	7 (juvenile)	29–47, 1.1–1.8	29–47, 1.1–1.8	10, 0.4
3	7 (juvenile)	48–68, 1.9–2.7	48–68, 1.9–2.7	36, 1.4
4	10 (juvenile)	6–25, 0.2–1.0	6–25, 0.2–1.0	50, 2.0
5	10 (juvenile)	26–50, 1.0–2.0	26–50, 1.0–2.0	63, 2.5
6	10 (juvenile)	51–73, 2.0–2.8	51–73, 2.0–2.8	70, 2.7
7	3 (1st winter)	33–81, 1.3–3.2	30–80, 1.2–3.1	76, 3.0
8	5 (1st winter)	0–37, 0.0–1.4	0–36, 0.0–1.4	89, 3.5
9	6 (1st winter)	0–31, 0.0–1.2	0–28, 0.0–1.1	101, 3.9
10	7 (1st winter)	0–24, 0.0–0.9	0–14, 0.0–0.5	101, 3.9
11	8 (1st winter)	0–5, 0.0–0.2	0–14, 0.0–0.5	140, 5.5
12	8 (1st winter)	6–47, 0.2–1.8	15–44, 0.6–1.7	165, 6.4
13	9 (1st winter)	0–4, 0.0–0.2	0–14, 0.0–0.5	203, 7.9
14	9 (1st winter)	5–36, 0.2–1.4	15–38, 0.6–1.5	241, 9.4
15	10 (1st winter)	0–9, 0.0–0.4	0–16, 0.0–0.6	279, 10.9
16	10 (1st winter)	10–49, 0.4–1.9	17–48, 0.7–1.9	292, 11.4
17	10 (1st winter)	50–82, 2.0–3.2	49–77, 1.9–3.0	305, 11.9
18	10 (1st winter)	83–107, 3.2–4.2	78–101, 3.0–3.9	330, 12.9
19	10 (1st winter)	108–129, 4.2–5.0	102–122, 4.0–4.7	355, 13.8
20	10 (1st winter)	130–144, 5.1–5.6	123–134, 4.8–5.2	380, 14.8
21	10 (1st winter)	145–174, 5.7–6.8	135–156, 5.3–6.1	406, 15.8
24	10 (1st winter)	174, 6.8	156, 6.1	≥500, ≥19.5

[a] Wing data mainly from Trautman (1950) and Etter et al. (1970); tail data from various sources.

[b] For tail length, the average of both sexes is reported here for 1–6 weeks of age; average of males is given for 7–24 weeks of age.

mid-September and its growth is completed by mid-November. Molt in the male rectrices begins in May, with new feathers of the second-winter plumage appearing between the end of May and early July, with the central rectrices appearing last. Similarly, the female rectrices appear between late June and early August, with rectrix growth being completed in females by late October and in males (which have appreciably longer central tail feathers) by early December (Mueller and Seibert, 1966).

In the ring-necked pheasant adult wing molt follows a similar pattern, with males beginning to molt their primaries a month earlier than females and completing molt earlier as well. Adult molt onset in females seems to be correlated with the date of hatching of their broods, whereas in males its earlier timing is seemingly associated with an earlier fall gonadal development cycle (Kabat et al., 1950). In the eared pheasants males also typically begin to molt 5–6 weeks earlier than do females. In these, wing molt likewise begins with the innermost primary, the third secondary, and the outermost rectrix. Adult tail molt in the eared pheasants is evidently essentially centripetal (Felix, 1964).

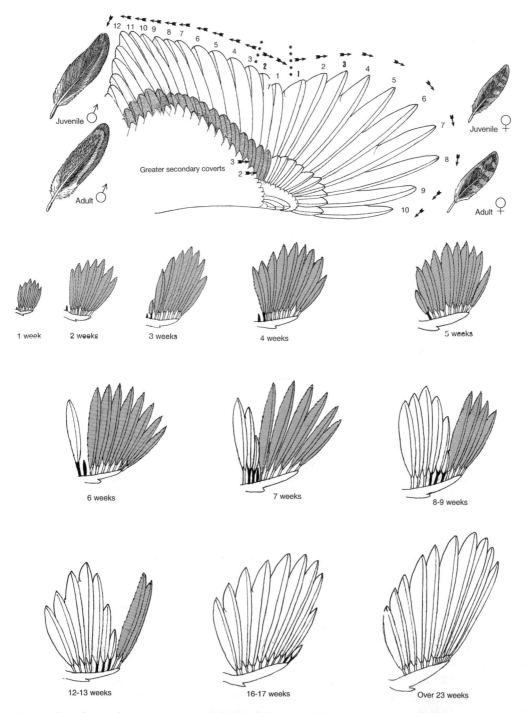

Figure 9. Dorsal view of common pheasant wing (top) showing numbering of primaries and secondaries, molt centers (asterisks), directions of molt in remiges (arrows), greater secondary coverts (shaded), and patterning of second and third greater secondary coverts in adults and juveniles of both sexes (after Sutter, 1971). Also shown (bottom) are selected stages of primary development in common pheasants during growth, with juvenile primaries indicated by shading and actively growing quills indicated by black bases.

Behavioral Development

The most thorough review of the ontogenetic development of behavior in any pheasant species is that of Kruijt (1964) for the Burmese red junglefowl, which may serve as a model for pheasants in general and is the primary basis for the following review. Nonsocial behavior that is typical of galliform species from hatching or shortly thereafter includes locomotory behavior (walking, running, jumping, hopping, and flying); behavior associated with feeding or drinking, defecation, and sleeping; and behavior that can be conveniently labeled as "comfort behavior," including such things as stretching, preening, and dust-bathing.

Walking, running, jumping, and hopping all occur in junglefowl from shortly after hatching, or by no later than the third day. Hopping differs from jumping only in that it consists of a series of consecutive jumps, with the wings being flapped with each hop. Thus, hopping gradually develops into flying behavior, which in junglefowl begins on the 12th day, when the wing feathers have grown sufficiently to support the weight of the body. As the bird develops it gradually loses its dependence upon hopping as a precursor to flight.

Feeding behavior develops around ground-pecking behavior, which is present immediately after hatching. Newly hatched birds typically peck at any small spots that contrast with the background, particularly those objects that are relatively more rounded than angular in shape. Ground-scratching, which is usually present from the third day, involves alternating series of backward strokes with each leg. Food-running, which is usually elicited when a chick seizes a large morsel of food (especially live food), is present from the second day after hatching. In this behavior, while running the bird also utters repeated peeping calls. The behavior's function seems to be that it attracts other members of the social group, thereby increasing the chances that the prey will be killed and made

edible. Later, the running behavior acquires a secondary function of preventing other chicks from stealing the prey. This behavior is the ontogenetic precursor of "tidbitting" behavior, one of the basic sexual displays of pheasants and many other galliform birds. A few other behaviors that are also associated with feeding activity are bill-wiping, bill-beating, head-scratching, and head-shaking movements, all of which appear by no later than the end of the first week.

Comfort activities (figure 10) of newly hatched chicks include stretching and wing-flapping movements. The wings may be stretched bilaterally, during which both wings are partially opened and stretched upwardly and forwardly, or unilaterally, in which a single wing is stretched to the side and rear, often simultaneously with the corresponding leg and foot. Wing-flapping consists of several wing-beating movements as the bird stands still, often on its toes. Wing-flapping, which occurs by the fourth day after hatching, is the precursor of display wing-flapping, a major sexual signal in many male pheasant species. However, wing-stretching does not appear to have been ritualized into a display in pheasants.

Preening behavior takes several forms. Generally it involves pecking, nibbling, and stroking or combing of the plumage, all of which occur by the fourth day after hatching. Preening rates vary greatly during growth of the chick, and probably are related to periods of growth and molt of particular feather areas. Related to preening is head-rubbing, head-scratching, and bill-wiping. Head-rubbing may rearrange the head plumage or remove materials from the head and is present in functional form only after at least 11 days. Head-rubbing with rotation, which is related to obtaining and spreading the secretions of the uropygial gland, is present from about the 11th day, when the gland first becomes functionally active.

Dust-bathing, which gradually develops into its full form, consists of several elements that

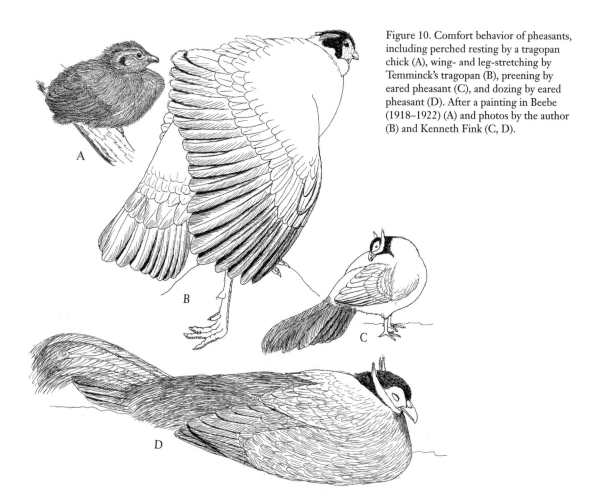

Figure 10. Comfort behavior of pheasants, including perched resting by a tragopan chick (A), wing- and leg-stretching by Temminck's tragopan (B), preening by eared pheasant (C), and dozing by eared pheasant (D). After a painting in Beebe (1918–1922) (A) and photos by the author (B) and Kenneth Fink (C, D).

may be present from the first day or appear after up to about 12 days. These include bill-raking, ground-scratching, wing-shaking (to sweep sand into the plumage), lying on the side, and head-rubbing. Dust-bathing is frequently performed throughout life, especially if the proper warm, dry, and loose substrate materials are present.

Various shaking movements, such as the previously mentioned head-shaking behavior, are commonly performed. They include shaking of the wings and body, the tail, the head and neck, and the legs, all of which often occur in various combinations with one another and with wing-flapping.

The sleeping posture of junglefowl and other galliform birds involves tucking the bill into the scapular feathers. In junglefowl this posture is not fully attained until about the 14th day, when feathers in this region have grown sufficiently to hold the bill in place.

Defecation in junglefowl is often associated with backward stepping and bilateral wing-stretching, and is present from the first day after hatching.

Early escape behavior includes the alert posture, head-shaking, running, and squatting. In the alert posture the bird stands still, with outstretched neck; in chicks this is often accompanied by calling, especially during the first few days. Head-shaking is also often performed in this posture. Squatting occurs from the first or second day after hatching and by the end of the second week is replaced by "freezing" without sitting. A variety of calls are associated with escape behavior, as will be noted later.

Early aggressive behavior also includes various components. Unoriented hopping, which develops out of simple hopping, may lead to approach or retreat from other individuals. By the time the chicks are about 1 week old, the hopping becomes directed toward other individuals, and at about the same time "frontal threatening" appears. This is a brief neck-stretching toward the other individual, and later this becomes more conspicuous as neck hackles appear and are erected during the behavior. Somewhat later, between the ninth and 12th day, leaping begins, and shortly thereafter aggressive pecking also appears. By the time the chicks are 3 weeks old the final component, kicking toward the opponent during leaping, is also beginning, although it does not generally occur until the birds are nearly twice that age.

From the age of 3 weeks, fighting behavior in young junglefowl begins to exhibit elements of escape, and from then on fighting becomes an ambivalent activity reflecting both attack and escape tendencies. Furthermore, irrelevant actions, such as ground-pecking, head-zigzagging behavior, and preening all begin to appear and seem to reflect ambivalent motivation. Social peck-orders, which start to develop during the first stage of aggressive behavior, become more fully established during the second stage. Young females (pullets) generally become lower in rank than young males (cockerels). Gradually a relatively linear social hierarchy develops, with the linearity probably resulting from differential rates of development in individual birds (Rushen, 1982).

An important social signal that first appears in the context of fighting is "waltzing." In its full form the displaying bird walks sideways around its opponent and holds its back and shoulders in an oblique manner, with the side nearest the opponent lower than the other one. Both wings are somewhat laterally extended, but the primaries of the outer wing are low-ered to the ground and pulled forward with their plane near the body and the longer primaries scraping the ground. The outer foot typically also makes scratching or stepping movements through the primaries. The tail is spread and turned toward the opponent, and the breast and belly feathers are also often spread. Waltzing may be virtually stationary, with foot movements limited to ground-scratching, or it may involve a circular movement around the other bird. Two displays are often associated with waltzing and contain similar components. The first of these is "side display," in which the outer wing is kept folded, but the same oblique body posture as occurs in waltzing is assumed. Small, deliberate steps are typically taken during side display. In "two-sided wing-lowering" a frontal rather than lateral orientation is assumed, and both wings are lowered toward the ground. In this display the ruff is frontally erected, and ground-scratching with both legs may be performed. All three of these displays may occur as early as 50 days after hatching, but are still rare at this time (figure 11).

As males become older, they exhibit a combination of sexual and aggressive patterns that become eventually integrated into courtship and copulatory behavior. Pure sexual behavior, or copulation, takes a similar form in all pheasants. In the junglefowl the male typically approaches the receptive female from behind, usually with the neck stretched upward but the bill pointed downward toward the female. As the female crouches he mounts, sits on his tarsi, which rest on her back, and often either pecks at her nape or grasps her head feathers. The male then begins an alternate trampling movement with his feet and begins to tilt his body backward. The tail is lifted, spread, and then brought downward to meet the female's cloacal area with rapid sideways movements. After the final cloacal contact the male releases the female and dismounts. No complex postcopulatory behavior is typical of pheasants.

Figure 11. Displays of male red junglefowl, including fighting by cockerels (A), lateral wing display by cockerel (B), running with bilateral wing-lowering (C), precopulatory approach (D), tidbitting (E), and waltzing (F). After various sources, including Kruijt (1964).

In junglefowl chicks between 30 and 80 days of age, copulatory behavior occurs more or less in behavioral isolation, without strong aggressive or escape elements. At later ages, these elements are more apparent, and at this stage differences in the behavior of males and females become important. Evidently the sitting (sexual crouching) position of a receptive female is especially important, because it is the only posture that does not elicit strong, competing agonistic tendencies. Because only females assume the sexual crouching position, this provides an automatic means of sexual identification and reduces the chances of males attempting to copulate with other males or to attack females. However, males may attempt to force females (or other males) into the crouching position and attempt to copulate with them.

Other than copulatory behavior, social responses of maturing male junglefowl toward other males or females consist of four more or less functional units or displays. The first two of these are waltzing and wing-flapping. Waltzing has been described in some detail above; it may be performed toward social partners of either sex. Waltzing may be followed by ground-pecking or pecking at the head of the other individual or may be followed by a conspicuous turning away of the body from the partner, so that the other side of the body is exhibited and the other wing lowered. Wing-flapping, in its social display context, appears between the ages of 80 and 120 days, and often precedes attack. A large variety of wing-flapping types occur, but in its most vigorous form the wings beat together and produce a clapping sound. The male may then fly into a high place and continue to beat the wings after landing. This is often followed by a crowing call. In many species of pheasants this becomes one of the major male advertising signals of sexually active birds. Wing-flapping may also occur when the male is in a submissive position, but in this case the tail is not erected, the wings are not raised high above the back, and the flapping is done in a weak and listless manner.

Tidbitting behavior, a term first used by Domm (1927) in describing the sexual activity of domestic fowl, is a basic form of sexual display in junglefowl and probably in all pheasant species. It consists of ground-pecking behavior that is directed toward edible or inedible objects, often with associated ground-scratching behavior, and is accompanied by high-pitched and rhythmically repeated calling. The ontogenetic precursor of the calling is the food-run call of chicks, and in adult females tidbitting calls are used to attract their chicks to food. Thus, tidbitting serves several roles, including parental bonding between females and chicks, social bonding between the chicks of a brood, and sexual bonding between adult males and females.

The last major social signal of male junglefowl is called "cornering" (Wood-Gush, 1954, 1956). In this behavior a male runs to a corner of its enclosure, stamps its feet, and lowers itself to the ground. The bird may remain silent or utter a repeated, low, purring note, especially after the male has settled and stopped moving its legs. Alternatively, the male may utter tidbit calls at this time. This behavior has a strong attractive effect on females, and perhaps helps the female decide on a particular nest-site location. It is possible that it also serves as a nest-building mechanism, because males often perform ground-pecking movements at this time and may hold bits of grass or straw in the bill. Both tidbitting and cornering behavior appear at about the same time that successful copulations begin. During all three behaviors similar leg movements occur. However, waltzing behavior most often occurs between these displays and copulation attempts. Thus, although waltzing is apparently derived from an ambivalence between aggressive and escape tendencies, it is actually more likely to occur in copulatory than in purely aggressive situations and so becomes a "courtship" display.

Female junglefowl behavior patterns develop along similar lines as those of the males, but fighting behavior is typically less prolonged and attempted mating behavior is extremely rare or perhaps absent, although it has been observed in domestic fowl. Although wing-flapping behavior is performed, the wing-clapping version is evidently lacking. Likewise, complete cornering behavior is apparently rare or absent. Sexual crouching behavior begins in females at 4 or 5 months of age. It is most often performed before males in response to waltzing, but may also occur as a response to attack behavior.

The effects of partial and complete social isolation during ontogeny have been reported by Kruijt (1964, 1966). Males raised in partial visual isolation for varying lengths of time tend to exhibit increased escape behavior that eventually disappears and is replaced by self-directed attack behavior toward their own tail as well as aggressive behavior toward humans. None of three males isolated from hatching until they were 15–16 months old performed copulatory behavior, and only two of eight that were isolated until 10–14 months old did so. Three males that were raised in total visual isolation (but able to hear one another) totally lacked normal social behavioral responses. Results of limited testing on females gave similar results to those found for males.

CHAPTER 4

General and Social Behavior

The behavior of any organism is controlled by its array of sensory input (e.g., vision, hearing, touch) and its motor output (e.g., reflexes, taxes, and more complex responses) as mediated by its central and peripheral neural network. Like other birds, the sensory abilities of pheasants are highly refined. An understanding of these abilities will help in any interpretation of their general and social behavioral attributes.

Based on reviews by Fischer (1975) and Wood-Gush (1955), a quick review of sensory capabilities of domestic fowl seems germane, inasmuch as it is probably relatively applicable to pheasants as a group. Vision in the domestic fowl is relatively acute, owing to the large number of cone cells present in the retina. This results in a great importance of color in the behavior of domestic fowl and undoubtedly most pheasants and helps account for the high incidence of bright male coloration as a social signaling device. The total range of color perception is probably very similar to that of humans. Unlearned color preferences of chicks are in the violet and orange regions of the spectrum, and there is a negative preference for green, which may aid in foraging by making other colors easier to find on a grassy substrate and thus more likely to be eaten. Domestic fowl have a total visual field of about 300°, with an area of binocular vision of about 26°. They apparently have an innate depth perception, and also an ability to perceive visual differences in size, shape, and pattern (Fischer, 1975). Domestic fowl evidently respond in the same way as do humans to optical illusions that subjectively differ in size. Shadows provide a very important cue to depth perception, which is a learned ability (Wood-Gush, 1955).

Hearing ranges in domestic fowl may be narrower than those typical of higher mammals. Studies with chicks suggest an auditory sensitivity centered between 100 and 2,800 Hz. Vocalizations also tend to fall within this range of frequencies, further suggesting its behavioral significance. Although the taste abilities of domestic fowl are generally considered to be quite poor, chicks do discriminate among various carbohydrates and are sensitive to salt and bitter-tasting substances. Sensitivity to flavors in water is greater than to flavors in solid foods. The sense of smell is believed to be poorly developed in domestic fowl.

Tactile senses in domestic fowl are not well studied, but are no doubt involved in eating, preening, and brooding behavior. Apparently tactile stimulation is especially important in developing and maintaining the brooding response (Fischer, 1975).

Intelligence, Memory, and Imprinting

Domestic fowl have been tested in a variety of ways for their learning abilities, only a few of which need to be mentioned here. A few of these studies suggest a learning ability equal to that of lower mammals, whereas others do not.

Both adult and young domestic fowl improve to about a 70 percent accuracy on successive discrimination-reversal learning tests, which is not much different from the performance of laboratory rats. Chicks can also learn to alternate turning responses in a temporal maze, and may be capable of latent learning. However, they are inferior to cats in active-avoidance conditioning tests, and they are generally very poor at learning detour problems (Fischer, 1975). Memory in domestic fowl is seemingly not well developed. Various studies suggest that hens may no longer be able to recognize members of their flock after 2 weeks in isolation, and that bitter or undesirable foods such as nettles or sour dock may be forgotten in no more than 14 days. One group of 5-month-old fowls remembered the feeding place of their former run after a 2-week period, but another group forgot the geography of their enclosure after 3 weeks. Learning abilities evidently improve from hatching to from 2 weeks to 2 months, perhaps depending on the complexity of the problem (Wood-Gush, 1955; Fischer, 1975).

Imprinting behavior, specifically filial imprinting and the following response, has been well studied in domestic fowl. Normally the critical period for filial imprinting begins almost immediately after hatching and lasts about 3 days, although isolated birds may imprint as late as 7–10 days after hatching. Chicks given a brief imprinting experience do not usually exhibit sexual preferences for the imprinting stimulus at maturity (Guiton, 1961). However, Lill and Wood-Gush (1965) found that in studies of breeds of chickens (white leghorn, brown leghorn, and two hybrid broiler strains) females of both brown and white leghorns exhibited intrabreed mating preferences, which were apparently based on plumage rather than on courtship display differences. In another experiment, males that had been raised with females of their own breed showed intrabreed preferences, whereas those raised in mixed breeds exhibited only weak intrabreed preferences, suggesting that imprinting effects may indeed influence adult mating preferences at least in males (Wood-Gush, 1971).

Peck-Orders and Social Dominance

The social hierarchy patterns, traditionally called "peck-orders," are extremely well studied in pheasants and were indeed discovered in domestic fowl. Peck-orders form the primary basis for social integration and group behavior patterns in pheasants and represent an important behavioral mechanism for group survival and social integration. Domestic fowl form individualized dominance-subordinate relationships soon after their first encounters, and, in this species at least, such relationships result in the formation of an essentially linear social hierarchy. Typically a mixed-sex flock has two peck-orders, each one unisexual, because males are more aggressive than females and the latter are normally passive in intersexual domination. The maintenance of such peck-orders obviously requires individualized memories and recognition of other flock members. With separations of 2 weeks or more, such peck-orders may be disrupted through forgetting. Evidently individual recognition depends on the overall appearance of a bird, rather than learning a single feature or a particular area. However, cues that are positively related to social dominance include the size of the comb, absence of molt, overall size, and relative threat posture. Because these factors are in part related to androgen levels, male hormones must also be considered important determinants of peck-order position. However, the role of estrogen in peck-order establishment and maintenance is apparently still unclear (Guhl, 1953; Wood-Gush, 1955; Fischer, 1975).

In addition to a very large number of studies of peck-orders in domestic fowl, a few studies have also been performed on captive red jungle-fowl (Banks, 1956) and with free-ranging ring-necked pheasants (Collias and Taber, 1951). Banks observed no significant differences between the social organization of red junglefowl females (in a confined environment) and those reported for female domestic fowl. He studied 26 females in four separate flocks. A well-developed social hierarchy was present in each flock, which was essentially linear but with minor geometrical complexities of organization. Only one observed case of dominance reversal between two females was found as a result of experimental transferring of birds from one flock to another. This occurred when two females that had been introduced into a new flock reversed their dominance positions relative to one another. Longtime members of a given pen enclosure were typically associated with higher social status following such experimental regroupings.

In a field study of ring-necked pheasants, Collias and Taber (1951) color-marked 38 male and 170 female birds and subsequently analyzed their dominance relationships. Among groups of three, nine, and eight males associated with specific feeding areas, they observed only two instances of deviation from a straight-line hierarchy arrangement. Both of these involved a male becoming territorial and thereby gaining at least temporary advantage over two other males. Among 14 females at a feeding station, there were only a few deviations from a straight linear social dominance pattern. Among 85 repeat observations of encounters between the same two females there were no cases of reversed social dominance. All males dominated all the females, but stopped pecking them soon after the breeding season began. Collias and Taber were unable to correlate body weight with dominance, although the trend was in a positive direction. At least among females, age may have been an important factor, because three of the seven most dominant females at one feeding station were at least 2 years old. Data on the males are inadequate for similar comparisons.

Dominance and Sexual Behavior

As may be expected, there is a positive relationship between the position of a male in the peck-order and his success in mating (Lill and Wood-Gush, 1965). Interestingly, the most dominant male may not perform courtship behavior as frequently as some of the less dominant birds or necessarily perform the most attempted matings, but he is nevertheless most successful in completed copulations. It is equally true that the most dominant females tend to be less sexually receptive than the most submissive ones, because the submissive crouching response of females facilitates copulation (Fischer, 1975). Some studies also indicate that males may vary in their rates of courtship, treading frequency, and ability to solicit crouching from females. There is evidence that these individual variations in sexual behavior may have a genetic basis and are to some extent sensitive to artificial selection pressures. In one study a negative correlation was actually found between comb size of males and mating frequency, suggesting that mating differences are not simply due to differences in testosterone levels. Differences in mating activities between the selected lines were apparently instead due to central nervous system factors rather than to endocrine or experimental differences. Other studies on heritability of mating tendencies (see Wood-Gush, 1971) have indicated a surprising negative correlation between semen volume (and sperm concentration) and mating rates. Evidently males that copulate at a high rate are more prone to ejaculate lower quantities of sperm. In another study it was concluded that there was in fact no correlation between mating frequency and number of males dominated in the flock or in male mating behavior and various indices of aggressiveness. It is thus possible that the most dominant male is not always the most sexually active one and that his presence might actually lower overall group fertility by hindering other males from mating (Wood-Gush, 1971).

Heterosexual Interactions

The major displays that occur in domestic fowl during heterosexual encounters were mentioned in chapter 3, and the major postural displays may be quickly summarized here. Some of these have associated vocalizations, which will be discussed separately. Major male displays of the domestic fowl, junglefowl, and indeed of most species of pheasants, include the following postures:

1. *Waltzing.* In this display one wing (the farther one in the junglefowl and domestic fowl) is lowered and the male advances past or around its partner. Frequently the primaries scrape against the ground, and the outer foot also scratches against the lowered primaries. In domestic fowl a lower intensity version of this display, without wing lowering, occurs and is called "circling." It probably corresponds to what has been described as "side display" in junglefowl.

2. *Wing-flapping.* A highly variable display in which the wings may be moved silently or flapped noisily, including clapping sounds made by the wings striking one another overhead. In many species the display is called "wing-whirring."

3. *Tidbitting.* The male pecks at the ground or scratches at the ground while giving food calls. In some species of pheasants actual items of food may be picked up and dropped, or may be held in the bill as the food call is uttered.

4. *Feather-ruffling.* In the domestic fowl the major feathers affected are the hackles of the neck. However, in many other pheasant species the crest, breast feathers, or body feathers in general may also be variably raised or ruffled. In peacock pheasants the entire dorsal plumage is often raised.

5. *Head-shaking.* In domestic fowl and junglefowl the head is vigorously shaken with circular movements. In other species the intensity or form of head-shaking may vary.

6. *Tail-tilting, tail-spreading, or tail-wagging.* Tail exhibition (or the exhibition of specialized tail-coverts, as in peafowl) are common forms of visual signaling in pheasants and usually occur in common with waltzing, side display, or frontal display.

7. *Frontal display with bilateral wing-lowering.* Although not well developed in junglefowl, this is a major display in many pheasants and reaches its peak in peafowl and peacock pheasants. Indeed in peafowl it has completely replaced lateral display posturing.

8. *Wattle, comb, or facial engorgement.* Nearly all pheasants utilize the exhibition

of temporarily enlarged areas of facial skin as sexual or aggressive signals. The wattled pheasant represents the culmination of this trend, whereas in others such as blood pheasants it is scarcely noticeable.

9. *Cornering.* This display, initially described by Wood-Gush for the red junglefowl and domestic fowl, is evidently an important precopulatory display in the genus *Gallus.* At least among red junglefowl it also serves in part as a nesting invitation display, and this may likewise be the case with feral domestic fowl. However, in confined domestic fowl it serves primarily as a preliminary to copulation. Another related display, the "rear approach," involves a direct male approach toward the female from behind, with his head high and his neck feathers variably ruffled.

10. *Crouching.* This female display, essentially a submissive posture, also serves as a specific invitation for copulation. In all pheasant species it takes a similar form, with the bird resting on her tarsi, her wings partially spread, and her head slightly raised. No specific copulatory or postcopulatory displays appear to be present among pheasants, although the usual postcopulatory feather adjustment and wing-flapping is common, especially among females.

Other species of pheasants possess certain additional displays that seem to be lacking in domestic fowl or, if present, have not yet been recognized as such. Thus, male tragopans evidently perform short display flights to elevated perches and of course they also have elaborate display postures associated with exhibition of their normally hidden gular lappets and "horns." Most, and perhaps all, male pheasants utter loud advertising calls, sometimes in conjunction with noisy wing-flapping displays or other posturing. Nearly all pheasant species so far studied perform a display similar to or equivalent to waltzing, although it takes greatly differing forms in different genera. Waltzing may, however, be lacking in peafowl, *Afropavo,* and some of the peacock pheasants, or perhaps it has been so greatly modified in them as to become unrecognizable.

Egg-Laying and Incubating Behavior

Nest-site selection and associated egg-laying behavior has been studied little in domestic fowl, although some aspects of nest-site selection have been studied in ring-necked pheasants. Among domestic fowl it has been learned that females typically visit several potential nest sites before selecting one, and that young females appear to be very nervous about choosing a site. If an egg is already present at a site, the site becomes more desirable, although china eggs are less attractive than real eggs and wooden or plaster eggs are sometimes superior even to real eggs. Likewise, concealed sites are more attractive than open sites, and certain materials such as straw, excelsior, or wood fibers have been found to be superior to other materials. Evidently the majority of domestic fowls leave their nests before 1:00 P.M., suggesting that even under domestication the usual galliform pattern of early morning egg-laying has been retained (Wood-Gush, 1955).

Nest-building behavior in other pheasants takes a similar form. As in all Galliformes, females pheasants are unable to carry nesting material about, and instead simply gather together what can be obtained by reaching out and tossing materials back toward the body. Nearly all pheasants construct simple nests in shallow scrapes in this manner. However, in tragopans, crested argus, and great argus the nests are often placed in elevated situations.

Domestic fowl and most pheasants other than a few tropical forms, such as *Polyplectron* spp., are indeterminant layers and continue to lay eggs if they are unable to complete a clutch. Likewise, if the eggs do not hatch or if artificial eggs are placed in the nest, the incubating behavior can be extended well beyond normal limits. The shift from incubating behavior to brooding behavior evidently requires both visual and auditory stimuli. The presence of downy chicks can induce broodiness even if the hen has not previously been incubating. These chicks must still be in the downy stage, and if they are replaced by young birds every few weeks the brooding period can also be extended well beyond its normal length (Fischer, 1975). Although caponized male domestic fowl can be readily induced to brood chicks and normal males can be stimulated to brood chicks by means of prolactin treatment, it is apparently not possible to stimulate males to sit on eggs (Wood-Gush, 1955).

The egg-laying rate of domestic fowl is one egg per day, whereas in the common pheasant the average egg-laying interval is about one egg per 1.4 days. In many pheasants continuous incubation behavior apparently begins with the last or penultimate egg, although some periodic incubation may occur during the egg-laying period, especially in later stages of laying.

Parental Behavior

Evidently chicks learn to recognize their mothers by various means. If chicks of different broods are placed together in the dark, there is a tendency for the young to find and go to their own hens, which suggests that acoustic clues may be useful in hen-chick recognition. However, visual clues may also be important: When chicks from hens of three different colors were placed together and in sight of hens with colors like their mothers, those from black hens went to the black female, those from red mothers went to the red hen, and those from white mothers went to the white female. However, some of the chicks did make incorrect choices. Studies on hen-chick interactions suggest that females learn to recognize their own young by a complex "recognition-impression"

process involving the entire brood. Thus a strange chick is likely to be accepted if it has the same general characteristics as the brood, whereas obviously different chicks may be attacked. Evidently, appropriate chick behavior is more important than chick coloration in this regard.

Domestic fowl chicks remain in close physical contact with their mothers for the first 10–12 days. They then enter a dispersal period in which they begin to feed independently but still sleep and brood under her. This stage lasts until the chicks are 6–8 weeks old, and the brood then gradually dissolves. Broods of pheasant chicks also tend to break up at this time. A few domestic fowl broods may last for as long as 12–16 weeks, but eventually the female loses her brooding tendencies and drives the chicks away (Wood-Gush, 1971). In peafowls and their relatives brooding takes an interesting form in that the female typically broods a single young under each wing (in most of these species two eggs are the normal clutch), with the chicks facing forward and their heads sometimes protruding from the front of the wing. Such brooding may occur on large branches of trees, as well as on the ground (figures 12, 13).

Chick Vocalizations

Three types of calls of chick ring-necked pheasants have been described by Heinz and Gysel (1970) as being peculiar to young up to 7 weeks of age. A "content" call is uttered by chicks when they are warm, resting with other young, or settling down for the night. This is a two-syllable call, with emphasis on the second syllable, *ter-rit* or *ter-wit*. A "caution" call is uttered by the chick when it is presented with a strange object. A third call, seemingly a general distress call, is uttered by birds separated from the hen or by isolated, hungry, or cold chicks.

Similarly, in the domestic fowl the chick utters a "trill" or prolonged "pleasure" call

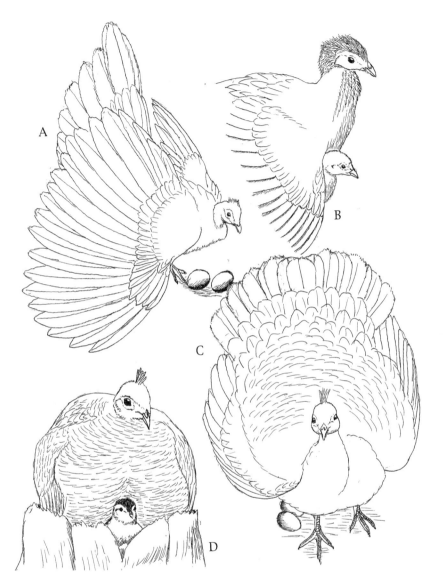

Figure 12. Maternal behavior of pheasants, including nest defense (A) and brooding (B) by female great argus, and nest defense (C) and brooding (D) by Congo peacock. After photos by David Rimlinger, Kenneth Fink, and Tim Greenwood.

when it is touched by a human or another chick and a shrill "distress" call, which is given as part of escape behavior and consists of repeated loud peeping notes (Wood-Gush, 1971). Judging from published sonograms (Heinz and Gysel, 1970; Wood-Gush, 1971), the distress calls of ring-necked pheasants and domestic fowl chicks are very similar acoustically. A third call, uttered by junglefowl chicks during food-running behavior, is similar to the distress call in that it consists of rapidly repeated peeping notes.

Adult Calls

Adult calls of the domestic fowl have been reviewed by Wood-Gush (1971), who has attempted to provide synonyms of various names that have been applied to them by earlier workers. "Crowing" is the best known of all male domestic fowl calls, although occasionally laying hens may also crow. The call may function as a territorial claim and attract females, and the crowing rate of a male seems to be positively related to his position in the

Figure 13. Maternal behavior of pheasants, including chicks feeding beneath tail (A) and brooding beneath wing (B) of Palawan peacock pheasant. After photos by the author (A) and J. del Hoyo (B).

peck-order. Crowing in male ring-necked pheasants has also been attributed to territoriality, although Heinz and Gysel (1970) found no evidence that it prevents other males from approaching nor did it obviously attract females. Crowing by pheasants is most often performed at dawn and dusk and especially during spring, but may occasionally be heard throughout the year. Temperature, humidity, fog, mild rain, dew, and other similar weather factors seem to have no significant effect on crowing rates in pheasants.

Warning or alarm notes take several forms in domestic fowl and pheasants. In domestic fowl the note uttered in response to hawks or other aerial predators is a loud, sustained, and raucous scream that stimulates chick to run and hide. According to Collias and Joos (1953), only adult males utter it. A "ground warning" call is an initially segmented and then sustained call that may last more than a second. A third type of warning call is a "fear squawk," which resembles the aerial predator alarm note in that it is sustained and has a mottled acoustic pattern. It is uttered by a female that is being held. In ring-necked pheasants, males utter a loud alarm call that somewhat resembles the crowing call, but is of varying numbers of syllables and is less well defined harmonically. Females utter a "brood caution" call similar to the ground-predator warning call of domestic fowl and also have a distress call that is rather high-pitched, descending in frequency, and similar to the fear squawks of domestic fowls in both structure and function.

Threat calls of male domestic fowls are fairly continuous, with an emphasis on lower frequencies; they are sometimes rather pulse-like. Likewise, male pheasants utter a similar "antagonistic" call that is a hoarse *kraah* and may be quite prolonged and pulsed. It is apparently often used by dominant males toward submissive ones (Heinz and Gysel, 1970).

An "alert" call is used by domestic fowl in response to the presence of a passing animal or a strange sound. It resembles the ground-predator call but has a more variable frequency pattern.

In pheasants, males often utter a "flight" call as they take wing to flee. Females sometimes do the same, especially those with broods. After taking flight, a female with a brood may land and begin to utter a "brood-gathering" call as she gradually returns to the point from which she took flight. This clucking or barking call evidently helps to reassemble the scattered brood.

Both sexes of pheasants utter hissing calls when they are intimidated, and "pecked" calls when they have been pecked. Female pheasants have also been heard to utter a high-pitched squeaking call when fleeing or when otherwise distressed by potential danger. Females also utter a special precopulatory call when crouching or at times also during copulation (Heinz and Gysel, 1970).

Special calls of the domestic fowl that are limited to the females include a *ku* call that is associated with feeding or mating, an aggressive call somewhat similar acoustically to part of the aggressive call of males, and a call that is given before egg-laying and in some other

situations. This same call may very occasionally be uttered by a defeated male. Brooding hens utter a considerable number of special calls. These include nest-defense calls, calls uttered before egg-laying, and those associated with agonistic encounters while sitting on the nest (Wood-Gush, 1971). Interindividual variations in calls might perhaps be especially associated with the crowing call, given its probable role in male advertisement and possible mate recognition. However, Heinz and Gysel (1970) reported more variation in male alarm calls than in crowing calls and a good deal of variation in chick calls as well. Nevertheless, female pheasants are evidently able to recognize their own mate's crowing call (Kozlowa, 1947).

It has recently been found that the male crowing calls of red junglefowl also exhibit individually distinctive acoustic characteristics. Miller (1978) reported considerable intraindividual consistency and interindividual variability in several aspects of red junglefowl crowing characteristics, including frequency modulation, number of discrete notes per call, call duration, duration of first and terminal notes, and dominant frequencies. He noted only minor differences in crowing characteristics between red junglefowl and domestic fowl, which supports the general notion of monophyletic origin of domestic fowl from the red junglefowl.

In all, Heinz and Gysel (1970) described a total of 15 different calls of ring-necked pheasants, including three characteristic of chicks, six limited to females, three limited to adult males, and two (hiss and "pecked") characteristic of adults of both sexes. Including a considerable number of calls that are limited to laying or brooding females, a total of 19 adult call types have been described for domestic fowl (Wood-Gush, 1971). Obviously, the classification and numbering of calls is greatly complicated by variations in their intensity and other difficulties of acoustical analysis. It is questionable whether such comparisons are useful at this stage. However, it is evident that some general coding characteristics of domestic fowl calls do exist. Thus, sounds that attract young chicks tend to be brief, repetitive, and contain an abundance of relatively low frequency notes. However, sounds that signal warnings to chicks tend to be of relatively long duration, have relatively little repetition or segmentation, and are weak or lacking in low frequencies (Collias and Joos, 1953). Similarly, in adult calls there may be both attracting or alerting qualities having similar acoustic characteristics; grading of acoustic signals can be attained by varying their duration, pulse rate, call loudness, and the duration of intervals between calls.

Feeding Behavior

Although it is obvious that pheasants vary individually, geographically, and taxonomically as to foods and foraging adaptations, it is difficult to generalize much on this activity. Probably all pheasants are predominantly vegetarian and the majority are adapted to seed-eating, although some species such as the koklass consume a surprisingly high proportion of green material, whereas others such as the monals obviously do a great deal of digging for subterranean vegetable matter. Studies of domestic fowl suggest that individual birds often have definite preferences for particular types of food. Evidently such individuals form preferences for foods on the basis of form and color, perhaps with the aid of tactile impressions, but with taste differences evidently of little significance. Foods with shiny or glittering surfaces may be preferred over less conspicuous types, and there seems to be an innate preference for grains having elongated rather than rounded forms. Wood-Gush (1955) has suggested that larger (unbroken) grains are preferred over smaller grain fragments and that the birds are also able to select foods that result in good diets over those that are inadequate in some dietary respect.

It is probably true that in all the Phasianidae there is a high incidence of live animal foods, especially insects, consumed during the first few weeks of life and this ratio of animal to plant foods rather quickly declines during ontogeny. To judge from information on a variety of mostly temperate-zone species of pheasants and partridges, by the time the birds are adults probably over 90 percent of food intake normally comes from plant sources; much less is known of tropical forest pheasant foods. A survey of wild pheasant diets (Dierenfeld et al., 1998) suggests that birds whose primary foods are mostly leaves (e.g., tragopans and eared pheasants) may ingest about 15 percent dietary protein, whereas highly insectivorous species (e.g., argus and peacock-pheasants) may consume up to 40 percent protein. A more detailed survey of foods and foraging adaptations of pheasants will be found in chapter 5, but it is apparent that feeding behavior in young chicks is largely related to the detection and capture of living prey, whereas in adults it is probably more dependent upon the recognition of inanimate food sources.

CHAPTER 5

Ecology and Population Biology

Relatively few detailed ecological studies have been carried out on wild pheasant species within their native ranges, and fewer still have attempted to deal with all of the pheasant species of a given region. Perhaps the best of these, and one that will provide a useful insight into such environmental controls as elevation, vegetational relationships, slope effects, and other local ecological factors, is that done by Gaston et al. (1981) in the western Himalayas of Himachal Pradesh, India. In this area of extreme northwestern India a large number of river systems flow westward out of the Himalayas, all of which eventually reach the Indus. There is also a substantial accumulation of snow, reaching as low as 1,600–1,900 m (5,250–6,235 ft) during midwinter. The mountain flora is primarily derived from the Sino-Japanese floral region, and the temperate fauna is also mainly eastern rather than western in its relationships.

The major montane vegetational types are zonally organized. The lowest of the montane forests is the subtropical pine forests, from 600 to 1,700 m (1,970 to 5,580 ft), and dominated by Chir pine (*Pinus roxburghii*). There is also a subtropical, dry evergreen forest from 500 to 1,000 m (1,640 to 3,280 ft) that is dominated by a mixture of several genera of tropical trees. The most important pheasant habitat zone is the Himalayan moist temperate forest between 1,500 and 3,000 m (4,920 and 9,845 ft), which is dominated by such temperate trees as oaks (*Quercus*), firs (*Abies*), pines (*Pinus*), horse chestnut (*Aesculus*), and deodar cedar (*Cedrus deodara*). There is also a dry phase of Himalayan temperate forest, dominated by Holm oak (*Q. ilex*) and edible pine (*P. gerardiana*), occurring at approximately the same altitudinal zone. Above these is the subalpine forest zone, from 3,000 to 3,400 m (9,845 to 11,155 ft), which is dominated by birches (*Betula*), firs (*Abies*), and pines (*Pinus*). There is also a subalpine scrub community of *Rhododendron campanulatum* at about 3,400–3,500 m (11,155–11,485 ft), finally grading to a dry alpine scrub of lower rhododendrons and junipers reaching up to approximately 3,800 m (12,470 ft).

Ecological Habitat Distributions and Population Densities

Seven species of pheasants occur within this general area (table 8), all of which are partially to totally herbivorous and variably important in ecosystem composition. The red junglefowl is largely limited to the subtropical zone below 1,200 m (3,940 ft), where it is most abundant in dry evergreen forest, especially where dense undergrowth also occurs. When it is locally protected, the Indian peafowl occurs widely on agricultural lands. Otherwise it is limited to similar habitats to those of red junglefowl, but extends locally into lower oak forests. The kalij pheasant occurs locally between about 1,200 and 2,500 m (3,940 and 8,200 ft) in forested areas along the front ranges of the Himalayas, often being most abundant in disturbed forests fairly close to human habitations. It is often common both in lower oak and lower conifer forests, with some apparent preference for the former. The cheer pheasant is typically associated with steep, grass-covered hillsides having scattered trees, especially where rocky crags are also present. The presence of dense grasses also appears to be an important habitat attribute. In some protected areas its density may be as high as 6 pairs per square kilometer (15.5 per square mile). The koklass occurs over a similar altitudinal range, but is associated with a variety of forests having well developed understories and is especially associated with oak forests. It locally reaches densities of as many as 17–25 pairs per square kilometer (44.0–64.8 per square mile). The fourth species associated with middle-altitude grasslands and temperate forests is the western tragopan, an endangered form now limited to relict areas of temperate and subalpine forests and especially to higher altitude conifer forests. The last of the Himalayan pheasants of this area is the Himalayan monal, which inhabits subalpine oak forests and reaches as high as subalpine meadows and scrub vegetation. This same zone is sometimes also used by the cheer pheasant, perhaps because of the subalpine zone's vegetational characteristics. The koklass, however, rarely reaches the subalpine zone. The major habitat difference between the monal and the koklass appears to be in the extent of ground and shrub cover needed, with most of the koklass records attained in areas having more than 70 percent ground cover, whereas less than 30 percent of the monal records occurred in such cover. Both of these species exhibit a strong seasonal shift to lower habitats between December and March, with a return upward movement in April and May. In this area slope and exposure, in so far as they affect snow cover, are important to pheasants because they not only influence food availability but also relative access by people and livestock. Thus, steep forest slopes usually support denser vegetation and perhaps also facilitate escape by downhill running. Therefore monals, for example, tend to inhabit valley slopes rather than valley bottoms (Gaston et al., 1981).

By comparison, Schäfer's (1934) description of the distribution of pheasants in western China indicates a similar vertical segregation of pheasant (and partridge) species and associated habitats, in this case involving six species of pheasants and three of partridges. Only the koklass was reported for both areas. In China that species was found in similar habitats but at approximately 1,000 m (3,280 ft) higher. Similarly, the Chinese monal occurs at a substantially higher altitude (approximately 2,000 m [6,560 ft]) in China than does the Himalayan monal in India. Interestingly, the major difference between these two species is a substantially larger body size in the Chinese form, which perhaps is an ecological adaptation for survival at these appreciably higher elevations and presumably colder temperatures.

Undoubtedly similar patterns of altitudinal zonation of pheasants occur throughout Asia,

Table 8

Comparative altitudinal distribution of Himalayan Phasianidae

Western Himalayas[a]		Elevation (m, ft)	Eastern Himalayas[b]	
Typical species and its altitude range (m, ft)	Vegetation zone		Vegetation zone	Typical species and its altitude range (m, ft)
			Alpine zone	Snow partridge (4,500–5,000, 14,765–16,405)
		4,500, 14,765		Chinese monal (4,500–5,000, 14,765–16,405)
			Subalpine scrub (Rhododendron)	Tibetan snowcock (4,500–5,000, 14,765–16,405)
		4,000, 13,125		Verreaux's monal partridge (4,500–5,000, 14,765–16,405)
				White eared pheasant (3,200–4,200, 10,500–13,780)
Himalayan monal (2,500–3,300, 8,205–10,830)	Dry alpine scrub (Juniperus)	3,500, 11,485	Holly oak, conifer forest	Blood pheasant (3,200–4,200, 10,500–13,780)
Western tragopan (2,500–2,800, 8,205–9,185)	Moist alpine scrub (Rhododendron)	3,000, 9,845		Koklass (3,000–3,200, 9,845–10,500)
Koklass (1,800–3,300, 5,905–10,825)	Subalpine forest	2,500, 8,205	Agricultural lands	Common pheasant (<3,000, <9,845)
Cheer pheasant (1,500–3,100, 4,920–10,170)	Himalayan temperate forest (moist and dry)	2,000, 6,560		Lady Amherst's pheasant (<3,000, <9,845)
Kalij pheasant (1,200–2,500, 3,935–8,205)				
Indian peafowl (500–1,900, 1,640–6,235)	Subtropical forest	1,500, 4,920		
Red junglefowl (700–1,400, 2,300–4,595)	Subtropical dry forest	1,000, 3,280		

[a] Data from Himachal Pradesh, India, adapted from Gaston et al. (1981).

[b] Data from western Szechwan, China, after Schäfer (1934).

Table 9

Comparative altitudinal distribution of pheasants on the Malay Peninsula[a]

	Elevation		Species and its altitude range (m, ft)	Estimated density (no./km², no./mi²)
	m	ft		
Scrub and disturbed habitats	Variable	Variable	Red junglefowl	Common
			Green peafowl	Extirpated
Forest habitats				
Lowland dipterocarp	0–200	0–655	Crested fireback (?–200, ?–655)	2.7–10.7, 6.9–27.6 (both sexes)
			Crestless fireback (?–200, ?–655)	0.6–6.0, 1.6–15.5 (both sexes)
			Malayan peacock pheasant (?–200, ?–655)	2.5–10.5, 6.5–27.2 (males)
Hill dipterocarp	200–1,000	655–3,280	Great argus (?–1,000, ?–3,280)	0.3–4.5, 0.8–11.7 (males)
Upper dipterocarp	1,000–1,200	3,280–3,935	Crested argus (790–1,080, 2,590–3,545)	1.9–3.0, 4.9–7.8 (males)
Oak–laurel	1,200–1,500	3,935–4,920	Rothschild's peacock pheasant (900–1,800, 2,955–5,905)	Common (?)
Montane ericaceous	>1,500	>4,920	—	—

[a] After data presented by Davison and Scriven (1987). Dashes indicate that no data were available (habitat not evaluated).

including tropical areas. One of the few studies that allows for a vegetational and altitudinal habitat analysis is that of Davison and Scriven (1987), who described the vegetational zonation and pheasant distributions typical of peninsular Malaysia. In that area eight species of pheasants and two additional partridges occur, and their altitudinal and vegetational habitat relationships are summarized in table 9. Two of the eight pheasants, the red junglefowl and the green peafowl, have generally broad distributions in Southeast Asia, but the green peafowl has apparently been extirpated from peninsular Malaysia. The climax forest vegetation of this area consists of tropical evergreen and semievergreen rainforests. The Malayan evergreen forest is characterized by an abundance of both types of shorea (*Anthoshorea* and *Rubroshorea*), high tree species diversity, and a tall forest profile with a multilayered vegetational canopy. The lowland forests consist of lowland and hill dipterocarp (Dipterocarpaceae) subtypes, the lower montane forest of upper dipterocarp and oak-laurel subtypes, and the upper montane forest of a montane ericaceous community. Three of the six species of forest-adapted pheasants are found in the lowland dipterocarp community type, including the crested fireback, the crestless fireback, and the Malayan peacock pheasant. The Malayan peacock pheasant is found on level or gently sloping ground in both logged and unlogged forests. The crested fireback is found only near moderate to large rivers in lowland dipterocarp forests developed over level alluvial riverine terraces. Thus, it has a much more restricted distribution than the peacock pheasant or the crestless fireback. The latter occurs on level, gently sloping, and steeply sloping habitats within the dipterocarp forest community. It has also been observed in both logged and unlogged sites, whereas the crested fireback has not been observed at the latter type of habitat. The great argus pheasant is found in most lowland and hill dipterocarp forest

sites, but is absent from heavily disturbed and fragmented sites. It reaches the boundary of the lower montane (upper dipterocarp) community only in one site where the crested argus is lacking, which suggests ecological exclusion. The crested argus has a much more restricted geographic distribution than does the great argus and also exhibits a curious habitat restriction to the hill dipterocarp and lower montane transitional forest, perhaps in both cases as a reflection of competition with the great argus. The Rothschild's peacock pheasant has been observed only in locations above 900 m (2,955 ft), typically on steep slopes with exposed rocks and some bamboo and climbing stemless palms. Rothschild's peacocks seem to be limited to ridge crests, where they appear to be fairly common, although no density estimates have been made.

Home Ranges and Movements

A considerable number of studies have been performed on the ring-necked pheasant as to its home range, all of which have indicated a remarkably small daily and seasonal mobility pattern (Olsen, 1977). Probably individual daily movements average no more than 0.80 km (0.50 mi) and during winter the daily movements between food sources and cover are usually less than 0.40 km (0.25 mi; Schick, 1952). Of 12 groups of wintering pheasants studied in Wisconsin by Gates and Hale (1974), only one group traveled as far as 0.80 km (0.50 mi) between food and cover, and the other 11 moved 0.40 km (0.25 mi) or less. These and other observations led the authors to believe that wintering birds rarely travel more than 0.40 km (0.25 mi) from day to day and probably never more than 0.80 km (0.50 mi). Indeed, the average distance of movement of all birds observed between 1 January and the winter breakup of flocks averaged only 0.60 km (0.40 mi; 405 observations), which closely compares to earlier studies on the species.

The onset of spring dispersal differed considerably with sex and age, with males tending to move away from wintering areas and establish breeding territories in late March or early April, whereas females remained on wintering areas until mid-April. Young females typically departed from wintering areas after the adults, whereas young males did not exhibit obvious differences from adult males. However, young males tended to disperse farther from winter cover than did adults (averaging 1.08 km [0.67 mi] versus 0.39 km [0.24 mi]), and 1 of 77 dispersed as far as 5.47 km (3.40 mi), whereas no adult male moved farther than 2.90 km (1.80 mi). Likewise, young females dispersed farther than adults (averaging 2.11 km [1.31 mi] versus 1.21 km [0.75 mi]), and 6 of 428 moved 6.44 km (4.00 mi) or more from their winter cover, whereas the maximum movement of adult females was one movement of 7.88 km (4.90 mi) from winter cover. Generally females were more prone to return to earlier nesting areas than to areas used the previous winter. Likewise, adult males remained faithful through their adult lives to the areas in which they first bred. Thus, the location of the breeding area exerted a major influence on male movements after their first spring of life, with the birds either occupying the same home range in winter or moving the minimum distance to suitable winter cover.

Spring activities in male ring-necked pheasants were found by Gates and Hale (1974) to center on very localized areas that typically were no more than about 1.30 km² (0.50 mi²) in size. More than 90 percent of their May–September observations of males were within 0.40 km (0.25 mi) of a central (16.2-ha, 40-acre) unit. Such areas have sometimes been called crowing territories or breeding territories, but few researchers have attempted to measure the degree of territorial defense or exclusion of males. Gates and Hale (1974) found that spring home ranges tended to be overlapping and daily movements of males were not confined to a particular area. No part of the home range actually seemed to qualify as a defined territory (i.e., a defended area), although the males' crowing areas most nearly approached this definition. In such areas the male did his most intense crowing, the harems were gathered within, and courtship and display were most prevalent. However, crowing areas seldom had a common boundary along which aggressive encounters between males occurred, and some males occasionally crowed when as far as 0.80 km (0.50 mi) from their regular activity center. In general, there seemed to be a moving zone of intolerance or individual distance limitations between individuals, rather than geographically definable territories. However, in this study the overall male densities were fairly low (0.003 males per hectare, 0.007 per acre) compared with earlier studies such as those of Taber (1949) and Burger (1966), in which more typical territoriality evidently occurred and higher densities were present (0.020–0.023 males per hectare, 0.050–0.058 per acre). Thus, Burger reported finding well-defined territories ranging in size from 0.5 to 4.0 ha (1.2 to 9.8 acres), and Taber concluded that territories in April were 4.9–5.3 ha (12.0–13.0 acres), although by mid-June they had compressed to about 2.4 ha (6.0 acres). Stokes (1954) reported similar male territories of 2.4–4.0 ha (6.0–10.0 acres) on Pelee Island. Although both Taber (1949) and Burger (1966) reported the presence of noncrowing and nonterritorial birds in the population, Gates and Hale (1974) did not observe any such males.

Spring and summer movements by females were similar to those of males, at least during the period May–August. Females showed no clear relationship between the location of their prenesting activities and their nest sites, and in fact the birds seemed to show some tendencies to spread out from the areas of harem activities toward more secluded sites. Females that had to nest twice as a result of nest failure remained in the same general area for their second nesting, with 11 such females moving an average of only 0.37 km (0.23 mi) between successive nestings. Likewise the average distances between successive recorded brood movements was only 0.43 km (0.27 mi), and generally the broods were raised within the home range occupied by the female during nesting and prenesting periods.

In England, Hill and Robertson (1988) found that during spring females gradually dispersed from winter flocks. Immature females dispersed farther than adults, namely an average of 309 m (1,013 ft) by nine young hens, as compared with 196 m (643 ft) by 11 older hens. Both sexes moved away from wintering habitats in woodlands into more open habitats for nesting, with the females establishing relatively small, nonoverlapping nesting ranges where they associated with males. Younger as well as older males also dispersed from their winter ranges, the younger ones moving farther than older, previously territorial males. Male breeding territories in Great Britain have been found to range from 1.8 to 4.5 ha (4.5 to 11.1 acres; unweighted mean of five studies, 2.7 ha [6.7 acres]), or comparable to the territories of 1.2–5.3 ha (3.0–13.1 acres; unweighted overall mean, 3.2 ha [7.9 acres]) observed in one Swedish and five North American investigations. In at least 8 of these 11 studies some nonterritorial males (typically younger birds) as well as predominantly territorial individuals were present in the potentially breeding male population (Hill and Robertson, 1988).

The studies of Göransson (1984) have illustrated that male common pheasants establish their breeding territories in Sweden during a 2-week spring period, and that older males return in the following year to their previous territories. Females were found to move about in the males' territories, eventually establishing a nesting range within a male's domain. Older females typically returned to their previous-year's mates and their respective territories.

About 20 percent of the yearling males failed to establish territories and constituted a "floating" male population having few breeding opportunities. Ridley and Hill (1987) found that although relative harem size was not positively correlated with the size of the male's territory, older males were more effective in attracting females than younger ones. Furthermore, females escorted by territorial males spent three times as much time foraging than did unescorted females and far less time running or being alert. The survival and reproductive advantages to females gained by associating with older, territorial males are thus apparent. Similarly, Papeschi et al. (1997) found that a male's tail length, spur length, and the vertical length of its wattle were significantly and positively correlated with its survival probabilities under natural conditions. These same traits were positively associated with a male's incidence of territorial behavior. However, the male's body weight, wing length, and tarsal length did not exhibit such correlations.

In a radio-telemetry study of the Cabot's tragopan, Ding (1995) found that home ranges increased in spring among the 11 birds he studied. Dominant males had stable home ranges, whereas females wandered over large areas, sometimes associating with a male within his range. Female home ranges were largest during the prebreeding season and decreased during nesting, only to increase again following hatching.

Somewhat comparable kinds of information are available for the red junglefowl, based on studies of free-ranging birds in the San Diego Zoo. Collias et al. (1966) reported that during their study there were in the zoo ten major roosts, each having from 6 to 30 birds, as well as 18 additional minor groups. These birds were highly fixed as to their home canyon, with only about 1 percent ever ranging out of the canyons in which they were hatched. Typically each flock might range 55–73 m (180–240 ft) from its roosting site up canyon slopes or into branch ravines, but only short distances of 30–46 m (100–150 ft) down the main canyon, which was in the direction of neighboring flock territories. Banded individuals moved an average distance of only 76 m (248 ft) over a 3-year period, with females not exhibiting obviously greater movements than males nor young birds substantially greater movements than adults. Within the flock's home range individual males were organized in dominance hierarchies. Dominant males tended to move outside the flock's overall range more than did subordinate ones. Females also exhibited a considerable degree of site specificity and reluctance to move much beyond their own flock's normal home range. Dominant males typically associated with one to several females, whereas subordinate males were often excluded from contact with the females. The diameter of the defended area where the birds were most often concentrated typically averaged only 30–46 m (100–150 ft). A few birds that were experimentally moved 305–427 m (1,000–1,400 ft) exhibited no indication of any homing abilities. This high level of locality fixation and territorial attachment is perhaps a hindrance to the species' capacity for range expansion. It is unknown whether significant seasonal movements occur among red junglefowl in their native habitats. Collias and Saichuae (1967) observed that in Thailand separate roosting flocks were usually situated from 0.25 to 0.50 km (0.16 to 0.31 mi) apart, but some limited evidence there suggests that seasonal movements of at least several kilometers might occur in southern Thailand. However, in west-central Thailand even greater seasonal movements of several miles might occur (Johnson, 1963). Studies in north-central India indicate very restricted daily movements of some flocks, which often restricted their activities to an area only about 137 m (450 ft) in diameter, although a few birds were observed to move as much as 275 m (900 ft)in a relatively short time (Collias and Collias, 1967).

Competition and Predators

Little can be said of competitive interactions between species of pheasants in the wild. The earlier discussion of habitat affinities in areas of high species density would indicate that even where six or more species of pheasant occur in geographic sympatry there is a high degree of habitat specificity and probably relatively few direct interspecific interactions. However, in some cases, such as where the great argus and crested argus occur sympatrically in Malaysia, there may indeed be some local interactions and competitive exclusion.

Similarly, the role of predation is extremely difficult to determine in influencing bird distribution and abundance. In spite of hundreds of field studies on ring-necked pheasants in North America, little can be said with certainty about the role of predation on pheasant populations. One of the best such studies was that of Wagner et al. (1965) using females that had been equipped with radio transmitters. Among 105 deaths of such birds, 76 (80.8 percent) of the total were attributed to predation. Of these 76 deaths, 46 (60.6 percent) were believed to be the result of mammalian predation, 19 (25.0 percent) caused by birds, and 11 (14.4 percent) were of undetermined cause. By comparison, Schick (1952) attributed 77 of 214 (36 percent) pheasant kills found during a 2-year period in Michigan to mammals and 81 (38 percent) to raptors. Gates (1971) attributed 95 of 194 (49 percent) cases of winter pheasant predation in Wisconsin to avian predators and only 66 (34 percent) to mammalian predators. Both Schick and Gates believed foxes to be responsible for the majority of mammalian predation cases, whereas smaller numbers were considered to result from cats, dogs, minks, and weasels. Generally, losses to foxes are believed to be heaviest during winter, although foxes also take some juveniles and adults in summer as well. Among avian predators, Wagner et al. (1965) implicated the great horned owl (*Bubo*

Table 10
Some estimates of major pheasant predators and their effects[a]

| | Results of stomach, pellet, or scat analysis | | | Population estimates | |
| | % of samples having pheasant remains | | % of studies listing pheasant remains | Predator density (no./km^2, no./mi^2) | Annual pheasant loss from predators |
	Range	Average[b]			
Mammals					
Red fox	0–65	8.9	72.7 (22 studies)	0–1.9,[c] 0–4.9[c]	8–14%[c]
Grey fox	0–13	4.0	62.5 (8 studies)		
Raptors					
Cooper's hawk	0–1	18.1	85.7 (7 studies)	—	1%
Great horned owl	0–41	10.5	81.8 (11 studies)	0.1–1.1, 0.2–2.8	—
Red-tailed hawk	0–31	5.3	57.1 (7 studies)	0.4–1.2, 1.0–3.0	—
Northern harrier	—	—	—	—	1–10%
Overall	—	—	—	—	9%

[a] Data derived from various studies cited by Wagner et al. (1965). Dashes indicate that no data were available.

[b] Unadjusted for sample size differences.

[c] Includes both species of fox.

virginianus) as a significant source of mortality, whereas Gates (1971) found that in his area the red-tailed hawk (*Buteo jamaicensis*) was responsible for the largest percentage of avian losses. Schick (1952) believed that in his study area the Cooper's hawk (*Accipiter cooperi*) and northern harrier (*Circus cyaneus*) were the most significant sources of predation. Based on information presented by Wagner et al. (1965), table 10 provides some estimates of major known predators of ring-necked pheasants in North America and their possible population effects. Nest predators are not included in this summary. Wagner et al. concluded that predation was the most important direct cause of adult pheasant mortality in their study and that it was likely to be a limiting factor for pheasant abundance in Wisconsin.

Mortality and Survival Rates

Only a few observations have been made on the mortality and survival rates of pheasants other than the ring-necked pheasant. In their study of free-ranging red junglefowl in the San Diego Zoo, Collias et al. (1966) observed a surprisingly high mortality rate. They observed that, of 42 birds banded as adults, 32 (76.2 percent) survived no more than 12 months and only 10 (23.8 percent) survived at least a year. Of 46 birds banded when 3–5 months of age, only 5 (10.9 percent) survived at least a year. Generally, of every 100 chicks that hatched, only 25 percent survived to the time of brood breakup and only about 6 percent survived to the end of their first year of life. Only three banded individuals were known to have survived more than 3 years after banding; these birds must have lived at least 4 years, because they were all banded as adults. Foxes, feral cats, and Cooper's hawks were all considered as possible causes of junglefowl mortality in that area.

A substantial amount of information is available on mortality rates in wild ring-necked pheasant populations, a sample of which is summarized here. Using a variety of methods of calculating mortality rates, Gates and Hale (1974) reported that adult survival rates among wild female ring-necked pheasants ranged between 11 and 33 percent, with fall-to-spring survival between 27 and 64 percent and spring-to-fall survival between 34 and 58 percent. Annual estimated survival for adult females was 24 percent. Male survival varied from 3 to 14 percent annually and averaged only 7 percent. There was little evidence of age-specific survival change; thus, life expectancies did not change after the first autumn of life. These mortality rates are higher than were estimated in various earlier Wisconsin studies (Buss, 1946; McCabe, 1949) and also tend to be somewhat higher than rates reported from other states.

Probably one of the best studies of ring-necked pheasant mortality is that of Dumke and Pils (1973), which is based on radio-tagged birds. Too few males were tagged to provide useful mortality data, but 137 mortality records for females were obtained. These data for females are summarized in table 11, which indicate an annual mortality rate of 73.9 percent (annual survival rate of 26.1 percent) or extremely close to the figures cited earlier by Gates and Hale. This study indicated that female mortality rates are higher in the spring-to-fall period than between fall and spring and that they are highest during the hunting season (females are illegal game) and during the winter period. Lowest seasonal mortality rates occur during the nesting, brood-rearing, and postbrood-rearing periods. However, during that same period, losses of brood members are relatively high. Baxter and Wolfe (1973) summarized data on brood losses during the first 6–10 weeks after hatching, and noted that chick mortality rates for eight different study areas ranged from 30 to 56 percent (unweighted average 38.2 percent), even though none of these studies was able to pinpoint the primary causes of such high chick

Table 11

Estimated annual and seasonal mortality rates of female ring-necked pheasants[a]

	Mortality rate (%)
Annual Mortality	
Overall female population	73.9
Radio-equipped females ($N = 201$)	79.7
Adult females ($N = 43$)	67.3
Juvenile females ($N = 132$)	82.4
Semiannual Mortality	
Fall to Spring (1 Oct–31 Mar)	
Overall female population	35.3
Radio-equipped females only	57.8
Spring to Fall (1 Apr–30 Sep)	
Overall female population	57.5
Radio-equipped females only	44.8
Seasonal Mortality of Radio-equipped Females	
Hunting Season (16 Oct–14 Dec)	
Juvenile females	34.6
Adult females	20.5
Early Winter (15 Dec–18 Feb)	
Juvenile females	36.0
Adult females	31.5
Late Winter (19 Feb–14 Apr)	
Juvenile females	9.6
Adult females	28.2
Nesting Season (15 Apr–28 Jun)	
Juvenile females (yearlings)	14.9
Adult females (>1 year old)	14.5
Brood-rearing Season (29 Jun–27 Aug)	
Juvenile females (yearlings)	4.7
Adult females	5.2
Post–Brood-rearing Season (28 Aug–15 Oct)	
Juvenile females (yearlings)	0.2
Adult females	0.0

[a] Data adapted from Dumke and Pils (1973).

mortality. Some studies have implicated high mortality rates with late hatching dates, but others (e.g., Baxter and Wolfe 1973) have suggested that late-hatched broods actually suffer somewhat lower mortality rates, perhaps because of the more favorable late summer weather and better brooding conditions at this time than earlier in the nesting season.

Wagner et al. (1965) extensively reviewed a variety of factors that might influence ring-necked pheasant reproductive behavior and reproductive success, including weather and climate, farming practices, predation, legal and illegal hunting of females and males, and population density. They concluded that, at least in Wisconsin, populations are generally in balance with the local temperature norm and that yearly springtime positive deviations from that norm tend to be accompanied by population increases. Prenesting temperatures are probably the most important single weather factor, but may not be the only control. In some parts of the range the severity of the winter may also play an important controlling role. Among farming practices, it was concluded that losses of both nests and nesting females from hayfield mowing are often important and locally may cause losses of at least one-fifth of the nests and nesting females. In areas where haying is important, access to available wetlands for nesting is especially important, because such areas tend to attract enough females to reduce significantly female mortality and increase nesting success. Information on predation effects was inconclusive, but it was suggested that at least in marginal range predators might influence pheasant populations significantly. Likewise, legal hunting of females was considered to exert an important effect on female losses in the population, and even those female losses resulting from illegal hunting might have some depressing effects. Normal hunting harvests of males were not believed to be significant in population control. The major miscellaneous mortality factor was apparently mortality caused by traffic accidents, which perhaps represented a loss of 3–4 percent of the total fall population. Much of this loss occurs in spring; thus, its effects are magnified by the loss of potentially breeding birds.

The effects of population density of females in influencing reproductive rates was reviewed at length by Wagner et al. (1965), who found evidence that in Wisconsin as well as in several other states the reproductive potential decreases as population densities increase. They suggested that this might result from competition for space, which was the only definite environmental population-control mechanism having density-dependent characteristics that they were able to identify. They concluded that density-dependent effects operate at all population levels, rather than only above certain threshold levels, but that they only influence actual population density in combination with other limiting factors such as weather. Thus, weather may well be the most important overall annual determinant of population productivity in Wisconsin pheasant populations. However, the local productivity tends to oscillate around a mean that is dependent upon local density and thus results in compensatory population adjustment. In the view of Wagner et al. (1965), mean densities in different areas are probably the result of variations in density-independent mortality factors, whereas temporal variations in a given area are the result of a combination of density-dependent and density-independent effects.

Foods and Foraging Ecology

Foraging by all pheasants is done primarily on the ground, and their stout toes, strong claws, and sharp bills are well adapted for digging and scratching. There is some variation in bill shape among the pheasants. The monals, for example, have unusually well-developed bills for digging out foods from below the ground surface. Roots, bulbs, tubers, subsoil insects, and other invertebrates all are likely to be excavated and consumed in this manner. A few pheasant species, such as the blood pheasants, have relatively weak bills and probably primarily consume greens and fruits, although the koklass has a considerably heavier bill but also evidently requires a high proportion of green materials. The tragopans similarly seem to be

Table 12

Major seasonal foods (percent composition of diet) of ring-necked pheasants in Missouri and South Dakota[a]

	Missouri[b]					South Dakota[c]				
	Autumn (N = 114)	Winter (N = 74)	Spring (N = 85)	Summer (N = 45)	Overall (N = 319)	Autumn (N = 206)	Winter (N = 213)	Spring (N = 288)	Summer (N = 167)	Overall (N = 874)
Farm crops										
Corn (*Zea*)	66.9	72.1	65.0	29.5	64.8	67.3	67.2	37.3	25.8	57.2
Wheat (*Triticum*)	1.4	—	tr.	19.0	2.0	4.2	6.9	16.4	27.9	10.7
Soybeans (*Glycine*)	10.8	1.5	—	—	—	—	—	—	—	—
Sorghum (*Sorjum*)	4.0	0.6	tr.	tr.	1.6	0.6	1.5	0.7	0.3	0.7
Rice (*Oryza*)	—	tr.	0.1	14.7	1.3	—	—	—	—	—
Barley (*Hordeum*)	tr.	tr.	0.3	—	0.1	4.4	5.4	11.4	11.9	6.6
Oats (*Avena*)	—	—	—	—	—	1.7	2.4	8.1	14.1	3.9
Flax (*Linum*)	—	—	—	—	—	1.0	1.5	0.7	0.3	1.2
Millet (*Setaria*)	—	—	—	—	—	0.1	0.3	1.2	1.6	0.7
Other crops	—	—	—	—	—	1.5	0.6	0.3	0.8	0.7
Overall	86.2	75.2	65.1	63.3	73.0	80.9	86.0	75.8	79.1	81.7
Noncrops										
Foxtail (*Setaria*)	3.6	0.9	0.9	21.5	3.4	3.4	0.9	1.3	4.1	2.7
Dayflower (*Commelina*)	0.7	7.9	17.5	—	8.4	—	—	—	—	—
Bur-cucumber (*Sicyos*)	0.4	4.3	2.6	0.3	2.4	—	—	—	—	—
False buckwheat (*Polygonum*)	0.2	2.6	1.5	tr.	1.4	—	—	—	—	—
Sunflower (*Helianthus*)	1.2	tr.	0.1	tr.	0.3	4.5	0.2	0.1	1.7	2.4
Miscellaneous seeds	5.8	8.4	9.8	8.0	8.3	1.9	2.6	0.7	1.5	2.0
Overall	4.9	15.6	22.5	21.8	15.6	5.2	1.8	12.8	9.3	5.4
Miscellaneous foliage	—	—	—	—	—	4.5	6.8	6.5	3.7	4.5
Animal materials										
Grasshoppers (Orthoptera)	1.9	tr.	tr.	1.0	0.6	2.6	1.2	1.1	4.7	2.3
Ground beetles (Carabidae)	0.6	tr.	tr.	0.3	0.2	0.1	0.1	2.3	0.7	0.5
Caterpillars (Lepidoptera)	—	tr.	0.1	0.2	0.1	—	—	4.2	0.2	0.7
Crickets (Gryllidae)	—	—	—	—	—	0.4	tr.	0.3	1.4	0.5
Ants (Formicidae)	tr.	tr.	0.4	1.2	0.2	tr.	—	1.7	0.4	0.3
Other animal materials	tr.	tr.	0.4	1.9	0.7	0.3	0.6	3.2	1.8	1.1
Overall	2.5	tr.	0.9	4.5	1.5	3.5	1.8	12.8	9.3	5.4
Mineral matter (grit)	0.1	0.2	0.4	1.5	0.4	0.5	1.6	2.6	0.5	1.1

[a] *N* is the number of birds sampled; a dash indicates that no data were available; "tr." denotes trace amounts.

[b] Data adapted from Korschgen (1964).

[c] Data adapted from Trautman (1952). Monthly means have been converted to seasonal means without adjustment for minor sample size variations.

relatively dependent upon fruits and berries, but the peacock pheasants appear to have a diet that is relatively high in insects.

Although a limited amount of information is available on the diets of junglefowl based on analysis of crops and gizzards, only the common pheasant has been extensively analyzed as to variations in diet associated with age, season, and geographic locality. Cramp and Simmons (1980) and Glutz (1973) have recently summarized the information on foods of pheasants in Europe. Their reports make it clear that

Table 13

Major foods (percent composition of diet) of juvenile ring-necked pheasants[a]

	Ontario sample, by age of birds (in weeks)[b]					South Dakota sample, by month[c]			
	1–3 wk	4–6 wk	7–9 wk	10–12 wk	Overall	Jul	Aug	Sep	Overall
	(N = 61)	(N = 82)	(N = 90)	(N = 18)	(N = 251)	(N = 57)	(N = 67)	(N = 72)	(N = 196)
Animal materials									
Arthropods									
Ephemerida	41.3	31.4	5.4	0.2	19.4	—	—	—	—
Orthoptera	0.6	1.2	1.7	1.9	1.3	36.3	27.3	18.8	23.3
Trichoptera	4.3	3.3	2.1	4.3	3.5	—	—	—	—
Lepidoptera	14.0	3.0	1.0	0.4	4.6	tr.	0.2	0.5	tr.
Coleoptera	1.2	0.4	0.4	0.4	1.8	tr.	1.0	3.0	1.9
Others	34.1	7.1	2.8	0.9	10.2	tr.	6.6	0.2	3.0
Molluscs	2.5	0.4	0.8	0.4	1.0	—	—	—	—
Overall	98.0	46.8	14.2	8.5	41.8	36.3	35.1	22.5	28.2
Plant materials									
Grains	1.8	35.1	39.7	45.6	30.5	50.7	48.8	51.3	50.3
Seeds	0.2	6.4	19.1	22.3	12.0	3.4	13.7	34.3	17.5
Fruits	—	9.4	19.5	16.8	11.4	—	—	—	—
Foliage	tr.	2.2	7.5	6.8	4.1	9.2	1.6	1.2	2.8
Overall[d]	2.0	53.1	85.8	91.5	58.1	63.1	64.1	76.8	70.6

[a] N is the number of birds sampled; a dash indicates that no data were available; "tr." denotes trace amounts.

[b] Data derived from weekly totals of Laughrey and Stinson (1955).

[c] Data adapted from Trautman (1952).

[d] Mineral matter (grit) not included.

cultivated grains (wheat, oats, barley, maize), seeds, nuts, and fruits are all favored year-long foods, whereas greens are often taken in spring and arthropods are especially important during summer months. During summer in Europe, up to half or even more of the diet may be composed of animal materials, but plant materials clearly dominate only during the winter months. Most studies in North America (e.g., Olsen, 1977) suggest that cultivated grains, primarily corn (maize) and secondarily wheat, oats, and barley, play a much higher role in the dietary economy of pheasants than in Europe; animal materials also seem to be consumed in appreciably lower amounts. Olsen (1977) has summarized a large number of food-intake studies of ring-necked pheasants in North America. He reported that there animal materials make up relatively small percentages of the annual diet, including an estimated 1.5 percent in Missouri (Korschgen, 1964), 3.2 percent in Michigan (Dalke, 1937), 3.9 percent in Minnesota (Fried, 1940), 5.0 percent in South Dakota (Trautman, 1952), and 14.5 percent in Utah (Cottam, 1929). In most of these studies grasshoppers (Locustidae) were found to be the most important single source of animal foods. Two studies on the seasonal foods of ring-necked pheasants in North America are summarized in table 12. Studies in all areas indicate that foods of chicks and juveniles are much higher in animal materials, particularly during the first few weeks of life. During the first week or so insects may represent as much as 80 or 90 percent of the chicks' food, but between 4 and 6 weeks after hatching their diet shifts strongly toward vegetable matter, with grain often becoming a major food item (table 13).

CHAPTER 6

Comparative Mating Systems and Social Signaling Devices

Like other vertebrate species, pheasants have evolved reproductive "strategies" that provide for efficient modes of reproduction under the diverse ecological and social conditions existing throughout their broad range. These include varied mating systems, such as differences in male participation in postfertilization aspects of reproduction, with consequent variations in pair-bonding strengths and the associated potential for nonmonogamous pairings. In turn, these variations have influenced the kinds and complexities of social signals that have evolved and are used to facilitate reproduction among pheasants.

Mating Systems

Reproductive strategies among the pheasants are in part a reflection of the relative antipredator importance of maintaining a gregarious social structure through most of the year (which is typically the case with smaller, more vulnerable species such as blood pheasants and the closely related partridges) versus the need or opportunities for maximum spacing owing to limited environmental resources and/or relative security against predators. Generally among pheasants there is a fairly close correlation between plumage dimorphism and reproductive strategy. Species that are sexually monomorphic tend to be monogamous and often have somewhat extended pair bonds. However, those that are strongly sexually dimorphic both in overall body size and plumage development tend to be polygynous or promiscuous, forming either male-dominated harems if maintaining their sociality through much of the year or establishing individual male display areas that individual reproductive females may visit.

In the pheasants, as in the quails and grouse (Johnsgard, 1973, 1983b), the general trends in reproductive strategies seem to have been ones that resulted in evolutionary shifts from earlier monogamous pair-bonding systems to polygynous or promiscuous ones whenever ecological conditions have allowed. The evolution of nonmonogamous mating has probably occurred more than once among pheasants, as I suggested earlier (Johnsgard, 1973, p. 122). The pheasant group may thus actually be of polyphyletic origin and derived from those partridge species that have for the most part abandoned monogamous mating tendencies for polygynous or promiscuous ones.

There is recent biochemical evidence supporting the idea of a polyphyletic origin of the pheasants from various perdicine ancestors. Fumihito et al. (1995) found mitochondrial DNA evidence supporting the proposed dendrogram of the pheasants shown in figure 1 (see chapter 1), namely that at least three distinct phyletic branches of pheasants exist. One such branch comprises the single junglefowl genus *Gallus,* one contains the peafowl and great argus (other potential group members such as peacock pheasants, crested argus, and Congo peafowl were not tested), and a third branch includes the genera *Phasianus, Chrysolophus,* and *Syrmaticus,* with *Lophura* being a more distant subsidiary branch. The remaining major phyletic group recognized in this book, including the tragopans, monals, and koklass, was not represented in the study. Fumihito et al. (1995) also biochemically linked *Gallus,* and less directly the peafowl, with the seemingly ancestral perdicine genus *Bambusicola.*

Ridley (1987) has summarized a variety of ecological aspects of pheasant relative to their mating systems and made a number of interesting observations and conclusions. He classified each of the pheasant genera as being either solitary or gregarious in social organization and their mating systems as monogamous, polygynous, or promiscuous. In his view, relative sociality in pheasants is largely related to habitat, because pheasants living or foraging in open habitats are gregarious due to higher predation risks. Such species do not compete strongly for food during the breeding season because open-country species tend to be mostly herbivorous, whereas forest-dwelling pheasants are typically solitary, apparently to avoid severe competition for limited food sources.

According to Ridley (1987), the habitat also plays a strong role in shaping pheasant mating systems. Among species that forage in open habitats the females are typically guarded by the males, thus such species tend toward the establishment of longer pair bonds. In edge habitats harem polygyny is characteristic, whereas in tropical forests the usual mating system is one of serial polygyny or promiscuity, the latter situation being promoted by the tendency toward asynchronous breeding cycles in tropical environments. Ridley considered the genera *Ithaginis, Catreus,* and *Crossoptilon* to be characteristic of scrub and open grassland. All of these exhibit whole-season pair-bonding and a lack of gregariousness among females. Genera he considered as forest-edge or scrub-edge species that forage in the open included *Gallus, Phasianus, Syrmaticus, Lophophorus,* and *Pavo.* In all of these genera the females tend to be gregarious and pair-bonding may last as long as the entire season (*Gallus*) until incubation begins (*Phasianus, Syrmaticus* and *Lophophorus*) or only until fertilization (*Pavo*). The montane forest pheasant genera include *Pucrasia, Tragopan,* and *Chrysolophus.* In all of these the females are solitary, with pair-bonding lasting only to incubation or (in *Chrysolophus*) only through fertilization. Lastly, the tropical forest genera include *Lophura, Polyplectron, Argusianus, Rheinartia,* and *Afropavo.* In all of these the females are nongregarious. In all but *Lophura* and probably *Afropavo* the mating systems are polygynous. *Afropavo* is exceptional in that apparently monogamous pair-bonding occurs and it evidently lasts through the breeding season. Because *Afropavo* is considered a generally very primitive form, this is presumably the result of a retention of this original trait, rather than a secondarily acquired characteristic.

The limited movements of most pheasants probably facilitate some remating with mates of previous years, although in at least some species (e.g., the common pheasant and the red junglefowl) the high mortality rates among adults, especially males, are seemingly so severe that in relatively few instances would the same mates be available in successive years. In the forest-dwelling species loud male calling and the use of traditional territorial display areas probably serve as important cues for females to locate sexually active males. Apparently only in peafowl (*Pavo*) has a trend toward social display evolved that is somewhat comparable to the lek display grounds of various grouse, in which several males gather in close proximity for display and the female may select a single male from the group for mating (Morris, 1957).

Social Signaling Systems and Sexual Dimorphism

It is generally true that sexual dimorphism in the broad sense (including color and plumage dimorphism, or "dichromatism," and behavioral dimorphism, or "diethism") tends to be highest in polygynous or promiscuous species of vertebrates. This is in line with generally accepted ideas on the effects of sexual selection on social signaling systems as related to their differential intersexual attraction and intrasexual dominance/fitness gradients. A summary (table 14) of the situation in the 16 pheasant genera provides a limited degree of support for this general position. For example, it is true that the three genera of pheasants believed to have season-long monogamous pair bonds all exhibit rather limited sexual dichromatism (and always are lacking in male iridescence) and have very low male : female adult weight dimorphism.

Among the polygynous and edge-adapted species the pattern is less clear. In this group the degree of sexual dichromatism is generally high, as might be expected both by the pair-bonding system and the ecological influences favoring bright male signal coloration in an open environment. However, the range of estimated male : female weight ratios is rather high, from 1.1 to 1.7, and indeed seems to be highest in the forms with the least dichromatism and lowest in those with the greatest dichromatism. Perhaps in some of these forms there has been sexual selection favoring large male size and associated territorial or peck-order dominance, whereas in others such as the monal the primary selective influence may have been on the development of sexually attractive male signaling devices and behavior.

The two apparently monogamous forest-dwelling pheasants, the Congo peacock and koklass, pose some uncertainties. Both show limited amounts of sexual dimorphism and both generally have dark plumage coloration, with no brilliant facial skin engorgements, combs, or wattles. Their degree of territoriality is unknown, but neither species is believed to defend and advertise large territorial areas. The length of the pair bond is still uncertain in the Congo peacock and information on its adult weight ratio is still inadequate.

Among the polygynous or promiscuous forest-dwellers there are a few surprises. The ruffed pheasants seem to have a surprisingly small disparity in male : female weight ratios, which appear to be no greater than those of the monogamous species. However, their actual degree of polygyny and length of contact with females under natural conditions is still uncertain. The peacock pheasants likewise have a considerable diversity in plumage dichromatism, being slight in the bronze-tailed pheasant in particular and perhaps highest in the Palawan peacock pheasant. Quite possibly this group also exhibits variations in pair-bonding tendencies that have yet to be determined in the field. Certainly the crested argus, great argus, and peafowl exhibit the combination of extreme male : female weight ratios, male

Table 14
Relationship of pair-bonding systems to sexual dimorphism in pheasants

	Extent of sexual dichromatism	Male : female weight ratio[a]	Length of pair bond[b]
Monogamous open-country or scrub species[b]			
Blood pheasant	Moderate	1.1	Entire breeding season
Eared pheasants	Slight	1.1–1.2	Entire breeding season
Cheer pheasant	Slight	1.2–1.3	Entire breeding season
Polygynous edge-adapted species			
Junglefowl	Moderate	1.4–1.7	Entire breeding season (harem polygyny)
Typical pheasants	Moderate	1.2–1.6	To incubation (harem polygyny?)
Monals	High	1.1	To incubation (territorial polygyny?)
Long-tailed pheasants	Moderate	1.3–1.5	To incubation (territorial polygyny)
Monogamous forest species			
Congo peacock	Slight	1.2	Entire breeding season
Koklass	Moderate	1.3	To incubation (?)
Polygynous/promiscuous forest species			
Tragopans	Moderate	1.0–1.7	To incubation (?)[c]
Peacock pheasants	Variable	1.3–1.4	To incubation (?)[c]
Gallopheasants	Moderate	1.1–1.4	To incubation (?)[c]
Ruffed pheasants	High	1.1	Mating only (?)
Crested argus	High	?	Mating only
Great argus	High	1.4–1.5	Mating only
Peafowl	High	1.5	Mating only

[a] Based on weight data in the species accounts.

[b] In part after classification of Ridley (1987).

[c] Some species are seemingly monogamous, and the tragopans are of uncertain status.

plumage elaboration, and highly elaborated male displays that are to be expected in promiscuous species. Adult weight data are still unavailable for the crested argus, but in the other species the females tend to be only about two-thirds the size of the males, or approaching the 2:1 ratio that is found in some large lek grouse such as sage grouse and capercaillies (Johnsgard, 1983*b*).

Evolution of Specific Male Signaling Devices

The pheasant group is relatively consistent in its use of only a relatively few types of signals for sexual attraction and/or male-male agonistic interactions. These signals have been briefly mentioned in earlier chapters, and at this point

it is perhaps of interest to concentrate on a few that appear to be of virtually universal occurrence and importance in the pheasants. These include tidbitting, waltzing, and double wing-lowering, or frontal display. Other displays, such as facial engorgement, tail-shaking or head-shaking, and wing-flapping, are of relatively similar form throughout the group and their differential development patterns are fairly self-evident.

Tidbitting is an extremely important social display in pheasants. As noted by Schenkel (1956–1958), it is of special interest because in most pheasants the same or similar behavior serves as an important food-showing device between females and their chicks, whereas in adults a ritualized version of this activity becomes a releaser for female sexual behavior or at least directs her attention to males that are performing it. In some species the original food-showing function completely disappears, as in *Pavo,* where the display occurs in the absence of actual food. There the function of the behavior seems to be largely concerned with "luring" the female into a favorable position so that the maximum visual effect of the males posturing can be projected toward the female. Tidbitting behavior is widespread in the Galliformes (Stokes and Williams, 1972).

The ambivalent motivational origin of the waltzing display has already been discussed, at least with reference to the red junglefowl; it is highly likely that this explanation is of general applicability among pheasants. The nearly universal occurrence of waltzing in pheasants helps to account for some aspects of their male plumage signals, such as the great visual importance of the upper back and rump coloration, which is typically exposed during waltzing. Likewise pheasants show a tendency toward lateral tail-spreading, but usually not the extreme tail-cocking that often occurs in grouse, which is associated with frontal rather than lateral display. However, underpart coloration in pheasants is less likely to be brilliant. Also, bright patterning or coloration is more likely to occur on the wing-coverts than on the remiges themselves, because typically it is only the primaries on the side opposite the "target" bird that droop fully to the ground and thus do not provide great opportunities for concentration of effective visual signals.

Although in the gallopheasants, ruffed pheasants, typical pheasants, and long-tailed pheasants the predominant male orientation during intense display is lateral, an interesting trend away from a lateral display and toward a frontal orientation becomes evident in the peacock pheasants and reaches its developmental peak in the great argus and typical peafowl. In the bronze-tailed pheasant, for example, the major male sexual display is a stationary lateral waltzinglike orientation with strong lowering of the near wing but almost no tilting of the tail and upper back toward the female. In that species iridescent color is wholly lacking on the back and only poorly developed on the tail. In the Rothschild's peacock pheasant a much stronger degree of back-tilting and tail-spreading is apparent and there is a correspondingly greater development of iridescence on the back, wing-coverts, and tail. In the grey and Germain's peacock pheasant a shift toward a frontal orientation during intense display is evident. In these species the tail is relatively more rounded than pointed and the entire dorsal body and wing-covert surface is studded with iridescent ocelli. The Bornean and Palawan peacock pheasants also exhibit rounded tails and somewhat intermediate postural tendencies. The Bornean species has an extremely well-developed lateral display and only rarely uses a full frontal display (judging from limited observations). Similarly the Palawan peacock pheasant has evolved a spectacular lateral display with extreme tail-spreading and tail-tilting that emphasizes both the dorsal body and tail coloration as well as retaining a clear view of the side of the head and eye, which form a point of strong visual focus for the overall display. In the great argus sexual display becomes fully frontal in its orientation. However, in this species the head is oriented laterally and the eye remains visible, again being situated in the optical center of a spectacular series of radiating "eyes" that extend radially outward along the ornamental secondary feathers. Finally, in the peafowl the role of the wings is taken over by the tail-coverts, which form a similar breathtaking array of "eyes" that radiate out from the head and iridescent mantle during full frontal display. The plain-colored remiges and rectrices are completely hidden between the train of iridescent tail-coverts, although periodic vibration of the rectrices produces a shimmering optical effect on the iridescent train and the primaries are rhythmically lowered to the ground and are sometimes scraped along the ground surface.

Several recent studies have dealt with the question of whether individual male traits can influence mating choices in females. Grahn (1992) studied mate selection in a group of free-ranging pheasants over a 4-year period. He found that male spur length, but not male dominance or relative territory quality, influenced mate choice by females. Males were divided into three experimental groups, those with their spurs shortened, those with artificially increased spur lengths, and a control group with normal, unmodified spurs. It was found that males with the greatest spur lengths not only had more surviving offspring (with increased spur length statistically accounting for 15 percent of overall female productivity), but also these same birds maintained their territories later in the breeding season, thereby attaining a longer potential reproductive period. Additionally, spur length in unmodified males in another study proved to be the best predictor of harem size and similarly was the best single predictor of overall male reproductive success (Göransson et al., 1990). Long-spurred males also survived better during the breeding season (Grahn, 1993). However, when short-spurred males were released and allowed to establish territories 3 weeks in advance of long-spurred males, fewer of the long-spurred males were able to establish territories than did short-spurred ones, suggesting the importance of early territorial establishment. The two groups also did not differ in their relative abilities to attract females (Grahn et al., 1993a), which suggests that perhaps females are assessing territorial quality as well as specific male attributes when selecting mates. Studies of differential male attractiveness to females (based on average harem size beginning with the latter half of the egg-laying period and extending through the subsequent breeding stages) indicated that the more attractive males exhibited an overall higher degree of territoriality, in that

they made shorter daily movements, had smaller home ranges, and crowed more frequently than did less attractive males (Grahn et al., 1993b). Continued work on this population using DNA fingerprinting has further established a positive relationship between long-spurred males and improved reproductive success among those females that mated with them (Schantz et al., 1994).

In a related Swedish study, Wittzell (1991) also manipulated spur length and found that it was the most important predictor of male success, as measured by average harem size and number of reared offspring, although it did not correlate significantly with relative male dominance. In contrast, using both captive birds and models, Mateos and Carranza (1995) reported that male tail length, length of ear tufts, and the presence of black spots on the wattle all positively influenced female choice of mates, but wattle size and a male's relative plumage brightness did not. Mateos and Carranza (1996) also reported that females showed no clear preference for those males having artificially lengthened spurs, although among adult males spur length was positively correlated with both wattle display performance and relative social dominance. Furthermore, in male-male hostile interactions, ear tufts and head wattles were the most important social signals in informing other males of an individual's relative willingness to fight, his fighting ability, and his resource-holding abilities. Mateos and Carranza (1997a,b) believe that, instead of being primarily used by females as a basis for mate choice, the bright and colorful aspects of male plumage allow males to assess the quality of their competitors.

In studies with red junglefowl, Sullivan (1992) reported that mate choice among females did not correlate with male morphology attributes or relative dominance. Rather he judged that the extended period of social contact between the sexes in this gregarious species allows females to make desirable mate choices independently of specific male traits. Similarly, Ligon and Zwartjes (1995*a*) found that in a group of red junglefowl where competing males had wild-type plumage or represented either of two other mutant types, the wild-type birds were not at a measurable sexual advantage. However, females did strongly prefer males with large combs. In a related study, Ligon and Zwartjes (1995*b*) reported that female red junglefowl chose to mate with both of two males when such were available, regardless of their possible plumage differences. This was true even when the males had comb sizes that were markedly different. Ligon et al. (1998) reported that female red junglefowl are insensitive to asymmetry of paired ornaments in males and that male comb size was the only trait significantly associated with mate choice.

Zuk et al. (1990*a,b*, 1995) have recently reported that the traits most highly correlated with male sexual signals and used by female red junglefowl in mate choice (comb length and color, eye color, and spur length) are ones that are highly condition dependent. Variations in the males' waltzing displays were the only behavioral variable that significantly helped to explain differing rates of individual male mating success; furthermore, chosen versus rejected individuals exhibited differing display rates even when females were absent. Surprisingly, differing plasma testosterone levels in males did not directly correlate with individual mating success, although it seems likely that some correlation between sex steroid levels and relative dominance must exist. For example, it has been recently reported (Leonard and Horn, 1995) that dominant males crow more often than subordinate ones and their crows are uttered at higher average frequencies; dominant individuals also tend to attack subordinates when the latter attempt to crow. Thus, social status influences both a male's crowing rate and sound quality and, unlike plumage traits, crowing also serves as an effec-

tive and temporally flexible status signal that can be broadcast at ecologically appropriate times but hidden at others.

The Indian peafowl also offers wonderful opportunities for studying the possible effects of male plumage variations and behavior on individual mating success. Manning and Hartley (1991) reported that relative symmetry and ornamentation of the male's ornamental train of long, iridescent, and ocelli-bearing feathers are positively correlated with mating success. Such differences in relative male ornamentation are age related (Manning, 1989). Petrie et al. (1991) similarly found that peahens prefer to mate with those males having a relatively large number of ocelli in their trains. Thus, the most successful males were those having the largest number of ocelli in all but one of 11 observed copulations, and over 50 percent of the variations in male mating success could be attributed to such differences in train morphology. Petrie and Halliday (1994) found that by experimentally reducing the number of ocelli in his train, a male peacock's mating success was reduced significantly. Studies by Petrie et al. (1992) suggested that female peahens will mate with the same male repeatedly and will mate with more than one male, especially if her first mating was with a nonpreferred male. Evidently there is a competition among females for preferred males and more dominant females tend to monopolize preferred males by initiating repeated sexual encounters with them.

All of these studies show to varying degrees that female phasianids are probably more closely attuned to minor differences in plumage, morphology, and behavior than has been generally appreciated in the past. Also, Darwinian sexual selection has very probably been a driving force in the evolution of elaborate plumages and sexual displays among male pheasants. In addition to their sensitivity to largely ontogenetic male variables such as tail length and spur length, it is known now that females can also detect and will differentially respond to a variety of condition-dependent variables among males. These include relative parasite loads (inasmuch as they might affect wattle color and size), iris color, or display rates among males. Thus, females adaptively prefer mating with those males having visible or behavioral traits that are correlated with low parasite loads (Hillgarth, 1990*a,b*).

CHAPTER 7

Reproductive Biology

In most species of pheasants the breeding season is distinctly seasonal. Breeding is typically associated with the spring months in the temperate regions or with the wet season in more tropical areas. Even in relatively temperate areas, such as the northern states and prairie provinces of North America, the nesting season of ring-necked pheasants is often prolonged. Hatching dates often extend over a span of 3 or 4 months, as a result of renestings by females that were unsuccessful with their first clutch. Wagner et al. (1965) reported that during the period 1946–1956 in Wisconsin the average annual hatching date varied only to a minor degree between years, always falling between 12 and 21 June. However, the total range of 5,985 estimated hatching dates varied from mid-April to the end of August, a 21-week span. They suggested that although the start of pheasant nesting varies somewhat from year to year, the onset of egg-laying is relatively fixed, and thus the incidence of "dump-nesting" (in which eggs may be deposited by several females, but incubated by none of them) may vary from year to year. Wagner et al. (1965) suggested that the phenology of egg-laying may be set by certain rather fixed environmental controls such as photoperiod, whereas nesting and brooding behavior may be influenced by more annually variable environmental factors such as springtime temperatures.

Average clutch sizes in the pheasants vary greatly in average number between species, with relatively large clutches being typical of grain-eating, edge or open-country adapted forms such as *Phasianus* and *Gallus,* and small clutches, of as few as only one or two eggs, being found in forest-dwellers such as *Polyplectron* and *Argusianus.* Evolved clutch-size is presumably largely a reflection of the average amounts of food available to the female around the time of nesting (Lack, 1968). In general, phasianids that have clutches of only two to four tend to lay eggs that are proportionately larger than the average for the family, whereas those with clutches of ten or more have proportionately smaller eggs than the average for the family. However, Lack also noted that although species laying actually larger eggs tend to have longer incubation periods, those that lay proportionately large eggs relative to their adult weight do not have noticeably longer incubation periods. At least in two genera (*Tragopan* and *Argusianus*) the evolution of a proportionately large egg seems to be correlated with the fact that the chicks are unusually precocial at the time of hatching, which may be of considerable ecological significance in some habitats.

Incubation periods among the pheasants range in length from 18 to 29 days, with the longer ones typical of such genera as *Argusianus, Pavo, Lophophorus,* and *Crossoptilon,* whereas periods of less than 23 days occur in forms such as *Pucrasia, Chrysolophus,* and *Polyplectron.* In most species incubation begins with the last egg, and thus hatching is essentially synchronous. In all cases only the female incubates; in a considerable number of species it is known that she will usually lay a second clutch following the loss of the first. Eggshells are left in the nest at the time of hatching, and the young are taken away from the nest site shortly thereafter. In a few species (*Tragopan, Rheinartia*) the young attain a limited flying ability when only a few days old, whereas in others the flight feather growth is relatively slower and fledging may require a week or more. Thus, in the common pheasant flight is attained initially at about 12 days, which is probably fairly typical for many species of pheasants (Cramp and Simmons, 1980). In most species adult plumage and sexual maturity is attained the first year, but in some genera two (e.g., *Tragopan, Lophura*) or even three (*Pavo, Rheinartia,* probably *Argusianus*) years may be required before the adult male plumage and maximum breeding capabilities are attained.

Reproductive Success and Recruitment Rates

Determination of the overall reproductive success of any animal species is a complex process. It is dependent upon a large number of variables that are not only difficult to measure, but also may vary from year to year or from place to place. Perhaps the simplest method of estimating overall annual reproductive success is to determine the fall or early winter recruitment rate (percentage of juvenile birds in the population), because this is the end-product of all the individual reproductive variables, such as clutch size, age of initial breeding, nesting success, hatching success, and rearing success. Furthermore, because mortality rates of pheasants are believed to be fairly constant with age by the end of their first year, the recruitment rate should closely approximate the annual adult mortality rate in an essentially stable population. Recruitment rates have the further advantage of being relatively easy to estimate, at least in those species of pheasants that can be legally hunted, because only a simple determination of the percentage of young birds in the autumn harvest is necessary for this determination.

In the North American ring-necked pheasant population the age ratios have been found to vary considerably, but average about 60 percent young of the year (table 15), which is reasonably close to the average estimated female adult mortality rate of 66 percent cited in chapter 5. Male mortality rates of pheasants in North America average considerably higher than this, because game management practices in North America have been to harvest the maximum number of males possible, in the belief that a relatively few males are necessary in the population to maintain fertility among females. Data from Denmark also suggest that male pheasants suffer the considerably higher mortality rate of approximately 78 percent annually, as compared to a female annual mortality rate of about 62 percent (Paludan, 1959).

Table 15 summarizes the major variables and parameters of reproduction in the North American ring-necked pheasant population,

Table 15

Some reproductive characteristics of wild ring-necked pheasant populations

	Range	Average[a]	Source
Overall nesting success	15–46%	26.4%	8 studies cited by Gates and Hale (1975)
% of females hatching broods	43–88%	57.6%	8 studies cited by Gates and Hale (1975)
% of eggs fertile	89–95%	92.9%	8 studies cited by Gates and Hale (1975)
% of fertile eggs hatching (in successful nests)	85–96%	90.7%	8 studies cited by Gates and Hale (1975)
% of all eggs hatching (in successful nests)	75–91%	84.8%	10 studies cited by Gates and Hale (1975)
% of nests destroyed by agricultural activities	0–74%	29.6%	16 studies cited by Wagner et al. (1965)
% of nests destroyed by predation	3–78%	31.1%	16 studies cited by Wagner et al. (1965)
No. of young females in fall relative to spring female population	1.3–1.9	1.5	6 studies cited by Gates and Hale (1975)
% increase in females from spring to fall	86–179%	115.7%	6 studies cited by Gates and Hale (1975)
% increase in total population from spring to fall	134–227%	187.6%	6 studies cited by Gates and Hale (1975)
% of young birds in fall population	47–83%	59.7%	4 studies cited by Hickey (1955)

[a] Unadjusted for sample size variations in some cases.

based on a variety of studies from nearby parts of the species' North American range. A limited amount of comparable information is available from other parts of the species' range as well. In England a 61-percent nesting success rate (of 345 nests) and an 83-percent hatching success (of eggs in 210 nests) were estimated (Cramp and Simmons, 1980). Both of these statistics are fairly close to the North American figures. There seems to be little published information on age ratios in wild European pheasant populations, but in Denmark the mean annual mortality of all birds was estimated to be 81.4 percent the first year and 58.4 percent in following years (Paludan, 1959). These data suggest that fall recruitment rates probably range between 60 and 80 percent in Denmark, or quite close to those reported for North America.

The data summarized in table 15 suggest that although the overall nesting success is typically fairly low, an average of more than half of the females in the studies cited eventually succeeded in hatching broods, which indicates a strong renesting capability in ring-necked pheasants. Indeed, on average it would seem that most unsuccessful females in these studies

must have renested at least once (2.2 estimated average nests per female) to account for the differences between overall nesting success and overall ultimate female success in bringing off broods. It is known that up to three renestings are possible in the wild following the loss of the first clutch (Dumke and Pils, 1979). Under captive circumstances individual females have been known to lay up to as many as 140 eggs, with second- and third-year females generally having higher reproductive capacities than first-year birds (Perkic and Leporati cited in Glutz, 1973). Gates and Hale (1975) indirectly estimated that 88 percent of the females on their Wisconsin study area that were initially unsuccessful in hatching broods made a second nesting effort, although in that area the farming practices caused high nest mortality rates in late-nesting birds. It would thus seem that the length of the optimum breeding season might thereby become an important local factor influencing overall reproductive success in this species.

Table 16 provides some specific examples from two relatively comparable and unusually complete field studies on the reproductive characteristics of ring-necked pheasants

Table 16

Summary of two field studies on reproduction of ring-necked pheasants[a]

	Wisconsin	Iowa
Clutch production		
Total no. of clutches produced	4,120	533
% of clutches incubated	81.1%	—
No. of successful clutches	1,235	136
% of clutches successful	29.9%	25.5%
Egg production		
No. of eggs per incubated clutch	11.2	9.5
No. of eggs per unincubated clutch	5.6	—
% of eggs randomly laid	3.3%	—
Chick production		
No. of eggs per successful clutch	11.1	9.5
% of eggs hatching	90.7%	82.8%
Production rates		
Nesting success (%)		
Of all clutches	30.0%	32.4%
Of incubated clutches	36.9%	—
Hatching success (%)		
Of all eggs laid	28.8%	—
Of all eggs in clutches	29.8%	—
No. of clutches produced per female	1.8	—
No. of clutches incubated per female	1.4	—
No. of eggs produced per female	18.6	—
% of females hatching chicks	52.8%	—
No. of chicks produced per female	5.3	2.7
Sources of nest failures		
Disruption by farm machinery	40.5%	7.2%
Predation	39.2%	37.6%
Nest abandonment	10.1%	50.8%
Other or unknown causes	10.2%	4.4%

[a] Wisconsin data derived from Gates and Hale (1975), and Iowa data from Baskett (1947). Dashes indicate that no data were available.

(Baskett, 1947; Gates and Hale, 1975). Although some aspects of reproduction, such as nesting and hatching success, are quite similar in these two studies, there are major differences in the individual sources of nest mortality, especially insofar as losses resulting from nest desertion and farm machinery are concerned. As a result of the renesting capabilities of the species, it seems that locally high rates of individual sources of nest losses, whether by predation, weather, or human-related factors, can probably be adjusted for and that areas having widely differing environmental characteristics might have remarkably similar overall hatching success.

Reproduction and Habitat Relationships

A large number of studies in North America have attempted to determine the kinds of nesting habitats most suitable for pheasants and differential rates of habitat use during the reproductive season. Generally these studies have attempted to compare reproductive success with particular "cover types" or physical aspects of cover, such as its height, density, or configuration.

One of the physical aspects of nesting that has received particular attention is that nesting density or nesting success might vary with the width or area of the particular nesting habitat. Gates and Hale (1975) observed in Wisconsin that nesting densities were greater in wider and denser strips of cover. Gates and Hale also reported that, although statistical significance fell somewhat short of acceptance, they believed that pheasants exhibited a probable preference for nesting in larger blocks of cover (4.1 ha [10 acres] or more) than in smaller ones. They found the lowest nesting densities in hayfields of less than 4.1 ha (10 acres) in area and the highest densities in fields of 12.6–16.2 ha (31–40 acres), with intermediate densities in hayfields of intermediate sizes.

An area of similar concern has been the possible role of habitat "edge" with respect to the placement and density of pheasant nests. Some early North American studies such as Hamerstrom's (1936) suggested that perhaps pheasants prefer to nest in locations fairly close to the edge of their chosen nest habitat, regardless of its actual area. However, more recent studies (see table 17) have failed to substantiate this view, and instead suggest that there is no direct relationship to the placement of a nest

Table 17

Results of two studies of ring-necked pheasant nest locations relative to habitat edge[a]

	Area studied		Nests found		
	No. of acres	% of total	No.	% of total	Nest density (no./100 acres)
Minnesota study area					
By distance from habitat edge:					
0–50 ft	200.3	31.3%	168	33.1%	84
51–100 ft	158.2	24.6%	117	23.0%	85
101–200 ft	198.1	30.8%	145	28.5%	73
>200 ft	87.3	14.4%	78	15.4%	89
Overall	643.9	100%	508	100%	79
Ohio study area					
By distance from habitat edge:					
0–50 ft	683	27.5%	124	27.5%	18
51–100 ft	601	24.2%	114	25.3%	19
101–200 ft	889	35.8%	157	34.8%	18
>200 ft	310	12.5%	56	12.4%	18
Overall	2,483	100%	451	100%	18

[a] Minnesota data from Nelson et al. (1960). Ohio data from Strode (1941); recalculated where necessary for comparability.

Table 18

Relations of habitats to hatching success and relative chick production of ring-necked pheasants[a]

	Nest distribution (% of total nests built)[b]	% hatching success within the habitat[c]	Relative chick production (% of total chicks produced)[d]
Hayfields	45.6	20.9	10.0
Strip cover	20.3	19.8	25.2
Small grains	13.7	38.5	45.0
Wetlands	5.5	32.9	10.4
Pastures	4.1	28.5	6.1
Woodlands	1.9	22.0	0.3
Other habitats	9.7	21.5	3.2
Overall	100	24.9	100

[a] Derived from tabular percentage data of 7–12 studies summarized by Olsen (1977); mean percentages were estimated without adjustment for sample size variations, which often were unspecified.

[b] Based on 12 studies, with some recalculations where required.

[c] Based on 11 studies. Overall hatching success was calculated from only those studies with indicated sample sizes.

[d] Based on 7 studies.

and its distance from the edge of the habitat. Judging from the data presented in table 17, this seems to be the case in areas of fairly high nesting densities, such as Minnesota, as well as in areas of much lower nesting density, such as California.

The physical aspects of the nesting site, in particular associated cover density and cover height, do seem to represent important aspects of nest-site selection in ring-necked pheasants. Hamerstrom (1936) reported that female ring-necked pheasants often situate their nests in cover that provides partial to complete over-head concealment. He also reported that hatching success of wholly concealed nests tended to be slightly higher than the success estimated for partially concealed or unconcealed nests. Hanson (1970) similarly reported that at least for early-nesting females, the vegetative life-form of the nesting cover seems to be critical, with the density or height of the plant positively influencing nest-site selection tendencies. Hanson judged that these factors might influence temperature and humidity characteristics of the nest itself, as well as its relative visibility to predators. Yearly averages of plant heights around the nest sites ranged from 24.9 to 68.1 cm (9.8 to 26.8 in), and average light penetration levels ranged annually from 9 to 30 percent.

Perhaps in part because of their widely differing height and density characteristics, there tends to be widely differing levels of use of various vegetation types by nesting ring-necked pheasants (see tables 18 and 19). As in many other North American studies, Gates and Hale (1975) found that in Wisconsin there were major differences in nesting densities and less conspicuous differences in nesting success associated with various nesting habitat types. They observed that wetlands, especially permanent wetlands, provided nesting cover that was associated with the highest average nesting success percentages. Wetland habitats were also associated with relatively high nesting densities, especially in the case of herbaceous

Table 19

Habitat influences on nest density and nest success in ring-necked pheasants[a]

	No. of nests	Nest density[b]	Nest success (%)
Hayfields			
Pastured hay	27	5	—
Mowed red clover hay	69	19	—
Mowed alfalfa hay	212	20	—
Unharvested hay	55	31	—
Overall	—	—	14
Wetlands			
Herbaceous	49	68	—
Canary grass (*Phalaris*)	62	33	—
Sedge meadow (*Carex*)	145	23	—
Shrub swamp	9	10	—
Cattail (*Typha*)	2	5	—
Temporary	24	29	—
Overall	—	27	46
Peas and small grains	—	1.5–9	31
Strip cover			
Roadsides	—	—	31
Grassy	50	1.1	—
Herbaceous	10	2.1	—
Sedge–canary grass	14	2.9	—
Woody	11	3.7	—
Overall	—	—	26
Fence lines			
Grassy	31	0.6	—
Herbaceous	15	1.1	—
Woody	28	2.2	—
Overall	—	—	24
Ditch banks			
Herbaceous	12	3.1	—
Canary grass	37	5.0	—
Woody	11	6.9	—
Overall	—	—	20

[a] Data derived from Gates and Hale (1975). A dash indicates that no data were available.

[b] Nest densities are given as number per 100 acres for hayfields, wetlands, and peas and small grains, and as number per mile for strip cover, fence lines, and ditch banks.

wetlands. "Strip cover," consisting of roadsides, fence lines, ditch-banks, and similar kinds of linear cover, also had fairly high nesting densities in some cases, particularly if measured on a somewhat misleading density-per-acre rather than nests-per-mile basis. Nesting density was found to be highest in woody and canary-grass types. Overall density apparently depended upon cover density and cover width more than upon other vegetational characteristics. However, hayfield use by pheasants was largely influenced by phenology; early-nesting years or early-nesting birds were associated with higher use of wetlands for nesting, but in later-nesting years and later in the nesting season hayfields and cultivated fields of peas and grain become much more important nesting habitats. Thus, nest-site selection in pheasants is apparently a complex interplay between cover preferences, cover availability, and varying cover conditions as the nesting season progresses. In general blocks of vegetation seem to be preferred for nesting over strips. Comparatively low and dense cover associated with residual plant material in wetlands or newly growing vegetation in hayfields or cultivated fields is evidently also important in nest-site selection by ring-necked pheasants.

Egg Size and Clutch Size

Comparative sizes of eggs to adults in birds allow for various ecological analyses, and some major points have already been made on this matter based on Lack's (1968) studies. However, Lack had to rely on limited information on egg weights and adult weights and did not discuss clutch weights as a significant ecological factor in reproductive biology of pheasants. Thus, I have recalculated relative egg weights for as many species of pheasant for which I could find suitable data. I have also calculated the average collective clutch weight relative to adult female weight as a measure of the physiological stress of reproduction in this group.

Table 20 summarizes the results, with clutch weights calculated on the basis of midpoints between the ranges indicated in table 21 and the species organized into mating system categories as used in table 14.

There are few obvious correlations between mating systems and relative female energy investment in eggs or clutches, although the lack of reliable information on both average adult female weights and average clutch sizes in the wild tend to obscure some of the trends that might have been expected. Generally, it seems that monogamous species of open habitats do not expend significantly different amounts of energy in individual eggs or collective clutches than do polygynous edge-adapted forms. Both groups typically lay eggs from 3 to 6 percent of the adult female's weight and produce clutches from about 15 to 50 percent of the female's weight. However, in the polygynous or promiscuous forest-adapted species there tend to be smaller clutch sizes (*Chrysolophus* being the major exception), relatively large egg sizes (especially in *Polyplectron*), and relatively low total energy investment in total clutches (particularly in *Polyplectron*, some *Lophura*, and *Argusianus*). These trends possibly are related to the desirability of producing a few, fairly precocial offspring in a woodland environment that has limited food resources for both the female and her chicks.

Brood-Rearing and Juvenile Mortality

The brood-rearing period is a critically important one in the lives of pheasants. During the first few weeks of life mortality rates are at their highest; thus, the success or failure of a species' local population may be determined by the weather and other environmental factors influencing mortality during these critical times of the year. Much has been made of these high juvenile mortality rates and of the difficulties of determining their causes under natural conditions. Stokes (1954) stated that

Table 20

Relationship of egg weight and clutch weight to adult female weight in pheasants[a]

	Estimated egg wt. (g, oz)	Estimated female wt. (g, lb)	Egg wt. as % of female wt.	Clutch wt. as % of female wt.
Monogamous open-country or scrub species				
Brown eared pheasant	44.5, 1.6	1,650, 3.6	2.6	16.9
White eared pheasant	58.4, 2.0	1,600, 3.5	3.6	19.8
Blue eared pheasant	52.1, 1.8	1,600, 3.5	3.2	20.8
Blood pheasant	28.8, 1.0	515, 1.1	5.6	47.6
Cheer pheasant	71.6, 2.5	1,200, 2.6	6.0	63.0
Overall			4.2	33.6
Polygynous edge-adapted species				
Ceylon junglefowl	30.4, 1.1	550, 1.2	5.5	16.5
Elliot's pheasant	25.2, 0.9	910, 2.0	2.8	19.6
Himalayan monal	70.7, 2.5	1,950, 4.3	3.6	21.6
Common pheasant	32.2, 1.1	1,025, 2.2	4.5	33.8
Mikado pheasant	46.2, 1.6	1,015, 2.2	4.5	33.8
Red junglefowl	29.6, 1.0	500, 1.1	5.9	35.4
Reeves' pheasant	34.8, 1.2	950, 2.1	3.7	37.8
Overall			4.1	27.9
Monogamous (?) forest species				
Blyth's tragopan	62.0, 2.2	1,250, 2.7	5.0	15.0
Western tragopan	61.3, 2.1	1,325, 2.9	4.6	16.1
Satyr tragopan	63.3, 2.2	1,100, 2.4	5.7	17.1
Temminck's tragopan	47.7, 1.7	1,035, 2.3	4.6	18.4
Koklass	40.0, 1.4	932, 2.0	4.3	25.8
Overall			4.8	18.5
Polygynous/promiscuous forest species				
Great argus	74.3, 2.6	1,600, 3.5	4.6	9.2
Germain's peacock pheasant	30.4, 1.1	400, 0.9	7.6	15.2
Grey peacock pheasant	37.3, 1.3	480, 1.1	7.8	15.6
Edwards' pheasant	32.3, 1.1	1,050, 2.3	3.1	17.1
Crestless fireback	33.2, 1.2	840, 1.8	3.9	17.5
Crested fireback (*ignita*)	47.6, 1.7	1,600, 3.5	3.0	18.0
Silver pheasant (*nycthemera*)	42.8, 1.5	1,150, 2.5	3.7	18.5
Indian peafowl	103.5, 3.6	3,375, 7.4	3.1	18.6
Palawan peacock pheasant	32.2, 1.1	320, 0.7	10.0	20.0
Wattled pheasant	45.3, 1.6	1,130, 2.5	4.0	22.0
Lady Amherst's pheasant	31.1, 1.1	700, 1.5	4.4	39.6
Golden pheasant	28.1, 1.0	600, 1.3	4.7	42.3
Kalij pheasant (*melanota*)	37.4, 1.3	890, 1.9	4.2	50.4
Overall			4.9	23.4

[a] Within mating ecology categories, species are listed by increasing relative clutch weight to female weight. Data adapted from Ridley (1987). The tragopans are probably at least occasionally polygynous.

Table 21

Some reproductive characteristics of pheasant species in captivity[a]

	Years to sexual maturity	Clutch size	Incubation period (days)	Birds in captivity[b]
Blood pheasant	1	5–12	27–28	5
Western tragopan	2	3–6	28	0
Satyr tragopan	2	2–4	28	583
Temminck's tragopan	2	2–6	28	562
Blyth's tragopan	2	2–4	28?	31
Cabot's tragopan	2	2–4	28	126
Koklass	1	5–7	26–27	207
Himalayan monal	2	4–8	28	798
Sclater's monal	?	?	?	0
Chinese monal	2?	3–6	28–29	0
Green junglefowl	1–2	4–6	21	225
Red junglefowl	1–2	4–8	19–21	651
Grey junglefowl	2	4–8	20–21	231
Ceylon junglefowl	2	2–4	20–21	77
White eared pheasant	2	4–7	24–25	355
Blue eared pheasant	2	5–8	26–28	439
Brown eared pheasant	2	5–8	26–27	399
Salvadori's pheasant	1–2	2	21–22	0
Imperial pheasant	2	5–7	25	4
Edwards' pheasant	1	5–6	24–25	418
Kalij pheasant	1	9–15	24–25	952
Silver pheasant	2	4–6	25–26	1,835
Swinhoe's pheasant	2	6–12	22	802
Crestless fireback	2	3–6	23–24	162
Crested fireback	2–3	4–8	24	429
Wattled pheasant	2	3–8	24–26	11
Cheer pheasant	1	9–12	26	363
Reeves' pheasant	1	7–14	25	908
Copper pheasant	1	6–12	25	208
Bar-tailed pheasant	1	6–11	27–28	340
Elliot's pheasant	1	6–8	25	482
Mikado pheasant	1	5–10	27	471
Common pheasant	1	8–12	24–25	millions
Green pheasant	1	8–12	24–25	277
Golden pheasant	2	6–12	22	7,488
Lady Amherst's pheasant	2	6–12	22	2,128
Bronze-tailed pheasant	2?	2	22	71
Rothschild's peacock pheasant	1–2	2	22	0
Grey peacock pheasant	1–2	2	22	729
Germain's peacock pheasant	1–2	2	22	194
Malayan peacock pheasant	1–2	1	22	102
Bornean peacock pheasant	1–2	1	22	0
Palawan peacock pheasant	1–2	2	18–19	429
Crested argus	3	2	25	0
Great argus	3	2	24–25	172
Indian peafowl	3	4–8	27–29	3,922
Green peafowl	3	4–6	28	869
Congo peacock	1–2	2–3	26–28	47

[a] Data on captive reproduction adapted primarily from information in Delacour (1977) and Howman (1979).

[b] Captive population data from the 1991 World Pheasant Association census.

"The disappearance of so many thousands of chicks in the short space of a summer almost beneath one's eyes and yet not noticed is a baffling experience and an enigma still to be solved." Estimates of the actual magnitude of these losses are quite variable and are made more difficult because of strong tendencies toward brood amalgamation toward the end of the summer. Gates and Hale (1975) concluded that so many factors influence the sizes of observed broods in the field that such data are highly suspect as indices to annual juvenile mortality. They judged that broods coming from the early stages of the hatching season showed rather little tendency to combine in later life, whereas those hatching later in the summer exhibited a greater tendency for amalgamation. Gates and Hale concluded that the brood sizes of 4–6-week-old age classes provided the best estimates of actual juvenile mortality rates. They judged that rates of juvenile mortality between hatching and 1 October averaged about 42 percent over a 6-year period. Most other field studies in North America (e.g., Stokes, 1954) have made similar estimates of juvenile mortality rates that have ranged from 35 to 56 percent. Most observers believe that

the majority of the losses actually occur within a few days after hatching (Baxter and Wolfe, 1973). During this vulnerable period the chicks are evidently highly sensitive to adverse weather effects such as chilling and wetting. Although some studies (e.g., Stokes, 1954) have positively correlated late-hatched broods with higher mortality rates, others have found that late-hatched broods instead tend to suffer lower mortality rates than do earlier ones (Baxter and Wolfe, 1973). Information on species other than the ring-necked pheasant is extremely limited. However, Collias et al. (1966) judged that only 25 percent of the chicks hatched among semiferal red junglefowl in the San Diego Zoo survived until the time of family breakup, or an approximate 75 percent mortality rate during the first 2 months of life. A further 76 percent mortality rate occurred from that point to the end of their first year of life in this population. This is not very different from the estimated 81.4 percent mortality rate of Danish pheasants in their first year of life (Paludan, 1959) or the estimated 82.4 percent annual mortality rate of juvenile radio-equipped pheasants in Wisconsin during their first year (Dumke and Pils, 1973).

CHAPTER 8

Aviculture and Conservation

There are so many excellent references on pheasant-keeping (Gerrits, 1974; Delacour, 1978; Howman, 1979, 1993; Roles, 1981) that it seems unnecessary to review detailed techniques of pheasant aviculture. Nevertheless the role of aviculture in the propagation and preservation of endangered species and the potential for aviculture to introduce or reintroduce pheasant populations is so great that no discussion of pheasant biology would be complete without some reference to these matters.

Pheasant-raising is perhaps the only branch of aviculture in which one can currently buy a pair of an endangered species for a very modest amount of money, breed it, and perhaps produce offspring (Howman, 1979). Many of the numerous rare, vulnerable, or endangered species of pheasants are fairly easily available from aviculturists, and nearly all of these can be bred under favorable conditions. It is indeed important that as many separate and dispersed collections of rare pheasants be kept as possible to reduce the danger of disease or other local calamity destroying the entire gene pool of a captive flock. Large numbers of birds in captivity also help to maintain a diversified gene pool, assuming these did not arise from a small founder stock. This is considerably important if fertility is to be maintained and genetic drift in lines, which is almost inevitable in captive populations, is to be avoided.

Pheasants as Avicultural Subjects

Keeping pheasants does require considerable aviary space if the birds are to be effectively maintained and bred. Howman (1979) suggests that about 60 m (200 ft) per pair is a reasonable aviary design, although some species may need much more space than this for optimum maintenance and breeding. For some highly herbivorous species, such as the koklass, a grassy floor may be desirable, but the dangers of such parasites as gapeworms (*Syngamus*) must be considered as well. Thus some aviculturalists may prefer wire-mesh floors over natural substrates. All aviaries require protection from terrestrial and aerial predators and depending upon the climate may also require protection from the elements. The latter is especially important for some delicate and tropical-adapted forms. Control for photoperiod may also be needed for species that are being maintained at latitudes well away from their normal range.

Most of the diseases of pheasants are the same ones that attack domestic fowl, turkeys, and similar poultry; thus a variety of treatments are available for most species. Similarly, the vast majority of pheasant species are able to consume the standard poultry rations that are widely available and these are generally of very high quality and provide well-balanced nutrition. However, some species that naturally consume large quantities of insects in the wild may require additional feeds of mealworms or other insects. Virtually all pheasant chicks can be stimulated to feed most effectively by providing such live food. Also, discarded eggshells are eagerly eaten by many captive birds, presumably for their calcium. Highly herbivorous species, such as the koklass and tragopans, greatly profit from such green foods as lucerne (alfalfa) or even fresh grass clippings. Fruits or nuts such as peanuts are often particularly attractive to most pheasant species as well. Finally, a source of grit is important, especially for the species in which grass or other cellulose-rich vegetation forms an important component of the diet.

Incubation of pheasant eggs can be done by bantam breeds of domestic fowl or by commercial incubators of the type used for domestic fowl and turkeys. Collecting the eggs as they are laid normally results in the laying of a second clutch, which may also be removed and incubated separately or left with the female to incubate on her own. Continued collection of eggs will often stimulate the laying of a much larger number of eggs than ever occurs in the wild, but brings with it some risk of reproduc-

tive stress on the female. However, because female ring-necked pheasants have been known to lay more than 100 eggs in a year, the dangers of this are probably not great provided that a fully adequate diet is maintained throughout the egg-laying period.

Table 21 (see chapter 7) summarizes the major reproductive parameters of all the pheasant species, based primarily on avicultural information. Frequently the indicated time to full reproductive maturity is longer than that required to attain the adult plumage. It is also true that species that normally breed only in their second year of life will occasionally attempt to breed their first year, albeit generally without high success. There is relatively little information available on the effective breeding lifetime of pheasants. However, in the wild common pheasants have been known to survive for almost 8 years and have survived even longer in captivity.

As an indication of the potential longevity of various pheasants under captive conditions, some examples of extreme longevity that were mentioned by Mitchell (1911) might be cited, remembering that these records were obtained well before modern methods of feeding and housing had been perfected. The maximum captive longevity that he cited for any species of pheasant was 179 months for the grey peacock pheasant, followed by 163 months for great argus, 161 months for kalij, 148 months for grey junglefowl, 133 months for Blyth's tragopan, 123 months for Temminck's tragopan and Himalayan monal, 116 months for crested fireback, 114 months for silver pheasant, and 109 months for crestless fireback. Several other species (golden pheasant, Lady Amherst's pheasant, copper pheasant) were also reported as surviving in excess of 90 months in captivity. Records of the New York Zoological Society indicate that two individuals of the Malayan peacock pheasant survived more than 228 and 300 months (Donald Bruning, pers. comm.). It is of interest that many of the species with unusually long captive life spans (i.e., great argus, peacock pheasants, and tragopans) are ones that have correspondingly low reproductive potentials in the wild. Quite possibly these long captive lifetimes are reflections of adaptations associated with a relatively long lifetime under wild conditions. Likewise, Delacour (1977) reported that one captive crested argus survived nearly 240 months in his collection, which is an example of another long-lived species having a very low reproductive potential in conjunction with a small clutch size and a delayed reproductive maturity.

Captive Status of Endangered and Rare Pheasants

Wayre (1969) has discussed the role of aviculture in ensuring the survival of rare endangered species of pheasants. In particular he described the role of the Norfolk Wildlife Park and Pheasant Trust in reintroducing the endangered Swinhoe's pheasant back to Taiwan, where it has been in the danger of extinction. After beginning a captive breeding program in 1959, more than 150 birds of this species were produced. In 1967 15 pairs were taken back to Taiwan and released in a protected area of forest. The following year six more pairs were released in the same area. A similar project was undertaken with the mikado pheasant, another rare endangered species from Taiwan. Finally, the trust sent 12 captive-bred pairs of cheer pheasants to India in 1971 for release near Simla in Himachal Pradesh. An additional group of 24 birds were sent to the same area in 1973, and the species now occurs in the wild not far from the original point of release (*Norfolk Wildlife Park and Pheasant Trust Annual Report*, 1980, p. 76).

Of similar interest are the efforts of the Jersey Wildlife Preservation Trust in breeding the vulnerable white eared pheasant from two pairs of this species obtained from China via Russia in 1966. After an initial breeding in 1969 by the single pair still surviving, which reared 13 birds, 22 more were raised during 1970 and 1971. More recent breedings have substantially increased this number (Mallinson and Taynton, 1978; Mallinson, 1979). As of 1979 the breeding program had resulted in a net population increase of 162 white eared pheasants. Similar efforts of the Jersey Wildlife Preservation Trust have been directed toward the Edwards' pheasant and the brown eared pheasant. These have resulted in more modest increases in the captive numbers of these two endangered or vulnerable species. Major captive breeding efforts are now underway involving the Malayan peacock pheasant and Rothschild's peacock pheasant. For both of these species, international studbooks are now being maintained by the Wildlife Conservation Society, a branch of the New York Zoological Society.

A major project involving the preservation and reintroduction of the endangered cheer pheasant has been undertaken jointly by the World Pheasant Association and the Game Conservancy; the project resulted in the production of 500 eggs of this species in 1978 and 1979 (Beer and Cox, 1981). Some cheer pheasants have been released in the Margala Range of Pakistan as a result of these activities (Mirza, 1981b), although the results of these efforts are still not very promising, owing to high predation levels (Roberts, 1991).

Since its inception in 1975 the World Pheasant Association has been concerned with all aspects of pheasant conservation, in particular the role of aviculture in maintaining breeding stocks of the world's rare and endangered species of Galliformes. At various intervals this group has surveyed the galliform holdings of private and institutional collections around the world and has published its results in the *World Pheasant Association Journal* and in its own newsletters. Table 21 (see chapter 7) provides a species-level summary of the association's findings for the most recent (1991) census, although this is admittedly a distinctly incomplete survey.

These very interesting data provide a valuable index to the captive status of all the species of rare pheasants and earlier surveys provide some comparative data. It is probable that captive numbers of pheasants are now at the highest level in history. Certainly avicultural methods for maintaining the breeding pheasants are constantly improving. Thus, a brief summary of the captive status of each of these species is provided here. A more general statement about the species' status in the wild and prospects for its conservation will be found in the individual species accounts in part two.

Endangered Species

Western Tragopan

Efforts are underway in both Pakistan and India for the propagation and eventual reintroduction of this endangered species. In the late 1970s facilities for hatching and rearing were established in Pakistan and field studies were also begun. A similar program was begun in Himachal Pradesh in 1978 (King, 1981). There are no breeding birds in Europe or North America at the present time. Sharma (1993) has summarized the history of this species in captivity, and Stewart (1994) reported its first captive breeding of this century.

Cabot's Tragopan

Although the Cabot's tragopan was once fairly common in captivity in Europe, its stocks declined after World War I (King, 1981). During the 1960s and 1970s a few birds reached England. Several have also recently reached the United States, where Charles Sivelle of Florida has made a special effort toward breeding them and building up a captive population. This species is now considered vulnerable rather than endangered and its known overall range

has been somewhat expanded. As of 1995 there were an estimated 125 captive individuals worldwide (McGowan and Garson, 1995). An international studbook is being maintained.

Sclater's Monal

This little-known and endangered Chinese endemic has never been kept in captivity by aviculturists outside China. However, efforts are currently underway to breed this species at Beijing's Breeding Center for Endangered Animals, where three pairs were present in 1997 (David Rimlinger, pers. comm.).

Chinese Monal

This species has been maintained in captivity only very rarely and has never been known to breed in the western world. A few were present in the Peking Zoo in the early 1970s, but all of these had died by 1976 (King, 1981). However, the Chinese obtained more wild stock; in the spring of 1983 a pair of these endangered birds were sent to the San Diego Zoo by the Chinese government in hopes that a captive breeding stock might be developed. Although several young were reared, these were eventually lost to disease.

White Eared Pheasant

White eared pheasants were not bred in captivity until 1938 and have always been relatively rare in collections. Two pairs were received by the Jersey Wildlife Preservation Trust in 1966, and others were sent to the Antwerp Zoo and the East Berlin Zoo (Tierpark Berlin). As mentioned earlier, two pairs were also received by the Norfolk Wildlife Park and Pheasant Trust in 1974. The current captive flocks are largely derived from these two sources, particularly those bred at the Jersey Wildlife Preservation Trust (King, 1981). Of the nine birds imported in 1966, more than 300 descendants had been bred as of 1980 (Grummt, 1980). A studbook has been organized to monitor the rate of inbreeding. The endangered *harmani* taxon of this pheasant (sometimes considered a full species) is not currently known to be represented in captivity, but there were perhaps 1,000 individuals of the other taxa (that collectively are considered vulnerable) in captivity in 1995 (McGowan and Garson, 1995). More effective captive management of this species is needed.

Brown Eared Pheasant

Although a rather large number of brown eared pheasants are in captivity, as of 1976 all of those outside China were descended from the original three importations. However, additional birds were sent from China to the Norfolk Wildlife Park and Pheasant Trust in 1976, from which offspring have been produced (Mallinson, 1979). More recently, wild-stock birds have also been sent to the San Diego Zoo from China, thus improving the genetic diversity of the captive population. This endangered species is fairly well represented in captivity, with about 1,000 captive birds worldwide as of 1995 (McGowan and Garson, 1995).

Cheer Pheasant

Like the Cabot's tragopan, this endangered species was common in captive collections until World War I, after which stocks declined. However, since 1933 the numbers have gradually grown, in large measure as a result of efforts of aviculturists via groups such as the Norfolk Wildlife Park and Pheasant Trust and the World Pheasant Association. As noted earlier, there have also been some relatively unsuccessful releases of captive-bred birds back into the wild.

Elliot's Pheasant

The Elliot's pheasant was first bred in captivity in 1880. Since then this species has always been

fairly common in captivity. However, since 1960 the only additions of wild-trapped birds to the stocks in captivity have been two males sent to the Norfolk Wildlife Park and Pheasant Trust in England in 1974 and a male in Hong Kong in 1976 (King, 1981). Since 1980 several additional birds have come to the San Diego Zoo from Chinese sources. As of 1995 there were probably about 1,000 birds in captivity worldwide (McGowan and Garson, 1995).

Edwards' Pheasant

This species was not introduced to western aviculture until 1924, but breeding began the following year. Since then, wild birds have occasionally been added to the captive gene pool and a moderately large number are now in captivity, primarily in the United States. With the discovery of an allopatric northern population (the Vietnamese or Vo Quy's pheasant), the known range of this species (as here recognized) is slightly expanded. However, both taxa are believed to be critically endangered (McGowan and Garson, 1995). A few Vietnamese pheasants were present in the Hanoi Zoo as of 1997. This group originated from two pairs captured in 1990 and numbered about 20 birds in 1995. Four of these birds were donated to the World Pheasant Association; the first successful breeding of these birds in Britain occurred in 1997. World Pheasant Association survey records indicate that 418 birds of the nominate Edwards' taxon were reported to be in captivity in 1991, which represents a substantial decline from earlier surveys. The only location where Edwards' pheasants were still known to be surviving in the wild as of 1997 is Bach Ma National Park, near Hue, Vietnam (Shamblin, 1997).

Imperial Pheasant

The imperial pheasants upon which the description of the species was initially based were brought into captivity in 1924 by Delacour and bred the following year. This critically endangered species is not currently represented in captivity. Earlier captives that were present in collections until as late as the 1980s were probably all of variably hybrid origin (with the silver pheasant). Thus, this stock has virtually died out.

Crestless Fireback

The Bornean race *pyronota* of this species is considered endangered, whereas the nominate Malayan race is classified as vulnerable. As of 1995 there were about 50 individuals of the Bornean race in captivity and about 250 of the Malayan taxon. About half those totals were present during a 1991 World Pheasant Association international census, and only a few dozen of each race were present during surveys of the 1970s.

Swinhoe's Pheasant

This Taiwan endemic has been maintained and bred in captivity since 1866 and immediately began to breed freely. However, no wild birds have apparently been imported from Taiwan for more than 70 years (Delacour, 1977). Thus the genetic variability of the captive population is probably relatively restricted. Nonetheless, it is still the most common captive species of all the forms listed in table 21 (see chapter 7).

Wattled Pheasant

Although this species was first brought into captivity in 1876, the first successful breeding did not occur until 1974 (King, 1981). Since then several breedings have occurred in the United States and Mexico, but the numbers in captivity are dangerously low. The large majority of the captive population is currently to be found in the United States. Although the World Pheasant Association listed only 11

birds in their international 1991 census, it has been estimated that as of 1995 there might be about 100 individuals in captivity worldwide (McGowan and Garson, 1995).

Mikado Pheasant

The mikado pheasant was first brought to England in 1912 and was raised successfully the following year. Additional wild stock was obtained during the 1920s by Delacour, and since then the total captive stock has gradually increased. There have evidently been no recent additions of wild stock to the captive population. This species was fairly common in private collections in the United States during the 1970s and 1980s, but has declined more recently because of restrictions on sales associated with its endangered species status (Donald Bruning, pers. comm.).

Bornean Peacock Pheasant

This critically endangered taxon has only infrequently been kept in captivity. Vernon Denton kept and successfully bred these birds in California during the 1970s and early 1980s, producing only hybrids of varying purity between the Bornean and Malayan peacock pheasants. No individuals are known to be surviving in captivity at present. Recent surveys in Kalimatan suggest it still survives in several areas, but there are no known major population centers.

Reeves' Pheasant

This species probably has more individuals in captivity than any of the endangered pheasants. Feral populations also exist in Europe and these are often supplemented by releases from captivity. In World Pheasant Association surveys held during the 1970s and 1980s at least 1,700 captive birds were represented. By 1991 this total had dropped to 908, but it is likely that about 3,000 individuals were in captivity throughout the world during the mid-1990s (McGowan and Garson, 1995).

Palawan Peacock Pheasant

The Palawan peacock pheasant was imported into the United States in 1929, and bred the following year. Since then there have been several importations of wild stock into North America and Europe, but the total captive population remains fairly low, with the largest numbers present in the United States. Although the World Pheasant Association listed 419 birds in captivity in 1991, there were perhaps about 1,000 captive individuals worldwide in 1995 (McGowan and Garson, 1995). Over 500 were reported in a recent unofficial studbook. Probably more than 1,000 were present as of 1998 in the United States (Donald Bruning, pers. comm.).

Green Peafowl

Green peafowl have been maintained in captive collections for a very long time and were present in the London Zoo as early as 1831 (Delacour, 1977). Although this species breeds well in captivity, it is less hardy and more aggressive than the more common Indian peafowl and the captive stocks are far smaller. There were about 600 green peafowl in captivity in 1995, most of which probably represented the inadequately studied subspecies *spicifer,* rather than the two endangered races *muticus* and *imperator* (McGowan and Garson, 1995). Hybrids with the common peafowl are frequent and pose a danger to the captive gene pool of the green peafowl.

Rare Species

Blyth's Tragopan

This tragopan was maintained in European collections as early as 1870, but did not breed

until 1884. After World War II the captive stocks gradually declined for lack of supplementation by wild stock. However, two pairs were imported into England in 1983 from Nagaland, which have subsequently been bred successfully. Captive numbers have thus increased, but the small founder population makes the captive flock highly inbred and vulnerable.

Bar-tailed Pheasant

This species was initially maintained in India as early as 1888, but did not reach Europe until 1961, when a pair was imported into England. These birds bred the following year and eventually were raised in fairly large numbers in Europe as well as North America. At present the largest numbers are in captivity in England, with somewhat smaller numbers present in North America. The western race *humiae* of this species is considered endangered and the eastern race *burmanicus* as vulnerable. In 1991 the World Pheasant Association census listed 340 individuals of *humiae* in captivity, but none of the eastern race. Earlier World Pheasant Association surveys listed 876 in 1971, 626 in 1979, and 532 in 1982, suggesting a progressive decline in captive stocks during recent decades. This decline in captive populations has largely been the result of permit requirements associated with the species' threatened status, causing most aviculturists to abandon their breeding efforts (Donald Bruning, pers. comm.). There may have been about 1,000 individuals in captivity worldwide as of 1995, all of which presumably represent the western race (McGowan and Garson, 1995). Evidently the eastern race has never been maintained in captivity outside China.

Crested Argus

This spectacular pheasant was first brought into Europe in 1924, and the birds were bred in France the following year. They proved to be very long-lived and bred rather easily. By 1940 they were established in both Europe and North America, but all the captive stock died out shortly after World War II. Of the two races of crested argus, the Malayan form *nigrescens* is considered as endangered and the Indochinese race *ocellata* is regarded as vulnerable. As of 1994, four individuals of the Malayan race were in captivity by the Malaysian Wildlife Department. There was also a lone male of *ocellata* in Bangkok and five birds were present in the Saigon Zoo (McGowan and Garson, 1995). In 1996 five wild-caught pairs were present at that zoo and two chicks were hatched successfully (Phan, 1996). By the autumn of 1997 a total of 16 chicks had been raised and nearly two dozen were present at the zoo (Phan Viet Lam, pers. comm.). During 1997 a few others were in captivity in Da Nang Botanic Gardens, the Hanoi Zoo, and at Bach Ma National Park, Vietnam.

Summary

This is a rather depressing review, because it seems unlikely that aviculture can do much to prevent the extinction of even a few of a group of species that are perhaps better adapted to captive breeding than almost any other taxonomic assemblage of wild birds. The selection pressures resulting from capturing, maintaining, and breeding a species in captivity are virtually the opposite from those that favor its survival in the wild. Therefore, it should not be surprising that captive-bred pheasants (or other birds) have essentially no chance of being successfully returned to the wild. There can be no question that watching a giant argus display in a zoo cage is a thrilling sight, but in effect we are seeing an incomplete bird performing on an artificial stage. We must always remain aware that the real stage is now quickly disappearing if not already gone forever. Nevertheless, zoos perform valuable services by educating the public as to the beauty of the world's birds, learning some of their basic biological attributes and requirements for survival, and perpetuating remnant stocks of species that are disappearing before our eyes throughout the world. Aviculturists can perform some of these same functions, but such persons often fail to preserve their data, observations, and experiences through publication. This book could not have been written without the direct cooperation and assistance of many zoos and aviculturists. To this end, I owe a great debt of gratitude.

PART TWO

TAXONOMIC KEYS
AND SPECIES ACCOUNTS

Keys to Classification of the Family Phasianidae

Key to Subfamilies and Tribes of Phasianidae

A. Head and upper neck virtually naked in adults, including the crown.
  - B. The tail short (under 200 mm [7.8 in]); 14 rectrices, mostly hidden by the tail-coverts: Numidinae (guineafowl, 7 spp.).
  - BB. The tail long (over 200 mm [7.8 in]) and rounded; 18 rectrices: Meleagridinae (turkeys, 2 spp.).
AA. Head and upper neck mostly or entirely feathered (if sparsely feathered, then the tail is greatly elongated and longer than the wing and the crown feathered).
  - B. Tarsus more or less completely feathered; nostrils densely feathered: Tetraoninae (grouse and ptarmigans, 16 spp.).[1]
  - BB. Tarsus unfeathered or at most only partially feathered; nostrils never completely hidden by feathers.
    - C. Lower mandible serrated toward the tip; males never with spurred tarsi: Odontophorinae (New World quails, 30 spp.).[2]
    - CC. Lower mandible completely smooth; males often with knobs or spurs on the tarsi.
      - D. Tail much shorter (maximum 200 mm [7.8 in]) than wing, flat and usually rounded; males similar to females and usually not iridescent or with extensive bare skin around eye: Perdicini (Old World quails, partridges, and francolins, 103 spp.).
      - DD. Tail usually longer (minimum 150 mm, [5.9 in]) than wing, sometimes vaulted and often graduated; males often dissimilar to females and variably iridescent, nearly always with bare skin around the eye: Phasianini (pheasants, junglefowl, and peafowl, 49 spp.).

[1]See Johnsgard (1983*b*) for key to species.
[2]See Johnsgard (1973) for key to species (exclusive of *Odontophorus*).

Key to Genera of Phasianini and Closely Related Perdicini[3]

A.   Wing usually over 300 mm (11.7 in; minimum 270 mm [10.5 in]); tarsus usually over 100 mm (3.9 in); both sexes crested or tufted as adults.
> B.   Tarsus unspurred; central rectrices greatly elongated.
> > C.   Tarsus grayish; innermost primary not the longest: *Rheinartia* (1 sp.).
> > CC. Tarsus reddish; innermost primary the longest: *Argusianus* (1 sp.).
>
> BB. Tarsus spurred and gray; tail-coverts long and iridescent.
> > C.   Wing over 400 mm (15.6 in); 20 rectrices: *Pavo* (2 spp.).
> > CC. Wing under 350 mm (13.7 in); 14 rectrices: *Afropavo* (1 sp.).

AA. Wing usually under 240 mm (9.4 in; maximum 330 mm [12.9 in]); tarsus usually under 100 mm (3.9 in); crested or tufted condition of adults variable.
> B.   Tail long and graduated; rectrices ocellated or partly iridescent; 4th primary (from inside) the longest: *Polyplectron* (7 spp.).
>
> BB. Tail variable but never ocellated to tipped with iridescence; 5th to 7th primary the longest.
> > C.   Wings usually under 200 mm (7.8 in) in females and 225 mm (8.8 in) in males; 10–14 rectrices, shorter than wing.
> > > D.   Ten rectrices; no spurs present in either sex: *Ophrysia* (Perdicini).
> > > DD. Fourteen rectrices; males or both sexes with up to 4 spurs.
> > > > E.   Tail rounded, less than three-quarters length of wing; claw of hind toe rudimentary: *Caloperdix* (Perdicini).
> > > > EE. Tail somewhat graduated, over three-quarters length of wing; claw normal.
> > > > > F.   Most body feathers lanceolate: *Ithaginus* (1 sp.).
> > > > > FF. Body feathers not lanceolate: *Bambusicola* (Perdicini).
> >
> > CC. Wing usually longer (over 200 mm [7.8 in] in females, 225 mm [8.8 in] in males); usually at least 16 rectrices (rarely 14); tail often longer than wing.
> > > D.   Body feathers lanceolate in both sexes; face fully feathered: *Pucrasia* (1 sp.).
> > > DD. Body feathers not lanceolate; males (and often females) variably naked around eye.
> > > > E.   Body and upper wing feathers with white to buffy rounded spots; tail brown, rounded, nearly flat, and no longer than wing; males with bare lappet and erectile "horns": *Tragopan* (5 spp.)
> > > > EE. Body and upper wing feather not as described above; tail often graduated and usually much longer than wing; males never with erectile horns.
> > > > > F.   Outermost primary longer than innermost; tail elongated and flat, with at least central pair of rectrices strongly barred.
> > > > > > G.   Central rectrices of males very long (400–1,600 mm, 15.6–62.4 in) and not fringed laterally; females with shorter and white-tipped rectrices: *Syrmaticus* (5 spp.)
> > > > > > GG. Central rectrices of males moderately long (400–500 mm, 15.6–19.5 in) and fringed laterally; females with shorter barred rectrices that lack white tips: *Phasianus* (2 spp.).

---

[3]Keys to species and subspecies of polytypic genera and species precede accounts of the genera in the text.

FF. Outermost primary shorter than innermost; tail variable, but if greatly elongated the central pair of rectrices enlarged and variably drooping.

    G. Twenty to 24 rectrices; feathers large and broad, with the central pair drooping and disintegrated; no iridescent feathers present: *Crossoptilon* (3 spp.).

    GG. Usually 14–18 rectrices (but 24–32 in one species), with the central pair sometimes drooping but not disintegrated; males often iridescent.

        H. Fourteen to 18 rectrices (rarely more), variably vaulted and compressed; bare facial area scarlet or blue, with lobes, wattles, or lappets in males.

            I. Smooth bare comb and throat lappets (reduced in females); ornamental and variably lanceolate neck feathers: *Gallus* (4 spp.).

            II. Papillose blue or red face wattles, but no comb or lappet; neck feathers not lanceolate: *Lophura* (10 spp.).

        HH. Eighteen rectrices; tail not strongly vaulted or compressed; no combs or lappets in males.

            I. Tail squarish and chestnut or iridescent and shorter than wings; both sexes with bright blue skin around eyes; males blackish below and extensively iridescent above: *Lophophorus* (3 spp.).

            II. Tail highly graduated, not iridescent or chestnut, and much longer than wing; bare area around eyes never bright blue.

                J. Red facial skin; grayish throat; straight brownish occipital crest: *Catreus* (1 sp.).

                JJ. Never with red facial skin or a grayish throat; decumbent colorful crest present in males: *Chrysolophus* (2 spp.).

Genus *Ithaginis* Wagler 1832

The blood pheasant is a small, partridgelike, montane species with a plumage that is very soft and with lanceolate feathers in males. The sexes are moderately dimorphic. A soft, short crest is present in both sexes, and males have a crimson feather coloration on the chin and the base of the tail. A small, bare orbital skin area is present. The wing is short and rounded, with the tenth (outermost) primary shorter than or nearly equal to the first. The tail is rounded and has 14 rectrices. The tail molt is perdicine, proceeding from the central feathers outwardly (centrifugal). The bill is short, stout, curved, and similar to that of some partridges and grouse. The tarsus is long and stout, with several spurs in males and knobs in females. A single species is recognized.

Key to Subspecies of Males of *Ithaginis cruentus* (in part after Delacour, 1977)

A. Throat red or tinged with red.
    B. Ear-coverts wholly black.
        C. Supercilium black: Kuser's blood pheasant (*kuseri*).
        CC. Supercilium red and black: Mrs. Vernay's blood pheasant (*marionae*).
    BB. Ear-coverts streaked with black and white.
        C. Supercilium red and black.
            D. Crest feathers disintegrated: Tibetan blood pheasant (*tibetanus*).
            DD. Crest feathers normal: Greenway's blood pheasant (*holoptilus*).[1]
        CC. Supercilium black.

---

[1]Doubtfully distinct from *rocki* (Cheng et al., 1978, p. 113).

D.   Median wing-coverts gray with a wide, pale green, central streak.

    E.   Crimson on crown: Himalayan blood pheasant (*cruentus*).

    EE. No crimson on crown: Sikkim blood pheasant (*affinis*).[2]

DD. Median wing-coverts entirely green.

    E.   Darker; crest shorter: Rock's blood pheasant (*rocki*).

    EE. Lighter; crest longer: Clarke's blood pheasant (*clarkei*).

AA. Throat gray.

  B.   Median wing-coverts wholly green: Geoffroy's blood pheasant (*geoffroyi*).

  BB. Median wing-coverts not wholly green.

    C.   Median wing-coverts brown and green.

      D.   Median wing-coverts marked brown and green: Bianchi's blood pheasant (*michaelis*).

      DD. Median wing-coverts brown, washed with green: Beick's blood pheasant (*beicki*).

    CC. Median wing-coverts entirely brown.

      D.   White shaft-lines on mantle narrow: David's blood pheasant (*sinensis*).

      DD. White shaft-lines on mantle wider.

        E.   Lighter; more reddish on cheeks, chin, and wings; tail pinker: Mrs. Sage's blood pheasant (*annae*).[3]

        EE. Darker; less reddish on cheeks, chin, and wings; tail more crimson: Berezowski's blood pheasant (*berezowskii*).

---

[2]Doubtfully distinct from *cruentus* (*Ibis* 1915, pp. 124–125).

[3]Doubtfully distinct from *berezowskii* (Cheng et al., 1978, p. 111).

BLOOD PHEASANT

Ithaginis cruentus (Hardwicke) 1822

Other Vernacular Names
None in general English use; ithagine ensanglantée (French); Blutfasan (German); semo (Tibetan); chilime (Nepalese).

Distribution of Species
Himalayas from Nepal eastward through Tibet and Sikang to the ranges of the Nan Shan in Tsinghai and Kansu and through northern Szechwan to southern Shensi and neighboring Honan, south from Sikang to northwestern Yunnan and neighboring northeastern Burma. Sedentary; breeding at altitudes varying from about 3,050 to 4,575 m (10,000 to 15,000 ft), but moving altitudinally down to about 2,135 m (7,000 ft) in the winter in some regions. Occurs in upper reaches of the coniferous forest and in rhododendrons or other scrub above it to the edge of the snow fields, receding or advancing with the snows on which it is often seen (Vaurie, 1965). See map 1.

Distribution of Subspecies
(after Vaurie, 1965; Wayre, 1969)
Ithaginis cruentus cruentus (Hardwicke): Himalayan blood pheasant (includes *affinis* Beebe). Resident from northern Nepal eastward to northwestern Bhutan, a population that has at times been recognized as a separate race (*affinis*). Ranges from subtropical pine to snowline, from 1,830 to 4,270 m (6,000 to 14,000 ft).

Ithaginis cruentus tibetanus Stuart Baker: Tibetan blood pheasant. Bhutan east of nominate *cruentus* and

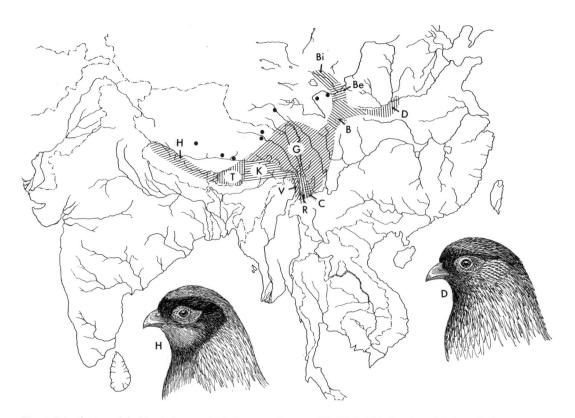

Map 1. Distribution of the blood pheasant, including races Berezowski's (B), Beick's (Be), Bianchi's (Bi), Clarke's (C), David's (D), Geoffroy's (G), Himalayan (H), Kuser's (K), Rock's (R), Tibetan (T), and Mrs. Vernay's (V). Some peripheral records of uncertain racial attribution are shown as solid circles.

southern Tibet, where it intergrades with *kuseri* between 92 and 93° E, in rhododendron scrub.

Ithaginis cruentus kuseri Beebe: Kuser's blood pheasant. Tibet and Himalayas, east of *tibetanus*, eastward to the upper Salween and Mekong, south to about 28° N. Intergrades with *marionae* and *rocki*. Occurs from 2,440 to 4,270 m (8,000 to 14,000 ft), in scrub and alpine forest.

Ithaginis cruentus marionae Mayr: Mrs. Vernay's blood pheasant. Mountains of northeastern Burma on the border of Yunnan east to the Shweli-Salween Divide. Intergrades with *kuseri*.

Ithaginis cruentus rocki Riley: Rock's blood pheasant. Northwestern Yunnan at about 27° N in the region between the Salween and Yangtze Rivers. Possibly an intergrade form between *kuseri* and *clarkei*. Includes *I. c. holoptilus* Greenway (1933).

Ithaginis cruentus clarkei Rothschild: Clarke's blood pheasant. Northwestern Yunnan in the Likiang Range.

Ithaginis cruentus geoffroyi Verreaux: Geoffroy's blood pheasant. Western Szechwan and eastern Tibet. Also reported from northern Yunnan and southern Tsinghai.

Ithaginis cruentus berezowskii Bianchi: Berezowski's blood pheasant. Mountains of western and north-western Szechwan north to the region of Sungpan and southern Kansu. Includes *annae* Mayr and Birckhead (1937).

Ithaginis cruentus beicki Mayr and Birckhead: Beick's blood pheasant. Northeastern Tsinghai in the region of the middle River Tatung to neighboring central Kansu, where it intergrades with *michaelis*.

Ithaginis cruentus michaelis Bianchi: Bianchi's blood pheasant. Northern and central Nan Shan Range (Kansu), grading southward into *beicki*.

Ithaginis cruentus sinensis David: David's blood pheasant. Mountains of southern Shensi (Tsinling Range) east to those of southwestern Honan. Also west to southern Kansu.

Measurements

Wing and tail lengths of the many subspecies are given in table 22. Ali and Ripley (1978) report the weights of males of *cruentus* to range from 482 to 568 g (1.1 to 1.3 lb). Cheng et al. (1978) reported the weights of five males of *berezowskii* as 520–600 g (1.2–1.3 lb) and of three females as 410–620 g (0.9–1.4 lb). A single male of *michaelis* weighed 655 g (1.4 lb). The eggs average 48 × 33 mm (1.9 × 1.3 in), and the estimated fresh weight is 28.8 g (1.0 oz).

Table 22
Ranges of wing and tail lengths of blood pheasants[a]

	Males		Females	
	Wing	Tail	Wing	Tail
cruentus	194–228, 7.6–8.9	164–178, 6.4–6.9	179–197, 7.0–7.7	140–154, 5.5–6.0
tibetanus	197, 7.7	177.5, 6.9	190–196, 7.4–7.6	—
berezowskii	188–204, 7.3–8.0	162–178, 6.3–6.9	178–184, 6.9–7.2	138–159, 5.4–6.2
marionae	190–200, 7.4–7.8	—	197, 7.7	130, 5.1
rocki	196–199, 7.6–7.8	—	180–186, 7.0–7.3	135, 5.3
clarkei	202–214, 7.9–8.3	162–176, 6.3–6.9	185–208, 7.2–8.1	141, 5.5
geoffroyi	208–223, 8.1–8.7	165–190, 6.4–6.7	205–210, 8.0–8.2	150–162, 5.9–6.3
beicki	204–225, 8.0–8.8	—	192–211, 7.5–8.2	—
kuseri	197–210, 7.7–8.2	147–171, 5.7–6.7	184–190, 7.2–7.4	—
sinensis	205–213, 8.0–8.3	165–180, 6.4–7.0	190–199, 7.4–7.8	132–150, 5.1–5.9
Overall	188–228, 7.3–8.9	162–190, 6.3–7.4	178–211, 6.9–8.2	130–162, 5.1–6.3

[a] Data from various sources, including personal observations. All measurements are reported in millimeters, followed by the equivalent in inches. A dash denotes no data.

Identification

In the Field (457 mm, 18 in)

This rather "chunky" species resembles a large partridge, but no partridge or other species of pheasant has crimson on the forehead, tail-coverts, and tail or pale green sides, underparts, and tail. Females are similar to males, but are uniformly reddish brown with a grayish to dull tawny head and throat that is sometimes tinged with crimson. Both sexes have red feet as well as a red area of bare skin around the eye. Both sexes are also quite vocal. The calls include a rallying call of scattered coveys, which is a long, high-pitched squeal resembling that of a kite. Short monosyllabic alarm notes are also used. Found in upper montane forest and subalpine scrub.

In the Hand

The distinctive crimson tinge on the face and tail of the male will immediately serve for identification, but females are less easily identified. The rather large area of bare red skin around the eye and the red legs, together with a relatively small size (wing 178–211 mm [6.9–8.2 in]) and a faintly vermiculated rufous to brownish plumage without strong patterning, should serve to separate females from all other pheasants and the few species of large partridges with which they might be confused.

Geographic Variation

Variation among the males of this species is very great and tends to be clinal. Two general groups are recognized, including a more southern and western series of populations (the nominate *cruentus* group), and a more northern and eastern group (the *sinensus* complex, including *michaelis, beichi, berezowskii, annae,* and *sinensis*). In the nominate group the greater upper wing-coverts are green or are at least well tinged with green, whereas in the *sinensis* group these feathers are reddish brown. The *sinensis* group males also have distinct "ear-tufts" and lack crimson pigments on the head and breast. However, these differences are not entirely clear-cut, because some individuals of the *sinensis* group may show crimson feathers on the forehead and chin, and some races (*beicki* and *michaelis*) of the *sinensis* group have wing-coverts that are somewhat tinged with green. The crimson pigments are increasingly disseminated on the head and breast from west to east in the nominate group (or from *cruentus* to *kuseri*), but decrease in northwestern Yunnan (from *rocki* to *clarkei*) and disappear farther east in *geoffroyi*. Furthermore, the ear-tufts may be seen in reduced form as far west as *kuseri*, and there is a clinal decrease in general color saturation from north to south among the races of the *sinensis* group. Females of the two groups are much more uniform, but in the *sinensis* group they also exhibit ear-tufts and are generally more grayish and less ochraceous on the face and throat than are those of the nominate group (Vaurie, 1965).

Ecology

Habitats and Population Densities

Beebe (1918–1922) described the habitats of *geoffroyi* as ranging from 3,660 to 4,575 m (12,000 to 15,000 ft) and including firs, larches, and oaks, plus the higher treelike and shrublike zones of rhododendrons, and finally the alpine meadow and grasses at elevations of about 4,875 m (16,000 ft). Beebe suggested that the seasonal movements may involve migrations of several thousand feet of altitude annually, although on more sheltered ranges blood pheasants might remain the entire year with much less seasonal movement. Of *kuseri* Beebe stated that the birds are usually found in bamboo jungle close to snowline, but in early mornings and late evenings they come out into the open. Where there was no snow, blood pheasants were very conspicuous in this habitat, which probably accounts for the restriction of their activity in such open areas to periods near sunrise and sunset.

According to Beebe, the summer period is spent hardly lower than the upper zone of small pines, and blood pheasants often range upward to the tundralike zone of alpine meadowland. Soon after the autumn establishment of coveys, blood pheasants begin to be forced downward by snowfall. They evidently show no preference for either northern or southern exposures. During this period they spend much of their life in the open, but always close to low, dense undergrowth. In the eastern Himalayas rhododendron forests and silver firs grow in this zone, whereas in parts of Nepal blood pheasants apparently prefer clumps of mountain bamboo. Where bamboo is abundant, blood pheasants seem to favor it for shelter, but prefer to roost in trees such as rhododendrons or firs. By winter, they are found among the open coniferous forests of fir and juniper nearer the lower elevations.

There is little direct information on population densities, but most accounts suggest that blood pheasants are sometimes fairly common, occurring in flocks of about 10–30 birds. There is an early account of an English sportsman who shot 36 birds on a single ridge during one day. Lelliott and Yonzon (1981) reported that in Nepal they observed a density of between six and nine pairs in a study area of 1.2 km² (0.5 mi²). Li (1996) summarized density data for China, which among four races varied greatly, from 0.7 to 66.0 individuals per square kilometer (1.8 to 170.1 per square mile), and averaging about 17 (44.0). Jia et al. (1997) reported densities of 11.3 (breeding season) to 22.6 (winter) individuals per square kilometer (29.3 to 58.5 per square mile) in Wolong Sanctuary, Szechwan. Home ranges were greatest among both sexes in February, smallest for females but very large for males during incubation, and intermediate for broods. The large home range of males during incubation might suggest that males give little or no nest protection. Some vertical migration occurred seasonally and even on a daily basis during some summer days. Home ranges of neighboring pairs overlapped. Similar studies in a Kansu sanctuary (Sun and Fang, 1997) indicated a prebreeding density of 17 birds per square kilometer (44.0 per square mile), and home ranges of 15.2–17.6 ha (37.5–43.5 acres) during that period. Brief monogamous pair bonds were formed, and both parents helped tend for the chicks. Home ranges of two broods were estimated at 24.7 and 37.2 ha (61.0 and 91.9 acres).

Competitors and Predators

Beebe (1918–1922) said that little information was available to him on competitors and predators. He judged that neither Himalayan monals nor snowcocks posed significant competitive threats. Of predators, he judged the grey fox (*Vulpes ferilata*?) to be probably the most serious threat and noted that it was one of the most common carnivores of the upper rhododendron zone. Beebe also reported that beech martens (*Martes flavigula*?) were sometimes in the same area and imagined that both dholes (*Cuon alpinus*) and Himalayan leopards (*Panthera pardus*) probably take some blood pheasants. The major large

raptors of the area are the golden eagle (*Aquila chrysaetos*) and Bonelli's eagle (*Hieraaetus fasciatus*). Beebe commented that the females more nearly match their rocky background than do males and that when squatting the males do not exhibit any of the crimson coloring of their underparts.

General Biology

Food and Foraging Behavior

Foods evidently vary greatly by season in blood pheasants. During the winter blood pheasants are believed to feed mostly on shoots of firs and junipers, as well as on berries, mosses, and bamboo leaves. Beebe (1918–1922) noted that birds shot during periods of heavy snow in fir forests were almost inedible because of their diet of fir and juniper sprouts. Earlier in the fall they apparently live on small fruits, leaves, seeds, and moss spore cases, as well as bamboo shoots, berries, and rose pips. In mid-April Beebe observed a flock foraging after a heavy snowfall on lily seed cases, which upon inspection proved not only to have seeds present but also various insects.

According to Beebe, there is apparently no specific foraging period. Blood pheasants sometimes may be found feeding at high noon, scratching for berries and insects among moss and coarse grasses. Unlike the Himalayan monals of the same area, blood pheasants are strong scratchers, but do not have favorite scratching sites.

In captivity, it has been found that blood pheasants do a great deal of grazing. In Major Iain Grahame's aviaries they kept a grassy area plucked so short that it resembled a lawn (Grahame, 1971, 1976). They also ate lucerne (alfalfa) pellets, turkey pellets, kibbled maize, and a mixture of such things as boiled eggs, diced apples, raisins, carrots, mealworms, and other foods used for "softbill" birds. Other aviculturists have noted that blood pheasants favor chopped lettuce, onions, and potatoes (Roles, 1981). At the Bronx Zoo blood pheasants fed largely on pelleted diet supplements with a few insects, fruit, and greens (Donald Bruning, pers. comm.).

Movements and Migrations

The seasonal altitudinal movements typical of blood pheasants were mentioned above in the Habitats and Population Densities section. These movements are evidently fairly leisurely ones, done on foot. Their movements are rather slow and fowl-like, but with the tail held low and partridgelike. When alarmed, blood pheasants arise in a quail-like covey, but their flights are brief and direct. On alighting they prefer to run uphill into the nearest cover, which is usually scrub rhododendrons. If trees are nearby and a dog flushes them, they may simply fly up to the lower branches. With lesser degrees of alarm, they may flee on foot, running with their necks outstretched and their heads and tails held rather high (Beebe, 1918–1922).

Daily Activities and Sociality

Throughout most of the day blood pheasants forage actively, typically in coveys or families, and the birds also roost together through the nighttime hours. At least at times, the same roosting site will be used on consecutive nights. Roosting may be done in trees, amid thickets, or on the ground. Occasionally when roosting on the ground blood pheasants will cluster in the shelter of a crag or boulder and orient themselves outward from the center, in typical quail-like manner (Beebe, 1918–1922). There is less evidence that blood pheasants regularly burrow into snow; although this was reported as typical in early literature, Beebe (1918–1922) suggested that the cold night fogs of the Himalayas often turn the snow crust to ice, and thus make such behavior extremely risky for birds. He believed that snow tunneling is thus either extremely rare or perhaps even of accidental occurrence owing to the possible snowing-in of a covey. Grahame (1971) likewise doubts that blood pheasants ever tunnel, for their habitats contain innumerable caves and overhanging boulders under which they might easily roost. In captivity they seem to be very vulnerable to excessive heat and winter rains (Donald Bruning, pers. comm.).

It is not yet known when the coveys dissolve prior to the breeding season, but this is likely to occur shortly prior to the nesting season; the coveys are certainly reestablished in early autumn by the amalgamation of family units. Beebe (1918–1922) states that this autumn period of covey formation occurs about October. A flock of several families is likely to consist of 15–40 birds. These groups remain established throughout the winter and roost and feed closely together. The sex ratio of such flocks is apparently nearly even or perhaps has a slight excess of males.

Social Behavior

Mating System and Territoriality

Nearly all writers have commented on the monogamous nature of this species. Beebe (1918–1922) reports that pairs of blood pheasants and their young remain together through the winter; thus, presumably an essentially permanent pair bond is maintained. This conclusion was partially based on Beebe's observations that the flocks he saw consisted of approximately equal numbers of males and females. Although there seems to be no good evidence to support this view, there is also little reason to doubt it, given the low degree of sexual dimorphism and the seemingly rare and simple display tendencies of blood pheasants. However, it has been suggested without specific supporting data that perhaps the birds may also be occasionally polyandrous or polygynous (Ludlow and Kinnear, 1944). There is likewise no good information on territoriality, although in captivity it has been noted that males tend to be quarrelsome with one another. Grahame (1971, 1976) judged that the species shows "loose monogamous pair-formations within coveys." In captivity, he found that when two pairs were placed in a group of three adjacent and connected aviaries (10 × 2 m, 33 × 7 ft), each of the two pairs occupied one of the end cages and the middle cage became a "no man's land" into which occasional sallies were made by the males or sometimes both members of a pair. This resulted in much chasing, either by male after male or male after female.

Voice and Display

The earliest notes on vocalizations come from Beebe (1918–1922). He observed that when threatened the flock members utter a series of sudden sharp notes, *seep! seep! seep!* After the scattered flock members have begun to reassemble, they utter a covey call, *see-e-e-elpe!*, which is simply a more drawn out alarm note, but with a snapping off at the end. Beebe also once heard a male within the flock uttering a repeated *silpe* note that somewhat approached a cackling call. Others have described the alarm note as a harsh *ship, ship*, and the presence of another note sounding something like the squeal of a kite (Baker, 1935). Beebe (1918–1922) described the call of *geoffroyi* as a single, long, drawn-out, wheezy whistle followed by several sharper notes and another call as loud and long-continued squealing sounds. Both of these descriptions were based on secondhand observations.

Recently, Lelliott (1981*b*) has observed the behavior of blood pheasants in some detail and has provided a good deal of new information on their vocalizations and other behavior. He recognized five types of calls of fairly definite meaning. The first is the *sree* call, which is a high-pitched squeal uttered about four to nine times per minute while the birds were foraging. The second is a high-pitched trill, lasting nearly a second and seemingly associated with mild alarm. The trill calls were uttered by both sexes and were used to maintain contact between pairs. A loud, piercing, muted squeal, *sree cheeu, cheeu, cheeu . . .* with a varying number (two to six) of *cheeu* notes was uttered by both sexes and evidently served to bring scattered individuals of a covey back together. There is some evidence that birds of a covey are able to recognize the call notes of other birds of that covey and approach such calls, whereas the playback of calls from another covey tended to repel them. This is the call that Beebe (1918–1922) referred to as a covey call; it has been described by others as resembling the squeal of a kite. Lelliott described a fourth call as a high-pitched, strident *chic* note of short duration and repeated at intervals of about 1 or 2 seconds. This call was uttered by both sexes and, at least in the case of males, was usually associated with alarm behavior. The last call, a buzzing note of about a second in duration, was heard only from males. Lelliott suspected that it may have been produced by the bird's feathers, because it was apparently associated with tail-fanning during display.

Lelliott observed blood pheasants primarily during the breeding season, during which 71 percent of the birds observed were in pairs or multiples of pairs, 20 percent were single males, and there was one observation each of a group of four males and of a single female. On this basis he judged the birds to be monogamous. Although he observed no courtship display, Lelliott did observe one copulation, which was preceded by the female standing on a raised rock and uttering the trill call prior to squatting down. The male then approached and the pair copulated for about 10–15 seconds while both birds called softly. Following treading the birds continued to feed quietly.

Almost nothing has been written on display in the blood pheasant, which presumably reflects its relative rarity and inconspicuous nature. Beebe (1918–1922) stated that during courtship the male "spreads its tail and wings, drooping the latter, raises the crest, swells out the breast feathers, and struts before the female, turning round and round." In the courtship period the males are said to fight fiercely with one another. Whether the wing-drooping and turning around the female Beebe described corresponds to the waltzing display, as seems likely, is not certain. It would also be of interest to learn if tidbitting behavior occurs in this species. Grahame (1976) stated that the only sound he heard from his captive blood pheasants during display was a *purrrh*, made by a sudden fanning of the tail feathers. He said that their display resembles that of the koklass, although it has more "urgency" of movement and is less noisy than for that species. In general the display consisted mostly of crest-raising and a good deal of chasing, but no evident vocalizations. Drawings of this species' general behavior pattern are shown in figure 14.

Reproductive Biology

Breeding Season and Nesting

The nesting season of the blood pheasant is only generally known. Reportedly *cruentus* breeds in April and May, and a nest containing three fresh eggs of *kuseri* were found in the beginning of May (Baker, 1935). Grahame (1976) reported that in Nepal the coveys have apparently broken up by late April, but laying has not yet begun. In Nepal laying apparently coincides with the blooming of the rhododendrons (*R. campanulatum*). A nest was later found on 13 May with seven eggs. One nest of *geoffroyi* was found at 4,115 m (13,500 ft) under brushwood in a forest and contained seven eggs. These were a narrow and elongated oval shape and were pale reddish buff with irregular blotches of several shades, the darkest being deep reddish brown (Beebe, 1918–1922).

The nest of *cruentus* is said to be constructed of grass and leaves on the ground among bushes and grass. Two nests found in Sikkim were at about 3,660 m (12,000 ft), and the eggs were located in a hollow scratched in a pile of loose fallen leaves at the foot of bushes in a pine forest. A nest of *kuseri* with two eggs was located at about 3,660 m (12,000 ft) under a clump of bamboos, with the surrounding area then still being largely under snow. In Bhutan a nest with four eggs was located on 30 May situated on a

Figure 14. Behavior of the blood pheasant, including wing- and leg-stretching (A), dust-bathing (B), alert walking (C), and relaxed standing (D). After photos by Jean Howman (D) and the author.

bank under an overhanging shrub. Similarly, a nest of *kuseri* was located on 29 May at 3,660 m (12,000 ft) near Lagong, Tibet. It contained six incubated eggs and was on a bank under a juniper (Ludlow and Kinnear, 1944).

Incubation and Brooding

There are no good accounts of incubation in the wild, but Grahame (1971, 1976) has provided a good deal of avicultural information on the species. Of two pairs that Grahame had in connected aviaries, the females laid a total of 15 eggs between 20 April and 16 May, with the females sometimes laying in one another's nest. The incubation period ranged from 26 to 29 days, averaging 27 days. Grahame obtained fertile eggs from a 1-year-old female mated to an older

male and from two females mated to a single male. Females seemingly preferred to nest in baskets located a meter or two above the ground. Their eggs were incubated artificially.

Growth and Development of the Young

Chicks tended to be very wild during the first few days and often hid in corners (Grahame, 1971, 1976). By the time they were 24 hours old, chicks were already grazing and made gooselike plucking movements. When they were only 5 days old, the males could be told from females by the first signs of gray on their shoulders. Full adult plumage was attained when the birds were 5 months old. By their first winter they differed from older birds only in having shorter spurs. There was much individual variation in the amount of red present on the breast.

Of 62 eggs laid, Grahame was able to hatch 29 young and rear 12 of them. The birds are apparently very susceptible to diseases such as aspergillosus and tuberculosis and to infections of the liver, kidneys, and caecum.

Evolutionary History and Relationships

There are still no good bases for classifying the blood pheasants in either the Perdicini or the Phasianini, although there is certainly little reason for believing that they are closer to the pheasants than to the Old World partridges. Apart from the monogamous mating system, the coveylike social organization and the majority of the plumage characteristics are certainly partridgelike. Delacour (1977) suggested that the lanceolate feathers, crest, and short beak all suggest a possible relationship to *Pucrasia*. Grahame (1971, 1976) also mentioned some similarities in the sexual displays of these two forms, both of which also seem to be highly herbivorous. In a study of 259 specimens from China, Yang et al. (1994) concluded that only a single species of blood pheasant should be recognized and that the forms *rocki*, *marionae*, and *holoptilus* are best considered synonyms of *kuseri*. Studies of the green pigment present on the underparts of the blood pheasant show that it is a copper-containing molecule (phasioverdin) similar or identical to that in the tropical partridge genus *Rullolus* as well as the comparable pigment found in turacos. This might argue that the blood pheasant is better placed taxo-nomically with the partridges than among the pheasants (Dyck, 1992).

Status and Conservation Outlook

The remote habitat of the blood pheasant is likely to keep it out of direct conflict with human activities for the foreseeable future. Also, it is too small and too inconspicuous to be overhunted for either sport or food. Nevertheless, it is perhaps the tamest of all the pheasants and thus is subject to local extirpation by natives if hunting is not controlled. Gaston (1981*a*) judged that the species is fairly common throughout its range in the central and eastern Himalayas and is still numerous in some places. He considered its habitat to probably be in no significant danger of reduction and its status there to be safe. The status and abundance of the Chinese populations is not as well known. However, they have evidently decreased, especially in Shansi and Honan. Among the races occurring in China, there are only two locality records for *michaelis* and *clarkei* and three for *marionae* and *rocki* (Li, 1996). As noted in the key above, the two latter forms are doubtfully valid. McGowan and Garson (1995) classified five subspecies (*kuseri*, *rocki*, *marionae*, *holoptilus*, and *clarkei*) as vulnerable. Del Hoyo et al. (1994) stated that in addition to occurring in China (70 sites, including four protected reserves), this species is known from more than 20 sites (including four protected reserves) in Nepal, three sites in India, and from many locations in Sikkim and Bhutan, where it is probably still fairly common. Its status in Afghanistan is unknown.

Genus *Tragopan* Cuvier 1829

The tragopans, or horned pheasants, are medium-sized montane pheasants in which the sexes are highly dimorphic. The coloration of males tends toward crimson on the head and sometimes elsewhere, with extensive white to buffy dorsal spotting. The wings are rounded, with the tenth primary the shortest, and the fifth and sixth the longest. The tail comprises 18 feathers, is rounded, and usually shorter than the wing. The tail molt is perdicine (centrifugal). The bill is short and stout, with the forehead feathers almost reaching the nostril. The tarsus is very stout, about as long as the middle toe and claw, and in males has a short spur. Males also have a short occipital crest, two erectile and brightly colored fleshy horns that are erected during courtship, and a brilliantly colored gular lappet or bib that can be expanded and exposed during display. The sides of the head and throat are naked or only thinly feathered and brightly colored. Five species are recognized.

Key to Species (and Subspecies of Males) of *Tragopan*

A. Head red and black, with a colorful face and lappet (males).
  B. Underparts spotted; red to bluish facial skin.
    C. Spots on underparts large, pearl gray, and edged with brown; face bluish: Temminck's tragopan.
    CC. Spots on underparts small and white, edged with black.
      D. Breast and upper parts mostly reddish brown; face dark bluish: satyr tragopan.
      DD. Breast and upper parts mostly gray to blackish; face reddish orange: western tragopan.
  BB. Underparts not spotted; yellowish to orange facial skin.
    C. Breast and underparts buffy; upper parts spotted with buffy: Cabot's tragopan.[1]

[1]See species account for subspecies determination.

CC. Breast and underparts reddish to grayish; upper parts spotted with white: Blyth's tragopan.

    D.   Paler; red of breast forming a broad band: eastern Blyth's tragopan (*blythi*).

    DD. Darker; red of breast forming a narrow band: western Blyth's tragopan (*molesworthi*).

AA. Head brownish or grayish; no bare lappet (females).

    B. Dorsal plumage streaked with white; orbital skin reddish to orange.

        C. Gray dominant in ventral plumage; small round white spots dorsally: western tragopan.

        CC. White or buff dominant in ventral plumage; whitish triangular markings dorsally: Cabot's tragopan.

    BB. Dorsal plumage not streaked with white; orbital skin yellowish to bluish.

        C. Bend of wing not orange-rufous; bluish orbital skin: Temminck's tragopan.

        CC. Bend of wing usually a rich orange-rufous; if not, then with yellowish orbital skin.

            D.   General plumage tone olive, with coarse markings; black ocelli dominant dorsally; yellowish orbital skin: Blyth's tragopan.

            DD. General plumage tone creamy buff, with fine markings; black ocelli not dominant in dorsal plumage; bluish orbital skin: satyr tragopan.

WESTERN TRAGOPAN

Tragopan melanocephalus (J. E. Gray) 1829

Other Vernacular Names
Western horned pheasant, black-headed tragopan; tragopan de Hastings (French); West-Satyrhuhn, Jewar (German); singmoonal (western Himalayan).

Distribution of Species
Western Himalayas, between about 2,440 and 3,660 m (8,000 and 12,000 ft), from Hazara eastward to Garhwal. This species also has been reported from Ladakh.

The Ganges seems to constitute the eastern limit of the range. The western tragopan moves altitudinally, wandering occasionally down to 1,220 m (4,000 ft) in winter. It inhabits dense coniferous or mixed mountain forests to rhododendrons or birch scrub at the upper edge of the forest (Vaurie, 1965). See map 2.

Distribution of Subspecies
None recognized by Delacour (1977).

Map 2. Distribution of tragopans, including Blyth's (B), eastern Cabot's (C), eastern Blyth's (E), satyr (S), Temminck's (T), western (We), western Blyth's (W), and western Cabot's (WC). Dotted lines show presumed original ranges of western and western Blyth's. Cross-hatching indicates the presumptive maximum range of eastern Blyth's. Blackened areas or locality circles indicate known current ranges of these forms. Upper inset maps detail known current distributions of the western tragopan (left) and of the Blyth's (right) in Nagaland and Arunachal Pradesh.

Measurements

Delacour (1977) reported that males have a wing of 255–290 mm (9.9–11.3 in) and a tail of 220–250 mm (8.6–9.8 in), whereas females have a wing of 225–250 mm (8.8–9.8 in) and a tail of 190–200 mm (7.4–7.8 in). Males weigh about 1,800–2,150 g (4.0–4.7 lb) and females 1,250–1,400 g (2.7–3.1 lb; Ali and Ripley, 1978). The eggs average about 63 × 42 mm (2.5 × 1.6 in) and have an estimated fresh weight of 61.3 g (2.1 oz). Measurements of wing chords obtained by Islam (1992) are the largest of any of the five tragopan species.

Identification

In the Field (635–686 mm, 25–27 in)

This species resembles a large but stout pheasant. It is usually seen singly, in pairs, or small coveys in fairly dense montane vegetation. The white circular spotting on the grayish upper parts, especially the lower back and upper tail-coverts is distinctive, but the spotting is less regular in females than in males. Both sexes have reddish faces, and the legs are usually pinkish. During the breeding season the male can be identified by its distinctive call, uttered most often at dusk and daybreak at intervals of about 5 or 10 minutes. It is a far-carrying *waa* note, resembling a goose calling or the bleating of a small goat. These call sequences are very extended, lasting about 45 seconds on average (Islam and Crawford, 1996). The only other tragopan possibly occurring within this species' range is the satyr.

In the Hand

Males may be easily distinguished by their predominantly gray upper parts and reddish facial skin. Females are more brownish gray, both above and below, and their distinctly grayish cast with black vermiculations will separate them from all other species of the genus. The extensive white spotting helps to separate female tragopans from those of other pheasants. The elongated white central spot on each feather is bordered with black. The male's lappet is bluish purple centrally, with four or five reddish and irregular to leaflike markings along each side, vertically arranged much like those of the satyr. They presumably likewise become crimson during display, but are usually described as pink, probably on the basis of dead or nonbreeding specimens. Museum specimens suggest that the shape of the lappet is more concave ventrally than in the other tragopans and is similar in length to that of the satyr, or about 100 mm (3.9 in). However, it is more densely covered by bristle-like feathers, especially on the rear surface.

Ecology

Habitats and Population Densities

Islam (1983) has summarized the general habitat characteristics of the western tragopan as consisting of a summer range in forests of spruce (*Picea smythiana*), deodar cedar (*Cedrus deodara*), and brown oak (*Quercus semicarpifolia*) at the upper edge of the treeline, from 2,500 to 3,600 m (8,200 to 11,810 ft) elevation. During winter they are found in midaltitudinal dense coniferous or mixed mountain forests with a northern aspect and between 2,000 and 2,800 m (6,565 and 9,185 ft). An undergrowth cover of rue (*Ruta*) and ringal bamboo (*Arundinaria*) provides dense cover in both summer and winter habitats in eastern parts of its range.

In Pakistan the western tragopan inhabits steep forested slopes in dense forests in a transition zone between moist and dry temperate climatic zones. Islam (1983) observed them in predominately coniferous, predominately deciduous, and mixed coniferous–deciduous forests, with the coniferous species consisting of fir (*Abies pindrow*), blue pine (*Pinus wallichiana*), spruce, and yew (*Taxus wallichiana*), and the deciduous species being brown oak, cherry (*Prunus padus*), walnut (*Juglans regia*), horse chestnut (*Aesculus indica*), maple (*Acer caesium*), and birch (*Betula utilis*). All forests where tragopans were observed were characterized by having thick undergrowths of *Viburnum nervosum*, with *Skimmia laureola* and bracken fern (*Pteridium* spp.) in some areas. Rue and ringal bamboo do not grow in Pakistan (Islam, 1983).

In Himachal Pradesh western tragopans primarily occur between 2,500 and 3,000 m (8,200 and 9,845 ft), but occasionally occur below 2,000 m (6,560 ft), presumably in winter. However, they have been observed as high as 2,900 m (9,515 ft) in November and January, suggesting that at least some of the birds are fairly sedentary (Gaston et al., 1981). There the species shows a preference for higher altitude coniferous forests, but Gaston et al. occasionally observed them in mixed deciduous forest and in higher altitude oak forests.

The species is now so rare that population densities are very difficult or impossible to estimate. However, Gaston et al. (1981) noted that in 400 party-hours spent in the proposed Manali National Park, only four sightings were obtained in the three main river valleys (Solang, Manalsu, and Hamta). They judged that the entire upper Beas Valley probably supports no more than about 1,000 birds. This area consists of approximately some 15,000 km^2 (5,790 mi^2). Mirza (1981*a*) judged that in the Neelum Valley of Pakistan there may be 0.4–0.8 pairs per square kilometer (1.0–2.0 per square mile) of habitat. Islam (1983) observed 29 birds in three areas of Pakistan, giving a density of 1.3 birds per square kilometer (3.4 per square mile). In 1977, in the Salkhala reserve 12 males were heard and six birds were seen in a 31-km^2 (12.0-mi^2) area, and nine were reported in a 26-km^2 (10.0-mi^2) area in Kuttan reserve in the Neelum Valley (Mirza, 1976). In an analysis of habitat requirements of this species in Pakistan, Islam and Crawford (1988) concluded that the presence of relatively shorter vegetation, including shrubs as well as young coniferous and deciduous trees, are crucial parts of its environment. Grazing and wood-collecting activities are thus deleterious to its survival.

Competitors and Predators

In Himachal Pradesh, this species occurs in company with at least four other species of pheasants, but only the koklass and Himalayan monal are relatively common in the ecological zones favored by the western tragopan. Neither one is likely to be a significant competitor. At the eastern edge of its range the western tragopan probably encounters the satyr tragopan. The two species seem to be ecological replacement forms, suggesting that competition between them does indeed exist. The Jumna Valley seems to represent the approximate point of geographic contact between them (Gaston, 1981*a*).

Islam (1983) did not mention any predators of significance to the status of the western tragopan in Pakistan. Gaston et al. (1981) list a substantial number of predatory mammals inhabiting Himachal Pradesh (three canids, two bears, a weasel [*Mustela sibirica*], a marten [*Martes flavicula*], a civet [*Paguma larvata*], a cat [*Felis chaus*], and a leopard [*Panthera pardus*]), most of which might represent varying levels of threat to tragopans, particularly the felids and mustelids. However, they found no direct evidence of tragopan predation for any of these species.

General Biology

Food and Foraging Behavior

Beebe (1918–1922) found western tragopans to be foraging on newly sprouted leaves. Based on the accounts of various sportsmen, he believed that such vegetable matter forms their principal diet. Beebe quoted an extensive account by a Mr. Wilson who stated that the birds forage on the leaves of trees such as box and oak and shrubs such as ringal bamboo and one something like a privet. Western tragopans were also said to eat roots, flowers, insects and their grubs, acorns, seeds, and berries of various kinds, but in small amounts as compared with leaves. In captivity western tragopans seem to be typical of their genus, consuming primarily vegetable materials, with emphasis on fruit and berries (Howman, 1979).

Movements and Migrations

Gaston et al. (1981) suggested that this species is relatively sedentary in Himachal Pradesh, showing little vertical movement with the seasons. However, Ali and Ripley (1978) give the species' elevations as breeding from 2,400 to 3,600 m (7,875 to 11,810 ft) and wintering at about 1,350 m (4,430 ft). Wilson (cited in Beebe, 1918–1922) also states that in winter western tragopans are found in the thickest parts of the oak, chestnut, and morenda pine forests having a dense undergrowth of ringal bamboo. However, during the breeding season, they are to be found in the higher parts of the forest, up to the zone of birches and white rhododendrons and almost up to the extreme limits of the forest.

Daily Activities and Sociality

Roosting by western tragopans is done in trees, preferentially low evergreens, where there are closely interwoven leaves and branches, rather than in taller trees. The birds were once normally found in groups from two or three to a dozen or so, which tended to be rather widely scattered over the forest. Such groups are apparently typical only where the birds are undisturbed; in such cases they tend to remain in ones and twos scattered over considerable distances. At least during the winter western tragopans seem to be quite sedentary, rarely moving far to forage. However, during conditions of heavy snowfall they may sometimes be found on bare, exposed hillsides, in narrow wooded ravines, patches of low brushwood and jungle, and other places where the ground is sheltered from the sun by trees and bushes.

When several are alarmed simultaneously, western tragopans all begin to cry at once and scatter in different directions, with some flying into trees while others flee on the ground. When first flushed they may simply alight in a nearby tree, but after a second flushing they often go some distance, almost invariably downhill (Wilson cited in Beebe, 1918–1922).

Like the other tragopans, western tragopans disperse in spring, with males establishing territorial calling perches that are well separated from one another. In early April the males are said to be moving a good deal, and they begin to call loudly, usually from a large stone or while perched on the thick branch of a tree or the trunk of one that has fallen to the ground. During autumn the families gradually begin their descent to wintering areas.

Social Behavior

Mating System and Territoriality
Ali and Ripley (1978) judged this species to be monogamous, with the male assisting in tending the chicks, even though the incubation was said to be done entirely by the female. This degree of monogamy seems as yet unproven, but it is also not established that the species is serially polygynous, as has been suggested by Ridley (1987). Given the relatively short advertisement and breeding season, it seems most likely that monogamy or simultaneous polygyny prevails.

Wilson (cited in Beebe, 1918–1922) reports that western tragopans begin to pair in early April, when the males are found well scattered and calling at intervals, sometimes all day long. The calls can be heard for upward of a mile, suggesting that relatively large territories are held by these birds. However, there are no available estimates of territory sizes.

Voice and Display
Wilson (cited in Beebe, 1918–1922) stated that the only calls of the species that he knew of were an alarm call consisting of a series of loud wailing cries sounding similar to the calls of a young lamb or kid, *waa, waa, waa . . .* , with each syllable uttered slowly and distinctly at first, but increasing as the bird is

hard pressed or about to take flight. The male's advertisement call is similar but much louder, with only a single note uttered each time, and similar to the bleating of a lost goat. It may be uttered every 5 or 10 minutes for hours on end or may be produced only two or three times during an entire day. Islam and Crawford (1996) found that the male's advertisement call has from 3 to 36 notes (average 11), uttered 2–6 seconds apart, the entire sequence lasting from 8 to 197 seconds (average 45 seconds). There is no change in pitch or volume during the series. The species' alarm call is a series of brief notes uttered at several-second intervals. This call is uttered progressively faster and higher in pitch among species in the sequence of satyr, Temminck's, and Cabot's tragopans, which probably reflects their graduated differences in average body size.

Beebe (1918–1922) heard a few other vocalizations of the western tragopan including a call similar to a "drowsy *waaa-waaaaaaaak!*" of a domestic hen, and a low chuckle uttered by a female approaching its nest.

The lappet color and pattern may be easily seen from study skins (figure 15), but there is almost no detailed information on the species' actual display. Delacour (1977) implies that all species of tragopans have essentially the same display sequence, but this is certainly not true, judging from the comparative studies of Kamal Islam. Nonetheless the display may be quite similar to that of the satyr, its apparent nearest relative.

Islam and Crawford (1998) have identified several major components of tragopan displays, based on studies of the other four species. It is likely that the western tragopan follows this general pattern. Lateral display by males of these four species consists of several elements: the wing nearer the female is lowered, the body is laterally compressed, the male makes half-turns toward the female, and performs side-stepping toward her. Males also make ground-pecking movements, similar to the tidbitting displays of typical pheasants. A strongly arched body posture during lateral display occurs in the Temminck's and satyr tragopans. The frontal display is more elaborate, prolonged, and complex, consisting of several stages. The male begins with initial peering and head-bobbing toward the female while hiding most of its body behind a solid object, such as a large rock or tree stump. Soon the male erects his "horns" and expands his lappet to form a shieldlike gorget. He

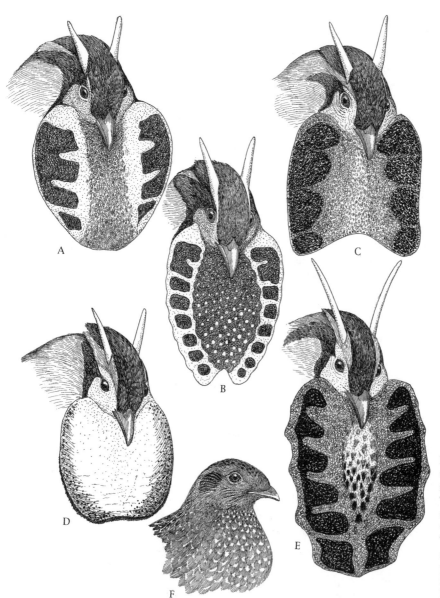

Figure 15. Lappet patterns of male tragopans, including satyr (A), Temminck's (B), western (C), Blyth's (D), and Cabot's (E). A female Temminck's tragopan is also shown (F). After photos of museum specimens (C) and of live birds.

also fans and depresses his tail and progressively synchronizes a long series of wing-flapping movements with repetitive clicking vocalizations, causing the horns to vibrate rapidly. The wings may be flapped either while they are outstretched or partially folded, and the often colorful anterior upper wing-coverts are variably raised. The drumroll-like series of rapid clicking calls and wing-flaps is followed suddenly by a rapid vertical rearing of the male on the tips of his toes while orienting his entire body toward the female and hissing loudly, his horns fully erect and his lappet maximally expanded. He then often rushes toward the female with his body and neck feathers ruffled and his wings drooping or held out some distance from the body. Typically he either climbs over or runs around the object behind which he had been hiding, but sometimes he simply "runs in place" while remaining behind the obstruction.

This entire frontal display sequence, one of the most spectacular of all pheasant displays, may require from about 30 seconds to several minutes to complete, depending on the duration of the preliminary stages. These in turn may depend on how long the female remains in optimum proximity and in full view, and the sequence may terminate if she moves out of sight. However, once the wing-flapping phase begins, the sequence is likely to continue to completion. It seems probable that the male's final rush may often lead to a copulation attempt. The sudden, bizarre appearance of the male as it rises from behind its hiding place perhaps serves to immobilize the female, causing her to remain in place long enough to be mounted. There are some visual similarities between this amazing display sequence and the equally incredible frontal display of the great argus. The very complexity of the tragopans' male displays and signaling structures would cause one to wonder if they are similarly polygynous.

Reproductive Biology

Breeding Season and Nesting
There are few exact dates of nesting for the western tragopan. However, one nest of six eggs was found on 25 May in Hazara Province (now a district of Pakistan), and another nest of three eggs was found on the ground, being carelessly constructed of grass, small sticks, and a few feathers. The second was on a slanting tree about 3 m (10 ft) aboveground in a hollow where a large branch had been torn away. The tree was a wild cherry and about 30 m (100 ft) or so above a stream on a well-wooded slope. A third nest was found in the Nila Valley of Garwhal and was located under the protection of a small bush in an open glade that was situated in very dense ringal jungle on a steep and rocky hillside. Only fragments of eggs were present. Also, a clutch of two eggs was found during June in a tree nest (Baker, 1935).

Besides these nests, Beebe (1918–1922) describes seeing a clutch of apparently three eggs collected 4 June in Pir Pangal, Kashmir. He also observed a female on a still-uncompleted nest 12 m (40 ft) up in a silver spruce tree in western Garwhal. This nest was evidently an old nest of a corvine bird, to which a lining of spruce twigs, oak leaves, and some weeds had been added.

Incubation and Brooding
There is no good information on this aspect of breeding biology because western tragopans have only very rarely been bred in captivity (Delacour,

1977). Very probably its incubation behavior and other aspects of the species' nesting biology are the same as those of other tragopans.

Growth and Development of the Young
Only a few people have successfully raised young of this species in captivity (Delacour, 1977), and there does not seem to be any detailed information on this subject.

Evolutionary History and Relationships
Although this species seems to be most closely related to the satyr tragopan on geographic grounds and perhaps on the basis of the plumage pattern of the male, there is more gray present on both sexes, perhaps as a reflection of the generally drier environment typical of the western tragopan as compared with the satyr. Generally, the satyr tragopan is largely limited to the watershed of the Ganges River, whereas the western is associated with the watershed of the Indus, although no major ecological or physiographic barrier separates the two at present. Islam (1992) likewise concluded that the western tragopan is most closely related to the satyr.

Status and Conservation Outlook
The known present range of the western tragopan is but a remnant of the original one, which perhaps included some 10,000 km^2 (3,860 mi^2) of forest habitat. The range is essentially entirely restricted to a small area of Pakistan and Himachal Pradesh, northern India (Gaston et al., 1983).

In Pakistan, the species is apparently largely restricted to the area between the Jhelum and Kunhar Rivers, of the Hazara District. There are no recent surveys in Swat to confirm its possible occurrence there, but skins have been brought out of the area (Mirza, 1981*a*). The western tragopan may also inhabit the Bichela and Bhunja forests of Kaghan Valley, Hazara District. However, the bulk of the population occurs in the Neelum and Jhelum Valleys of Azad Kashmir. It apparently still occurs from Reshna to Bor, from Bugina to Phalakan along the ceasefire line, and from Kuttan to Machiara in the Neelum Valley. It has also been reported from Pir Chinase, Pir Hasi Mar, Leepa, and Chinari in the Jhelum Valley, and may still occur locally in the Murree Hills and Hunza (Islam, 1983). Gaston et al. (1983) indicated the range in Pakistan as extending beyond the

Kunhar River into the Kanghan Valley, based on earlier surveys in Pakistan by Mirza et al. (1978), and west to the Jhelum River in the vicinity of Chinari.

In westernmost India western tragopans may occur locally from the historic ceasefire line of Jammu and Kashmir, Pakistan, east perhaps to somewhere in Tehri-Garwhal, western Uttar Pradesh, although there have been no sightings in Uttar Pradesh for at least 20 years. In Jammu and Kashmir the birds probably occur very locally along the ceasefire line (Jhelum Valley of Azad Kashmir), and the species still probably occurs locally in Kishtwar (Gaston et al., 1983). In Himachal Pradesh the western tragopan is known to still occur in the upper Beas watershed, especially near Manali, including Solang Nalla, the core area for a potential national park (Gaston et al., 1981). The species was observed once during 1978–1979 in the adjacent Ravi Valley to the north, and the Ravi Valley population is probably distinct from the one in the upper Beas Valley. Further east it has been seen in the Simla District, but the upper Beas Valley is probably the species stronghold in Himachal Pradesh.

Based on their studies in this region, Gaston et al. (1981) suggested that the area of the proposed Manali National Park might support about 50 birds, and that the entire upper Beas Valley might hold no more than 1,000. There is no information for the eastern parts of Jammu and Kashmir, the Dhaula Dhar Range in Kandra District, or western Uttar Pradesh. They judged that unless these areas support unknown populations of birds, the world population of western tragopans must be less than 5,000 individuals. More recently Gaston et al. (1983) estimated that only 2,000–3,000 km^2 (770–1,160 mi^2) of habitat still remain that are suitable for the species, and that the world population is perhaps between 1,600 and 4,800 individuals. This would seem to be an objective assessment, but the actual number may be even lower than that. Islam (1983) mentioned that the skin of a

male is worth from 100 to 150 rupees (about 15 dollars), or much more than its value as meat. According to Islam, grazing, logging, and gathering of branches for firewood all occur in the habitat of the western tragopan as well and add to the disturbance it now receives. Gaston et al. (1983) suggested that three preserves should be established for the species, one each in the main habitat blocks still used by the birds. These include the Neelam Valley, the Inner Seraj area, and a site somewhere in the Ravi-Chenab area still to be determined. During the past decade much work has been done to establish the status of the western tragopan in Pakistan and Himachal Pradesh, India. In Himachal Pradesh, where the birds were once distributed from Chamba in the west to the Pamber Valley in the east, they have disappeared from some ranges (Choor, Kufri, Hatu, and Dharamsala) and are now limited to small and isolated populations (Chauhan and Sharma, 1991). The new Great Himalayan National Park at Kullu, Himachal Pradesh, is the only Indian national park that protects a population of this species. However, the western tragopan is also potentially protected in eight additional sanctuaries, from Limber in Jammu and Kashmir to Darangati in Himachal Pradesh (McGowan and Garson, 1995). Its possible survival in western Uttar Pradesh needs to be established.

In Pakistan the species' population may number up to 900 birds, and it is protected at Machiara National Park, in Azad Jammu and Kashmir (Islam and Crawford, 1988). Perhaps the species' greatest surviving population is in the Palas Valley, Hazara District, in the Indus Kohistan region of Northwest Frontier Province, where perhaps as many as 330 birds were present in 1994 (Whittington et al., 1994). Smaller populations exist along several Pakistani valleys, including the Duber, Patan, Kayal, Khagan, Nhelum, and Machiara watersheds (Roberts, 1991). The species is listed as vulnerable by McGowan and Garson (1995) and by Collar et al. (1994).

SATYR TRAGOPAN

Tragopan satyr Linné 1758

Other Vernacular Names

Crimson horned pheasant, Indian tragopan; tragopan satyre (French); Rot-Satyr-huhn (German); monal (Nepalese); see-a-gea (Chinese Tibetan).

Distribution of Species

Himalayas, between about 2,440 and 3,660 m (8,000 and 12,000 ft) and occasionally 4,270 m (14,000 ft), from Garwhal eastward to Bhutan and forests of neighboring southern Tibet to Monyul to about 95° E. This species moves altitudinally, down to occasionally 1,830 m (6,000 ft) in winter. The satyr tragopan inhabits moist and dry temperate coniferous forests (Vaurie, 1965.) See map 2.

Distribution of Subspecies

None recognized by Delacour (1977).

Measurements

Delacour (1977) reported that males have a wing of 245–285 mm (9.6–11.1 in) and a tail of 250–345 mm (9.8–13.5 in), and females have a wing of 215–235 mm (8.4–9.2 in) and a tail of 195 mm (7.6 in). Fifteen adult males had a wing range of 260–277 mm (10.1–10.8 in; average 268.5 mm [10.5 in]), and seven adult females had a wing range of 216–245 mm (8.4–9.6 in; average 229.5 [9.0 in]). Males weigh from 1,600 to 2,100 g (3.5 to 4.6 lb), and females from about 1,000 to 1,200 g (2.3 to 2.6 lb; Ali and Ripley, 1978). The eggs average 65 × 42 mm (2.5 × 1.6 in), and have an estimated average fresh weight of 63.3 g (2.2 oz).

Identification

In the Field (635–686 mm, 25–27 in)

This species is most likely to be found in steep hillsides with scrubby undergrowth and bamboo or in forests of oak, deodar, and rhododendron. Males are a nearly blood red throughout, with contrasting white markings on the back and underparts. They most resemble the Temminck's tragopan, but in that species the pale underpart spotting is not distinctly bounded or edged with black. Females are generally rufous brown, with sandy and blackish mottling

above and large buffy white spots below that are not distinctly black edged. The male's courtship call is a loud *wak* or *kya* that is repeated several times in succession; such sequences of this bleating call are repeated at regular intervals. No other species of tragopan is found in the area of the satyr.

In the Hand

The male's strongly reddish plumage, with a bluish face and throat, is distinctive. Females are less grayish than the western tragopan and very closely resemble the Temminck's, but the tail length is slightly longer (over 180 mm [7.0 in]). The satyr's underpart markings tend to be more lanceolate, and it tends to be more yellowish orange at the bend of the wing. The male's orbital skin and lappet are Prussian blue, edged with four of five leaflike to irregularly shaped patches of brilliant scarlet, with the intensity of the colors probably varying with age and season. At maximum extent the lappet is 90–100 mm (3.5–3.9 in) long.

Ecology

Habitats and Population Densities

Beebe (1918–1922) reported that in Sikkim and eastern Nepal the favorite habitats of this species are narrow side gorges with tiny streamlets flowing down them. In such areas there are overarched tangles of broken bamboo stems, outjutting boulders, forests of rhododendrons and magnolias, and undergrowths of lilacs, primroses, violets, strawberries, and forget-me-nots. Beebe never encountered satyr tragopans where the adjacent valleys were dry, and he believed the birds to be highly sedentary, probably spending their entire lives on a single range.

Ali and Ripley (1978) stated that satyr tragopans are associated with oak, deodar cedar, and rhododendrons on *khuds* and steep hillsides with scrubby undergrowth and ringal bamboo (*Arundinaria*), primarily between about 2,400 and 4,250 m (7,875 and 13,945 ft) elevation, but moving down to about 1,800 m (5,905 ft) during severe winters. Studies in Singhalila National Park, Darjeeling District of western Bengal, indicate that the satyr tragopan occurs there in steep, densely forested slopes between 2,400 and 4,300 m (7,875 and 14,110 ft) elevation, but it mainly occurs in the wet temperate forest zone between 2,600 and 3,100 m (8,530 and 10,170 ft). Most birds observed during a recently initiated study were singles or in pairs, but sometimes males were seen with two females. Maximum calling frequency was in April, with extremes of late March and May (Khaling, 1997).

Relatively little information is available on population densities, but Lelliott and Yonzon (1981) reported that in a census area of approximately 1.2 km² (0.5 mi²) of Annapurna Himal, Nepal, a total of ten calling male tragopans were recorded in late May. These birds were found there in thick rhododendron mixed forest on both gentle and steep slopes, and were widespread in such habitats between 2,750 and 3,345 m (9,025 and 10,975 ft) elevation. During winter satyr tragopans moved downward to as low as 2,300 m (7,545 ft).

Competitors and Predators

Beebe (1918–1922) suggested that although man is certainly the major enemy of satyr tragopans, because of the species' arboreal nature it probably has several important terrestrial predators, including perhaps leopards, foxes, jungle cats, and the larger species of civets. He mentioned a leopard having been shot that had just killed a tragopan. Eagles were also mentioned as possible predators, although in heavy forest cover these would seem unlikely to be efficient hunters. Relatively large numbers of birds are sometimes taken by natives using noosetraps, a device that tends to capture about four or five times as many males as females, according to Beebe.

Because the satyr overlaps slightly with the western tragopan on the western edge of its range, there are probably no significant competitors except other tragopan species. The satyr tragopan apparently also comes into contact with the Blyth's tragopan at the extreme eastern edge of its range.

General Biology

Food and Foraging Behavior

Beebe (1918–1922) stated that tragopans are omnivorous foragers, but tend to specialize in leaves and buds. Of the crops that he examined, two held many torn leaves and flowers of the paper laurel and one of

the birds had packed its crop with the petals of rhododendrons mixed with a few laurel leaves. One of the birds had also eaten a considerable number of insects, including small earwigs, ants, a few spiders, a cockroach, and a centipede. Others have suggested that in addition to insects satyr tragopans also eat bamboo shoots, onionlike bulbs, wild fruits, rhododendron seeds, and the leaves of daphne and bastard cinnamon. Thirteen birds from Sikkim collected in December all contained leaves and fern materials believed to be of *Diplazium* and *Polypodium* (Ali and Ripley, 1978). Lelliott (1981b) suggested that tragopans tend to favor scarce but potentially high-quality foods such as insects or fallen fruits. In captivity the birds are largely vegetarians, with an emphasis on fruits and berries (Howman, 1979).

Beebe (1918–1922) reported that like most forest and low-country pheasants satyr tragopans usually confine their foraging activities to early morning and late afternoon hours. However, on dull and cloudy days they may forage at more irregular intervals. Satyrs typically forage on the open edges of the forest, scratch deep among its undergrowth, or they may feed in low trees and bushes to obtain petals, buds, and berries. When feeding in jungle undergrowth satyr tragopans apparently concentrate their scratching in a few likely spots, rather than scratching superficially and randomly over a wide area, as kalij pheasants are inclined to do.

Movements and Migrations

Beebe (1918–1922) stated that these birds seemed to feed on the upper slopes for about 2 hours during the early morning hours and then descend again to spend the rest of the morning in the lower valley. He said that at times a bird would make its way up a ridge for 150 m (500 ft) or more to its crest.

There are of course much longer seasonal movements typical of this species. As noted above satyr tragopans may winter as much as 1,000–2,000 m (3,280–6,560 ft) lower than where they occur during the breeding season. Undoubtedly this is a leisurely and gradual migration, the timing of which is certainly dictated by local snow conditions.

Daily Activities and Sociality

Beebe (1918–1922) stated that, compared with blood pheasants, tragopans are highly solitary birds. He was never able to determine just when the birds went to roost or when they left their nocturnal perches. However, he did locate one such roost of a male. This was some 3.7 m (12 ft) aboveground in a moderate-sized magnolia tree and only a few yards from a gushing spring. During both morning and evening the birds would forage well up on the slope sides, indicating that normally two trips a day would be taken up and down the mountainside. Even during the winter there is little evidence of gregariousness in these birds, although Beebe believed from various reports that females and one or two offspring normally remain together for the greater part of the winter.

Social Behavior

Mating System and Territoriality
Although there is little direct evidence from the wild, it is believed that tragopans are in the group of pheasant species in which the male remains with a single female only until egg-laying or early incubation and then may remate with a different female (Lelliott, 1981*b*; Ridley, 1987). Regardless of this uncertainty, there can be little doubt that satyr tragopans are highly territorial. During their pairing period of a few weeks in the spring the males call from their favorite perch day after day, especially in early morning. Because satyr tragopans will often respond to imitations of their call and approach such sounds, it is clear that the call is at least in part a challenge to rivals and not simply a sexual attraction call to females. Beebe (1918–1922) stated that the call is heard only during April and May. This suggests that perhaps there is normally only a single mating per year rather than successive matings with different females, in which case one would have expected a much longer or at least interrupted periods of calling. Males sometimes also respond to other loud noises, such as shouts or gun discharges, further suggesting the probability that the call is as much a challenge as it is a sexually motivated signal.

Territorial sizes have not been estimated. However, if population densities approximate 10 males per square kilometer (25.9 per square mile), the area defended might approach about 10 ha (25 acres) per male.

Voice and Display
Beebe's (1918–1922) description of the male's "challenge" call is excellent. He describes it as a weird,

full-throated cry that is deep, half-booming, and half-bleating. At a distance of 45 m (150 ft) or more it sounds like *wah, waah! oo-ah! oo-aaaaa!*, with the last tone sometimes drawn out into a heartrending crescendo wail. The syllables usually run together and seem to be a single utterance. Wayre (1969) stated that the male utters 12–14 such notes, the series gradually rising in volume until it becomes almost a shriek. The entire sequence lasts for 20–25 seconds or more. Unlike the Temminck's tragopan this call has no terminal growl or croak, according to Wayre. Islam and Crawford (1996) reported that their challenge-call sample had 3–17 notes (average 8.5), uttered at intervals of 1–4 seconds, the entire sequence lasting 8–55 seconds (average 24), and gradually rising in pitch and volume.

Lelliott (1981*b*) distinguished four types of call, which he considered to be an incomplete description of the species' vocalizations. The first is the *wah, wah* call, which is a repeated monosyllabic note uttered by both sexes at any time of the day during spring and autumn. The function of the call is uncertain, although it may relate in part to courtship and have other functions, including possible male-to-male aggressive signaling. The second call, the *wak, wak* call, is similar to the first, but is of lower amplitude and thus less audible. It was observed to be uttered by both sexes when alarmed after disturbance by humans and possibly other enemies. Most commonly it is uttered during flushing; more than half the birds flushed by Lelliott were heard to utter this call. The third type of call recognized by Lelliott is the "bleat" note. This is a short, monosyllabic note similar to the bleat of a sheep or goat, which often precedes wailing. The call was heard only during the breeding season, suggesting that it functions in pair formation or courtship. The "wail" call, the vocalization that corresponds to the "challenge" call of Beebe, is a drawn out mammal-like call made by the male only. It consists of the repeated series of notes mentioned earlier and ranges up to 33 seconds in length. "Wail" calling is greatest at dawn; its onset is closely related to the time of sunrise. Lelliott judged its function to be uncertain, but that it might be related to territorial advertisement.

Lelliott (1981*b*) observed display in wild birds on only one occasion, when bleating and wailing calls were heard in conjunction with wing-whirring and with an observation of copulation. During the breeding season he observed the birds most often as single males or single females (58 percent), less often in pairs (17 percent), and otherwise in unknown social combinations. He judged that the mating system is still uncertain, but perhaps the satyr tragopan is normally a monogamous species.

There are relatively few detailed descriptions of the displays of the satyr tragopan, but Beebe (1918–1922) quoted an account provided to him by Mr. Barnby Smith that is worth quoting in full:

The lateral display of a cock Tragopan in good plumage is interesting; that is, he presents one side of the head, body, and tail to the hen, and lowers one wing and raises the other until he almost looks like the mere skin of a bird stretched flat on a wall. This pose is constantly assumed during the breeding season (from February onwards), the cock taking up a position about a yard distant from the hen and repeatedly assuming a new position if she moves off.

This lateral display, however, is as nothing compared to the frontal display, which I usually notice some three or four times each season. In this case the cock faces the hen (about two or three yards distant) and commences by crouching down slightly, ruffling his feathers and spreading his wings, which are slowly flapped on the ground. The head is nodded repeatedly with increasing

Figure 16. Display postures of male satyr tragopan, including wing-whirring (A), wattle engorgement (B), frontal display with expanded lappet (C), and calling posture (D). After photos from various sources.

speed and the brilliant light blue horns gradually become inflated and extend forwards from the black feathers of the head, whilst the bib (or gular wattle), which is also blue with pink side stripes, is gradually let down to its full length. Whilst this is being done the shivering and the rustling of the feathers have increased to an alarming extent, the body of the bird has been lowered quite near the ground, the wings are extended sometimes almost to their full width, and the whole business is preceded and accompanied (particularly in the early stages) by a curious noise like the "clacking" of two bones together, but how this noise is made I have never found out, though I should much like to know. When the bib has been extended to full length for a few moments the bird gathers himself together, moves forwards about a yard, draws

himself up to his full height (and it is surprising how high he can reach), keeps the bib fully extended in front of the hen for one moment, and then, within half a minute, horns and bib have entirely vanished and the cock is strolling about pecking grass as if nothing unusual had happened.

Likewise, various other observers and personal observations indicate that the major display is a forward run with the body held erect while the male emerges from the darkness of some undergrowth and spreads the breast and flank feathers like a skirt while exposing the large blue and red lappet (figures 16 and 17). The bright reddish upper wing-coverts are erected conspicuously during the wing-flapping and climax

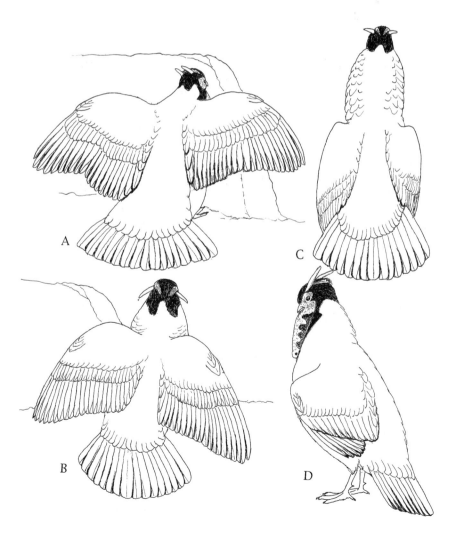

Figure 17. Frontal display sequence of satyr tragopan, including wing-flapping stages (A, B), erect rush toward female (C), and frontal display to female (D). After a video sequence by David Rimlinger.

rearing phases. In Islam and Crawford's (1996) samples of both satyr and Cabot's tragopans, the clicking phase lasted an average of about 11 seconds as compared with 5 seconds in Blyth's and 22 seconds in Temminck's tragopan.

Wing-flapping is apparently also part of this frontal display (figure 12, see chapter 4) of the satyr tragopan. As with the other tragopans, the lappet is displayed for only a few seconds and shortly thereafter the horns retract, the lappet is contracted, and both disappear again into the crown and throat feathers. Although the male's calling is evidently normally done from an elevated perch, full male display is seemingly performed on the ground, presumably after a female has been attracted to the male's vicinity by his calling. Quite probably copulation attempts normally follow such a running approach to the female. Lelliott and Yonzon (1980) observed copulation in a pair of wild satyr tragopans in mid-May, after a male had "fanned its tail, opened its wings slightly, and quivered violently while lowering its wings to the ground and raising its tail."

Reproductive Biology

Breeding Season and Nesting
Little is known of the breeding season of wild satyr tragopans, but it evidently occurs during May and June or about a month later than the peak of male calling. Baker (1935) has provided the best summary of nesting. He described a clutch of two eggs found in the Chambi Valley near the Tibetan border, in a rhododendron-oak forest. The nest was located in a tree about 6 m (20 ft) aboveground and was well hidden from view. The forest was very thick but stunted and the ground greatly broken up with huge rocks covered by mosses and ferns. The nest was a jumble of old, dead twigs and branches that were mostly rotten. Baker judged that perhaps an old mass of sticks had been found and converted into a nest. Another nest in the same area was similar, but was empty and closer to the ground. Natives of the area stated that two eggs were usual in the clutch and that never more than four are present. Baker believed that in the wild the nesting season begins in early May and extends through June, with some birds in the highest elevations not laying until July.

Nests made by captive birds in W. Shore Baily's collection were typically sticklike and placed in open

baskets put up in apple trees at heights of 3–6 m (10–20 ft) aboveground. The nesting season in captivity is more extended than in the wild. Frequently eggs will be laid on the ground and subsequently ignored. The eggs are rather variable in color, but typically are reddish buff with a freckling of deeper brick red over the larger end or the entire surface. Delacour (1977) stated that old nests of crows or other birds may be used as nest sites, although female tragopans do a good deal of nest-building and nest-lining themselves.

In captivity, the eggs are typically laid on alternate days, until the female has a clutch of four to six eggs. If the eggs are taken away from the female immediately, she will often lay a replacement clutch. This may be repeated three or four times in a single season (Delacour, 1977) or perhaps until the female's reserves are exhausted.

Incubation and Brooding
The incubation period is typically 28 days. When the chicks emerge they are able to survive without any food for their first 48 hours. However, they do need water about 12 hours after hatching. Satyr tragopans are highly precocial at hatching. Within 2 or 3 days they are even able to fly up to an elevated perch, where they typically roost under their mother's wings (Wayre, 1969).

When rearing satyr chicks in captivity, their early food should consist of bread and milk, ant eggs, various insects, hairless caterpillars, and small worms. Later they can be fed finely chopped greens, and gradually they may be provided a mixture of crushed hemp, small groats, and broken wheat. Apparently they should not be fed seeds, starchy foods, or boiled eggs until they are at least 12 days old. Because female tragopans feed their chicks only one or two insects at a time, after having carefully softened them with their bills, the young birds should only be fed soft-bodied insects (Delacour, 1977). Sivelle (1979) has provided additional information on feeding young tragopans in captivity.

Growth and Development of the Young
In the wild the brood probably remains with its mother for most or all of their first year; likewise, in captivity the brood may be kept together for their first year. Males typically will not develop their full plumage in their first year, although some will de-

velop more quickly and rarely may be able to breed when only 1 year old.

Evolutionary History and Relationships

There seems to be little doubt that the nearest relative of this species is Temminck's tragopan because the male plumages of these two species are extremely similar, although their lappet patterns are quite different. The satyr tragopan is also somewhat intermediate between Temminck's tragopan and the western tragopan, both in its geographic distribution and in the intensity of black on the male's underparts. These three species seem to me to represent one superspecies group in the genus *Tragopan*, whereas the Blyth's and Cabot's appear to comprise a second group. However, Islam (1992) associated the Temminck's tragopan with the Cabot's and Blyth's as a central phyletic assemblage, based on his analysis of behavioral and morphological data.

Relationships of *Tragopan* to other genera of pheasants are far less certain. There is indeed some question as to whether the tragopans should be included with the pheasants or the partridges (Boetticher, 1939). At minimum it seems clear that this is a rather isolated genus, without any close relatives except perhaps the blood pheasants (Cheng et al., 1978), which probably also should be transferred to the partridge group.

Status and Conservation Outlook

The satyr tragopan is a favored food of the local inhabitants in many areas, and is heavily trapped by them in Nepal by noosetraps (Beebe, 1918–1922). Yonzon and Lelliott (1981) reported more recently that in their studies the second highest trapping toll of pheasants in the Annapuma Himal area of Nepal was of satyr tragopans, which comprised about 36 per-

cent of the trapping take. Likewise, among pheasants shot with guns the tragopan kill was second highest, and the altitudinal zones in which satyr tragopans occur are those favored by both professional and amateur hunters. Because of hunting pressures the pheasant populations are declining. Because the hunting pressures increase areas of previously nearly untouched steep terrains, the favorite habitats of tragopans are increasingly being exploited for hunting purposes. Hunting in Nepal is carried on throughout the year, with no restrictions on the age or sex of the game. Thus, in Nepal as well as in India, it is likely that the satyr tragopan will continue to decline in numbers unless better conservation measures are instituted. The satyr tragopan is one of four species of birds now accorded legal protection in Nepal (the Himalayan monal being the other pheasant), but this does not actually provide much help toward its survival (Roberts, 1981).

This species was listed as vulnerable by McGowan and Garson (1995) and as near-threatened by Collar et al. (1994). Most of the six Chinese locality records come from Tibet, but there is one Yunnan record (Li, 1996). An additional Tibetan locality is Mount Jumulang Ma National Nature Reserve. Also, there are records from four Nepalese national parks (Khaptad, Langtang, Sagamatha, and Malaku Barun), the Anapurna Conservation Area, and Dhorpatan Hunting Reserve (Inskipp, 1989). The satyr tragopan extends from Garhwal east to western Arunachal Pradesh in India and has been seen in two protected areas, Kedarnath Wildlife Sanctuary in Uttar Pradesh and Singhalila National Park in Darjeeling (McGowan and Garson, 1995). The species is well established in Bhutan, where the population is fairly common and believed stable (del Hoyo et al., 1994). It also occurs in Sikkim.

TEMMINCK'S TRAGOPAN

Tragopan temmincki (J. E. Gray) 1831

Other Vernacular Names
Chinese crimson horned pheasant; tragopan de Temminck (French); Horn-huhn, Temminck-Satyrhuhn (German); kiao-ky (Chinese).

Distribution of Species
Eastern Himalayas, at about the same elevations as *T. satyra*, and forests of neighboring southern Tibet from about 93°30' E (and probably farther west to the upper reaches of the Subansiri River) eastward through southeastern Tibet to the mountains of western and northern Szechwan, southern Kansu, southern Shensi, and northern Hupeh southward to northeastern Burma and Yunnan (where it breeds up to 4,575 m [15,000 ft]) to extreme northwestern Tonkin. This species inhabits dense evergreen or mixed mountain forests and dense rhododendrons and bamboos (Vaurie, 1965). See map 2.

Distribution of Subspecies
None recognized by Delacour (1977).

Measurements
Delacour (1977) reported that males have wing lengths of 225–265 mm (8.8–10.3 in) and a tail length of 185–230 mm (7.2–9.0 in), whereas females have wing lengths of 220–225 mm (8.6–8.8 in) and a tail length of 175 mm (6.8 in). Cheng et al. (1978) report five males with wing lengths of 210–242 mm (8.1–9.4 in) and tail lengths of 200–232 mm (8.6–9.0 in) and five females with wing lengths of 202–218 mm (7.9–8.5 in) and tail lengths of 158–178 mm (6.2–6.9 in). The males ranged in weight from 980 to 1,120 g (2.1 to 2.5 lb) and the females from 970 to 1,100 g (2.1 to 2.4 lb). Two captive males weighed 1,362 and 1,447 g (3.0 and 3.2 lb) and two females 907 and 1,021 g (2.0 and 2.2 lb; D. Rimlinger, pers. comm.). The eggs average 54 × 40 mm (2.1 × 1.6 in) and have an estimated fresh weight of 47.7 g (1.7 oz). Li (1991) reported a mean weight of 52.9 g (1.9 oz) for 27 eggs.

Identification

In the Field (635–686 mm, 25–27 in)
This species is found in the thick undergrowth of dense montane forests and like the other tragopans is highly arboreal. Males resemble those of the satyr tragopan, but the pale spotting on the flanks and underparts is grayish and lacks the black edging of the satyr, producing a less contrasting plumage pattern. Females cannot be readily separated in the field from those of the satyr. The male's courtship call is a plaintive series of increasingly louder *waaa* notes, lasting from 6 to 27 seconds (Islam and Crawford, 1996).

In the Hand
Males are easily separated from the similar satyr tragopan by their grayish rather than white underpart spotting, which lacks black borders. In females, the light spots are large and oval rather than lanceolate and have a border of yellowish buff. Females also resemble Cabot's tragopans, but are generally more fulvous than grayish, and typically exhibit bluish orbital skin rather than being reddish to orange in this area. The male's orbital skin is light blue, the lappet is also blue, mottled with lighter blue spots centrally, and has about nine bright red markings of irregular or leaflike shape extending in vertical series along each side.

Ecology

Habitats and Population Densities
Beebe (1918–1922) has summarized the relatively small amount of information he could learn about the habitats of this species. In Szechwan Temminck's tragopan can be found in heavy hardwood forests that are rich in undergrowth and occur between 915 and 2,740 m (3,000 and 9,000 ft). It prefers steep mountain slopes that are rich in arborescent vegetation. However, in Yunnan, Beebe observed the species in an area of stunted rhododendrons and bamboo stubble on a steep mountainside. In Tibet, Ludlow and Kinnear (1944) found this species to be abundant in Packakshiri between 2,135 and 3,660 m (7,000 and 12,000 ft), inhabiting thick rhododendron and bamboo undergrowth in the densest forests. Delacour (1977) stated that in Tonkin he never found them below 2,440 m (8,000 ft) and that they seemed to prefer hardwood forests along sharp ridges and

slopes with thick undergrowth and moss-covered trees.

Li (1996) reported density range estimates of 1.1–9.6 individuals per square kilometer (2.8–24.9 per square mile) in four different localities (means of individual locality range from 3.9 to 5.1 per square kilometer [10.1 to 13.2 per square mile]). Earlier Li (1991) reported that the highest estimated densities (13 individuals per square kilometer [33.7 per square mile]) occurred in mixed deciduous and coniferous woodlands, especially subtropical forests having dense undergrowths of shrubs and bamboos, and that these woodlands represent ideal habitats. Seasonal differences also exist. In spring, Temminck's tragopan prefers forest-edge habitats rich in grasses and bushes, in summer it moves to areas having good cover and abundant bushes; in general it prefers mixed broadleaf forest and bamboos forests rich in fruits of trees, and in winter it uses similar habitats but expands its range size (Shi et al., 1996).

Competitors and Predators

So far as is known, there is almost no contact with any other tragopan in this species range. Cheng (1979) states that Temminck's tragopan occurs in company with Cabot's tragopan at the extreme western limit of that species' range in northeastern Kwangsi, but it is the rarer of the two forms there. It also closely approaches the range of Blyth's tragopan in southern Tibet; Cheng et al. (1978) show locality records of the two species occurring less than 100 km (62 mi) apart. Ludlow and Kinnear (1944) judged that the Subansiri-Manas watershed (Dafla Hills) may be the dividing line between the satyr and Temminck's tragopans in the vicinity of the Bhutan border.

Nothing specifically has been noted of this species' predators, which probably include the usual large raptors and various predatory mammals of eastern Asia. Li (1991) mentioned several raptors and mammalian predators as possible enemies.

General Biology

Food and Foraging Behavior

Beebe (1918–1922) has summarized what little is known of foods in the wild. Reports from China indicate that Temminck's tragopans there eat frozen fruits and berries in the winter, and more generally consume grain, berries, fruits of *Cotoneaster* and allied shrubs, and maize. Beebe noted that a bird he shot in late fall had a crop that was filled with a mass of comminuted leaves and an almost equal amount of macerated insects. The insects were unrecognizable, but two small spiders that had just been swallowed were identified. A Temminck's tragopan that was shot during May in Tibet had been feeding on the unopened flower-buds of *Berberis nipalenesis* (Ludlow and Kinnear, 1944). Li (1991) listed a large variety of plant taxa (mostly grasses) and the buds, leaves, and berries of various shrubs and trees as primary foods.

In captivity Temminck's tragopans are primarily vegetarian, with a special fondness for fruits and berries (Howman, 1979) as well as peanuts (Sivelle, 1979).

Movements and Migrations

Nothing specific seems to have been written on this, although undoubtedly the same kinds of vertical migration occur in this species as in other tragopans and high-montane pheasants.

Daily Activities and Sociality

The few observations of this species in the wild suggest that Temminck's tragopans occur singly or at most in very small groups, perhaps pairs or family units. Their daily activity cycles are probably much like that described for the western and other tragopans.

Social Behavior

Mating System and Territoriality

Presumably this species shows the same kind of relatively monogamous mating system that apparently is typical of the other tragopans. Males utter a very loud cry during the spring, as do the other species. Beebe (1918–1922) judged that males spend the winter solitarily and that their calling serves both as a challenge to other males and as an attraction to females. He noted that in captivity a male and female will call to each other periodically during the day if they are separated, but that the female always remains silent when she is in the same enclosure as the male. However, captive females have been observed calling while in the same cage as the male (Kenneth Fink, pers. comm.).

Voice and Display

The male's challenge call has been described as an *ona*, which is repeated twice (Beebe, 1918–1922). It has also, and probably more accurately, been characterized as a single plaintive note, *waaa*, repeated four to eight or more times, rising in pitch and becoming louder and more insistent, like someone crying for help. The sequence may end with a croak on a lower note or a disyllabic growl of far less carrying power. The entire sequence may take from 8 to 16 seconds (Wayre, 1969). Islam and Crawford (1996) found the male's challenge call to consist of 4–12 notes (average 7.4) uttered at intervals of 1–2 seconds; the entire sequence lasts 6–27 seconds (average 14.5), with the notes gradually rising in pitch and volume.

Beebe's (1918–1922) description of the display, as consisting of three separate phases, has been widely quoted, but does not seem to agree well with recent observations by myself or by David Rimlinger (1984), curator of birds at the San Diego Zoo. Based on our individual observations, as well as analysis of videotapes of a captive pair of birds, the major phases of display are summarized in some detail here as a basis for future comparison with other tragopan species.

Rimlinger (1984) classified the male tragopan displays he observed into two major categories, lateral and frontal. During lateral display (figure 18), the male flattens his body feathers and raises the wing away from the female, while slightly lowering the nearer one. In an upright posture, the male thus slowly walks around the female, occasionally "freezing." Alternatively, the male may walk up to a female and arch his head down toward the ground, slowly side-stepping toward the female. Although the posture is similar to tidbitting in typical pheasants, the male neither offers the female food in this posture nor takes any for himself. Lateral displays may also be used as threats toward other males or toward reflections in a mirror.

Frontal display, by comparison, is much more complex and is highly stereotyped (figure 19). It usually begins from behind a solid object, such as a large rock. Typically, the male starts by walking behind the rock and peering over it toward the female. He may stand there for a time or immediately begin the display sequence by twitching his head vertically, gradually exposing the colorful throat lappet and expanding his "horns." The tail is then spread and the wings begin to beat. As the wings are beaten the horns vibrate and the orange-colored upper wing-coverts become visible. This first phase lasts about 19 seconds

and is terminated by the onset of calling and head-jerking. The calling phase lasts only about 14 seconds, and during it the wings are beaten but the body is held rather still, with the head lowered and the bill resting on the lappet. As calling begins, the male crouches down until he is nearly out of sight of the female momentarily. The calls are repeated clicking or gasping sounds. The wing beats, which begin as fairly slow shallow movements, are synchronous with the calls, both of which speed up to a rate of about five calls and wing beats per second. The third and final phase consists of the bird suddenly hissing, rising up as high on tiptoes as it can reach, pushing the wings downward, pointing the beak downward, and spreading the lappet maximally as the horns are held erect. This fantastic posture is held only momentarily, after which the bird sinks back down, the lappet retracts, and a normal posture is assumed. This climax phase lasts only about 3 seconds. In a small proportion of the sequence (10 of 35 sequences observed by Rimlinger), the male may complete the sequence by running over the top of the rock with his lappet and horns extended and his wings and tail spread and scraping the ground, accompanied by a hissing sound. The male may thus chase the female a short distance or may simply stop where she had been standing. It is possible that copulation normally follows such a rush, but the behavior has not yet been described in the Temminck's tragopan. It is known, however, that copulation in the closely related satyr tragopan may occur after a very similar display sequence and rush (Lelliott and Yonzon, 1980).

This bizarre display is totally different from that known for any of the true pheasants, and is also distinct from that of any known partridge species. This therefore supports the view that the genus *Tragopan* is very isolated in the pheasant family. Rimlinger (1984) has also observed a few other apparent displays, including a short display flight to an elevated perch. This also is unusual behavior for pheasants, but is the presumed evolutionary precursor to stationary wing-whirring, which has also been observed in this species.

Reproductive Biology

Breeding Season and Nesting

Little is known of the nesting season in the wild, but young have been collected in Tibet during early July

Figure 18. Postures of male Temminck's tragopan, including normal resting posture (A), partial lappet engorgement (B), upright lateral display (C), display flight (D), and arched lateral display toward a female (E). After photos by the author and David Rimlinger.

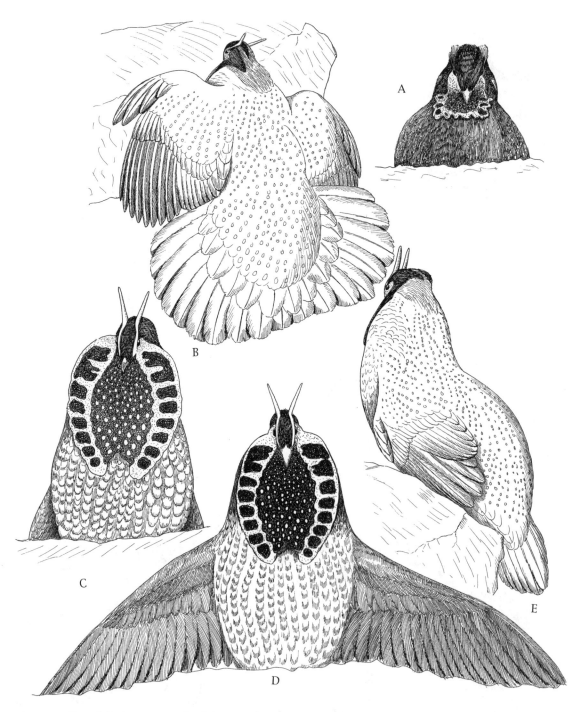

Figure 19. Frontal display sequence of male Temminck's tragopan, including preliminary hiding posture behind rock (A), wing-flapping phase from behind (B), rearing phase (C, D), and rush toward female (E). After photos by David Rimlinger (D) and the author.

(Ludlow and Kinnear, 1944), suggesting that incubation there began about 1 June. In Britain, laying by captives typically begins in early April and continues until early May, with later layings in June and July the result of earlier nesting failures or removal of earlier clutches. Nests are probably normally elevated, although the single nest of a wild Temminck's tragopan that has been described from China was a ground nest containing an unusually large clutch of six eggs (Beebe, 1918–1922). Li (1991) stated that eight nests of wild birds were elevated 0.5–8.0 m (1.6–26.2 ft) aboveground and had from two (an incomplete clutch) to five eggs, typically (in three cases) four eggs. These nests were found between 28 April and 20 June, with five of them being found during May. No participation by the male in incubation or brood-protection was detected. At least five of the eight nests under observation hatched.

Beebe (1918–1922) made the interesting point that when his females were provided with elevated nesting sites they laid clutches that averaged smaller than when they laid eggs on the ground. In seven cases females laid three to six eggs in ground nests, whereas the same birds laid only one or two eggs in elevated nests. He judged that the small clutch of tragopans was perhaps adaptively related to the small available platform of most elevated nests. Beebe also mentioned that tragopan eggs are unusually large for the size of the female, which he attributed to the advantages of having the young hatch at a highly advanced state so that they can flutter down from a tree nest or up to elevated perches shortly after hatching. Beebe considered this precocial behavior typical of tragopans, peafowl, and monals, all conspicuous species that he believed might have a special need to be able to escape by flight.

Incubation and Brooding

Incubation is done by the female alone. It requires 26–28 days, rarely 30 days, one of the longest incubation periods in the pheasant group. Within 2 or 3 days after hatching the young are easily able to fly up to elevated perches, where they typically roost beneath their mother's wings in the manner of peafowl and argus pheasants.

Growth and Development of the Young

Although able to fly within a few days of hatching, the chicks actually grow relatively slowly and are quite sensitive to disease and chilling. The full adult male plumage is not attained until the second year (Delacour, 1977).

Evolutionary History and Relationships

There seems little doubt that the nearest relative to the Temminck's tragopan is the satyr. The birds hybridize in captivity extremely freely, and the offspring are fully fertile (Sivelle, 1979). The two species now very closely approach one another geographically, but are not known to overlap in distribution. Islam (1992) considered the Cabot's tragopan to be the nearest relative of Temminck's, based on his behavioral and morphological studies.

Status and Conservation Outlook

This species has by a considerable measure the largest range of any tragopan, and thus its safety seems to be the most secure. However, it is extensively trapped or shot for its feathers and flesh or for keeping in captivity. More importantly its habitat of mature montane forests is in many areas being cut for timber or razed for agricultural purposes. Thus, like the other tragopans, Temminck's is quite vulnerable to population destruction over large areas of its range. In China 33 nature reserves occur within this species' range, although the level of protection in many such reserves is minimal at best (Li, 1991). Li (1996) listed 97 Chinese locality records, primarily in Szechwan. There are no recent records from Burma, Vietnam, or Tibetan China and only two from Bhutan. The species has also been seen in Mehao sanctuary, Arunachal Pradesh, but it is of marginal occurrence in India (del Hoyo et al., 1994). McGowan and Garson (1995) considered the species as secure, but Collar et al. (1994) classified it as near-threatened.

BLYTH'S TRAGOPAN

Tragopan blythi (Jerdon) 1870

Other Vernacular Names
Grey-bellied tragopan; tragopan de Blyth (French); Blyth-Satyrhuhn (German).

Distribution of Species
The mountains of Assam and northwestern Burma; the hills south of the Brahmaputra from the Barail Range and the Naga Hills eastward through the Patkoi Range into northern Burma and southeastward through Manipur into the Chin Hills; also eastern Bhutan and southeastern Tibet. This species inhabits forests between 1,830 and 3,660 m (6,000 and 12,000 ft). See map 2.

Distribution of Subspecies (after Ripley, 1961)[1]
Tragopan blythi blythi (Jerdon): eastern Blyth's tragopan. Assam south of the Brahmaputra River in the Patkoi, Naga, and Barail Ranges south to Manipur, Lushai Hills, and the adjacent hills of Burma, from 1,830 m (6,000 ft) up; in moist temperate forest. One record from Yunnan. Now largely limited to Nagaland and surrounding areas.

Tragopan blythi molesworthi Stuart Baker: western Blyth's tragopan. Southeastern Tibet, Bhutan, and the upper ranges of the Assam Hills east to the Mishmi Hills, from 1,830 to 3,660 m (6,000 to 12,000 ft); in moist temperate and coniferous forest. Current distributional and numerical status uncertain; the populations in Aranachal Pradesh are probably of this race.

Measurements
Delacour (1977) reported that males of *blythi* have a wing length of 260–265 mm (10.1–10.3 in) and a tail of 180–220 mm (7.0–8.9 in), whereas females have a wing of 230–245 mm (9.0–9.6 in) and a tail of 170–175 mm (6.6–6.8 in). Adult males are reported to weigh 1,930 g (4.2 lb) and females 1,000–1,500 g (2.2–3.3. lb; Zeliang, 1981). The eggs average 58.5 × 44 mm (2.3 × 1.7 in), and their estimated fresh weight is 62 g (2.2 oz). The two known male specimens of *molesworthi* have wing lengths of 250 and 260 mm (9.8 and 10.1 in) and tails of 180 and 195 mm (7.0

and 7.6 in; Delacour, 1977; Johnsgard, pers. obs.). A single female had a wing of 232 mm (9.0 in) and a tail of at least 155 mm (6.0 in; Biswas, 1968).

Identification

In the Field (635–686 mm, 25–27 in)
This species is found in the thick undergrowth of dense evergreen montane forest. Males are distinctive in being almost uniformly gray underneath, whereas the neck, chest, and head are mostly orange red. The male's courtship call is a sonorous *wak*, sometimes lengthened to *wa ak-ak*, which produces a sort of two-toned effect. Females are also distinctly grayish on the undersides and are dark olive brown above, with a whitish chin and throat. Blyth's are probably not separable from female Temminck's tragopans in the field, but it is unlikely that any other species of tragopan occurs in the area occupied by this species.

In the Hand
The uniformly gray underpart coloration of males and the yellowish facial skin provide for immediate identification. Females are most similar to those of the satyr tragopan. However, they are paler, have yellowish orbital skin, a more grayish underpart coloration with white ocelli on the feathers of the midbelly, and the dorsal feathers have distinct black lateral ocelli. The male's lappet is small (length 50–75 mm [2.0–2.9 in]), rounded, and mostly yellow, bordered with pale blue. The bare facial skin is golden yellow, and the erectile horns are pale blue.

Geographic Variation
The westernmost population of Blyth's tragopan is apparently recognizably distinct from birds in the remainder of the species' range, but is so far known only by three specimens. The type specimen of *molesworthi* was described as being darker on the upper parts, with browner red ocelli, narrower buff vermiculations, smaller white spots, and generally paler on the underparts than in the nominate race. Apparently only two additional specimens of the race have

[1]Koelz (1954) has described an additional race (*rupchandi*) that was not recognized by Ripley.

since been obtained. One of these, a female, is darker than those of the nominate form and is generally more grayish and less reddish on the upper parts (Biswas, 1968).

Ecology

Habitats and Population Densities
Very little has been written of this species in its natural habitat. However, Beebe (1918–1922) quoted extensively from unpublished notes of E. C. Stuart Baker, who observed Blyth's tragopans in North Cachar, in an area of the Barail Range adjoining the Naga Hills. The area where the birds were seen was at 1,830 m (6,000 ft), in densely wooded valleys occurring below such peaks as Mahadeo, Hengmai, and Hungrum. The vegetation at the bases of these mountains is luxurious and massive, with some trees over 30 m (100 ft) tall. Above 1,220 m (4,000 ft) the vegetation begins to become more scant, and from 1,525 m (5,000 ft) upward the major trees are stunted oaks rarely more than 9 m (30 ft) tall. However, a variety of epiphytes drape the trees, including mosses and orchids such as *Coelogyne*, *Dendrobium chrysotoxum*, and *D. densiflorum*. The forest floor is covered by bracken and other ferns of all kinds from maidenhair to palm ferns, as well as various begonias and wild jasmine. The environment is quite moist, with many seasonally turbulent mountain streams.

The nominate subspecies has been recorded as low as about 1,525 m (5,000 ft) during the winter dry season and apparently summers in the moist temperate montane forest zone between about 1,830 and 3,050 m (6,000 and 10,000 ft). The western race *molesworthi* has been locally recorded in southeastern Tibet (in mountains north of Dibrugarh in Assam) and probably extends southward in these mountains through eastern Bhutan and northern Assam at elevations between 1,830 and 3,660 m (6,000 and 12,000 ft) in moist temperate and coniferous habitats comparable to those of the nominate form.

There are no studies of population densities, but Blyth's tragopans are now rare everywhere. In Pulibadze sanctuary in Nagaland, an area of 9.2 km^2 (3.6 mi^2), about 40 birds were present (Zeliang, 1981), suggesting a density of about 4.3 birds per square kilometer (11.1 per square mile).

Competitors and Predators
There is no specific information available on predators and competitors. The forests of Nagaland support a wide array of feline predators, including tigers, clouded leopards, and panthers, as well as other smaller cats and various mustelids, all of which might be potential predators.

Blyth's tragopan may well encounter and compete with the satyr tragopan at the extreme western edge of its range in Tibet and in Bhutan or adjoining Assam, but there is no direct evidence of this. It also closely approaches or may actually encounter the Temminck's tragopan in southeastern Tibet (Cheng et al., 1978); the nominate form likewise approaches this species near the Assam–Yunnan border. Delacour (1977) reported that there is a possible hybrid specimen apparently involving these two species from the Shweli–Salween Divide area of western Yunnan that is in the collection of the American Museum of Natural History.

General Biology

Food and Foraging Behavior
Baker (1935) states that in the wild Blyth's tragopans feed on all kinds of seeds, berries, fruits, and buds; his captive birds greedily ate worms, insects, and even small frogs. Beebe (1918–1922) states that captive birds will eat berries, worms, and unhulled rice. Howman (1979) indicates that in captivity the species is primarily vegetarian, but with an emphasis on fruits and berries. The only known female specimen of *molesworthi* was collected while foraging in the undergrowth of a rhododendron forest. Her crop was filled with freshly consumed plant materials including the leaves and fronds of ferns, leaves of *Spiraea*, leaves of *Herpetospermum caudigerum* (Cucurbitaceae), the leaves of a species of Ranunculaceae, and a variety of other leaf fragments, shoots, petioles, and leafbuds, but no animal materials (Biswas, 1968).

Apparently Blyth's tragopans make daily movements up and down the mountain slopes in trips between roosting sites and foraging areas, following paths along which natives often set snares (Beebe, 1918–1922).

Movements and Migrations
Apparently the seasonal movements of this species are often not very great, because where they occur

the winters are not severe. Blyth's tragopans are more prone to move down the mountainside during the winter to avoid the drying out of vegetation associated with the dry season than to escape cold weather. Beebe (1918–1922) suggested that during the breeding season the rains are so intense that nesting aboveground is advantageous inasmuch as seasonal rains may cause flooding or even sweep clean many of the slopes in the areas where these birds breed.

Daily Activities and Sociality

There is no specific information on this, but activities and sociality are probably no different from that described for the other better-known tragopans. Baker (1935) stated that Blyth's tragopans move about in small parties of four or five birds, although natives of Nagaland reported that the birds are always to be seen in pairs, both in and out of the breeding season.

Social Behavior

Mating System and Territoriality

The Blyth's tragopan male advertises in the usual tragopan manner. It is assumed that one or more females are attracted to individual males when mating. Zeliang (1981) suggested that the males are "selective in their choice of mates," and that the male guards the female while she is within his territory. These comments would suggest that a monogamous pair-bonding system might exist, but tragopan mating systems are still very uncertain.

Voice and Display

The territorial advertisement call of the male has been described in various ways. Thus, it has often been called a fine, sonorous *wak*, sometimes lengthened into a *wak-ak-ak* reminiscent of that of a peacock, although much less harsh. Zeliang (1981) described it as a deep bass sound, *mao, mao*, uttered in early morning and again in the evening. The only sound attributed by Zeliang to the female was a sharp, musical, quacking call. He said that the mating season is in March. Islam and Crawford (1996) found that the male's advertisement call ranged from 3 to 11 notes (average 7.4), uttered at intervals of 2–5 seconds. The entire sequence lasted 9–45 seconds (average 30) in their sample, with no apparent changes in pitch or volume.

E. C. Stuart Baker (quoted in Beebe, 1918–1922) has provided the only detailed description of the frontal sexual display of the male, based on the observation of a wild bird. It is worth quoting in full:

> For a few minutes the two birds, male and female, scratched about the hill just like a pair of barn-door fowls, now and then picking up an insect disturbed from under the pebbles, or seizing a grasshopper from the scraps of herbage scattered about over the bare ground. But presently, ceasing to take any interest in the abundant food all about him, the cock bird began to attempt to attract the attention of the hen by all sorts of antics and displays. At first he merely came up to her and bowed and scraped with his wings slightly raised, and his purple blue horns fully dilated and projecting forward. Then, seeing that she took no notice, he depressed his wings and walked slowly around her, nodding violently as he walked and swelling out his throat and breast, the feathers of which were ruffled and standing almost on end. After a short time of this ineffectual display he once more stopped in front of the hen, and standing still, leaned forward until his breast almost, or quite, rested on the ground; he then extended both his wings, so that their upper portions faced the same way as his head, and stood thus for some seconds—a blaze of deep crimson, with his weirdly shaped horns quivering with excitement, and his wattles displayed to the fullest possible extent. Then suddenly his feathers collapsed, his horns nearly disappeared; he held himself erect, and once more quietly commenced to scratch and feed until he and his mate shortly disappeared into the adjoining forest.
>
> As far as I could see, the hen bird took little or no interest in the display of the male, and continued serenely feeding all the time it was going on, but this was perhaps only a ladylike way of inducing him to exert himself to the uttermost. Both birds constantly uttered a soft, chuckling note, and now and then the cock bird gave a rather loud *quawk*.

Islam (1992) has recently analyzed lateral and frontal display in this species. During lateral display the male does not perform obvious head-lowering but does exhibit strong wing-lowering (figure 20). During the wing-flapping phase of frontal display, the wing beats are slower and more labored than in the other three

Figure 20. Posture of male Blyth's tragopan, including lateral display to female (A), erect posture (B), crouching during lateral display (C), and head detail (D). After photos from Kamal Islam (B), the Zoological Society of London (C), and Jean Howman (D).

tragopans so far described, and the lappet that is maximally expanded during the rearing phase is relatively small and not very colorful as compared with the other species. In a video sequence filmed by Islam (figure 21), the male held his wings well out to the side as he rushed toward the female and his neck feathers were widely erected. At the end of the sequence the male repeatedly bowed his head forward and downward somewhat. As compared with the other tragopans, this species' entire frontal display sequence is relatively primitive in Islam's view.

Reproductive Biology

Breeding Season and Nesting

According to Baker (1935), the breeding season begins in early April and lasts well into May. Ghose and Thanga (1998) found a nest in Blue Mountain National Park, Mizoram, that was situated 7.6 m (24.9 ft) up in a tree. No other nests have been definitely found in the wild, but natives in Nagaland stated that they are always placed in trees, stumps, or small bushes and are never on the ground. Their heights are

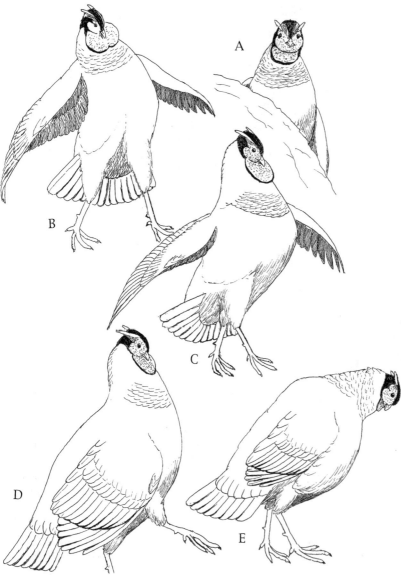

Figure 21. Frontal display sequence of Blyth's tragopan, including head-bobbing behind rock (A), rush toward female (B–D), and final head-jerking (E). After a video sequence by Kamal Islam.

said to range from 2 to 6 m (6 to 20 ft) aboveground. Nests are made of good-sized sticks, with a lining of smaller twigs and sometimes grass or weed stems. Favored nesting sites are said to be hidden among masses of creepers growing over dead trees. At times the nest may be placed in the fork of a leafless sapling and thus quite visible (Baker quoted in Beebe, 1918–1922). As in other tragopans, Blyth's probably normally use old nests of other birds; thus their actual locations largely depend on the availability of such sites.

The clutch size in the wild has been reported as consisting of two to five eggs, usually three or four.

Baker (quoted in Beebe, 1918–1922) confirmed this as typical also of his captive birds. Zeliang (1981) stated that in captivity females lay from two to six eggs. He noted that, over various years, a total of 25 females laid 76 eggs, or an average of 3.1 eggs per female.

Incubation and Brooding
Zeliang (1981) reported the estimated incubation period in captive birds as ranging from 36 to 45 days, which is far greater than that reported for any other pheasant, and is therefore probably in error. More

probably it is closer to the 28 days, which is considered typical of tragopans in general. Howman (1993) reported an incubation period of 28–30 days.

Growth and Development of the Young
The only quantitative data on the avicultural aspects of hatching and survival of young are those of Zeliang (1981). From his data it would appear that hatching success was approximately 70 percent (29 chicks from 41 fertile eggs), and rearing success was about 34 percent (10 of 29 chicks). No specific information was provided on growth rates or rates of attaining reproductive maturity, which is evidently 2 years.

Evolutionary History and Relationships
From a zoogeographic standpoint, this species' nearest relative should be either Temminck's tragopan or Cabot's tragopan. Blyth's tragopan exhibits the grayish bases to the feathers of the flanks and underparts that occur in Temminck's and has the rather yellowish facial coloration and pale buffy to rusty markings on the upper tail-coverts and mantle that are so conspicuous on Cabot's tragopan. It thus seems to be somewhat transitional in plumage between Cabot's tragopan and the three more western forms (satyr, Temminck's, and western). Islam (1992) grouped the Blyth's, Cabot's, and Temminck's tragopans in a single phyletic assemblage, using a variety of behavioral and morphological traits.

Status and Conservation Outlook
The status of this species is quite difficult to assess. It was considered "rare" by the International Council for Bird Preservation (King, 1981). In China, the nominate race of this species is known from a single Yunnan site, and there are three older records of *molesworthi* in Tibet (Li, 1996). This subspecies may be present in the Mount Jumulang Ma National Nature Reserve in Tibet (McGowan and Garson, 1995), although there are no recent definite records of Blyth's tragopans in Tibet. It has also been reported from Maenam sanctuary in Sikkim (Ghose and Sumner, 1997). In India Blyth's tragopans have been reported from five states. In Arunachal Pradesh they occur at least at Mouling, Mehao Wildlife Sanctuary, Kamlang, and Namdapha (Choudhury, 1996); these populations possibly represent *molesworthi*. In Nagaland the species is known from three small sanc-

tuaries totaling 71 km² (27.4 mi²; Fakim Rang tanki, and Pulebadze) and from four of seven ts (Kohima, Phek, Zunheboto, and Tuensang; C hury, 1997). It has been reported from a local (Shiroi Ridge) in northeastern Manipur, a potly protected area, and in a very small area of east Assam state, near the Manipur-Nagaland bor Blyth's tragopan has also been found well to t south in Blue Mountain National Park, the hi point in Mizoram (Choudhury, 1996), and on nt Victoria in the southern Chin Hills of adjoinir northwestern Burma (Wheatley, 1996). McGo and Garson (1995) suggested a total potential an population of nominate *blythi* in the range of 5 5,000 birds. Besides its possibly continued prese in Tibet and Yunnan, the species (presumably *es*-*worthi*) is also present in Bhutan in unknown b perhaps considerable numbers.

In India the nominate subspecies has been kwn mainly in the past few decades from the Naga hge of Nagaland (Pauna Mountain on the west to Fim, near the Burmese border, on the east). There it as recorded on Pauna Mountain; Dzukou, Kipamzu, and Phekekedzumi Mountains; at Fakim; and in he Pulebadze sanctuary. In all these areas the total sti-mated population was about 400 birds as of 198 (Zeliang, 1981). Its status in the adjoining Patko range of Burma is unknown, although as noted it has been reported from the Chin Hills.

The status of the more westerly race *molesworthi* is even more uncertain. It is known from only two males and a female, all taken in eastern or southeastern Bhutan (Biswas, 1968). However, the race apparently also once extended eastward through the Dafla Hills across the Brahmaputra to the Mishmi Hills of extreme northeastern Assam (Arunachal Pradesh). In recent years it has been reported only from eastern Bhutan and the Dafla Hills of Arunachal Pradesh (King, 1981). Cheng et al. (1978) list a locality record for southern Tibet, directly north of the Bhutan–Arunachal Pradesh border. Indeed Etchécopar and Hüe (1978) indicate a rather extensive range for the Blyth's tragopan in Tibet. Yet there is no good evidence that such a large Tibetan range exists, for suitable forested habitat would be extremely rare in that area. In fact, this race is likely to be even more threatened with extinction than is the nominate form, based on the little available information.

CAT'S TRAGOPAN

Tragcaboti (Gould) 1857

Othrnacular Names
Yellollied tragopan; tragopan de Cabot (French); Cabqyrhuhn, gelbäuchige Hornhuhn (German).

Disttion of Species
This sies inhabits areas in Chekiang, Fukien, nortl Kwangtung, and southern Hunan west to norttern Guangxi. See map 2.

Disttion of Subspecies
Trpan caboti caboti Gould: eastern Cabot's tragopa Range as indicated above excepting Guangxi.

Trpan caboti guangxiensis Cheng Tso-hsin: western (ot's tragopan. Endemic to Guangxi (Cheng, 1979

Mearements
Delaur (1977) reported that *caboti* males have wing lengs of 210–225 mm (8.2–8.8 in) and tail lengths of 21–215 mm (8.2–8.4 in), whereas females have wingengths of 200–214 mm (7.8–8.3 in) and tail lengs of 160–168 mm (6.2–6.6 in). Islam (1992) repcted that the Cabot's tragopan has the shortest wing chord measurements of any tragopan. Cheng et al. (1978) noted that six males had wing lengths of 225–233 mm (8.8–9.1 in) and tails of 207–230 mm (8.1–9.0 in), whereas six females had wings of 192–213 mm (7.5–8.3 in) and tails of 151–171 mm (5.9–6.7 in). One male weighed 1,400 g (3.1 lb), and females weigh about 900 g (2.0 lb). The eggs average 50 × 40 mm (2.0 × 1.6 in) and their estimated fresh weight is 44.2 g (1.5 oz).

Identification

In the Field (635–686 mm, 25–27 in)
This species is found in dense montane forests and, like the other tragopans, is highly arboreal. Unlike the other tragopans, the male has plain buff underparts without any ocelli; the upper parts have only indistinct buffy spots that are flanked by black and russet, but show little development of ocelli-like markings. Females are rufous brown above and grayish brown below, with whitish triangular or linear markings on the dorsal surface and larger whitish

underpart spotting. The male's advertisement or challenge call is very brief, high pitched, and rapidly repeated, with the entire sequence usually lasting under 4 seconds (Islam and Crawford, 1988).

In the Hand
The uniform buffy underparts and buffy rather than white spotting provide for ready identification of males. Females have reddish to orange orbital skin and thus closely resemble the western tragopan. However, females are more buffy to whitish on the underparts, rather than grayish, and the spotting on the dorsal plumage tends to be linear to triangular rather than rounded. The male's throat is pale blue, and the center of the extremely large lappet (length up to 150 mm [5.9 in]) is orange with bristly purple spots of increasing size on the lower half. This area is surrounded by cobalt blue, interrupted by up to nine wedge-shaped scarlet patches along each margin. The lower edge is distinctly bilobed. The facial skin is orange yellow, and the erectile horns are pale blue and up to 50 mm (2.0 in) long.

Geographic Variation
The recently described subspecies *guangxiensis* differs from the nominate form in being more maroon red on the back and rump of males and generally deeper maroon red throughout, but with paler yellowish white ovoid spots near the feather tips. Females of this race are much darker brown, with more black streaking throughout. The whitish spotting is also better developed in females of this race than in the nominate form (Cheng, 1979).

Ecology

Habitats and Population Densities
One of the few descriptions of the habitat of Cabot's tragopan is by Beebe (1918–1922), who observed the birds in west-central Fokien. This was in heavy forest, which had a dense undergrowth dripping with moisture and included azalea-like plants. The elevation was no greater than 855 m (2,800 ft). The species has also been collected at higher elevations of 915–1,220 m (3,000–4,000 ft) in the vicinity of

Kuatun, near the Fokien–Kiangsi border. This species was apparently once common in the range of mountains separating these two provinces, but now appears to be very rare there (King, 1981).

According to Cheng (1979), the race *guangxiensis* occurs as elevations of 700–1,100 m (2,300–3,610 ft), or somewhat lower than that reported for the nominate race. The birds inhabit mixed forests with heavy undergrowths of bushes, dwarf bamboos, and ferns. Population densities of 7.1–21.4 individuals per square kilometer (18.4–55.4 per square mile) have been reported in two areas, with means of the individual areas being 7.48 and 17.48 per square kilometer (19.4 and 45.3 per square mile; Li, 1996). Ding (1995) reported an average density of 14.65 birds per square kilometer (37.9 per square mile) in another study area.

Competitors and Predators

This species is isolated from most of the other tragopans by a considerable distance, but according to Cheng (1979) the race *guangxiensis* is often found with the Temminck's tragopan in the forested areas of northeastern Guangxi. No hybrids have been reported thus far. Nothing has been written of the possible predators of Cabot's tragopan.

General Biology

Food and Foraging Behavior

La Touche (1900) stated that the crop of an immature male he examined contained young leaves, whereas that of a female contained acorns. Beebe (1918–1922) stated that the crop of an adult male was crammed with laurel-like leaves having a strongly aromatic odor. There were also two small land molluscs in the crop and some grit in the gizzard. Cheng (1979) stated that birds of the newly described race *guangxiensis* feed mostly on seeds (e.g., *Castanopsis, Moghanis*) and red beans. Winter and early spring foods include a strong preference for the leaves of *Daphniphyllum macropodum*, and its fruits were highly favored during autumn. This species is part of the original forest flora, which may help explain the species' close association with old evergreen broadleaf forests (Sun, 1995).

As other tragopans, in captivity this species thrives on growing or freshly cut lucerne (alfalfa), grasses, squash, cucumbers, apples, raspberries, mulberries, raisins, grapes, and peanuts. Live insects are rarely fed, except to young birds, but Sivelle (1979) provided peanuts to breeding birds daily.

Movements and Migrations

Daily or seasonal movements are undescribed. However, they are likely to be very small because the mountains on which this species occurs are not very high and the seasonal changes are not likely to be severe.

Daily Activities and Sociality

No specific information has been recorded in the wild except that Cheng (1979) stated that Cabot's tragopans would gather in small groups for foraging and for going to water at about 10:00 A.M. and again about 3:00 P.M. In captivity the Cabot's tragopan seems to be at least as arboreal as the others of its genus, which is perhaps a reflection of the dense forests in which it lives. It is probably no more social than the other tragopans, which at most seem to occur in family units or pairs.

Social Behavior

Mating System and Territoriality

The mating system and territoriality of Cabot's tragopan has not been described for wild birds. Sivelle (1979) kept a group of three males and three females in adjoining aviaries. By late March one of the males appeared to have attracted one of the hens, and the birds were allowed together. The same was tried with a second pair, but the resulting eggs were infertile. Therefore the female was allowed to mate with the third male, which also resulted in infertile eggs. The third female, a first-year bird, showed no interest in mating and never laid any eggs. These albeit artificial conditions would suggest an essentially monogamous mating system in Cabot's tragopan. However, Ding (1995) observed matings with multiple females by a wild male and no male participation in brood-rearing.

Voice and Display

Islam and Crawford (1996) found that the male's advertisement or challenge call to consist of rapid bursts of 5–25 notes (average about 9), uttered at intervals of 0.1–1.0 seconds, the entire sequence lasting 2–8 seconds (average about 4). Each sequence is of diminishing pitch and volume, with intervals of about 6 seconds between bursts. Both sexes utter "summons" calls, which are sharp repeated notes that may be continued for several minutes. Males, and sometimes females, produce threat notes while performing lateral displays toward other males. Males also produce repeated *chi* notes during the frontal display sequence.

Virtually nothing has been written on the sexual behavior of Cabot's tragopan in the wild and relatively little on the species in captivity. However, Wayre (1969) has provided a fairly detailed account of the male's displays, which is quoted here:

The male Cabot's tragopan starts his display by standing in a very upright position so that his body and tail are in a vertical line; only his bill is not quite vertical. His crest is raised and the feathers round his thighs are puffed out like pantaloons. Though not extended, the blue border at the top of the lappet is strikingly visible and it is vivid cobalt blue in color. He then lowers the primaries of the wing nearest the hen, at the same time raising the shoulder on the opposite side, giving the appearance of flattening his body and exposing as much of his colorful plumage to the hen as possible. Still remaining very erect he circles round her.

The display may cease at this stage, or the bird may suddenly stop, facing the hen and go into the final grotesque climax, in which, with lower plumage fluffed out and wings half spread, he shakes his head and neck until his fleshy blue horns and elaborately patterned orange and blue lappet are extended to their utmost. At this moment the lappet or bib hangs down in front of the bird's breast and is spread out to display its pattern. It is all over in a matter of seconds; the horns and lappet are quickly retracted and the bird assumes its normal shape. We have also watched a cock Cabot's tragopan stand up and whir his wings like a silver pheasant, but in his case the wings are not vibrated so powerfully and the noise is not so loud. As with other pheasants, this display is probably in the nature of a challenge to other cocks.

The frontal display has also been described and illustrated by Zheng et al. (1989), and has been videotaped by Kamal Islam. Islam (1992) observed that the frontal display sequence (figure 22) is similar to those of the other tragopans, but the male's wings are held

more outstretched during the flapping phase. In one video sequence that Islam filmed, there is a strong, repeated head-lowering component following the climax phase of the frontal display. During lateral display there is no apparent head-lowering and body-arching component.

Zheng et al. (1989) reported that male displays extend seasonally from December through April, reaching a peak in early April. Copulation was observed only at the end of March. Lateral display was observed only when the male was approached by a female or another male. Zheng Guangmei et al. also noted that frontal display is performed most often during afternoon and evening hours, and each sequence lasts 45–58 seconds. Briefly, the frontal display sequence (figure 23) consists sequentially of preliminary lappet-spreading and head-nodding (about 26 seconds), wing extension with associated head-nodding and synchronous wing-flapping (about 8 seconds), which soon becomes faster and more violent (about 6 seconds). The male then utters about 15–16 *chi* notes in synchrony with his most violent wing movements, accompanied with maximum tail-spreading and tail-vibration. This phase lasts about 9 seconds and is abruptly followed by rearing the head, neck, and body upward to an erect or climax posture while simultaneously lowering the beak and uttering a loud and long *chi* call. This climax phase, which lasts only 2 seconds, is accompanied by quick alternate foot movements (probably ritualized or inhibited running), and the entire sequence is terminated by about six rapid vertical head-nods.

Although it is assumed that only adult males perform this display, it is of interest that La Touche (1900) described the lappet of an immature male collected in March as apparently fully developed in color. He described it as having a pink wattle, banded with pale cobalt, and the bands tinged in the center with silvery green. The "hairy" part of the wattle was livid purple spotted with orange vermilion. This description differs somewhat from that presented in the description of the species; perhaps there is some age or individual variation present, as well as probable seasonal variations.

Reproductive Biology

Breeding Season and Nesting
Not many nests of Cabot's tragopan have been described from nature. The first reported nest was

Figure 22. Frontal display sequence of male Cabot's tragopan, including erect lateral display (A), frontal posture before female (B), and normal resting posture (C). After photos by Kamal Islam (A, B) and Kenneth Fink (C).

found at Kuatan in northwestern Fokien near the Fokien–Kiangsi border in mid-May. It was some 9 m (30 ft) aboveground, in an old squirrel nest of the previous year. The nest contained four eggs, of which two were ready to hatch and the other two were addled. The embryonic young birds already had wing quills over an inch long at this stage of development (La Touche, 1900). Ding and Zheng (1997) reported finding 15 nests in mixed broadleaf-conifer forests and by radio-tagging females learned that renesting sometimes also occurs. These nests were mostly (80 percent) located in commercially planted conifer (*Cryptomeria*) woodlands, thus proving that Cabot's tragopan can breed in areas where the original forest has been logged. Tree characteristics, terrain slope and direction, elevation, and distance to ridgetop were all found to be significant factors in nest-site selection.

Beebe (1918–1922) believed that the four eggs discovered by La Touche represented two separate clutches of two each. However, it is now known that in captivity clutches of three or four eggs are the usual number. Sivelle (1979) stated that his birds averaged three to four clutches per season, totaling 12–14 eggs, assuming that clutches were removed as they were completed.

Wayre (1969) stated that in the collection of the Norfolk Wildlife Park and Pheasant Trust Cabot's tragopans do not always select an elevated nest site, but instead often lay their eggs in a scrape beneath the cover of an evergreen shrub. However, Sivelle (1979) reported that although about one in ten of his female

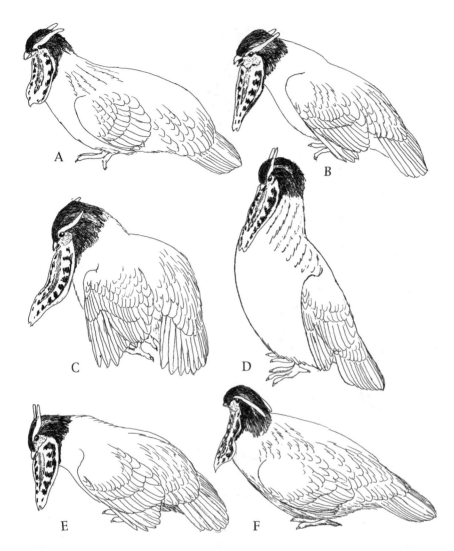

Figure 23. Postures of male Cabot's tragopan display sequence. After photos in Zheng et al. (1989).

tragopans (of various species) will not use one of his elevated nest boxes for laying an egg, their following clutch will invariably be in one of the boxes, which are from 0.6 to 1.8 m (2 to 6 ft) aboveground. W. H. St. Quintin (quoted in Beebe, 1918–1922) bred this species at least twice in captivity, with the birds once laying in an old woodpigeon's nest about 3 m (10 ft) from the ground, and the following year nesting in the same nest, then 4.3 m (14 ft) aboveground. A second female nested in a pigeon's nest about 2.4 m (8 ft) up, in a spruce tree, after lining the nest with dead twigs that she broke off from neighboring branches.

Incubation and Brooding

All that is known of incubation and brooding in this species comes from observations in captivity. The in-cubation period is 28 days, as in the other tragopans. Sivelle (1979) reported that he hatches tragopan eggs with an incubator temperature of 99.5°F and relative humidity of 84 percent, turning the eggs three times daily. The pipped eggs are removed to a hatcher, which is maintained at 88–94 percent relative humidity.

Growth and Development of the Young

St. Quintin (quoted in Beebe, 1918–1922) reported that the chicks are hatched with a thick coating of coarse, shaggy down. The primaries were so far developed that on the first day the chicks were able to flutter up and perch on the side of the foster mother. Their abilities at perching and climbing developed rapidly, and they would fly between their various perches as well as any young passerine bird. If left in their cage a little longer than usual at roosting time, the birds would fall asleep side by side on a perch.

Sivelle (1979) feeds his young tragopans greens and alfalfa supplemented by commercial turkey starter feed and grains (keeping protein levels below 20 percent), after starting them on mixed cracked grains and medicated water. After 2 weeks the birds are moved outside and given a variety of greens, fruits, and peanuts. Dierenfeld et al. (1998) suggested that a protein level of 15 percent is typical of adult tragopans.

Evolutionary History and Relationships

As suggested earlier, I believe that Blyth's tragopan is this species' nearest relative, even though at the present its range is much closer to that of the Temminck's tragopan. Like Blyth's, Cabot's appears to be adapted to very moist, temperate montane forests. Presumably the ancestral area of separation of these two forms was in the mountains of southern or southwestern China. Islam (1992) concluded that the Temminck's tragopan is Cabot's nearest relative.

Status and Conservation Outlook

This species is now considered endangered and is believed to be very rare throughout its range, largely as a result of habitat destruction associated with agricultural activities (King, 1981). The recent description of the western race *guangxiensis* extends the species' known range somewhat and provides additional hopes for the birds' survival.

All of the current captive stocks are derived from birds sent out of China since 1960, when five males and two females were received by the Norfolk Wildlife Park and Pheasant Trust. About 15 birds were imported over about a 10-year period, and quite a number of young were raised. In 1978 and 1979 the trust sold their stock to various private breeders, including C. Sivelle in the United States, who has had considerable success with breeding them (Sivelle, 1979).

Several nature reserves have recently been established by the Chinese within the historic range of this species, the largest of which is a 5,000-ha (12,350-acre) reserve (Luyuan Wuzhishan Chingjendong) in Kwangtung Province (Wang, 1980). At least three other smaller reserves also exist within the possible range of this species. Current evidence suggests a major decline in this species since 1980, especially in Fukien Province, but there are some small and isolated preserves in Kiangsi, Chekiang, Hunan, and Kwangsi Provinces that should help protect this species. Li (1996) listed 42 localities for nominate *caboti* and 12 localities for *guangxiensis* in China. Many of these localities were from Chekiang Province, where Wuyanling Natural Reserve is located and where several long-term studies have been conducted (Ding, 1995; Sun, 1995). McGowan and Garson (1995) considered the nominate race as vulnerable and *guangxiensis* as endangered; Collar et al. (1994) regarded the species collectively as vulnerable.

Genus *Pucrasia* G. R. Gray 1841

The koklass is a medium-sized montane pheasant in which the sexes are moderately dimorphic. Males have a well-developed occipital crest and lateral erectile feather tufts, whereas females have a shorter crest and lack the tufts. In both sexes the head is entirely feathered and many of the body feathers are lanceolate. The wing is somewhat rounded, with the tenth primary considerably shorter than the ninth, and the seventh primary slightly the longest. The tail is highly graduated; of 16 feathers, the middle pair of rectrices are about twice as long as the outermost pair. The upper tail-coverts are greatly elongated and almost as long as the tail. The tarsus is longer than the middle toe and claw and is spurred in males. A single species is recognized.

Key to Subspecies of Males of *Pucrasia macrolopha*
(in part after Delacour, 1977)

A.   Mantle feathers with single black streak on the shaft.
  B.   Sides and flanks mostly gray.
    C.   Chestnut collar present on hindneck: Kashmir koklass (*biddulphi*).
    CC. No distinct collar, although sides and red of neck sometimes buffy yellow.
      D.   Brighter and paler throughout: Indian koklass (*macrolopha*).
      DD. Duller and darker throughout: Punjab koklass (*bethelae*).
  BB. Sides and flanks mostly black or chestnut.
    C.   Sides and flanks black, edged with gray: Nepal koklass (*nipalensis*).
    CC. Sides and flanks chestnut: western koklass (*castanea*).

AA. Mantle feathers with two or more black streaks.
    B.   Mantle feathers with four black streaks, separated by three white wedges: Darwin's koklass (*darwini*).
    BB. Mantle feathers with two black streaks, separated by a white wedge.
        C.   No distinct nuchal collar present: Joret's koklass (*joretiana*).
        CC. A distinct yellowish nuchal collar present.
            D.   Nuchal collar orange: orange-collared koklass (*ruficollis*).
            DD. Nuchal collar yellow.
                E.   Basal portions of outer rectrices mostly gray: yellow-necked koklass (*xanthospila*).
                EE. Basal portions of outer rectrices mostly yellow: Meyer's koklass (*meyeri*).

KOKLASS

Pucrasia macrolopha (Lesson) 1829

Other Vernacular Names
None in general English use; eulophe macrolophe (French); Koklas-fasan (German); sung chi (Chinese).

Distribution of Species
Discontinuous, from eastern Afghanistan eastward through the Himalayas, normally from about 1,830 to 3,660 m (6,000 to 12,000 ft), to west-central Nepal; again from eastern Tibet and northwestern Yunnan northeastward through the mountains of western China to southwestern Manchuria; and again in the mountains of eastern and southeastern China from Hupeh and Anhwei south to northern Kwangtung. Sedentary, but moves altitudinally with the season. This species inhabits steep, broken, or rocky

mountain slopes in coniferous or mixed forests or sites grown with dense brush and bamboos (Vaurie, 1965). See map 3.

Distribution of Subspecies (after Ripley, 1961; Vaurie, 1965; Wayre, 1969)
Pucrasia macrolopha macrolopha (Lesson): Indian koklass. Resident in Lahul and northeastern Jammu north of the range of Goddulphi from 1,830 to 4,270 m (6,000 to 14,000 ft); in dry temperate forest.
Pucrasia macrolopha castanea Gould: western koklass. Resident in the mountains of Nuristan in eastern Afghanistan to Chitral in Pakistan at 2,135 m (7,000 ft).
Pucrasia macrolopha biddulphi Marshall: Kashmir koklass. Resident from Gilgit and Kashmir eastward

Map 3. Distribution of the koklass, including Darwin's (D), Indian (I), Joret's (J), Kashmir (K), Meyer's (M), Nepal (N), orange-collared (O), western (W), and yellow-necked (Y) races.

to Ladakh and to Kulu in northern Punjab, where it intergrades with the nominate race *macrolopha*, from 1,980 to 3,355 m (6,500 to 11,000 ft). The birds of Kulu (*bethelae* Fleming 1947) occur between the population of Kashmir and that of Garhwal farther to the east and appear to be a clinal form.

Pucrasia macrolopha nipalensis Gould: Nepal koklass. Resident in western Nepal, east to about 83°40′ E, above 1,830 m (6,000 ft).

Pucrasia macrolopha meyeri Madarasz: Meyer's koklass. Resident in western and southwestern Szechwan (regions of Bateng and Baurang) to northwestern Yunnan to the Likiang Range and Shweli-Salween Divide and eastern Tibet.

Pucrasia macrolopha ruficollis David and Oustalet: orange-collared koklass. Mountains of western Szechwan north to southern Kansu, Shensi, and Hopeh, possibly intergrading with *xanthospila* in the north.

Pucrasia macrolopha xanthospila G. R. Gray: yellow-necked koklass. Resident in northern Shensi and the mountains of Inner Mongolia, northeast to the mountains of western Hopeh.

Pucrasia macrolopha joretiana Heude: Joret's koklass. Resident in the mountains of southwestern Anhwei and eastern Hopeh between 610 and 1,525 m (2,000 and 5,000 ft).

Pucrasia macrolopha darwini Swinhoe: Darwin's koklass. Resident in the mountains near Itchang (Hupeh), southeastern Szechwan, south and east to Kweichow, Kwangsi, Chekiang, Fukien, and northern Kwangtung.

Measurements

Cheng et al. (1978) reported average male wing and tail lengths (all subspecies) of 212.4 and 206.8 mm (8.3 and 8.1 in), respectively, and average female wing and tail lengths of 196.7 and 148.8 mm (7.7 and 5.8 in), respectively (10 specimens of each sex). Delacour (1977) reported that *macrolopha* males have wing lengths of 215–245 mm (8.4–9.6 in), whereas females have wing lengths of 180–218 mm (7.0–8.5 in) and tail lengths of 172–195 mm (6.7–7.6 in). Ali and Ripley (1978) report *castanea* male wing lengths of 240–252 mm (9.4–9.8 in) and tail lengths of 178–252 mm (6.9–9.8 in). A female had a wing length of 228 mm (8.9 in). Males of *biddulphi* had reported wing lengths of 233–249 mm (9.1–9.7 in). Males of *macrolopha* were reported to weight from 1,135 to 1,415 g (2.3 to

3.1 lb) and females from 1,025 to 1,135 g (2.2 to 2.5 lb). Cheng et al. (1978) reported an average weight for 10 males of 1,184 g (2.6 lb) and for 10 females of 932 g (2.0 lb). The eggs average 51 × 37.5 mm (2.0 × 1.5 in) and have an estimated fresh weight 40 g (1.4 oz).

Identification

In the Field (533–610 mm, 21–24 in)

This is a medium-sized pheasant associated with montane woodlands, including both hardwood and coniferous forests. The chestnut breast, black head, and white patches on the sides of the neck are unique. The female also has a distinctive whitish patch along the side of the neck. Both sexes have somewhat elongated blackish to brownish tails with paler tips and tapering occipital crests, which in the male are sometimes raised into earlike display structures. The male's call in spring and summer is a loud *pok-pok-pok . . . pokras*, uttered mainly during morning and evening hours.

In the Hand

In both sexes the white to creamy white neck markings and a somewhat elongated tail (170–300 mm, 6.6–11.7 in) with a paler tip are distinctive. In both sexes the flank feathers are rather lanceolate, with dark brown centers and buffy edges. Iridescence is limited to the dark greenish black head of the male.

Geographic Variation

Geographic variation in this widely distributed species is very great and tends to be clinal in some respects. There are three groups of subspecies that apparently are completely isolated from one another. The westernmost or nominate group extends from *castanea* on the western edge of the species' range through *nipalensis* in central Nepal. The second (*xanthospila*) group consists of several widely dispersed and possibly isolated races, from *meyeri* in eastern Tibet to *xanthospila* in northern China. The third (*darwini*) group of eastern China consists of only two races, the widely distributed *darwini* and the localized *joretiana*.

Much of the plumage variation in these groups involves male upper-part plumage color and tail pat-terning. In the westernmost or *macrolopha* gr[...] feathers of the upper parts, breast, and fl[...] uniformly black areas along their sha[...]

those of the hindneck are dark chestnut to silvery gray, sometimes tinged with buff. In the intermediate (*xanthospila*) group the hindneck is golden yellow, which extends laterally around the sides of the neck to form an incomplete collar, and the feathers of the upper parts have gray shaft-streaks that divide the black area of each feather into two components. In the third (*darwini*) group the males lack the distinct golden collar and the feathers of the upper parts have four rather than two lateral black streaks. Additionally the lateral tail feathers vary from more or less chestnut colored in the *macrolopha* group (and *meyeri* of the *xanthospila* group) to patterned black and silvery gray in the remaining subspecies. This same plumage variation in the tail feathers also exists in females, which otherwise tend to be quite similar throughout the species' range (Vaurie, 1965).

Ecology

Habitats and Population Densities
In Himachal Pradesh the koklass has a relatively wide ecological distribution, reaching from the lower altitude oak forests to the alpine meadow or scrub zone, but being most commonly observed in the higher altitude oak and coniferous forests and fairly widely occurring through the several temperate forest types. The koklass seems to be most common where the undercover is well developed. It occurs only sporadically below 2,000 m (6,560 ft), but it shows a seasonal shift to the lower altitude coniferous forests between December and March and back to the higher altitude oak forests in April and May (Gaston et al., 1981). In Kashmir the birds are widely distributed, occurring from about 1,830 m (6,000 ft) up to the limits of the pines, but probably are most common above 2,135 m (7,000 ft). The koklass seems to be particularly associated with broken ground having good cover and avoids more open slopes and flats (Bates and Lowther, 1952). In Himachal Pradesh over 50 percent of the records made by Gaston et al. (1981) were in habitats having more than 70 percent ground cover. Baker (1930) stated that the birds are partial to forest of cypress, paludna, and other pines and prefer broken ground to other topographic sites. Wherever they occur, they are always found where there is a good deal of undergrowth such as ringal bamboo as well as trees.

Farther east the ecological distribution is well known, but Schäfer (1934) indicated a rather restricted vertical range in China for *meyeri* in the transition zone between agricultural lands and montane forests, at an altitude of somewhat above 3,050 m (10,000 ft). Cheng (1963) stated that in China the birds occur between 2,000 and 4,000 m (6,560 and 13,125 ft) in pine forests located on rocky slopes of tall mountains. The Nepal race *nipalensis* also occurs between about 2,440 and 3,355 m (8,000 and 11,000 ft), mainly in oak, conifer, and bamboo forests of western Nepal, west of the Kali Gandaki River (Roberts, 1981).

Because of its regular morning calling behavior, censusing of the koklass can be done fairly easily. In Pakistan, for example, the density in favorable habitats such as the Murree Hills was estimated at 1.9 pairs per square kilometer (4.9 per square mile; Mirza, 1981a). Severinghaus (1979) estimated that in one study area of Pakistan there were 11 males (pairs) in an area of 48.6 ha (120 acres), or about 23.0 pairs per square kilometer (58.9 per square mile). He cited a variety of earlier studies suggesting densities ranging from 1.5 to 10.4 pairs per square kilometer (4.0 to 26.9 per square mile).

In Himachal Pradesh a density of 17–25 pairs per square kilometer (44.0–64.8 per square mile) has been estimated in the Simla area, and many forest reserve areas support densities of at least 5 pairs per square kilometer (13.0 per square mile; Gaston et al., 1981). In China, population densities of 1.0–23.3 individuals per square kilometer (2.59–60.3 per square mile) have been observed (Li, 1996). Han (1990) estimated a density of 22 birds per square kilometer (57.0 per square mile) at Mount Da-Bie, Anhwei, in mixed evergreen and deciduous forest, with females outnumbering males by a ratio of 1.72:1.00. Because of this unbalanced sex ratio, Han De-min doubted that a strictly monogamous mating system prevails in the wild, although there is no evidence for a non-monogamous mating system.

Competitors and Predators
This species is highly herbivorous. Although it occurs in much the same habitats as the monal, the two species forage in very different ways and almost certainly do not seriously compete with one another.

Presumably the usual array of Himalayan predators affect the koklass. Baker (1928) cites an observation by C. Whymper in which he observed a female koklass hit in full flight by a crested eagle (*Spizaetus nepalensis*). Whymper ran up and caught the downed bird, which he was later able to release.

General Biology

Food and Foraging Behavior

Baker (1930) stated that this species feeds on all kinds of grain, grass seeds, acorns, berries, and buds and on insects and worms, although he believed it is probably much more vegetarian than insectivorous. He mentioned that one bird that had been examined had been eating almost nothing but coarse grass, with a little maidenhair fern and moss. Cheng (1963) reported that two birds collected in China had eaten ferns (*Selaginella*), maize, the seeds and fruits of solanaceous plants, and the seeds and tender needles of pines, spruce, and other plants. Examination of 22 crops from wild birds indicate that the koklass is strongly vegetarian, consuming a wide variety of grasses, sedges, herbs, and shrubs (Han, 1990).

Observations in captivity confirm the fact that this is a highly herbivorous species, eating large amounts of green food, particularly grass and lucerne (Howman, 1979). Dierenfeld et al. (1998) suggested that about 20 percent of their diet consists of protein.

Movements and Migrations

At least in Himachal Pradesh there is a seasonal movement downward of approximately 1,000 m (3,280 ft), so that by February the birds are concentrated at about 2,200–2,500 m (7,220–8,205 ft). In most locations where koklass were observed by Gaston et al. (1981) there were some areas of suitable habitat extending below 2,500 m (8,205 ft), suggesting the importance of snow-free winter habitat. In Nepal, the amount of snowfall is much less in the east than in the western Himalayas; thus, the birds may not have to descend so low in winter (Roberts, 1981).

Daily Activities and Sociality

This species apparently keeps very close to the same quarters and may be found morning after morning and evening after evening in the same open glades searching for food. The birds are monogamous and apparently remain in pairs throughout much of the year. Gaston (1981a) says that the birds are usually solitary or at most in pairs.

Roosting is done in trees, and Severinghaus (1979) reported that one such roost that he observed con-

sisted of pines, with the roost sites 6–9 m (20–30 ft) high. These pines were on a south-facing slope and were about 30–35 years old, with few shrubs or herbs directly below them. The Chinese name for this species, "sung chi," meaning pine chicken, apparently refers to the tendency of the koklass to roost in pines and to consume its needles (Cheng, 1963).

Social Behavior

Mating System and Territoriality

Baker (1930) stated that it is "almost certain" that these birds are monogamous and that the male may be found in the close vicinity when the female is incubating. Further, once the chicks have hatched the male participates in rearing and protecting the brood. Later writers have generally confirmed the view that an extended monogamous mating system prevails in the koklass.

Territoriality is well developed in the koklass, judging from the high level of male calling typical of the species. Gaston (1981a) indicated that the calling season lasts from at least November through to May, or longer than any other Himalayan pheasant species he listed except for the cheer. In January, the peak of male calling lasts about 15 minutes each morning, with a maximum of calling at intervals of about two calls per minute, gradually tapering off to about one call per minute. Severinghaus (1979) found that calling begins about 30–45 minutes before sunrise and is greatest during the first 20–30 minutes, declining thereafter. Calling in the late afternoon is less frequent than during early morning.

Territorial sizes have not been directly estimated, but Lelliott and Yonzon (1981) noted that 11 male koklass were heard calling in a 1.2-km² (0.5-mi²) study area in May, suggesting a maximum territorial size of about 10 ha (25 acres). As noted earlier, in some areas the density may even reach more than 20 pairs per square kilometer (51.8 per square mile), thus requiring territories of about half this size.

Voice and Display

The male's territorial call is a distinctive *kok-kok-kok . . . kokras*, or sometimes given as *khwa-kakak*. Severinghaus (1979) has listed 14 phonetic renderings of this call and concluded that there may be major geographic variations in its sound, at least in different subspecies. The call is uttered mainly during morning and evening hours, but also throughout the day in cloudy weather. Frequently several males will respond simultaneously to the sound of a gun or thunder (Ali and Ripley, 1978).

Besides this call, Severinghaus (1979) has described several additional call types, all of which seem to be variants of the usual crowing call. However, they differ in numbers of syllables, relative emphasis of the individual syllables, timing of the syllables, total call duration, and the duration of individual syllables.

More recently, Lelliott (1981*b*) has studied the vocalizations of the koklass. The first call is a harsh, rapid and staccato *kuk-kuk-kuk-kuk* . . . uttered during flushing in alarm. A second call is a repeated *aw-cuk* note, which is produced by both sexes and may be uttered for as long as 15 minutes, often when the bird is confronted by an unfamiliar object. A third, not heard by Lelliott, is a rather melodious clucking and six-syllable call *chuk-cher-ra-ka-pa-tcha*, associated with frontal threat display by males.

The fourth call (Lelliott 1981*b*) is the familiar crowing call of males, which occurs in many variant forms, as Severinghaus (1979) has previously noted. However, Lelliott noted that in spite of all these variations individual birds cannot be recognized on the basis of general call type, as males often shift their call types in the course of a morning's calling activity. Lelliott noted that the call type used in his area of study (*nipalensis*) differed from that reported as most common by Severinghaus for *castanea*. However, certain individuals could be recognized by their distinctive calling patterns by Lelliott. He judged that the function of the crowing call is uncertain but probably territorial, inasmuch as tape-recordings tended to cause males to approach the tape recorder or at least to answer the crow with calling of their own.

The last call type noted by Lelliott (1981*b*) consisted of a soft female call uttered in response to male crowing, which he described as a soft *oowow* or *kerwakow*. He judged that it might serve as a contact note, informing the male that the female was close by. Lelliott noted that the birds are highly solitary, even after the breeding season; he found them to be extremely shy and difficult to observe.

Displays of the koklass have been described by Wayre (1964) and by Harrison and Wayre (1969). The forward threat display is performed relatively quietly. In situations of threat the male typically faces the threatening individual with the ear-tufts held flat,

plumage sleek, neck and head extended forward to the level of the body, and the tail slightly cocked. This posture is accompanied by threatening lunges and a continually repeated, subdued, and somewhat melodious chuckling call, *chu-cher-ra-ka-pat-tcha*.

A second posture, oriented laterally, is in some ways the antithesis to the forward threat. The stance is relatively erect, with the bird at right angles to and sometimes leaning away slightly from the focus of its attention. The ear-tufts are erected vertically like rabbit ears and the white cheek patches are fluffed up, especially toward their lower edges (figure 24). The neck feathers are also fluffed, making the neck appear short and thick. The body is held in a slanting manner so as to expose the maximum of plumage toward the other individual. The plumage, including the upper tail-coverts, is ruffled, as are the flank and belly feathers, while the wing on the displayed side is drooped slightly. The tail is spread fanlike and is twisted sufficiently strongly toward the target of the display that it follows the same plane of the body slant. In this posture the male circles around its partner, sometimes making a sudden run of a few yards in intense lateral display, with the farther wing strongly drooped and the primaries scraping along the ground, which produces a rustling sound. This display is otherwise relatively silent, although the male may also assume an upright posture and utter the typical loud crowing call. When the female is the object of this attention, she may remain indifferent or stand still, with her neck stretched parallel to the ground and with her cheek feathers and short ear-tufts erected. Alternatively, she may suddenly crouch; the male will then immediately mount her.

Reproductive Biology

Breeding Season and Nesting

In various parts of India the koklass breeds from April to June (Ali and Ripley, 1978). In Kashmir, nests with eggs have been found as late as 15 July (Bates and Lowther, 1952), but probably over most of the country earlier nesting is typical. Baker (1930) says that the nominate race begins laying about the end of April and continues on well into June. Evidently most birds lay from the middle of May to about the end of June in this race. Most nests are placed under thick bushes, usually of evergreens, on the sides of hills in coniferous forests. The nest is

Figure 24. Postures of
male koklass, including
normal (A) and erected
pinnae (B), and waltzing
from farther (C) and
nearer (D) sides. After
photos of live birds.

sometimes hidden among bracken, but may also be placed in tangles of briars, raspberries, or other canes and is invariably well hidden from view. Sometimes the nest is wedged among the roots of a tree, and in such cases may be in a hole or hollow virtually out of sight. According to Baker (1930), the presence of thick undergrowth and perhaps a proximity to water appear to be the major requisites for nesting in this species.

The normal clutch is probably five to seven, with Baker (1930) recording one clutch of nine. Clutches of eight have also been recorded, yet these large clutch sizes are unusual. Sometimes full clutches of only four eggs also have been found. Most probably six is the commonest clutch size, judging from records in the wild, although Howman (1993) suggests that 9–12 eggs are typical in the artificial conditions of captivity.

Incubation and Brooding
Incubation takes 26–27 days and is performed by the female, with the male apparently remaining close at hand. The young are highly precocial and are able to fly well within only a very few days (Baker, 1930). Like all pheasants, young birds consume a high incidence of live insects.

Growth and Development of the Young
There is no specific information on growth rates and periods of dependency of young koklass. Maturity occurs during the first year, and probably young males become territorial the spring following hatching. Young birds raised in captivity may be fed the usual pheasant diet, but probably should be shifted to a diet of greens as early as possible. They are also highly sensitive to infections, and thus sometimes are best kept on wire netting frames off the ground, provided that access to green foods can be maintained (Delacour, 1977).

Evolutionary History and Relationships
The genus *Pucrasia* is apparently relatively isolated and has no close relatives. Delacour (1977) placed it between *Tragopan* and *Lophophorus*, mainly because of known hybrids with these two genera as well as with *Catreus*. However, he noted that the lanceolate plumage of the males is similar to that of *Ithaginis*. Similarities in social displays and mating systems between these two genera have previously been noted here. The downy young are distinctively ruffed in the occipital region, but Delacour (1977) states that they resemble those of tragopans in shape and behavior. I believe that the genus is fairly close to the partridge group, and thus probably should for the present be maintained in the relative linear position accorded it by Delacour, namely close to *Ithaginis* and the tragopans.

Status and Conservation Outlook
Over most of its Himalayan range this species seems to be fairly secure, although it is vulnerable to destruction of mature middle-altitude forests with thick undergrowth, its prime habitat (Gaston, 1981a). In Pakistan the koklass' population size is still quite favorable, but overgrazing and agricultural encroaching do pose some threats (Mirza, 1981a). The species is not highly prized for its plumage and it is apparently less prone to being trapped than are some of the other pheasants of the area. The koklass is considered as very secretive and ultrawary by Yonzon and Lelliott (1981). Thus its mortality rate as a direct result of human activities is still fairly low.

McGowan and Garson (1995) listed the Chinese races *darwini, joretiana, meyeri, ruficollis*, and *xanthospila* as vulnerable. The race *joretiana* occurs in a protected area of Anhwei, and *darwini* occurs in three such preserves. The race *ruficollis* also occurs in several preserves. Li (1996) generally regarded the koklass as uncommon to common in China, with the largest number of available locality records for *darwini*. The fewest locality records (three) were for *jeretiana*. Roberts (1991) questioned the validity of several of the reputed Pakistan subspecies, based on their plumage variability. He considered the koklass to be the least endangered of the country's pheasants.

Plate 1 *(top)*. Himalayan blood pheasant, male. Photo by the author.
Plate 2 *(bottom)*. Satyr tragopan, male. Photo by the author.

Plate 3. Western tragopans, two males *(foreground and background)* and a female *(middle)*. Plate by Joseph Wolf and J. Smit; photographed by John Weinstein. © 1998 The Field Museum, Chicago (negative A113384c).

Plate 4. Blyth's tragopan, male. Photo by the author.

Plate 5 *(top)*. Temminck's tragopan, male. Photo by the author.
Plate 6 *(bottom)*. Cabot's tragopan, male. Photo by Kenneth W. Fink.

Plate 7 *(top)*. Nepal koklass, male. Photo by the author.
Plate 8 *(bottom)*. Himalayan monal, male. Photo by Kenneth W. Fink.

Plate 9. Sclater's monal, male. Plate by Joseph Wolf and J. Smit; photographed by John Weinstein. © 1998 The Field Museum, Chicago (negative 113383c).

Plate 10 *(top)*. Chinese monal, male. Photo by the author.
Plate 11 *(bottom)*. Green junglefowl, male. Photo by Kenneth W. Fink.

Plate 12. Red junglefowl, an adult male *(foreground)* and adult females or young. Plate by Joseph Wolf.

Plate 13. Ceylon junglefowl, an adult male *(middle)*, a female *(above)*, and a young male *(below)*. Plate by Joseph Wolf.

Plate 14 *(top)*. Grey (Sonnerat's) junglefowl, male. Photo by Kenneth W. Fink
Plate 15 *(bottom)*. Tibetan white eared pheasant, male. Photo by Kenneth W. Fink.

Plate 16 *(top)*. Brown eared pheasant, male. Photo by the author.
Plate 17 *(bottom)*. Blue eared pheasants, a male *(foreground)* and a female. Photo by the author.

Plate 18 *(top)*. Salvadori's pheasant, male. Photo by Kenneth W. Fink.
Plate 19 *(bottom)*. Vo Quy's (Vietnamese) pheasant, male. Photo by Han A. Assink.

Plate 20. Imperial pheasant, male. Photo by the author.

Plate 21 *(top)*. Black-breasted kalij pheasant, male. Photo by Kenneth W. Fink.
Plate 22 *(bottom)*. Lineated kalij pheasant, male. Photo by Kenneth W. Fink.

Plate 23 *(top)*. True silver pheasants, a female *(foreground)* and a male. Photo by the author.
Plate 24 *(bottom)*. Swinhoe's pheasant, male. Photo by the author.

Plate 25 *(top)*. Siamese fireback, male. Photo by Kenneth W. Fink.
Plate 26 *(bottom)*. Malay crestless fireback, male. Photo by the author.

Plate 27 *(top)*. Vieillot's crested fireback, male. Photo by the author.
Plate 28 *(bottom)*. Wattled pheasant, displaying male. Photo by the author.

Plate 29. Cheer pheasant, male. Photo by the author.

Plate 30. Elliot's pheasant, male. Photo by the author.

Plate 31. Copper pheasant, male. Photo by the author.

Plate 32 *(top)*. Mikado pheasant, displaying male. Photo by Kenneth W. Fink.
Plate 33 *(bottom)*. Hume's bar-tailed pheasant, male. Photo by Kenneth W. Fink.

Plate 34 *(top)*. Reeves' pheasant, male. Photo by Kenneth W. Fink.
Plate 35 *(bottom)*. Golden pheasant, male. Photo by Kenneth W. Fink.

Plate 36. Kweichow common pheasants, males. Plate by Joseph Wolf.

Plate 37. Green pheasants, a male *(foreground)* and a female. Plate by Joseph Wolf.

Plate 38 *(top)*. Lady Amherst's pheasant, male. Photo by the author.
Plate 39 *(bottom)*. Sumatran bronze-tailed pheasant, male. Photo by Kenneth W. Fink.

Plate 40 *(top)*. Rothschild's (mountain) peacock pheasant, male. Photo by the author.
Plate 41 *(bottom)*. Grey peacock pheasant, male. Photo by the author.

Plate 42 *(top)*. Germain's peacock pheasant, male. Photo by the author.
Plate 43 *(bottom)*. Malayan peacock pheasant, displaying male. Photo by the author.

Plate 44 *(top)*. Bornean peacock pheasant, displaying male. Photo by the author.
Plate 45 *(bottom)*. Palawan peacock pheasant, male. Photo by the author.

Plate 46. Crested argus, male. Photo by Dang Gia Tung.

Plate 47 *(top)*. Malay great argus, a male displaying to a female. Photo by the author.
Plate 48 *(bottom)*. Congo peacock, male. Photo by Kenneth W. Fink.

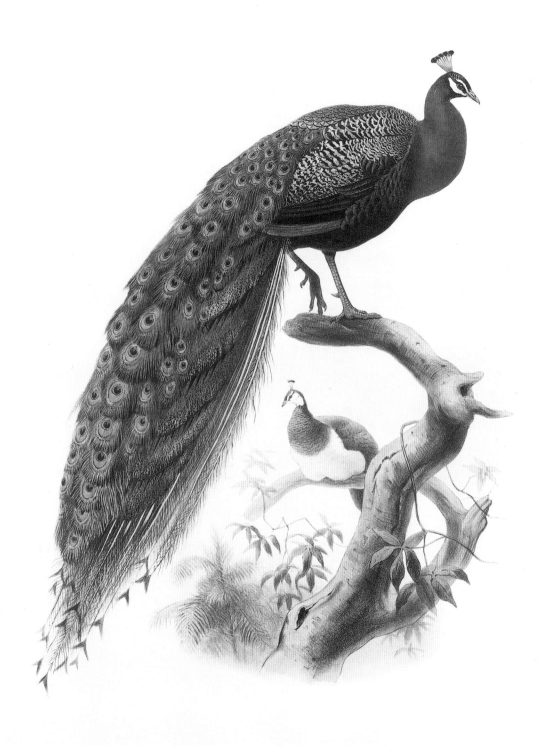

Plate 49. Indian peafowl, a male *(foreground)* and a female. Plate by Joseph Wolf.

Plate 50. Green peafowl, a male *(standing)* and a female. Plate by Joseph Wolf.

Genus *Lophophorus* Temminck 1813

The monals are large montane pheasants in which the sexes are highly dimorphic. Iridescent plumage is extensive in males excepting the underparts, which are velvety black. Males also have bare, bright blue orbital skin and crests of varying size and shape. The bill is long and highly curved, with the upper mandible strongly overlapping the lower one. The wing is rounded, with the tenth primary being the shortest and the fifth and sixth the longest. The tail is flat, broad, and shorter than the wing, and has 18 rectrices. The tarsus is stout and shorter than the middle toe and is spurred in males. The tail molt is phasianine (centripetal). First-year males resemble the females, which are dark brown with rufous and whitish markings. Three species are recognized.

Key to Species (and Subspecies of Males) of *Lophophorus*

A. Upper parts mostly iridescent (males).
   B. Tail bluish green: Chinese monal.
   BB. Tail mostly chestnut and less than 250 mm (9.8 in) in length.
      C. Tail entirely chestnut; crest feathers with naked shafts: Himalayan monal.
      CC. Tail white and chestnut; crest feathers short and curly: Sclater's monal.
         D. White tail-band 20–28 mm (0.8–1.1 in) wide: western Sclater's monal (*sclateri*).
         DD. White tail-band 10–20 mm (0.4–0.8 in) wide: eastern Sclater's monal (*orientalis*).
AA. Upper parts brownish and noniridescent females.
   B. Lower back pure white; wing at least 315 mm (12.3 in): Chinese monal.

BB. Lower back barred with black or brown; wing less than 300 mm (11.7 in).
  C. Crested; underparts coarsely marked; lower back barred with buff and black: Himalayan monal.
  CC. Uncrested; underparts finely speckled; lower back barred with brown and white:
  .   Sclater's monal.

HIMALAYAN MONAL

Lophophorus impeyanus (Latham) 1790

Other Vernacular Names
Impeyan pheasant, monaul; lophophore resplendissant (French); Himalaya Glanzfasan, Konigs-Glansfasan (German); moonal (central Himalayan vernacular).

Distribution of Species
Eastern Afghanistan (Nuristan and Safed Koh), Pakistan's Northwest Frontier Province eastward through the Himalayas to Bhutan and northeastern Assam (Mishmi Hills), and neighboring southern Tibet to area east of Lhasa (Pome and southern Chamdo to the valleys of the Po Yigrong and Brahmaputra, to at least the region of Showa Dzong, or to about 95°30′ E). This species is also reported from Burma (Yin, 1970).

It breeds from 2,745 to 3,660 m (9,000 to 12,000 ft) in the Himalayas, but it has been reported at 4,575 m (15,000 ft) and found breeding at 2,440 m (8,000 ft) in Garhwal and 2,135 m (7,000 ft) in Kashmir. The Himalayan monal is sedentary, but slight altitudinal movements have been reported. This species inhabits relatively open coniferous, mixed, or deciduous forests and rhododendrons, usually on rocky, broken, and precipitous slopes and in gorges (Vaurie, 1965). See map 4.

Distribution of Subspecies
None recognized by Delacour (1977), although females from eastern areas are said to be more rufous and richer in tone.

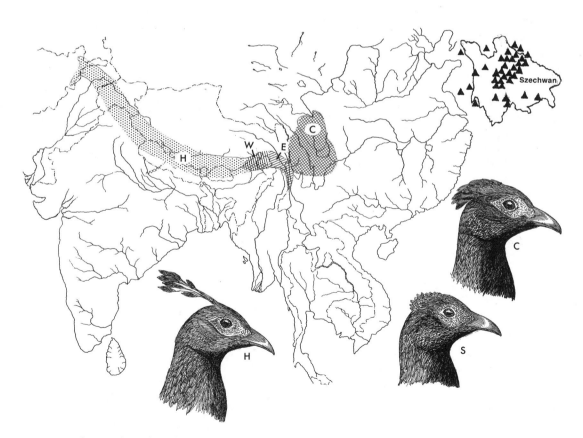

Map 4. Distribution of monals, including Chinese (C), Himalayan (H), and Sclater's (S), the last including its eastern (E) and western (W) races. Locality records of the Chinese monal are shown in the inset.

Measurements

Ali and Ripley (1978) report male wing lengths of
289–320 mm (11.3–12.9 in) and tail lengths of 215–
235 mm (8.4–9.2 in), whereas female wing lengths
were reported as 259–287 mm (10.1–11.2 in). Delacour (1977) reported a female tail length of 200 mm
(7.8 in), and Ali (1962) indicated a female tail length
of 189 mm (7.4 in). Three females measured by me
had tail lengths of 182–189 mm (7.1–7.4 in). The
weight of males ranges from 1,980 to 2,380 g (4.3 to
5.2 lb), whereas females range from 1,800 to 2,150 g
(3.9 to 4.7 lb). Lack (1968) reported the average adult
weight as 2,000 g (4.4 lb). The eggs average 63.5 ×
44.9 mm (2.5 × 1.8 in) and the estimated fresh weight
is 70.7 g (2.5 oz).

Identification

In the Field (559–635 mm, 22–25 in)

This montane species is found on steep slopes, especially where the ground is greatly broken and where
there are occasional grassy areas interspersed with
woods. In flight, the white rump of the male is conspicuous and the blackish underparts are also distinctive. The male's tail is entirely chestnut, and there is
also chestnut present on the wings. The female is
mostly a rich chestnut brown, with a rather short,
squarish tail; buffy body streaking; and a bluish area
of skin around the eye. Unlike the similar Sclater's
monal, the female also has an erectile crest and a
narrower white tail-band. The usual call is a wild,
ringing whistle, which is often used as an alarm note
and is similar to that of a curlew (*Numenius*). A
chuck-chuck call is also used by disturbed birds.

In the Hand

The long, decurved bill and the flat, fairly short tail
(under 250 mm [9.8 in]) identifies this as a monal.
In males the entirely chestnut tail and narrow, erect
plume feathers are unique. In females the short crest
(of about 25 mm [1.0 in]) and near-absence of a
white tail-band separates the species from the similar
Sclater's monal, whereas the absence of white on the
lower back and its smaller size separates it from the
Chinese monal.

Ecology

Habitats and Population Densities

In Pakistan this species occurs between 2,440 and
3,660 m (8,000 and 12,000 ft), generally, as elsewhere

in the Himalayas, in rocky crags near treeline, but
also in various valleys including those having birch
trees (Mirza, 1981*a*).

In Himachal Pradesh the birds occur between
3,000 and 3,800 m (9,845 and 12,470 ft), mainly in
low forests and scrub close to the treeline. However,
they descend to as low as 2,400 m (7,875 ft) in winter,
and then are found in dense forests that have good
undergrowth and usually on fairly steep slopes (Gaston, 1981*a*). Monals have been recorded in a variety
of habitat types, but especially in higher altitude oak
forests of Himachal Pradesh, especially during spring
and autumn. In the winter months they are more
generally distributed through various forest types,
including those dominated by pines and firs (Gaston
et al., 1981). Himalayan monals move up into subalpine meadow areas in September and October. A
minority of the records obtained by Gaston et al.
(1981) were in habitats with at least 70 percent ground
cover. The largest number of observations are made
in relatively dense canopy cover (over 50 percent)
and shrub cover ranging from 0 to 90 percent.

In Nepal the species inhabits open, rocky, grass-
covered slopes and the adjacent birch and rhododendron forests, mainly between about 3,000 and
4,100 m (9,845 and 13,450 ft). In one such area about
8 pairs per kilometer (20.7 per mile) of ridge were estimated to be present (Yonzon and Lelliott, 1981).

Other population density estimates are few, but
in some areas of Pakistan the density may be about
1.9 pairs per square kilometer (5.0 per square mile)
according to Mirza (1981*a*). Gaston et al. (1981) cen-
sused a variety of habitats in Himachal Pradesh and
found densities as high as 4–8 males in areas of 1 km²
or less. However, for larger areas the densities were
much lower, on the order of about 2.5–7.0 pairs per
square kilometer (6.5–18.1 per square mile).

Studies at the Great Himalayan National Park
(Yahya, 1993) in Himachal Pradesh found monals
to occur between 2,700 and 3,400 m (8,860 and
11,155 ft) during September, with seven sightings in
high-elevation oak forest, four in high conifer forest,
and one in mixed deciduous forest. In the Garhwal
Himalayas of Uttar Pradesh, Bisht et al. (1990) found
that during the summer the birds use alpine and subalpine areas as high as 3,600 m (11,810 ft) and during
winter they occupy subalpine and temperate forests.
In the Kedarnath Wildlife Sanctuary of the same
general region, monals were found to favor habitats
with low trees, high shrubs, and medium grasses in

oak-fir forests and temperate tree-scrub habitats between 2,600 and 3,000 m (8,530 and 9,845 ft). Population densities ranged from 5 to 20 pairs per square kilometer (13.0 to 51.8 per square mile; Sathyakumar et al., 1993).

Competitors and Predators

The unique foraging method of the Himalayan monal probably places it well out of competition with other genera of pheasants, as well as other birds of similar size.

Baker (1930) judged that most of the "troubles" of this species come from eagles because he observed that the birds seemed to repeatedly gaze up into the sky while they were foraging in exposed areas. Various predatory mammals undoubtedly affect the monal as well, but specific information is lacking. Beebe (1918–1922) found the carcass of a male that had apparently been killed by "some great bird of prey," and close by saw a golden eagle that he believed had been responsible for the kill.

General Biology

Food and Foraging Behavior

Beebe (1918–1922) has summarized the foods of Himalayan monals well. He suggests that terrestrial insects and tubers form their chief foods, but the specific foods vary greatly by locality. Wherever snow does not cover the ground, the birds spend a great deal of time digging with their beaks, apparently for tubers, roots, and subterranean insects. In autumn they are said to forage largely on insect larvae that they find under decaying leaves and at other times of the year on roots, leaves, and young shoots of various shrubs and grasses, as well as acorns, seeds, and berries.

Although in winter Himalayan monals may be seen in wheat and barley fields, they seek roots and maggots rather than the grain. Edible mushrooms, wild strawberries, currants, and the roots of ferns have also been mentioned as local foods. Beebe noted that the crops of the birds he examined were filled with hard tuber fragments, and he judged that the sharp edges of the bird's mandible were important in cutting and splitting plant tissues of such firm consistency as these.

The foraging behavior of the monal is very distinctive. The birds do very little digging with their feet, but instead pick at the earth with their shovel-like beaks, sometimes digging holes as deep as a foot. When a large tuft of grass or bamboo is encountered, the birds will dig around it until it is left supported only by its bare roots or it may actually be toppled over. Himalayan monals typically forage in small groups, but do not usually fight over foods that are excavated in this manner (Baker, 1930).

Movements and Migrations

Gaston et al. (1981) noted that this species seemed to exhibit the greatest altitudinal movements of the pheasants they observed in Himachal Pradesh, concentrating mainly between 2,000 and 3,000 m (6,560 and 9,845 ft) during January–March and mostly above 3,000 m (9,845 ft) during September–October.

In Nepal there is an altitudinal movement range from 3,200 to 4,350 m (10,500 to 14,270 ft; Lelliott and Yonzon, 1981; Roberts, 1981). Compared with the western Himalayas, the snowfall in Nepal is not so great toward the east; thus seasonal movements are probably not so great.

Daily Activities and Sociality

Beebe (1918–1922) reported that monals choose protected raised ledges on southern or southeastern slopes of steep cliffs or outjutting masses of boulders for their roosting sites. Because the sites were typically 3–5 m (10–15 ft) above level ground, Beebe imagined that this represented a site safe from beech martins, grey foxes, and wild dogs and protection from all but the worst storms. He observed three males, including one immature bird, at one such roosting site, as well as feathers of a female or immature bird. In another site Beebe observed the birds arrive at a favorite foraging area one morning. The first male arrived alone, apparently shortly after dawn. Within about the next hour more adult males arrived, singly, in pairs, or in trios, until finally 14 males were present. These males seemingly came from different directions. They fed for about half an hour, after which they were apparently disturbed and began to leave.

Himalayan monals are somewhat gregarious, but the ties between flock members seem to be quite loose. Probably only during the winter, when the birds are forced into restricted habitats, are real flocks formed. In the autumn females wander down the slopes with their offspring. During winter rather large flocks of 20–30 birds may gather in chestnut forests (Beebe, 1918–1922). During this time an

adult male may associate with flocks consisting of females and their young, but typically groups of three
or four males associate during the nonbreeding season (Baker, 1930).

Social Behavior

Mating System and Territoriality

Ridley et al. (1984) judged this species to be polygynous, a condition facilitated by the tendency for females to be gregarious. He judged the pair bond to
last from mating to incubation. Baker (1935) also
questioned whether the species is monogamous. Certainly the high degree of sexual dimorphism suggests
a polygynous or promiscuous mating system.

Locations of individual males are apparently ad-
vertised by loud daily calling. Gaston (1981a) reported the calling period in Himachal Pradesh to
last from March to June, the same period as he also
reported for the western tragopan and kalij. Gaston
et al. (1982) suggested that although dispersion of
males in spring gives some suggestion of territoriality,
14.5 percent of the males seen then were in groups
of two, and 13 percent were parts of larger groups
of birds (both sexes). Although aggressive behavior
between males was observed in May, these authors
believed that strictly observed territorial boundaries
did not seem to be present. Loosely defined home
ranges seemed to be a better description of dispersion
characteristics. The highest densities occurred in
areas close to precipitous crags, which probably provide both safe roosting sites and favorable launching
sites for display flights.

Voice and Display

The male's call is a shrill, loud, and curlewlike whistle.
Beebe (1918–1922) said that the same call is used by
the female and her young, although in the case of the
young birds it is shriller and higher than in adults.
He described the adult call as a high, reverberating
whistle, bringing to mind the beat of a dove's wings
in flight (see figure 28).

Lelliott (1981b) has distinguished three call types
for this species, based on fieldwork with wild birds.
The first is a succession of high-pitched piping notes
that begin very rapidly, but become more spaced out
toward the end of the call, which lasts up to 10 seconds. Each note lasts only about 0.14 seconds, but
the intervals between the notes gradually lengthen to

about 1 second as the call progresses. This call is uttered by alarmed birds of both sexes, but especially
females, as they are flushed. It was also uttered by
birds on the ground, in which case it had a seemingly
different and unknown function. For example, this
call might accompany the dawn crescendo call. This
call consists of a piping whistle of two to five notes,
with each set repeated faster and with increasing frequency and amplitude per set, which is followed by a
slowing and diminuendo of sets. The dawn crescendo
call was found to be uttered by both sexes at least
during spring and autumn, and probably is uttered
throughout the year except during the monsoon. The
functions of the crescendo call are more uncertain
than, for example, the dawn calling of tragopans and
the koklass pheasants. Lelliott (1981b) judged the call
to be possibly territorial in function, but inasmuch as
it is also uttered outside the breeding season this explanation is not altogether suitable. According to
Beebe (1918–1922), it may also be simply a concomi-
tant of the nervous excitement of awakening.

The third call observed by Lelliott is the high-
pitched whistling note described by various observers as being similar to that of a curlew (Numenius arquata). This call was sometimes found as an
extension of the piping call, with the ultimate note
of that call being extended into a pure tone and repeated at intervals from 1 to 5 seconds to form the
whistle call. This call might then be uttered for up
to 5 minutes while the bird was perched on the
ground or in a tree. It was never heard from a flying
bird, and was noted only in females or immatures.
Lelliott (1981b) judged that its function may be
that of indicating "anxiety," and on hearing it other
monals would sometimes become alert and utter the
same call.

The male's display has been described by a variety
of writers (Roden, 1899; Wayre, 1969; Catlow, 1982).
Catlow's recent description is most complete and is
based on observation of captive birds. Early stages of
male-male display establish dominance. Males stand
very tall, with the feathers of the neck, the mantle,
and the abdomen strongly fluffed. The beak is held
upright and the birds step slowly and elaborately,
occasionally lowering the head, causing the crest to
vibrate and shimmer. When males are directing the
display toward females, they may perform the same
ceremony, but without feather fluffing. Males also
suddenly arch the neck, bringing the crest forward,

Figure 25. Postures of male Himalayan monal, including normal appearance (A), lateral display to female (B), approaching female (C), and frontal display with tail-fanning (D). After photos and drawings in various sources, including Schenkel (1956–1958).

then call while suddenly raising the head and quickly lowering it again, causing the crest to vibrate violently. Alternatively, the male may puff out the feathers from the lower mandible around the ear-coverts and, finally, the feathers along the top of the head, making them appear black and increasing the width of the blue skin around the eye. The male then pulls his head in toward the shoulders, with the beak against the upper breast, compresses the neck feathers, and lowers the nearer wing (figure 25). Then, with the primaries scraping the ground, he circles the female, leaning toward her and raising the farther wing so that it is visible across his back. He may also tidbit

with food items or small stones. The next stage is the direct frontal display, in which the male faces the female and lowers its head while standing erect, half-opening the wings, fluffing the neck feathers, and holding the tail high and fanned. The wings are slowly opened and partially closed, and the tail is slowly flicked up and down. In this posture the white back patch is exposed and the head is oriented so that the crest directly points toward the female (Wayre, 1969). The display is normally silent, but the head is sometimes shaken, causing the crest to vibrate and shimmer. This part of the display may last up to 2 minutes and is usually the climax. On a

Figure 26. Display postures of male Himalayan monal, including wing-flapping and bounding phase (A), wing-spreading and tail-fanning (B), tail-cocking (C), and retreat with wing-spreading (D). After scenes in an Anglia Survival Film, *The Bird of Nine Colours.*

few occasions the male may follow the frontal display with other postures and rarely may call during the climax phase, uttering a very loud, drawn-out, churring call of 3–4 seconds duration, with the head thrown back (Catlow, 1982).

Additional visual information on male displays has been derived from nine sequences in a BBC Anglia Survival Film, *The Bird of Nine Colours.* Several sequences show a male approaching a female in a bouncing manner, his wings mostly open and flopping up and down in awkward manner, the bird somewhat resembling an enormous, inebriated fruit bat trying unsuccessfully to take flight (figure 26A). At the end of this saltatory approach sequence the

male would suddenly stop, lower his wings, and tilt his fanned tail forward, while facing the female in a bowing frontal display posture (figure 26B). Alternatively, he would quickly stop, raise his head, and tilt his tail forward (figure 26C). He would then turn about and retreat in a rather erect posture, with both wings extended and his white back pattern conspicuously visible (figure 26D).

Besides these ground displays, a display flight also occurs in wild Himalayan monals, which consists of an extended gliding, with the tail fanned and the wings held well above the level of the body, and with a repeated piping call that varies from fairly soft to strident (Gaston et al., 1982).

Reproductive Biology

Breeding Season and Nesting

In India Himalayan monals begin nesting primarily in May, but egg records extend from 20 April to 27 June (Baker, 1935). In northeastern Afghanistan and Tibet the season evidently lasts from April to July (Hüe and Etchécopar, 1970; Etchécopar and Hüe, 1978), although there seem to be no actual nest records for these areas. The altitudinal range during the breeding season is very great, with some breeding in India as low as 2,440 m (8,000 ft) and rarely even to 2,135 m (7,000 ft). However, Himalayan monals also may be seen at 4,270–4,575 m (14,000–15,000 ft) during the same season.

Nests are invariably placed in wooded habitats, typically in forests having large trees but not very thick undergrowth. The nest is a simple scrape, often under the shelter of a bush, a rock, or in the hole of some large tree. The nest is often unlined, although leaves may collect in the hollow and thus form a lining.

Incubation and Brooding

Clutch sizes in the wild are most commonly four or five eggs, with three-egg clutches also fairly common; sometimes only two eggs are present. The largest reputed clutch of a wild bird was apparently of eight eggs (Baker, 1930, 1935; Ali and Ripley, 1978), although such clutch sizes are highly suspect (Bates and Lowther, 1952). There are a few suggestions in the literature that the male helps in caring for the young, however, this is certainly not the typical situation.

The incubation period lasts about 26–29 days, usually 28 days. In captivity females will often lay a second replacement clutch if the first is removed or unsuccessful in hatching.

Growth and Development of the Young

Not much has been written on this phase. However, Wayre (1969) stated that in captivity the chicks are not difficult to rear on starter crumbs to which has been added live food (maggots and mealworms) for the first few weeks. Sexual maturity is not attained until the second year of life.

Evolutionary History and Relationships

Delacour (1977) suggested that *Lophophorus* is an ancient and long-established genus, with no apparent phyletic links to other genera. I agree with this point, and can suggest no special relationships within the pheasant group. Within the genus *Lophophorus*, the three species form a geographical replacement series that seems to reflect closely their origins and relative phyletic relationships to one another.

Status and Conservation Outlook

The status of this beautiful pheasant is still fairly secure in many areas. In some areas of Pakistan the Himalayan monal is still fairly common at elevations between 2,440 and 3,660 m (8,000 and 12,000 ft), at least in some valleys (Mirza, 1981a). In Nepal it is also still locally common. When afforded protection, as in Sagarmatha National Park, the Himalayan monal is quite tame (Roberts, 1981). However, where it is hunted it becomes extremely wary and difficult to shoot, although traps and snares take many young birds (Yonzon and Lelliott, 1981). In Himachal Pradesh the species has disappeared from some areas, but in others it is still widespread and certainly numerous in a few places (Gaston et al., 1981). Thus, in spite of the large size and valuable plumage, the species seems to be able to cope with humans to a surprising degree. It is considered endangered by McGowan and Garson (1995) and as vulnerable by Collar et al. (1994). Li (1996) listed only three locality records for Tibet and four for Yunnan. However, the Himalayan monal occurs in two Yunnan sanctuaries and is fairly common in one of them (Gaoligong Shan), as well as in Dibang Wildlife Sanctuary, Arunachal Pradesh (del Hoyo et al., 1994; McGowan and Garson, 1995).

SCLATER'S MONAL

Lophophorus sclateri Jerdon 1870

Other Vernacular Names
Crestless monal; lophophore de Sclater (French); Weisschwanz Glanzfasan, Stalhuhn (German).

Distribution of Species
Eastern Himalayas and neighboring southern Tibet, at about the same elevations as *L. impeyanus*, from about 93° E east to about 97° E and probably the Salween River in Tibet, north to at least 30°20′ N in the Po Yigrong Valley and south to northern Burma (Adung Valley and the region of Hpimaw, or about 26° N in the east), east to the mountains of northwestern Yunnan on the border of Burma, and south to the Shweli–Salween Divide or about 25°30′ N. This species is sedentary and inhabits ravines and rocky slopes in mountain forest in habitat more or less similar to that of *L. impeyanus* (Vaurie, 1965). See map 4.

Distribution of Subspecies
None recognized by Delacour (1977). Davison (1974) has described an eastern race (*orientalis*) in which the tail-band is somewhat narrower (10.5–20.0 mm, 0.4–0.8 in) than in the nominate western form (20–28 mm, 0.8–1.1 in). Davison considers the range of *orientalis* to extend from the upper Irrawaddy River eastward to northwestern Yunnan and south along the Shweli–Salween watershed.

Measurements
Ali and Ripley (1978) reported that males have wing lengths of 298–303 mm (11.6–11.8 in) and females 285–287 mm (11.1–11.2 in). Tails of males range from 194 to 206 mm (7.6 to 8.0 in) and one female's was 193 mm (7.5 in; Baker, 1930). Cheng et al. (1978) noted that three males had wing lengths of 290–310 mm (11.3–12.1 in) and tail lengths of 195–212 mm (7.6–8.3 in), whereas a female had a wing length of 266 mm (10.4 in) and a tail length of 180 mm (7.0 in). Two males weighed 2,500 g (5.5 lb). Baker (1928, 1930) reported male weights of 2,267–2,948 g (5.0–6.4 lb) and female weights of 2,126–2,267 g (4.7–5.0 lb). The eggs average 63.2 × 45.4 mm (2.5 × 1.8 in) and their estimated fresh weight is 70.5 g (2.5 oz).

Identification

In the Field (533–610 mm, 21–24 in)
This species is found in similar montane habitats as the Himalayan monal (but only locally overlapping with it in range). Sclater's monal usually occurs above 2,745 m (9,000 ft) in dense forests having local grassy openings or above the treeline. Males closely resemble those of the Himalayan monal, but have much more white on the lower back, rump, and upper tail-coverts; their chestnut tail is also white tipped. There is only a very short and curly crest, but the rest of the male's plumage is extremely similar to that of the Himalayan species. Females have a paler lower black, rump, and upper tail-coverts than do those of the Himalayan species. Additionally they have a distinct buffy tail-band. The calls are evidently very similar to those of the Himalayan monal and include a shrill, harsh whistle reminiscent of peacocks or guineafowl. Sclater's monals are often found in dense fir forests with a rhododendron understory.

In the Hand
Males can be readily distinguished from the Himalayan monal by their unconstricted, short, curly crest feathers and the presence of a white tip of at least 10 mm (0.4 in) on their tail feathers. Females lack the short crest typical of female Himalayan monals and are much paler on the lower back and rump. Female Sclater's monals are also somewhat less coarsely patterned on the underparts than are females of that species.

Ecology

Habitats and Population Densities
Almost nothing is known of this species in its natural environment. Ludlow and Kinnear (1944) described the breeding habitat in Tibet as consisting of silver fir (*Abies*) with dense rhododendron undergrowth. Sclater's monal is found at elevations of between 3,000 and 4,000 m (9,845 and 13,125 ft), moving up into the alpine zone during summer. Most probably it also occurs in forest openings and subalpine meadows. In various seasons its habitats are probably

much like those of the Himalayan monal (Ludlow and Kinnear, 1944).

There are no estimates of population densities.

Competitors and Predators

Nothing has been written on this for the Sclater's monal. The species is not known to be in contact with the Himalayan monal, although Ludlow and Kinnear (1944) stated that somewhere between 92 and 93° E these two species meet.

Predators probably include such large raptors as the golden eagle and various predatory mammals.

General Biology

Food and Foraging Behavior

According to Ludlow and Kinnear (1944), Sclater's monals feed during mornings and evenings in small forest openings. Few actual crop or gizzard contents have been analyzed, but among the contents specifically mentioned are *Polygonum* seeds and the heads of thistles or hard-headed flowers (Ali and Ripley, 1978).

Foraging behavior has not yet been specifically described. However, the shape of the beak would lead one to believe that it is essentially the same as has been described for the Himalayan monal.

Movements and Migrations

This has not been specifically described. However, downward movements to snow-free areas undoubtedly occur in winter, while in summer Sclater's monals are known to move up to alpine meadow.

Daily Activities and Sociality

Because they flushed three of them from a rhododendron brake in late May, Ludlow and Kinnear (1944) suggest that even during the breeding season males may remain fairly social. Birds collected at that time had enlarged gonads.

As mentioned above, Sclater's monals feed in the morning and evening hours, in the typical fashion of pheasants in this general region. They probably also roost in small groups, like the Himalayan monal.

Social Behavior

Mating System and Territoriality

Although nothing has been written on this, mating system and territoriality presumably resemble the situation in the Himalayan monal.

Voice and Display

The alarm call has been described as shrill, hoarse, and rather plaintive, sounding intermediate between that of a peacock and a guineafowl. This call may be uttered repeatedly by a "sentinel" from an overlooking rock on a steep hillside. The call has also been described as a wild, ringing whistle quite similar to that of the Himalayan monal, but distinct in tone (Ali and Ripley, 1978).

Reproductive Biology

Breeding Season and Nesting

Sclater's monals collected by Ludlow and Kinnear (1944) in Tibet during mid-May were all in breeding condition, with one female having an incubation patch and the other having an unlaid egg in the reproductive tract.

No nests have been described from the wild, and the birds are not known to have bred in captivity. Baker (1935) mentioned a clutch of five eggs that had been brought to him and had been collected between 1 and 3 June at an elevation of about 2,745 m (9,000 ft) on a peak north of Sadiya.

Incubation and Brooding

The clutch size, incubation period, and other aspects of breeding are unknown for the Sclater's monal, but are probably similar to those of the Himalayan monal.

Growth and Development of the Young

There is no information on this subject.

Evolutionary History and Relationships

There can be no doubt that this species is a very close relative of the Himalayan monal and probably has not been isolated from it for long. Davison (1978b) suggested a Pleistocene separation. He commented that the greater width and increased conspicuousness of the white tail-band in western populations of the Sclater's monal may be associated with increased possibilities of hybridization with the Himalayan monal in that area and needs for more effective visual isolation. Western males also tend to have more fulvous rather than blackish underparts, and Davison added that this too may be an important species-specific signaling device during frontal display in the two species.

Status and Conservation Outlook

The Sclater's monal is currently considered as rare (King, 1981). It may already be gone from Nagaland in India, but until at least in 1968 it was still fairly common in some parts of northern Burma, especially at elevations between 2,440 and 2,745 m (8,005 and 9,005 ft; King, 1981). In 1936 Ludlow and Kinnear (1944) observed a considerable number in Pachak-shiri, in southeastern Tibet, but there is no recent information on the status of the species in Tibet.

CHINESE MONAL

Lophophorus ihuysi Geoffrey St. Hilaire 1866

Other Vernacular Names

Chinese impeyan or monal; lophophore de Lhuys (French); Grünschwanz-glanzfasan, Schanzschwänzige Stalhuhn (German); koa-loong (Tibetan Chinese).

Distribution of Species

Mountains of northeastern Sikang and northwestern Szechwan north to those southeast of the Koko Nor in Tsinghai, and southern Kansu south to eastern Tibet and northern Yunnan, at elevations varying from about 3,050 to 4,875 m (10,000 to 16,000 ft). This species is sedentary, but it undoubtedly moves altitudinally with the season and has been reported at 2,745 m (9,000 ft) in the winter. It occurs from the upper limit of the coniferous forest and rhododendrons to the rocky alpine meadows and tundras above the forest and the zone of arborescent scrub (Vaurie, 1965). See map 4.

Distribution of Subspecies

None recognized by Delacour (1977).

Measurements

Delacour (1977) reported that a single male had a wing length of 345 mm (13.5 in) and a tail of 305 mm (11.9 in), whereas a female had a wing of 320 mm (12.9 in) and a tail of 270 mm (10.5 in). Two females measured by me had wing lengths of 285 and 317 mm (11.1 and 12.4 in) and tails of 185 and 228 mm (7.2 and 8.9 in). An adult male weighed 2,837 g (6.2 lb) and a female 3,178 g (7.0 lb; David Rimlinger, pers. comm.). Five eggs (from a captive female) averaged 71.2 × 50.4 mm (2.8 × 2.0 in) and had an estimated fresh weight of 99.4 g (3.5 oz). The average fresh weight of 13 eggs was 97.1 g (3.4 oz; Rimlinger and Whitman, 1986).

Identification

In the Field (762–813 mm, 30–32 in)

This is the only species of monal that is found in China, and thus identification is simplified. Like the other monals, males have velvety black underparts and a white rump area. In this species there is no chestnut on the tail, and the birds are appreciably larger than other monals. Females resemble the other monals, but are larger and generally more contrastingly patterned with dark brown and whitish. The species' calls include a clear and quadrisyllabic vocalization, presumably of the male, that is usually heard in early morning and sounds something like the native name, *koa-loong*. Rimlinger and Whitman (1986) reported that the male's usual advertisement or challenge call consists of one to eight notes, which are rapidly repeated in bursts up to 2 seconds apart. Double-note calls seem to serve as alarm notes.

In the Hand

The large size (wing at least 320 mm [12.5 in]) and absence of chestnut on the tail of males provide for ready separation from other monal males. Females are larger than in the other two monal species. They also have completely white backs, as well as white to grayish mottling or shaft-streaking elsewhere on the body.

Ecology

Habitats and Population Densities

This species inhabits high montane coniferous forests having a rhododendron undergrowth, which presumably serves as breeding habitat. The Chinese monal also extends during summer into alpine meadows and tundra areas as high as 4,880 m (16,010 ft). It has also been observed in a rocky scrub margin of alpine larches at 3,660 m (12,000 ft) and probably like the other monals is associated with rocky outcrops that provide hiding places and roosting sites.

A density of 0.1–13.0 individuals per square kilometer (0.3–33.7 per square mile) has been reported in Szechwan. Four total sites had an mean density of about 4.5 per square kilometer (11.7 per square mile; Li, 1996). Lu and Liu (1986) reported a density of 1.32–1.58 birds per square kilometer (3.4–4.1 per square mile), averaging 1.45 birds per square kilometer (3.8 per square mile) during breeding. Rimlinger et al. (1997) estimated densities of 4 (spring) to 6 (winter) birds per square kilometer (10.4 and 15.5 per square mile) and home ranges of 0.005–0.810 km^2 (0.003–0.502 mi^2), averaging 0.132 km^2

(0.08 mi²) for four birds. Altitudinal ranges varied from 2,927 to 3,593 m (9,605 to 11,790 ft) in winter to 2,945–3,900 m (9,665–12,795 ft) in spring.

Competitors and Predators
Schäfer (1934) suggested that the golden eagle is this species' only natural enemy. Nothing is known of possible competitors, which are probably few, if any.

General Biology

Food and Foraging Behavior
Practically nothing is known of this. Schäfer (1934) and others have said that the bulbs of an alpine *Fritillaria*, locally called "pei-mu" or "be-mu," is the major food of this species. Thus, a local Chinese name for the bird is "bemu-chi," according to Schäfer. Recent studies by Lu and Liu (1986) have added to our knowledge of this species' foods. They found that Chinese monals eat flowers, tender leaves, roots, and sprouts, with the enlarged roots and bulbs of plants such as onion and fritillary being dug out of the ground with their powerful bills.

Movements and Migrations
This species is sedentary, but it moves altitudinally within the season.

Daily Activities and Sociality
Little has been written of this, but probably what has been said of the Himalayan monal applies equally well to this species. Beebe (1918–1922) stated that Chinese monals occur in small parties during the day, sometimes with males and females in separate groups, foraging in open habitats. At night they descend to perch in dense, scrubby, and stunted rhododendrons or in the sheltering branches of pines farther down the mountainsides.

Social Behavior

Mating System and Territoriality
Schäfer (1934) stated that calling by males occurs regularly during spring and summer mornings. The few descriptions of the species would suggest that the calls might be louder and carry further than is true of the Himalayan monal. Whether these calls are indications of territorial defense or simply are used by males to attract females to them is somewhat uncertain even in the Himalayan species. The Chinese monal is believed by some Chinese biologists to be monogamous (Lu and Liu, 1986), in spite of its high level of sexual dimorphism and elaborate male displays.

Voice and Display
Most descriptions of the male's call indicate that it is distinctly multisyllabic, consisting of one to eight distinct and separate notes. One Tibetan name, "koa-loong," is suggestive of the sound that is produced (Beebe, 1918–1922). Schäfer (1934) states that during the call the tail is half-spread in a fan, both wings droop gently, and the upper part of the body, neck, and head are all strained forward. The bird thus stands with ruffled neck feathers, runs a few steps forward, turns in a circle, stretches its neck, utters the call again, and vaunts its beautiful plumage. The whistled note is uttered every few minutes, starting with a high tone that lasts unchanged for about 3 minutes, and then gradually shifts to a more mournful tone of deeper range, which eventually fades away. When alarmed, the Chinese monal takes flight with a whir of wings and utters repeatedly a loud, gabbling whistle.

Rimlinger and Whitman (1986) as well as Ruan et al. (1993) have provided new information on the displays of this species in captivity. The male's courtship or advertisement call is uttered from elevated sites, in an outstretched-neck posture (figure 27) that is similar to that of the Himalayan monal (see figure 28). The calls average about 1.7 seconds in length, with a sequence of 20 such calls lasting about 40 seconds. During spring, females utter similar calls that are either single-note or double-note. The male typically approaches the female in a hopping or stiffly walking posture, with lowered head, apparently much like the bouncing approach of the Himalayan monal described earlier. This approach usually is accompanied with ground-pecking, but no calls. Alternatively, the male may assume a more erect posture, with his head raised, his wings slightly lowered, and his white back feathers exposed, as he walks around the female (figure 27C). A very similar posture of the Himalayan monal is illustrated for comparison (figure 27D). The Chinese monal's frontal display consists of facing the female with the tail spread, initially assuming a rather erect posture prior to wing-spreading (figure 27E). The tail is then widely spread while it is alternately

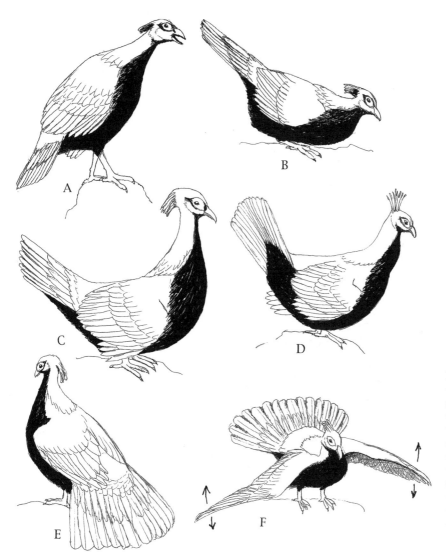

Figure 27. Display postures of male Chinese monal, including calling (A), crouching (B), erect walking (C), first phase of frontal display (E), and second phase of frontal display (F). After sketches by Ruan et al. (1993). A calling posture of the male Himalayan monal is also shown (D) for comparison with sketch A.

raised and lowered, the wings are spread and flapped, and simultaneously the head and neck are stretched forward and downward (figure 27F). Copulation attempts sometimes follow this nearly prostrate posture. A display flight, involving gliding down from the heights or circling around a valley with the tail spread and the white upper parts fully exposed, has also been observed in wild birds.

Reproductive Biology

Breeding Season and Nesting
Few nests of this species have yet been found in the wild. Some reputed eggs of this species are probably actually those of other species, judging from their relatively small sizes. Cheng et al. (1978) describe three eggs as averaging 73.7×51 mm (2.9×2.0 in), which closely approximates the size of eggs laid by a captive female at the San Diego Zoo (David Rimlinger, pers. comm.). Cheng et al. judged the clutch size to be from three to five eggs, which are laid in April and May (captive birds). Thirteen eggs were also laid by a captive female in San Diego between mid-April and mid-June during 1984 and 1985 (Rimlinger and Whitman, 1986). Among five nests found in nature there were usually 3 or 4 eggs; four nests had an average clutch of 3.25 eggs. However, up to 11 eggs have been seen in nests that presumably

Figure 28. Crowing postures of male junglefowl, including grey junglefowl (A), green junglefowl (B), Ceylon or Lafayette's junglefowl (C), and red junglefowl (D). After photos by Kenneth Fink. Calling by the male Himalayan monal (E) is shown for comparison. After a photo by Lincoln Allen.

involved laying by multiple females. Nests were always located in recesses or narrow crevices of steep cliffs, and were always at least 180 m (590 ft) apart (Lu and Liu, 1986). Nests of wild birds have been found in late April and early May (C. Zhang et al., 1997).

Incubation and Brooding
Studies on wild birds by Lu and Liu (1986) indicate that only the female plays a role in incubation, nest protection, and care of the young. Incubation lasted 28 days, with all 13 eggs hatching in four observed nests. Roosting of the young at night with their "parents" (presumably only their mothers) was observed in chicks as old as 2 months. Studies in captivity (Rimlinger and Whitman, 1986) indicate that mean weight at hatching is about 68 g (2.4 oz). Dark underpart feathers began to appear in captive males within 8 weeks, and they began to acquire iridescent feathers as early as 15–17 weeks. An average adult weight of about 3,200 g (7.0 lb) in males and 2,900 g (6.3 lb) in females was attained by 9–10 months. Similar studies by Cheng et al. (1996) indicate that males attain sexual maturity at 3 years of age, when they weigh about 3,500 g (7.7 lb), and females at 2 years, when they weigh about 3,100 g (6.8 lb).

Growth and Development of the Young
Schäfer (1934) stated that in May, when the young have reached the size of a thrush and can fly well, they begin to appear with their mothers. He mentioned seeing a brood of five young out looking for food on a misty morning. When alarmed, the young would hide under their mother's wings.

Evolutionary History and Relationships
This species is geographically separated from the Sclater's monal by the drainages of the Yangtze, Mekong, and Salween Rivers, but certainly the two populations evolved from a common ancestral type in rather recent times.

Status and Conservation Outlook
The Chinese monal is classified as an endangered species (King, 1981) and is believed to be rare throughout its Chinese range. Due to its relatively large size, it is sought out by hunters and trappers. It is believed that hunting rather than habitat destruction is mainly responsible for its rarity (King, 1981). Very few birds are in captivity and even fewer have bred under such conditions, making the prospects for saving the species by avicultural methods a daunting prospect. However, several nature reserves have been established by the Chinese government within the range of this species, which occurs in much the same area as does the giant panda. Thus the outlook seems more promising than previously (Wang, 1980). Among these is the 200,000-ha (494,000-acre) Wenchuan Wolong Nature Reserve in Szechwan, and a 95,000-ha (23,465-acre) reserve (Baishuaijiang) in Kansu, both of which are of subalpine coniferous forest types. Several other similar smaller reserves (ranging to 40,000 ha [98,800 acres]) in Szechwan are also located within this subalpine ecosystem. The Chinese monal is considered vulnerable by McGowan and Garson (1995) and by Collar et al. (1994). Li (1996) listed 39 locality records, with 30 of these from Szechwan. It has been recorded from reserves in seven Szechwan counties (del Hoyo et al., 1994).

Genus *Gallus* Brisson 1760

The junglefowl are small to medium-sized tropical pheasants in which the sexes are highly dimorphic. Males have a unique fleshy comb on the top of the head, one or two lappets below the bill, and an almost entirely naked face and throat. Males also have iridescent plumage, and the neck and rump feathers (hackles) are variably lanceolate and ornamental. The wings are rounded, with the tenth primary shorter than the first and the sixth the longest. The tail is strongly compressed laterally and vaulted, with the central rectrices greatly lengthened and sickle shaped. The tarsus is longer than the middle toe and is spurred in males. In females the plumage is duller, the combs and lappets are greatly reduced, and the spurs are lacking. Four species are recognized.

Key to Species (and Subspecies of Males) of *Gallus* (after Delacour, 1977)

A. Comb, spurs, and hackles rudimentary (females).
    B. Breast almost uniform in color.
        C. Breast rufous; upper parts finely vermiculated: red junglefowl.
        CC. Breast dull pale brown; upper parts strongly marked with black: green junglefowl.
    BB. Breast strongly patterned.
        C. Breast mottled with pale buff and dark brown; wings and tail strongly barred: Ceylon (Lafayette's) junglefowl.
        CC. Breast with brown-edged white spots; wings and tail finely vermiculated: grey (Sonnerat's) junglefowl.
AA. Comb, spurs, and hackles well developed (males).
    B. Neck feathers short, with broad tips of iridescent green: green junglefowl.
    BB. Neck feathers dull black or long and narrow.
        C. Breast striped.

D.　Breast striped with orange, red, and brown: Ceylon (Lafayette's) junglefowl.

DD. Breast striped with gray, black, and white: grey (Sonnerat's) junglefowl.

CC. Breast black: red junglefowl.

D.　Neck feathers with rather broad, blunt tips: Javan red junglefowl (*bankvia*).

DD. Neck feathers with narrow, pointed tips, forming hackles.

E.　Hackles rather short and dark red; comb small; earlobe small and usually red: Tonkinese red junglefowl (*jabouillei*).

EE. Hackles long and golden red.

F.　Hackles yellow at tip, with large and blackish central stripe: Indian red junglefowl (*murghi*).

FF. Hackles orange at tip, central stripe brown and narrower.

G.　Hackles very long; earlobes white and large: Cochin-Chinese red junglefowl (*gallus*).

GG. Hackles moderately long; earlobes small and usually red: Burmese red junglefowl (*spadiceus*).

GREEN JUNGLEFOWL

Gallus varius (Shaw and Nodder) 1798

Other Vernacular Names

Javan junglefowl; coq de Java, coq à queue fourchue (French); Gabelschwanzhuhn (German).

Distribution of Species

Java and neighboring islands (Madura, Kangean, Bawean, Bali, Lombok, Sumbawa, Flores, Alor) at low elevations (Delacour, 1977). Introduced but probably now extirpated from the Cocos-Keeling Islands. See map 6.

Distribution of Subspecies

None recognized by Delacour (1977).

Measurements

Delacour (1977) reported that males have wing lengths of 220–245 mm (8.6–9.6 in) and tail lengths of 320–330 mm (12.5–12.9 in), whereas one female had a wing length of 195 mm (7.6 in) and a tail of 115 mm (4.5 in). Three females measured by me had wing lengths of 180–195 mm (7.0–7.6 in) and tail lengths of 121–126 mm (4.7–4.9 in). Dunning (1993) reported that the mean of an adult sample (unspecified by sex or sample size) was 620 g (1.4 lb; range 454–795 g [1.0–1.7 lb]). The eggs average 44.5 × 34.5 mm (1.7 × 1.3 in) and have an estimated fresh weight of 29.2 g (1.0 oz).

Identification

In the Field (406–711 mm, 16–28 in)

This species is limited to Java and its immediate vicinity, where it is unlikely to be confused with any species except for the introduced red junglefowl. It is the only junglefowl with a rounded green and purplish comb in males, and has a generally greenish to blackish coloration over most of its body plumage, wings, and tail, except for the yellow-edged rump feathers and the brown-edged wing-coverts. The crowing of the male is a shrill, crisp, and continuous *chaw-aw-awk*, whereas the female has a slow, repeated, cackling *wuk*. Other notes include a repeated *chop*, *chak*, or *kowak*. Females are best identified in the field by their very dark tail and by their spotted or "scaly" back pattern.

In the Hand

This is the only junglefowl in which the male has broadly rounded and iridescent green neck feathers and a small, rounded comb that is greenish to purplish. Except for her rounded tail, the female's plumage is reminiscent of *Phasianus*, but the tail feathers are black with an iridescent green sheen and buffy edging.

Ecology

Habitats and Population Densities

The habitats of the green junglefowl are reputed to include seashores, coastal valleys, and to some extent inland forests, at least those of low altitude. Dry, rocky scrub around cultivated fields and bamboo or brush thickets are also among their favored habitats (Delacour, 1977).

There are no estimates available for population densities.

Competitors and Predators

This species overlaps with the red junglefowl over nearly all of its range, and undoubtedly the two must compete rather strongly. Because the red junglefowl is considerably larger than the Javan species, the latter must be at a considerable competitive disadvantage. However, there is no evidence of interbreeding between the two in the wild, so some kind of ecological isolation seems to keep them apart.

There is no specific information on possible predators.

General Biology

Food and Foraging Behavior

In the wild green junglefowl are believed to forage mainly on noxious insects and weed seeds and to regularly visit grain fields. In captivity they consume seeds, greens, berries, and other fruit, and such animal materials as insects and worms. Termites are important foods for young birds (Beebe, 1918–1922).

Movements and Migrations

There are probably no movements of significant distance in green junglefowl because the seasonal temperature changes are limited. Also, there is no prolonged and severe dry season for the birds to contend with. Indeed, nesting generally corresponds with the dry season, lasting from June to November, when there is an abundance of insect life (Beebe, 1918–1922).

Daily Activities and Sociality

Observers of green junglefowl in the wild suggest that they are usually to be found singly, in pairs, or at most in family units of no more than about six birds. During the largest part of the day they are said to remain hidden in stunted undergrowth, not venturing out into clearings and fields to forage. At other times they may even go out onto roadways to feed or perhaps obtain grit, and generally remain together in small numbers. At such times the birds can be captured by placing nets at certain openings in the bushes that lead to rice fields. Then, just before dusk, natives of the island rush through the cover while beating sticks. Birds already in the fields rush back to the cover provided by the bushes, and in so doing are captured (Beebe, 1918–1922).

Social Behavior

Mating System and Territoriality

In contrast to the other species of junglefowl, it has been suggested that the mating system of this species is probably monogamous. Beebe (1918–1922) reported seeing birds only in pairs and said that observations of semicaptive birds also support the idea of a monogamous pair-bonding that lasts well past the time of hatching. Beebe said that the young may remain with their parents for 6 or 7 months and that he repeatedly saw both parents leading their broods. However, some other earlier observers (e.g., Beebe 1918–1922) had suggested that the birds might be polygynous, with males having harems of as many as four females, which Beebe could not verify.

There does not seem to be any information on territoriality in this species.

Voice and Display

The male's challenge call (see figure 18, page 119) has been described as consisting of three sharp and shrill syllables, which sound like *cha-aw-awk*. The male also utters a slow and cackling repeated *wok* note, as well as a repeated *chop* alarm call. The female utters a fast repeated *kok* call, as well as a loud *kowak, kowak* (Delacour, 1977). David Rimlinger (pers. comm.) has observed lateral waltzing and a frontal display by males. He has further observed that during the last note of the crowing call the male will occasionally clap his wings together overhead, producing a single sharp clapping sound.

Reproductive Biology

Breeding Season and Nesting

Beebe (1918–1922) indicated that the breeding season is long and rather variable in Java, with as much as several months sometimes intervening between the laying of neighboring pairs. The breeding season generally occurs between June and November, most often occurring during the first month or two of the eastern monsoon. According to Beebe this period is one characterized by an abundance of insects and sprouting plant life.

Hoogerwerf (1949) provided nesting records for Java that extended from March to November, with the largest numbers in May and June (8 of 16). Green junglefowl from eastern Java seemingly breed somewhat earlier and over a shorter period than those from western Java (4 of 7 in April and May).

Relatively few nests have been described in the wild. Beebe (1918–1922) stated that the nest usually consists of a hollow in the ground hidden among dense vegetation. However, several have been found in ferns growing at some height against a tree trunk or even in the heart of a tree fern's top. One nest that Beebe found was among some oak-leaf ferns on a ledge that was breast high in an area of dense cacti and briers. The ledge itself was covered by fire ants; Beebe believed that such ants provided protection for the sitting bird, which he thought to be immune to their effects.

Beebe (1918–1922) judged the average clutch size to be from 6 to 12 eggs, averaging about 8. However, this is evidently an overestimate, and the average wild clutch is probably only of 3 or 4 eggs (Hoogerwerf, 1949). However, in captivity the females lay quite freely, sometimes as many as 40 eggs in a season, which is probably related to a high potential for renesting.

Incubation and Brooding

The incubation period of the green junglefowl is 21 days under captive conditions, which is typical of the genus *Gallus*.

Growth and Development of the Young
Nothing specific has been written on this subject.
The young are reportedly quite delicate and susceptible to infections as well as to cold. Two years are
required for full attainment of sexual maturity and
development of the fully adult male plumage.

Evolutionary History and Relationships
In all respects this species seems to be the most generalized or "primitive" of all the junglefowl, although
its seemingly monogamous mating system has perhaps contributed to its lack of elaborate male displays
and plumage specializations. The colors of the females and the shape of the tail show some similarities
to those of the genus *Lophura*, but the absence of fertile intergeneric hybrids with any other group of
pheasants would argue for a relatively isolated position of the genus *Gallus*. Mainardi (1963) suggested
that *Gallus* may be derived from a *Phasianus*-like ancestor, but its immunological distances as well as
chromosomal differences indicate differentiation
from this group. Relatively close immunological distances were obtained with both *Phasianus colchicus*
and *Lophura nycthemera*. Delacour (1977) judged the
genus *Gallus* to be isolated and strongly specialized,
without a clear appropriate location in the sequence
of pheasant genera. Within the genus *Gallus* the
green junglefowl is apparently the most isolated form
(Morejohn, 1968*a*). Although hybridization with
G. gallus is possible, the F$_1$ females show reduced fertility and produce young only when backcrossed to
gallus (Steiner, 1945). I have placed *Gallus* earlier in
the generic sequence than did Delacour, primarily to
emphasize its isolation rather than to suggest specific
affinities. However, it may be that such Old World
partridge genera as *Galloperdix* are actually more
closely related to *Gallus* than are any of the typical
pheasant genera. Stock and Bunch (1982) recently
compared chromosome morphology. They reported
that *Gallus* most nearly retains the apparent ancestral
phasianid karyotype (as compared with *Crax*, which
they considered nearest to the original galliform
state) and that it is fairly close to the partridge type
found in *Coturnix*, but more distantly removed from
Phasianus. Stock and Bunch also reported that both
Numida and *Pavo* are relatively primitive in their
chromosome structures, supporting the general notion that the guineafowl are descendants of an early
cracidlike stock from which the partridges and pheasants were both subsequently derived.

Status and Conservation Outlook
Little can be said about the status of the green junglefowl. However, because it thrives in edge and broken
environments and can adapt to the presence of humans, it is likely to be able to survive quite well in
spite of its limited geographic range. This species is
known to occur in high densities in Java's Baluran
National Park and to occur in Ujung Kulon National Park (Sumardja, 1981). In 1990 the "bekisar"
(a sterile male hybrid genotype produced by mating a
male green junglefowl with a female domestic fowl)
was named as the official mascot of eastern Java because of its notable vocal behavior. Owing to the
resulting pet market for these birds, the capture rate
of green junglefowl has greatly increased, and thus
increased pressures have been placed on its survival
in the wild.

RED JUNGLEFOWL

Gallus gallus (Linné) 1758

Other Vernacular Names
Wild junglefowl; coq bankiva (French); Bankivahuhn, Wilduhn (German).

Distribution of Species
Widespread in Southeast Asia, north to the lower ranges of the Himalayas (up to about 2,010 m [6,600 ft]) from northeastern Pakistan eastward to Assam and southward to at least 17° N on both banks of the Godavari River, where it encounters *sonnerati* and locally hybridizes with it. This species also extends east through Burma, across much of Indochina, and on the islands of Java, Sumatra, and Bali. The red junglefowl has been introduced widely and occurs as domesticated varieties almost worldwide. See maps 5 and 6.

Distribution of Subspecies (after Wayre, 1969)
Gallus gallus gallus (L.): Cochin-Chinese red junglefowl. Resident in Cochin-China, Cambodia and nearby islands, Vietnam (except extreme north), central and lower Laos, and eastern Thailand. It intergrades with *jabouillei* in North Vietnam.

Gallus gallus spadiceus (Bonnaterre): Burmese red junglefowl. Resident in southwestern Yunnan, Burma, Thailand (except extreme east), northern

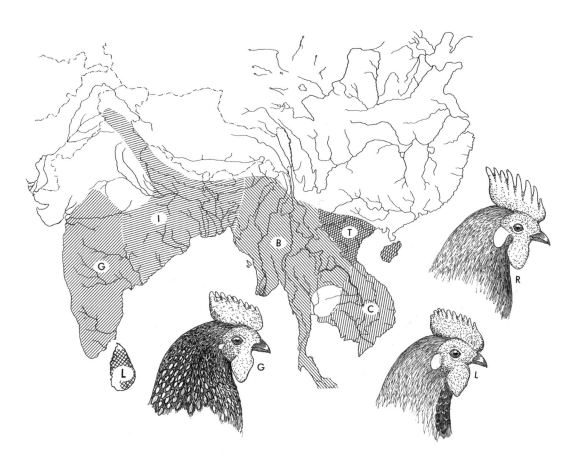

Map 5. Distribution of junglefowl, including Burmese (B), Cochin-Chinese (C), Indian (I), and Tonkinese (T) races of red junglefowl, the grey or Sonnerat's junglefowl (G), and the Ceylonese or Lafayette's junglefowl (L). See map 6 for remainder of red junglefowl's range. The introduced range of red junglefowl is not indicated.

Map 6. Distribution of Burmese (B) and Javan (J) races of red junglefowl (R; hatching) and green or Javan junglefowl (Ja; shading). Both species occur natively on Java, and feral red junglefowl also occur elsewhere in the region (Sulawesi, Lombok, Timor).

Laos, Malaya, and northern Sumatra. It intergrades with *jabouillei, gallus, murghi,* and *bankiva* near their respective boundaries.

Gallus gallus jabouillei Delacour and Kinnear: Tonkinese red junglefowl. Resident in Tonkin (North Vietnam) extreme southeast of Yunnan, Kwangsi, Kwangtung, and Hainan. It intergrades with *spadiceus* in northwestern Tonkin.

Gallus gallus murghi Robinson and Kloss: Indian red junglefowl. Resident in northern and northeastern India, the lower ranges of the Himalayas from southern Kashmir to Assam, and northern and east-central India. Rarely from the plains to 2,135 m (7,000 ft), normally not above the foothills; in sal forest and adjacent scrub. Koelz (1954) has described a new race (*gallina*) that Ripley (1961) considers a synonym of *murghi.*

Gallus gallus bankiva Temminck: Javan red junglefowl. Resident in the southern half of Sumatra, Java, and Bali.

Measurements

Delacour (1977) reported that males of *gallus* have wing lengths of 230–250 mm (9.0–9.8 in) and tail lengths of 260–275 mm (10.1–10.7 in), whereas females have wing lengths of 185–200 mm (7.2–7.8 in) and tail lengths of 140–155 mm (5.5–6.0 in). Ali and Ripley (1978) indicate that males of *murghi* have wings of 203–244 mm (7.9–9.5 in) and tail lengths of 300–380 mm (11.7–14.8 in), whereas females have wing lengths of 177–196 mm (6.9–7.6 in) and tail lengths of 145–165 mm (5.7–6.4 in). Weights of male *murghi* range from 800 to 1,020 g (1.8 to 2.2 lb) and females from 485 to 740 g (1.1 to 1.6 lb). Cheng et al. (1978) reported that ten males of *jabouillei* ranged in weight from 672 to 1,020 g (1.5 to 2.2 lb), averaging 844 g (1.8 lb), and a single female weighed 500 g (1.1 lb). They also gave the average male wing length of this race as 216 mm (8.4 in; range 207–223 mm, [8.1–8.7 in]) and the average tail length as 207 mm (8.4 in; range 167–340 mm, [6.5–13.3 in]) for ten males. The eggs average 45.3×34.4 mm (1.8×1.3 in) and their estimated fresh weight is 29.6 g (1.0 oz).

Identification

In the Field (432–762 mm, 17–30 in)
This species, which is the wild ancestor of the domestic fowl, is familiar to nearly everyone, but its plumages vary considerably according to the degree of domestic or feral influence. However, wild-type males have uniformly golden brown to reddish ornamental neck hackles, a serrated scarlet comb, and a generally blackish green tail and underpart coloration. The male's crowing call is very much like the familiar call of the domestic form, but the last note is typically more abbreviated. Females are much duller than males and are generally patterned with dull brown, yellowish, and blackish. Where grey junglefowl also occur, the females of the red junglefowl are separable by their paler and more yellowish brown overall coloration and their rather uniformly rufous-tinted breast, with no indication of the black-and-white spotting that is typical of female grey junglefowl.

In the Hand
Red junglefowl are most likely to be confused with Ceylon junglefowl. Males of the former differ mainly in their unstriped neck hackles, their more distinctly serrated comb (which also lacks a yellow interior), their brownish rather than purplish black wings, and their blackish rather than brownish underparts. Females are more difficult to separate. However, the breast of the red junglefowl is more rufous and less distinctly patterned with black and white, the upper neck is more yellowish, and the wing feathers are not so distinctly barred. Female green junglefowl are also rather similar, but are distinctly more "scaly" in their dorsal plumage pattern and show some iridescence on their tail feathers.

Geographic Variation
Geographic variation is marked and clinal in this species, particularly among males. Males of the most northern and westerly form, *murghi*, have unusually long and golden yellow neck hackles during the breeding season. Those of the most southern and easterly form, *bankiva*, have relatively short and rounded neck feathers. Fairly short and dark reddish neck hackles also occur in the easternmost mainland race, *jabouillei*. Males of this race also have small combs and very small earlobes, but those of the race *gallus* immediately to the south have large combs and lappets, as well as large white earlobes. Females have neck hackles that vary in length and color in a similar manner to those of males, but these feathers are uniformly shorter (Delacour, 1977).

Ecology

Habitats and Population Densities
The red junglefowl occupies a wide range of habitats throughout its broad range. It ranges altitudinally from sea level to at least 1,830 m (6,000 ft) and is found in nearly all tropical to subtropical habitats. In general the red junglefowl prefers flat or rolling terrain to steep country and second-growth or edge habitats to heavily forested habitats. In India the birds occur in forests ranging from fairly mature second-growth and spot-lumbered forests to more open, mixed hardwoods, tropical thorn forests, and the jungles of the alluvial plains. Typically the undergrowth is fairly dense; if it is open, it includes such clumped species as bamboo and lantana. In some areas scrub jungle ravines are used, as are deciduous forest slopes and flats. In yet other areas sal (*Shorea*) and teak-bamboo forests are also used. In Nepal red junglefowl typically occur in scrub at forest edges, especially among *Zizyphus* bushes and bamboo groves. In Assam, mixed evergreen and deciduous forests are used. Pure coniferous forests and wet or boggy habitats are apparently avoided (Bump and Bohl, 1961).

In Thailand and Malaya similar habitats are used, including secondary forests associated with abandoned clearings, the edges of bamboo forests (which often sprout after fire and provide an excellent food source for the birds), and similar successional habitats in forested regions. Red junglefowl apparently do not occur in primal undisturbed forest, at least in Malaya, and in Thailand apparently do not occur at higher elevations than those supporting bamboo growths (Collias and Saichuae, 1967). However, red junglefowl have been recorded at elevations as high as 1,830 m (6,000 ft) in Malaya (Medway and Wells, 1976).

In southern China the species reaches about 2,000-m (6,560-ft) elevations and occurs in a wide variety of tropical to temperate habitats, such as chestnut forests, secondary bamboo growths, mixed broadleaf forests, sparse forests, and shrubby areas (Cheng et al., 1978).

The species is apparently native to Sumatra, but has perhaps been only rather recently introduced into Java; it has become widespread in Indonesia, the Philippines, Micronesia, Melanesia, and Polynesia through introductions (Long, 1981).

There are surprisingly few estimates of population densities for the red junglefowl. However, Bump and Bohl (1961) thought that in the Siwalik foothills area of India a winter density of about 25–50 birds per square kilometer (64.8–129.5 per square mile) might be a conservative estimate. On the basis of average flock size and average distances between roosts, Collias and Collias (1967) estimated that a population density of about 100 birds per square kilometer (259.0 per square mile) probably existed there. Li (1996) reported a density of 1.1–1.6 individuals per square kilometer (2.8–4.1 per square mile) in Yunnan. In a feral but protected flock living within the San Diego Zoo, Collias and Collias (1996) observed that the mean density of their study population was 13.4 individuals per hectare (5.4 per acre), representing a density far greater than reported for any observed wild pheasant population.

Competitors and Predators

Except along the limited area where this species is in contact with the grey junglefowl, it is probably not in serious competition with any other pheasants. Bump and Bohl (1961) judged that the kalij pheasant is probably not a serious competitor with the red junglefowl, although the males sometimes fight to the death during the breeding season.

These same authors mentioned several potential predators of the red junglefowl in India, including the mongoose, jungle cats, various hawks, and great horned owls, but found no evidence that any of these posed serious problems to the birds. Collias and Collias (1967) listed a considerable number of potential predators in India, but found no direct evidence for any of these being important. Collias and Saichuae (1967) also mentioned that various hawks, eagles, and a variety of small cats reportedly prey on red junglefowl in Thailand and Malaya, but had little direct evidence of this. Additionally, a fishing cat (*Prionailurus vivernnus*) and a palm civet (*Paradoxurus hermaphroditus*) were reported as predators by a native guide, and

there was a similar report of a yellow-throated marten (*Martes charronia*) trailing a flock of red junglefowl. Goshawks (*Accipiter badius*) and serpent eagles (*Spilornis cheela*) have also been seen near flocks of junglefowl in Thailand. Various snakes and lizards may be important egg predators in the area. In China, reported predators include jungle cats (*Felis chaus*), yellow martens (*Mustela sibirica*), sparrow hawks (*Accipiter nisus*), various falcons, and owls. Other predators such as snakes, lizards, rodents, and small carnivores probably consume eggs and chicks (Cheng et al., 1978).

General Biology

Food and Foraging Behavior

Most evidence indicates that red junglefowl are highly opportunistic and omnivorous. In India, a sample of 37 crops revealed no less than 30 kinds of seed and many types of insects. Some of the plant genera represented in the crops of two or more birds were *Trichosanthes*, *Rubus*, *Carissa*, *Zizyphus*, *Shorea*, *Digitaria*, *Stellaria*, and *Oryza*, of which only *Oryza* (rice) was a cultivated grain. The insects included representatives of ants, beetles, termites, true bugs, and flies, and there were other invertebrates such as snails, spiders, and millipedes as well.

Collias and Saichuae (1967) reported on the food contents of 23 adult junglefowl collected in Thailand, and these also consisted of a large mixture of plant and animal materials. Important plant contents included the fruits of various Euphorbiaceae, seeds of bamboo, fruits of *Zizyphus* and Acanthaceae, and materials from such crops as maize, rice, beans, and tapioca. Animal foods present were primarily insects, especially termites and ants, with as many as nearly 1,000 termites counted in the crop of a single male.

In China, the foods are similarly diverse and include nuts, seeds, bamboo shoots, leaves, petals, cultivated grains, drupes, and berries. Termites, their eggs, and other insects are also eaten in quantity (Cheng et al., 1978).

Collias and Collias (1967) stated that junglefowl seem to eat a wide variety of foods as they become seasonally available, and typically scratch about in the leaf litter individually. An apparent peck-order relationship determines access to favorable foraging locations.

Movements and Migrations

There is little evidence of seasonal migrations in this species, although Bump and Bohl (1961) suggest that in the winter birds from the higher areas are forced into lowlands. There are also some unsubstantiated reports of seasonal migrations to rainforests in Thailand during the rainy season (Johnson, 1963). Most observers have found red junglefowl to be relatively sedentary, with limited home ranges (Collias et al., 1966; Collias and Collias, 1967).

Daily Activities and Sociality

The daily activities of red junglefowl center around their roosts, which tend to remain constant unless they are disturbed. The birds roost socially in trees and forage from dawn to about 9:00 A.M. and again from about 3:00 or 4:00 P.M. until dusk. During the dry season the birds also regularly visit water holes and typically drink early in the morning and again in the evening. Red junglefowl move about as they forage, but at least some have rather restricted daily movements of as little as about 137 m (450 ft) in diameter. During the hottest part of the day the birds rest in or near the roosting trees (Collias and Collias, 1967). Although Collias and Collias (1967) found the average number of birds using a single roost to be only about 5, there is an early account of as many as 30 birds seated side by side on a single bamboo (Beebe, 1918–1922).

Social Behavior

Mating System and Territoriality

Although a few earlier writers believed this species might be monogamous, most recent studies have supported the idea that it is regularly polygynous (Bump and Bohl, 1961; Collias and Collias, 1967). Collias and Collias found single males to be in association with zero to four females. However, early in the breeding season there were fewer polygynous pairings, whereas later there were cases of males in company with single females after some had gone off to incubate their eggs. Several single unmated males or groups of males were also observed, presumably reflecting first-year birds or other older males who were unsuccessful in obtaining mates.

Territories are presumably established and proclaimed by crowing (figure 28). Collias and Collias (1967) believed that crowing helps to reinforce the territorial relationships, facilitates spacing, and reinforces dominance relationships within a flock. Crowing by one male tends to stimulate crowing in others and sometimes also attracts others. Much of the crowing is apparently associated with territorial patrolling; subordinate males seem to crow much less than do dominant birds.

Collias and Collias (1996) estimated that male territories at the San Diego Zoo were 50–75 m (164–246 ft) in diameter and irregular in shape. The most dominant males tended to have large territories. One male remained in control of his flock for almost 4 years, eventually dying when over 5 years old. Territorial boundaries were primarily defended by crowing, with dominant or "despotic" males doing most of the crowing. Eight dominant males performed 60 percent of 135 observed copulations, and a single dominant male performed 38 percent of the total copulations. Four of the most dominant females (of 28 hens) successfully raised half of the flock's chicks. Only a small proportion of the most dominant males and females were responsible for generating most of the adults in succeeding generations. Dominant females not only raised most of the flock's chicks, but also had longer mean life spans than did subordinate hens (Collias et al., 1994).

Voice and Display

The crowing call of the male is the familiar four-noted *cock-a-doodle-doo* of the domestic fowl, but the individual notes tend to run together. The third note is the most sustained and has the highest amplitude (Collias and Collias, 1967). Individual differences in crowing characteristics occur and may allow for individual acoustic recognition (Miller, 1978).

Crowing is most intense during early morning hours, being slightly earlier each morning as the breeding season progresses. Typically a sharp peak in crowing occurs shortly after dawn and there is often a secondary peak before sunrise, probably coinciding with the birds' initial movement away from the roost to foraging and drinking areas. There is usually little crowing during the day, but a minor peak occurs before sunset as the birds go to roost (Collias and Collias, 1967).

The social displays of the red junglefowl have been discussed in chapter 4 and need not be repeated here. In the wild there seem to be relatively few actual fights among males, although subordinate males may

Figure 29. Maternal behavior of red junglefowl, including brooding (A) and showing food to chicks (B). After photos by the author.

persistently follow a male and his harem. Males attract their harem by uttering their distinctive food call associated with tidbitting behavior. This behavior (see figure 11 in chapter 3) as well as copulatory behavior is apparently the same in wild junglefowl as in domesticated forms (Collias and Collias, 1967).

Collias (1987) analyzed and classified the red junglefowl's vocal repertoire. He established ten general call categories within which the approximately 24 different calls that he detected could be grouped. These included five calls uttered by chicks, several calls uttered only by hens toward chicks or while protecting the nest or brood, and two calls (courtship and crowing) that were uttered only by males. The remainder were uttered by both sexes. Collias concluded that the male's crowing advertises and identifies the species, identifies the bird's sex and location, and proclaims territorial ownership and social dominance. Crowing calls of the three other junglefowl differ from the red junglefowl's in call duration, number and accenting of notes, pitch and acoustic structure of notes, and length of intervals between notes. Contrary to the findings of Leonard and Horn (1995), Sullivan (1992) found no statistical correlation between rate of crowing and relative dominance among male red junglefowl, although the presence of a dominant male had a suppressive effect on crowing by nearby subordinate males. Dominant males also had larger combs and tended to spend more time in vigilant activities. Thus, females might respond positively to more vigilant males when choosing mates and may also favor those males that exhibit their dominance in more obvious ways, such as crowing and fighting. Studies under natural conditions in Malaysia suggest that most males are seen with single females, but as many as three females may be present (Arshad and Zakaria, 1997). Additional information on mate-choice tendencies in this species is provided in chapter 4.

Reproductive Biology

Breeding Season and Nesting
In India the breeding season of this species corresponds with the dry season in spring and is centered on the period March–May. This also corresponds to the cycle of crowing, which in northern India peaks in early May and declines by early June (Collias and Collias, 1967). However, for India as a whole, eggs have been found from January to October (Bump and Bohl, 1961). On the Malay Peninsula the nesting season is apparently quite seasonal, with nests having been found from December to May (Medway and Wells, 1976). The peak of the nesting season there seems to be in January and February (Glenister, 1951). In China, breeding begins as early as February in southern Yunnan and peaks between March and May. Apparently some nests have been reported as late as October (Cheng et al., 1978).

Nests are found in a wide variety of locations, but red junglefowl prefer dense secondary growth, bamboo forests, and other dense growths. Often the nests are placed under a bush or in a bamboo clump; when placed in the middle of a bamboo clump they may be elevated a few feet above the ground.

Clutch sizes are apparently normally of five or six eggs, with four also frequently found and rarely as many as nine present. Estimates of clutches larger than nine are probably in error (Baker, 1930).

Incubation and Brooding
Incubation is done by the female alone. Under natural tropical conditions it takes about 20 days (Baker, 1930). Cheng et al. (1978) report the period as be-tween 18 and 20 days. Although only one brood is produced per year, renesting is typical when the eggs have been destroyed before hatching. Renesting efforts typically have only three or four eggs (Bump and Bohl, 1961).

Growth and Development of the Young
The mother and the chicks leave the nest site as soon as the last young is dried; thereafter the female leads them about (figure 29). The chicks are able to fly at a surprisingly early age and when they are no more than a quarter grown. It is uncertain as to how long the chicks remain with their mother in nature, but in captivity the period of association lasts for at least 3 months (Collias and Collias, 1967). There have been some observations of males attending females with broods, although it has been asserted that the male does not roost with them while they are growing (Cheng et al., 1978). Broods often stay together well into the autumn (Bump and Bohl, 1961). Sexual maturity is attained by the end of the first year. However, the males are at that age not in full plumage and probably remain sexually subordinate to older and more experienced males for some time.

Evolutionary History and Relationships
Comments have been made in the section above on the green junglefowl as to the evolutionary affinities of the genus *Gallus*. The red junglefowl seems to have a somewhat central position in the genus, with plumage similarities to several species. Thus, Beebe (1918–1922) mentions that it resembles *varius* in the ventral plumage of males and *lafayetti* in male dorsal plumage. However, females show a similarity to *sonnerati* in female dorsal plumage. The relationship with *sonnerati* is apparently close because there seems to be complete hybrid fertility between these two species (Danforth, 1958) and there is natural hybridization known in the limited zone of sympatry between them.

Status and Conservation Outlook
The red junglefowl is not a problem for conservationists because it breeds more commonly in disturbed second-growth habitats than in undisturbed forests. In its domesticated form it is probably the most abundant bird in the world, with populations numbering in the several billions.

GREY (SONNERAT'S) JUNGLEFOWL

Gallus sonnerati Temminck 1813

Other Vernacular Names
None in general English use; coq de Sonnerat (French); Sonnerathuhn (German).

Distribution of Species (after Ripley, 1961)
Peninsular India north to southern Rajasthan (Mount Abu), Gujarat, Madhya Pradesh, and Andhra Pradesh to Polavaram. This species inhabits evergreen as well as scrub, bamboo, teak, and mixed deciduous forest from plains level to 1,525 m (5,000 ft). See map 5.

Distribution of Subspecies
None recognized by Delacour (1977). Koelz (1954) has described a race (*wangyeli*) that was not accepted by Ripley (1961).

Measurements
Delacour (1977) reported that males have wing lengths of 220–255 mm (8.6–9.9 in) and tail lengths of 330–390 mm (12.9–15.2 in), whereas females having wing lengths of 200–215 mm (7.8–8.4 in) and tail lengths of 130–170 mm (5.1–6.6 in). Ali and Ripley (1978) reported male wing lengths of 220–255 mm (8.6–9.9 in) and tail lengths of 314–390 mm (12.2–15.2 in) and females wing lengths of 190–215 mm (7.4–8.4 in) and tail lengths of 100–170 mm (3.9–6.6 in). Adult males weigh from 790 to 1,136 g (1.7 to 2.5 lb) and females from 705 to 790 g (1.5 to 1.7 lb). Eggs average 46.3 × 36.5 mm (1.8 × 1.4 in) and their estimated fresh weight is 33.4 g (1.2 oz).

Identification

In the Field (381–762 mm, 15–30 in)
Although limited to southern India, in some areas this species may be found where the red junglefowl exists in wild or feral form and might be mistaken for it. The grey junglefowl is usually found only in pairs or small family groups. Also, its usual call is quite different, sounding like *kuck-kaya-kaya-kuck*, ending with a low *kyukun, kyukun*, which is repeated slowly. The male's crow has also been described as *kuk-ka-kurra-kuk* and as unmusical and curiously grating and halting in quality. A clicking sound also is uttered. Males are almost uniformly grayish in appearance,

lacking the red junglefowl's yellow to brownish neck, rump, and wing coloration. Females are generally darker brown and less yellowish in the neck region than those of the red junglefowl. Their underparts are also more distinctly spotted and mottled with black and white, rather than tending to uniform reddish brown to buffy.

In the Hand
Male grey junglefowl are the only species having grayish neck hackles ornamented with golden yellow spotting; such spotting also occurs on the rump and the wing-coverts. Females are best identified by their vermiculated dark brownish and blackish tail and their mottled black and brown wing feathers, which differ from the distinctly barred condition of green and Ceylon junglefowl or the brown and mottled rufous wing and tail coloration of the red junglefowl. Males in breeding plumage are unique in having the tips of the neck hackles, median wing-coverts, and scapulars with shafts that are expanded terminally into flat plates of orange yellow, most of which are fringed outwardly with deep yellow. These tips lack the "eclipse" plumage, but young males show them to a limited extent.

Ecology

Habitats and Population Densities
The grey junglefowl inhabits a wide variety of habitats, from secondary dry deciduous forest to moist evergreen forests, but is especially common in mixed bamboo jungle, the edges of forest villages, around cultivated fields, and in abandoned clearings or neglected plantations. It is primarily associated with broken foothill country, but also occurs up to the highest peaks of about 2,400 m (7,875 ft) in the Nilgiris and Kerala Ranges (Ali and Ripley, 1978). In southern India it extends to the coastline, but in the north it encounters the red junglefowl in the vicinity of the Godavari River. The grey junglefowl only locally occurs north of this river, its tributary the Indravati, or the Nerbudda to the west, where the species' northern limits are reached in the vicinity of the Avaralli Hills (Beebe, 1918–1922).

Collias and Collias (1967) reported that the habitat of this species is very similar to that of the red junglefowl, although it is typically more open and rocky. The more grayish color of the males seems to match closely the rocky background color. When in vegetation, grey junglefowl seem to prefer areas covered with shrubs, small trees, and euphorbias over grassy clearings.

There are no available estimates of population densities in this species. However, Baker (1930) reported that along quiet jungle roads as many as 15 birds have been seen within the space of a few miles.

Competitors and Predators
Certainly at the northern edge of its range the red junglefowl must be a significant competitor. Beebe (1918–1922) suggests that grey junglefowl also feed in the same areas as spurfowl, bustard-quail, and other small granivorous birds, which probably are not significant competitors. However, grey junglefowl often associate with various species of babblers (Timaliidae) and the two groups apparently mutually benefit. The babblers catch insects that are flushed by the junglefowl and the junglefowl use the babblers as effective lookouts for possible danger.

Beebe believed that several raptors, such as the Bonelli eagle (*Hieraeetus fasciatus*) and the crested hawk-eagle (*Spizaetus nipalensis*), are significant predators, particularly the latter. Beebe also believed that such mammalian predators as leopards, various other cats, and especially mongooses are undoubtedly important enemies. He also mentioned pythons as a possible source of danger.

General Biology

Food and Foraging Behavior
Foods of the grey junglefowl are very diverse and include grain; shoots of grass and crops; tubers; berries of *Zizyphus, Lantana, Streblus*; windfallen figs (*Ficus* spp.); and the seeds of bamboo and *Strobilanthes*, the latter of which are especially favored by the birds. Animal materials include insects, especially grasshoppers and termites, and even small reptiles (Ali and Ripley, 1978). After fields have been burned over grey junglefowl seem to enjoy the tender, juicy sprouts of freshly growing grasses. A young chick was found to have filled its crop with soft vegetable matter, some tiny beetles, and a small moth (Beebe, 1918–1922).

When foraging, grey junglefowl do not wander far from cover and at the first indication of danger quickly dart back into heavy brush. However, when surprised in the open they typically fly, rather than flee on foot, and head for the nearest cover. They typically forage day after day in the same area at about the same time each day (Baker, 1930).

Movements and Migrations
According to Beebe (1918–1922) there are no seasonal migrations in this species, although there may be a considerable amount of wandering for food during different seasons as particular food sources become locally available. When the seeds of *Strobilanthes* become locally available, hundreds of individuals may gather for a few weeks until the food supply has disappeared.

Baker (1930) mentioned an individually recognizable male that regularly foraged each morning and evening in two localities nearly 0.5 km (0.3 mi) apart for an extended period.

Daily Activities and Sociality
Like other tropical pheasants, grey junglefowl forage only in the mornings and evenings. They retire to heavy cover during the middle of the day, except on cloudy days when they may remain out for most or all of the day (Baker, 1930).

Grey junglefowl are not highly social. The birds usually occur singly, in pairs, or at most in family units of up to about five birds. Roosting is also apparently done nonsocially, even though several birds may be found in the same tree or in neighboring ones (Beebe, 1918–1922).

Social Behavior

Mating System and Territoriality
Ali and Ripley (1978) contend that the grey junglefowl has a mating system of serial polygyny, with males pairing with individual hens as they become physiologically mature. However, Beebe (1918–1922) believed that pairs are associated throughout the year. Although under certain circumstances a male may mate with two or three hens, Beebe considered that monogamy is the normal condition. Certainly in captivity the birds are facultatively polygynous, with three or four hens apparently the ideal number to mate with a male (Johnson, 1964).

Like the other junglefowl, males of this species seem to be seasonally territorial. The crowing season seems to last longer than the actual breeding season, from about October or November until May. The calls are typically uttered from trees in very early morning, but sometimes also at night, especially when there is moonlight. Territorial or home range sizes have not been estimated for this species.

Voice and Display
A good deal has been written on the male calls of the grey junglefowl, which are rather distinct from those of the red junglefowl. A loud wing-flapping that produces a clapping sound typically precedes the crowing call. As in the red junglefowl, a single crow (see figure 28, page 158) consists of four component notes and the duration of the call is about the same. However, the individual notes are more discrete and each of the notes is more varied in pitch. The second note appears to the human ear to be the most strongly emphasized note, although all four are actually of about the same amplitude (Collias and Collias, 1967).

Display behavior in this species is not well described, but it has sometimes been mentioned as similar to that of the red junglefowl. Morejohn (1968b) stated that males of both species perform tidbitting behavior, but that the associated call is markedly different in the two, with that of the grey junglefowl sounding similar to the nasal mewing of a cat. Morejohn did not describe waltzing behavior in this species. However, he found that after 4 or 5 months of contact with a male grey junglefowl a female red junglefowl would respond to his tidbitting call and presumably to his other courtship displays as well. Fertility and hatchability of these matings were even higher than those of intraspecific matings of either species. The resulting hybrid males courted all three types of female (grey, red, and F_1 hybrids), with backcross young being produced in both directions. However, no F_2 young were successfully raised, which suggests some genetic isolation between the species that apparently supplements geographic isolation and behavioral isolating mechanisms.

Reproductive Biology

Breeding Season and Nesting
The breeding season of the grey junglefowl is quite prolonged and probably conforms to the period of male nuptial plumage. Most breeding records are for the period February–May, beginning somewhat earlier in the south, but locally extending more or less throughout the year (Ali and Ripley, 1978). In the western Nilgiris the main breeding months may be from October to December, whereas in Travancore breeding occurs from March to July and sometimes until August (Baker, 1930).

The nests are located in small hollows, often in the shade of a bush, a bamboo clump, or rarely elevated on a dead tree or stump. Most records of wild clutches are of no more than four or five eggs and sometimes only three. However, occasionally clutches of six or seven have been recorded, perhaps as a result of the efforts of two females (Baker, 1930; Ali and Ripley, 1978). Beebe (1918–1922) has suggested that clutch sizes are larger at the southern end of the range than they are in more northern areas.

Incubation and Brooding
Incubation is performed by the female alone and lasts about 20–21 days. However, when the young are hatched the male apparently returns to the brood and is often seen with them and takes his part in finding food (Beebe, 1918–1922). If general, this behavior would certainly argue for a normally monogamous mating system.

Growth and Development of the Young
The young are quite precocial and are able to fly a few days after hatching. Even before that stage is reached, the chicks will clamber up into shrubs and bushes and then leap off and flutter a short distance before falling to the ground. The notes of both the female and the chicks are very much like those of the domestic fowl. The chicks remain with their parents at least until they have attained their "adult" (presumably first-year) plumages and perhaps for longer (Beebe, 1918–1922). First-year males are usually infertile. The full-adult male plumage is not attained until the second year.

Evolutionary History and Relationships
Morejohn (1968a) compared the plumages of the four species of Gallus. He concluded that gallus and sonnerati may have differentiated into separate forms northeast and southwest of the Godaveri River, respectively, with the grey junglefowl becoming adapted to a generally more xeric, scrub-jungle environment and the red adapting to an area of mixed jungle and

grasslands in a more mesic climatic region. Wild hybridization does not occur in the area of contact, but hybridization between the grey junglefowl and domestic fowl in the vicinity of villages has been reported. The two forms have certainly attained the level of full species. Their geographic isolation is supplemented with differences in plumage, behavior, and intrinsic genetic differences that result in reduced hybrid fertility and viability (Morejohn, 1968*b*).

Status and Conservation Outlook

The grey junglefowl was listed as vulnerable by McGowan and Garson (1995) and as near-threatened by Collar et al. (1994). Although still fairly common in many areas, the genetic integrity of the species has been greatly affected by influx of genes associated with domestication. In that sense, the grey junglefowl is one of the most critically endangered of all galliform birds (Brisbin and Peterson, 1997).

CEYLON JUNGLEFOWL

Gallus lafayettei Lesson 1831

Other Vernacular Names
Lafayette's junglefowl, Cingalese junglefowl; coq de Lafayette (French); Lafayette-Huhn (German); wali-kukula (Native Ceylonese).

Distribution of Species
This species is limited to Ceylon (Sri Lanka). It inhabits coastal scrub to damp mountain forests between sea level and 1,830 m (6,000 ft). See map 5.

Distribution of Subspecies
None recognized by Delacour (1977). Deraniyagala (1957) has described a race (*xanthimaculatus*) that was not accepted by Ripley (1961).

Measurements
Delacour (1977) reported that males have wing lengths of 216–240 mm (8.4–9.4 in) and tail lengths of 230–400 mm (9.0–15.6 in), whereas females have wing lengths of 170–180 mm (6.6–7.0 in) and a tail length of 110 mm (4.3 in). Ali and Ripley (1978) reported that eight males had wing lengths of 228–239 mm (8.9–9.3 in) and tail lengths of 290–338 mm (11.3–13.2 in), whereas three females had wing lengths of 187–195 mm (7.3–7.6 in) and tail lengths of 108–118 mm (4.2–4.6 in). Males weigh from 790 to 1,140 g (1.7 to 2.5 lb), whereas females weigh from 510 to 625 g (1.1 to 1.4 lb). The eggs average 46.3 × 34.5 mm (1.8 × 1.3 in) and their estimated fresh weight is 30.4 g (1.1 oz).

Identification

In the Field (356–711 mm, 14–28 in)
The limited Ceylonese range of this species eases identification, although feral red junglefowl might make the problem greater. Males differ from those of red junglefowl in having reddish brown underparts, brown streaking through the yellow neck hackles, and the interior portion of the comb tending toward yellow. Also, the wings are dark and iridescent bluish black rather than reddish brown. The male's crow is a distinctive three-syllable, musical, and ringing *chick, chaw-choyik* (or *tsek . . . George Joyce*), with the beak jerked sharply upward with

each syllable. Crowing is often preceded by a vigorous wing-clapping of three or four beats. Females have a more strongly barred wing and tail patterning than do female red junglefowl and lack the yellow neck tones of that species.

In the Hand
The male's brown-streaked neck hackles and brownish underpart coloration allow for easy identification from all other junglefowl, whereas the female is best identified by the heavy brown, blackish, and buffy barring on the wings and the brown and black markings on the tail. The comb of the male is less fully serrated than that of the male red junglefowl, the iris is yellow rather than red, as is the interior of the comb, and there are no distinct earlobes. The legs are reddish with sharp brownish to blackish spurs, rather than grayish to brownish spurs and legs as found in the red junglefowl. Hybrids with domestic fowl sometimes occur in the vicinity of villages and might pose identification problems.

Ecology

Habitats and Population Densities
The Ceylon junglefowl is ecologically quite widespread and occurs from tall forests that originally covered the sides of hills and mountains to low *Euphorbia* and other types of scrub jungle typical of the coastline and elsewhere. It also commonly inhabits bamboo thickets, brush, semicultivated areas, and secondary growth following old cultivation or other disturbance (Baker, 1930). Its vertical distribution ranges from the sea coast up to at least 1,830 m (6,000 ft). The Ceylon junglefowl is more common in the eastern and drier areas of Sri Lanka than in the damper areas to the west and southwest (Beebe, 1918–1922).

There are no estimates of population density. However, pairs have been known to breed as close as 183 m (600 ft) apart (Beebe, 1918–1922), which suggests a relatively high density in favored habitats. Collias and Collias (1967) counted 24 males crowing within earshot along a 1.6-km (1-mi) stretch of road in Wilpattus National Park in an area known to have one of the most dense populations of junglefowl in the park. This area was dense enough to provide good cover, but not too dense to walk through easily. Further, there were many termite nests, as well as an abundance of fruiting trees and shrubs, which seem to be important food sources for the young and adults, respectively.

Competitors and Predators

This is the only species of junglefowl native to Sri Lanka, although the red junglefowl is common as a domestic bird. Therefore, feral flocks might locally exist in competition with the native species.

Perhaps the most serious native predator is the mongoose (*Herpestes* sp.) according to Beebe (1918–1922), who on several occasions observed these animals stalking junglefowl. He also mentioned the jungle cat (*Felis chaus*) as a potential predator, and commented that ticks cause a greater number of deaths than might be appreciated.

General Biology

Food and Foraging Behavior

Beebe (1918–1922) reported on the crop contents of eight Ceylon junglefowl shot in March. Among the plant contents were grass seeds, seed pods, berries, and flower petals and the animal materials included scarab beetles, termites, molluscs, wood lice, ticks, centipedes, hemipterids, ants, a wood roach, and a grasshopper. Termites were found in four of the eight birds, and one of the crops contained several hundred termites. Four of the crops contained grass seeds and green seed pods were present in three of them. Beebe believed that termites are the most important part of the animal diet of junglefowl. He confirmed that one of their favorite plant foods is the berries and seeds of the *nilloo* or cone-head plant (a variety or species of *Strobilanthes* and *Stenosiphonium* that occur from about 1,525 m [5,000 ft] upward and flower only infrequently). The seeds of these plants are reputed to cause the birds to become temporarily intoxicated and relatively helpless, although Baker (1930) questioned this belief.

Males and females forage in grassy strips along jungle roads in the mornings and evenings, especially after rains. They do not remain out very long, however, when the weather is wet and cool they may remain out foraging all day long (Baker, 1930).

Movements and Migrations

There are apparently few movements of any great length in the Ceylon junglefowl. Beebe (1918–1922) considered the birds to be highly sedentary, with little seasonal shifting. He judged that two pairs inhabited a very small area of semidesert scrub for 1.5 years, without ever leaving this location. However, when *Strobilanthes* is in fruit the birds are attracted from lower areas "far and near," according to Beebe.

Daily Activities and Sociality

As mentioned above, there are apparently well-marked daily patterns of foraging, with morning and evening periods of activity. During the middle of the day Ceylon junglefowl seek out shady areas in which to roost; they also roost at night. Beebe (1918–1922) stated that he never saw more than five birds together at one time and judged that group to represent a family unit. He believed that the species is relatively nonsocial, except at times when several females and their broods may be found flocking, such as during the early life of the chicks.

Social Behavior

Mating System and Territoriality

Beebe (1918–1922) judged Ceylon junglefowl to be facultatively monogamous or polygamous, with no special predominance of one over the other. He knew of several pairs of monogamously mated birds as well as other cases where two hens were undoubtedly mated to the same male. Three females were the largest number that he knew of as being mated to a single male. (Baker, 1930) considered the species to be "apparently polygamous," with no indication of the male taking any interest in caring for his chicks.

Territorial proclamation, or at least male sexual advertisement, occurs in the usual junglefowl manner, with calling done through most or all of the year. Beebe (1918–1922) stated that males have definite territories that they announce daily by crowing. Calling begins on the roost itself, and the bird then moves lower in the roosting tree and continues crowing for a variable period. Calling begins at the first hint of dawn, about 5:15 A.M., and is at its peak just before sunrise, about 5:30 A.M. Calling begins to diminish about 6:00 A.M. and is usually over by 9:00 A.M. In the evening there is a second period of calling, but it is not so enthusiastic as the morning calling. On cloudy days calling may occur periodically throughout the day.

Voice and Display

The crow of the Ceylon junglefowl has three discrete and well-separated notes. However, its total duration is slightly shorter than that of the red or grey junglefowl, in spite of a marked interval between the first and second notes. There is little difference in the amplitude of the three components, but there is a marked variation of the frequency of the individual notes (Collias and Collias, 1967).

During calling (see figure 28, page 158) the male stands on its toes and often flaps its wings. Wing-clapping is apparently an important aspect of the display because males can be readily "called up" by imitating the clapping, whereas imitation of the crowing call is less effective (Beebe, 1918–1922). Detailed descriptions of male posturing are not available, but it evidently differs little from that of the red junglefowl.

Reproductive Biology

Breeding Season and Nesting

According to Beebe (1918–1922), eggs are laid during almost every month of the year, with February–May being the most usual period for the island as a whole. In the northern parts of the island February–August seems to be the favored breeding period, whereas in the Batticaloa District of eastern Sri Lanka eggs are apparently laid during every month except the period from November to January.

Nests are situated in a variety of locations, such as on the ground near a tree, under a bush, beneath a fallen log, or among the roots of a tropical tree. Several elevated nesting sites have been reported, such as on the top of a decayed stump about 2 m (7 ft) aboveground, in deserted squirrel nests, and in the old nest of a crow or hawk at about 9 m (30 ft) aboveground (Beebe, 1918–1922).

The clutch size is usually of only two eggs. However, sometimes three are present and very rarely four eggs may be laid (Baker, 1930). Larger clutches mentioned by Beebe (1918–1922) of up to eight eggs were evidently multiple clutches or artificially supplemented ones.

Incubation and Brooding

The incubation period, at least under artificial conditions, seems to last 20–21 days. Apparently two or three young are the usual brood size under wild conditions. One brood of seven young has been reported (Beebe, 1918–1922), although this brood size might well reflect the efforts of two females.

Growth and Development of the Young

Beebe (1918–1922) reported that for the first 2 weeks after hatching the young depend on concealment for protection, even though their flight feathers develop rapidly. When the young are about three-quarters grown they begin to escape by fleeing rather than by hiding. The food of the chicks is reported to consist mostly of termites. Two years are required for attainment of full adult plumage and sexual maturity.

Evolutionary History and Relationships

There seems little doubt that this species evolved from a mainland ancestral junglefowl type that probably generally resembled the modern Ceylon junglefowl (Morejohn, 1968a) because hybrids between *gallus* and *sonnerati* resemble pure Ceylon junglefowl, especially in the case of females. However, hybrids between the Ceylon and red junglefowl are not fertile when bred *inter se*, although backcrossing to either of the two parental types sometimes results in successful breeding (Beebe, 1918–1922).

Status and Conservation Outlook

There seems little reason to be concerned about the future of the Ceylon junglefowl because it survives well under conditions of habitat disturbance and human activities (Henry, 1955). Reduced hybrid fertility also probably prevents any significant infusion of red junglefowl genes into the population.

Genus *Crossoptilon* Hodgson 1838

The eared pheasants are medium-sized montane pheasants in which sexual dimorphism is virtually lacking. Both sexes have variably elongated ear-coverts that form distinct tufts or "horns" on the sides of the head. The rest of the plumage is also dense, long, and somewhat hair-like. The tail is variably compressed, with the feathers relatively wide, often disintegrated, and vaulted. The wings are rounded, with the tenth primary much shorter than the ninth and the fifth or sixth the longest. The tarsus is relatively long and is spurred in males. Iridescent plumage is virtually lacking in both sexes, but the rectrices are somewhat iridescent in one species. In all species the ear-coverts are white, and white is also often present on the tail-coverts and rectrices. There are 20–24 rectrices, which are somewhat graduated. The central pair are variably disintegrated, longer than the wing, and about twice as long as the outermost pair. The tail molt is phasianine (centripetal). Three species are recognized.

Key to Species and Subspecies of *Crossoptilon* (after Delacour, 1977)

A. Ear-tufts visible above the nape.
    B. General plumage color bluish gray: blue eared pheasant.
    BB. General plumage color brown and white: brown eared pheasant.
AA. Ear-tufts not visible above the nape: white eared pheasant.
    B. Breast gray.
        C. Breast dark gray: Harman's white eared pheasant (*harmani*).
        CC. Breast pale gray: Dolan's white eared pheasant (*dolani*).

BB. Breast white.
    C.   Wings almost pure white: Tibetan white eared pheasant (*drouyni*).
    CC. Wings gray.
        D.   Wings pale gray: Yunnan white eared pheasant (*lichiangense*).
        DD. Wings dark gray: Szechwan white eared pheasant (*crossoptilon*).

WHITE EARED PHEASANT

Crossoptilon crossoptilon (Hodgson) 1838

Other Vernacular Names
Tibetan eared pheasant; hoki blanc, faisan oreillard blanc (French); weisser Ohrfasan (German); sharkar (Tibetan).

Distribution of Species
Eastern Tibet from about 90° E eastward to western Szechwan north to southern Tsinghai and south to northwestern Yunnan. This species is sedentary. It ranges from the upper limit of the coniferous and mixed forests and rhododendrons and juniper scrub above the forest to open grassy hill slopes, occasionally up to the snowline (Vaurie, 1965). See map 7.

Distribution of Subspecies
(after Vaurie, 1965; Wayre, 1969)
Crossoptilon crossoptilon crossoptilon (Hodgson): Szechwan white eared pheasant. Southern Kansu, Szechwan, Yunnan, and adjacent parts of southeastern Tibet. It intergrades with *drouyni* and probably with *lichiangense*.

Crossoptilon crossoptilon lichiangense Delacour: Yunnan white eared pheasant. Northwestern Yunnan and southern Szechwan. It probably intergrades with nominate *crossoptilon*.

Crossoptilon crossoptilon drouyni J. Verreaux: Tibetan white eared pheasant. Western Szechwan, southern Tsinghai, and Tibet between the Yangtze and Salween Rivers on the mountains that divide these rivers from the Mekong Valley between 30 and 32° N. It intergrades with *harmani* and produces unstable gray forms.

Crossoptilon crossoptilon dolani de Schauensee: Dolan's eared pheasant. Reported only from southern Szechwan. Total range unknown. It probably intergrades with *drouyni*.

Map 7. Distribution of eared pheasants, including blue (Bl); brown (B); and Dolan's (D), Harman's (H), Szechwan (S), Tibetan (T), and Yunnan (Y) races of white eared pheasant (W). The shaded area represents the region of probable sympatry between the white and blue eared pheasants.

Table 23

Ranges of wing and tail lengths of white eared pheasants[a]

	Males		Females	
	Wing	Tail	Wing	Tail
drouyni	300–340, 11.7–13.3	310–365, 12.1–14.2	271–308, 10.6–12.0	280–419, 10.9–16.3
doulani	328, 12.8	388, 15.1	295, 11.5	465, 18.1
crossoptilon	297–395, 11.6–15.4	—	290–302, 11.3–11.8	—
harmani	272–306, 10.6–11.9	—	265–282, 10.3–11.0	—
lichiangense	318, 12.4	498–560, 19.4–21.8	290, 11.3	425–440, 16.6–17.2
Overall	272–395, 10.6–15.4	310–560, 12.1–21.8	265–308, 10.3–12.0	280–465, 10.9–18.1

[a] Data from Vaurie (1972) and Cheng et al. (1978). All measurements are reported in millimeters, followed by the equivalent in inches. A dash denotes no data.

Crossoptilon crossoptilon harmani Elwes: Harman's eared pheasant. Limited to southeastern Tibet north of the main Himalayan axis west to 90° E and east to 96° E in the Tsangpo Valley, where it intergrades with *drouyni*. This is sometimes considered a distinct species, the Tibetan or Elwe's eared pheasant (Ludlow, 1951; Sibley and Monroe, 1990).

Measurements

Wing and tail lengths of the subspecies are given in table 23. Cheng et al. (1978) reported that three males of *drouyni* ranged from 2,350 to 2,750 g (5.1 to 6.0 lb) and that seven females ranged from 1,400 to 2,050 g (3.1 to 4.5 lb), whereas two males of *lichiangense* were 1,017 and 2,010 g (2.2 and 4.4 lb) and two females were 1,410 and 1,450 g (3.1 and 3.2 lb). Felix (1964) reported male weights of 1,800–2,200 g (3.9–4.8 lb) and female weights of 1,550–1,800 g (3.4–3.9 lb). The eggs average 60 × 42 mm (2.3 × 1.6 in) and their estimated fresh weight is 58.4 g (2.0 oz).

Identification

In the Field (914 mm, 36 in)

The large size, white body plumage pattern, and drooping darker tail plumage are unique and easily identifiable. The white eared pheasant does not overlap with any of the other eared pheasants, and cannot be confused with other types of pheasants. Calls of eared pheasants include an alarm note that is a sharp, repeated *wrack;* a conversational cackling that terminates on a high *cuco, cuco;* and a breeding call often uttered in unison by both sexes and sounding like a repeated *trip-crrra-ah*, becoming progressively louder and lasting up to about 30 seconds. Wing-whirring is lacking in the eared pheasants.

In the Hand

The white eared pheasant has shorter ear-tufts than the others. Also, it is the only eared pheasant that is white over most of the head and body, or at least on the head and neck.

Geographic Variation

Geographic variation is apparently very well marked. However, information on distribution is still highly incomplete and the taxonomic situation remains unsettled. The easternmost nominate race is extremely white, whereas the westernmost race (*harmani*) is dark slaty gray (and sometimes considered a separate species). However, these two extremes are somewhat bridged by the northern race *dolani*, which is pale ashy gray. Although individually apparently constant, these three forms are geographically connected by the highly variable *drouyni*, considered by some to be the result of hybridization among the extreme types. The little-studied race *lichiangense* shows some tendency toward the plumage condition of *drouyni*, but is not known to be in contact with it (Delacour, 1977). Li (1996) indicated that the eastern race *crossoptilon* overlaps the range of *auritum* in at least six districts of northern Szechwan, but there are no known hybrids between these rather distinctively plumaged species. At present there does not seem to be any clear pattern for understanding the geographic variations in the plumages of this highly variable species.

Ecology

Habitats and Population Densities
The white eared pheasant typically lives on subalpine forests near the snowline, where there are thickets of rhododendrons present. In Szechwan these forests are of spruce, birch, and holly oak at 3,660–4,270 m (12,000–14,000 ft) elevation (Beebe, 1918–1922). It typically forages in open grassy hills near rhododendron thickets and when disturbed runs to the nearest thicket for protection. The white eared pheasant ranges in elevation from about 3,000 m (9,845 ft) upward to snowline, occasionally being recorded as high as 4,600 m (15,095 ft). In winter it evidently moves into montane subalpine forests, but even then it is rarely observed at elevation of less than 3,000 m (9,845 ft; Felix, 1964).

Estimated densities in China ranged from 1.5 to 4.0 individuals per square kilometer (3.9 to 10.4 per square mile) in two study areas of southern Szechwan (Li, 1996).

Competitors and Predators
In a similar manner to the monal, white eared pheasants sometimes dig for roots and bulbs, but they also consume an array of aboveground plant materials. Thus, these two types of alpine pheasant probably do not compete strongly. According to Schäfer (1934) the eared pheasants occupy a somewhat lower altitudinal zone than does the Chinese monal. He listed the blood pheasant as the only other major pheasant species using this subalpine altitudinal zone. However, these two species would be unlikely competitors.

Felix (1964) listed a considerable number of probable predators of eared pheasants, including the leopard (*Panthera pardus*), red fox (*Vulpes vulpes*), wild canids (*Canus lupus* and *Cuon alpinus*), marten (*Martes flavigula*), golden eagle (*Aquila chrysaetos*), falcons (*Falco peregrinus* and *F. cherrug*), goshawk (*Accipiter gentilis*), and eagle owl (*Bubo bubo tibetanus*).

General Biology

Food and Foraging Behavior
Although data are lacking, it is believed that the foods of the white eared pheasant are mainly plant bulbs of the lily family, as well as the tuberous stems and bulbs of wild onions, the latter sometimes giving their flesh a strong taste and odor. Digging behavior is highly developed in the eared pheasants, and their beaks are distinctly sharp and elongated. However, Beebe (1918–1922) stated that a Captain H. Bower reported seeing them feeding on juniper bushes in December and the crop of a bird that he shot there was full of juniper berries. Juniper berries are probably important autumn and winter food sources, according to Felix (1964). In the summer other berries, such as cranberries and strawberries, may also be consumed. Ludlow and Kinnear (1944) reported white eared pheasants feeding on the berries of a mountain ash during autumn.

Movements and Migrations
Apparently there is no marked movement to lower elevations by white eared pheasants. Captain Bower observed them in Tibet during mid-January at a height of more than 4,575 m (15,000 ft). Later the birds were also seen at elevations of as low as 2,745 m (9,000 ft). They probably remain close to the snowline through the winter, and their white plumage pattern would certainly be most effective in such an environment. Early observers have noted that white eared pheasants are often actually found in the snow (Beebe, 1918–1922).

Daily Activities and Sociality
White eared pheasants show typical daytime activity cycles. They feed early in the morning until about 10:30 A.M., rest during midday hours, and in the afternoon visit springs or brooks to drink. At least during the winter period the birds are highly social and have been seen in flocks numbering as many as 250 (Schäfer, 1934). More often, however, white eared pheasants move about during winter in smaller flocks of about 30 birds (Beebe, 1918–1922). During much of the rest of the year the birds undoubtedly break up into much smaller groups, and are said to associate in groups that range from two to five pairs (Cheng, 1963). There is probably a further dispersion during the actual breeding period as well.

Social Behavior

Mating System and Territoriality
Observers seem agreed that all the eared pheasants are monogamous. This impression is reinforced by

the virtual absence of sexual dimorphism in plumage and in the relatively simple courtship displays typical of the entire genus.

There is no definite information on territoriality, other than the fact that males regularly utter loud calls during the breeding season. These may be heard at distances described as "up to a mile away" (Ludlow and Kinnear, 1944) or even up to 3,500 m (11,485 ft) away (Cheng, 1963). The calls are uttered during early morning and late evening and sound something like *krrah-krrah*. They reportedly are not so full and sharp as are those of the blue and brown eared pheasants. (Felix, 1964).

In a study of the distinctive race *harmani* (which is sometimes accorded species recognition), Lu (1997) reported that these birds had a maximum home range of 12.6–58.7 ha (31.1–145.0 acres), a nuclear range of 1.2–8.0 ha (3.0–19.8 acres), and a regularly used range of 0.2–4.5 ha (0.5–11.1 acres). The birds spent most of their time in foraging and resting, with pair-formation occurring by mid-April. No changing of partners thereafter was observed in the birds, which formed monogamous pair bonds. However, after females began incubation, the subadult and adult males merged into small social groups with a linear dominance hierarchy.

Voice and Display

The male's crowing call is extremely loud and carries great distances. As in the other eared pheasants, it is typically uttered with the neck and head vertically stretched and the tail somewhat raised (see figure 30, page 188).

Calling by paired birds is sometimes done in unison, which is perhaps unique among pheasants. The two birds utter a resounding, repeated *trip-crraah*, which grows louder and louder and is often kept up for 30 seconds. When foraging they call almost constantly, uttering a mewing, drawn-out cackle that ends on a high note, *cuco, cuco* (Delacour, 1977).

The display of eared pheasants is relatively simple. Typically it consists of a sideways parade (waltzing) with the tail cocked, the wing closer to the female lowered to the ground enough to scrape the substrate, the tail tilted so that half of it is lowered and the other half raised, the facial wattle distended, and the neck somewhat arched. The male attracts the hen by uttering a repeated *kak* note, presumably the food call, and the female may respond with somewhat similar

repeated *krkrkr-krkrkr* notes. Typical tidbitting behavior, in which the male holds a bit of food in its beak while attracting the female with calls, also occurs and is apparently characteristic of birds that are forming pair bonds for the first time. Wing-flapping during display in the eared pheasants is apparently unknown (Felix, 1964).

Reproductive Biology

Breeding Season and Nesting

Rather little is known of nesting by this species in the wild. Ludlow and Kinnear (1944) reported locating three nests in Tibet. One of these contained nine well-incubated eggs and was found under a fallen fir tree on 23 May. Baker (1930) reported a nest from the Arbor Hills of northeastern Assam with four eggs, which was found on 26 May between 3,355 and 3,660 m (11,000 and 12,000 ft). This nest was located on the ground in deep forest. From this record and sightings of chicks in July and August, Baker judged that May and June are the months during which eggs are laid. Ludlow and Kinnear (1944) questioned the Assam origin of this clutch, thinking that it might have instead come from Tibet, but at least the date would seem to be authentic. A group of wild caught birds obtained in 1966 and moved to Tierpark Berlin initially began laying on 30 May (in 1968), but in subsequent years began laying earlier. After 11 years the initial laying date occurred as early as 6 May. The entire laying period lasted about 2 months for these birds, with egg-laying intervals of 2–3 days typical, with rarely a 4-day interval between successive eggs (Grummt, 1980).

In a study in western Szechwan, Wu and Peng (1996) reported that white eared pheasants live at elevations of 3,300–4,050 m (10,825–13,290 ft). Nests are usually located under subalpine spruces or protruding rocks, and six to nine eggs are laid. Only the female was observed to incubate over the 28–29-day period. One of four observed nests survived predation by crows, hawks, or other small carnivores.

Incubation and Brooding

Most authorities such as Felix (1964) and Grummt (1980) give the incubation period of white eared pheasants as 24 days. Mallet (1973) stated that the usual period for all three species of eared pheasants is 27–28 days, although Felix indicated that these longer periods are typical only of brown and blue eared pheasants. Mallet also stated that the average clutch consists of up to 15 eggs per pair, presumably reflecting a removal of the eggs as they are laid. Felix (1964) stated that the clutch size ranges from 4 to 14 eggs under natural conditions. In spite of the monogamous pair-bonding, all incubation is performed by the female.

Growth and Development of the Young
Felix (1964) stated that at the time of hatching young eared pheasants weigh approximately 40 g (1.4 oz), but by 10 days they average 85 g (3.0 oz) and at 50 days about 600 g (21.0 oz), with the weight of females about 50–70 g (1.8–2.5 oz) lower than the males at 50 days. By the time they are 100 days old males weigh about 1,500 g (3.3 lb) and females about 1,350 g (3.0 lb), and at 150 days they have attained adult weight. Feather growth occurs at a correspondingly fast rate; adult plumage development is completed by 150–170 days.

Grummt (1980) noted that by the time the birds are 3.5 months old the sexes can be readily distinguished by the presence of a 4–5 mm (0.2 in) tarsal spur in males. Sexual maturity is normally attained at the end of 2 years, but in some cases hens begin to lay at the age of 1 year.

Evolutionary History and Relationships
All three species of eared pheasant are geographic replacement forms and clearly constitute a superspecies. The white and blue forms are seemingly closely related, with such transitional forms as *harmani* confounding the problem of species limits. Vaurie (1972) believed *harmani* and *crossoptilon* to be borderline cases in speciation, and thought that to treat them simply as subspecies was to obscure the situation.

In any event, it is quite easy to imagine a pattern of possible speciation in this group, with an early separation of ancestral *crossoptilon* stock and a much more recent separation of *mantchuricum* and *auritum*. *Crossoptilon auritum* and *crossoptilon* are now mostly associated with the montane headwaters of the Yangtze and the Hwang Rivers, respectively. Cheng et al. (1978) also suggest an early separation of *crossoptilon* stock and a later division of *auritum* and *mantchuricum*. Delacour (1945) suggested that *crossop-*

tilon is the least morphologically specialized of the three species, *auritum* is the most highly specialized form, and *mantchuricum* is intermediate. Hybridization combinations within the genus apparently produce fully fertile offspring; natural hybridization with *auritum* has been suggested (Gray, 1958), but is apparently unproven.

Relationships of the genus *Crossoptilon* to other pheasants are rather uncertain, although Delacour (1977) suggests that limited hybrid fertility with *Lophura* suggests a fairly close relationship between these genera. He also believed that the genus *Catreus* might be closely related, which seems less likely to me than an affinity with *Lophura*.

Status and Conservation Outlook
The white eared pheasant is currently considered vulnerable (King, 1981) and is believed to be threatened by forest destruction and excessive hunting. There is a proposed reserve in the Mishmi Hills of northern Assam that should protect the very small Indian range of the species if the sanctuary materializes. The situation in Tibet is uncertain at present. In China there are a few relatively small sanctuaries that are barely within the probable range of the species (Wang, 1980), but most of its known range is unprotected.

There is now an intensive program in the captive breeding of the white eared pheasant (Mallinson, 1979; Grummt, 1980), and a studbook has been developed for facilitating the most effective breeding program (Mallinson and Taynton, 1978).

McGowan and Garson (1995) considered the possibly specifically recognizable taxon *harmani* to be endangered and the remaining forms as vulnerable. Collar et al. (1994) classified all of these taxa as vulnerable. There are no known protected areas within the species' range, although in Tibet there is little disturbance to *harmani* because of the widespread practice of the Buddhist faith. Li (1996) listed 13 localities for *harmani* in Tibet, plus 32 Chinese localities for nominate *crossoptilon*, 10 for *drouynii*, 4 for *lichiangense*, and 1 for *dolani*. The sole locality for *dolani* (Yushu, Tsinghai) was also listed for *drouynii*, which might cause one to question its racial validity. Cheng et al. (1978) similarly listed only two locality records for *dolani*, one of which was within the apparent range limits of *drouynii*.

BLUE EARED PHEASANT

Crossoptilon auritum (Pallas) 1811

Other Vernacular Names
Pallas' eared pheasant, Mongolian eared pheasant; hoki bleu (French); blauer Ohrfasan (German); ho-ki (Chinese).

Distribution of Species
Western and central China in the mountains of Inner Mongolia (Ala Shan) west to those of Kansu and eastern Tsinghai and south to those of northwestern Szechwan (region of Sungpan and north of Mowhsien). This species is sedentary and inhabits coniferous and mixed forests, junipers, and bushy sites on alpine meadows above the forest (Vaurie, 1965). See map 7.

Distribution of Subspecies
None recognized by Delacour (1977).

Measurements
Cheng et al. (1978) reported that five males had wing lengths of 285–314 mm (11.1–12.2 in) and tail lengths of 477–570 mm (18.6–22.2 in), whereas five females had wing lengths of 283–311 mm (11.0–12.1 in) and tail lengths of 470–510 mm (18.3–19.9 in). The males ranged in weight from 1,735 to 2,110 g (3.8 to 4.6 lb) and the females from 1,820 to 1,880 g (4.0 to 4.1 lb). Felix (1964) reported male weights of 1,700–2,050 g (3.7–4.5 lb) and female weights of 1,450–1,750 g (3.2–3.8 lb). The eggs average 59 × 40 mm (2.3 × 1.6 in) and have an estimated fresh weight of 52.1 g (1.8 oz). However, David Rimlinger (pers. comm.) estimated the average fresh weight of 31 eggs as 63 g (2.2 oz).

Identification

In the Field (965 mm, 38 in)
This forest-dwelling pheasant of western China is very rare and is likely to be confused with no other species of the area. It is mostly smoke gray, with paler areas on the rump and white on the bases of the outer tail feathers. Its calls are very similar to those described for the white eared pheasant. Blue eared pheasants are found in wooded mountain country and are likely to escape by running uphill and then flying off or dashing into heavy cover.

In the Hand
Like the brown eared pheasant, this species has long white ear-tufts, but otherwise is much more grayish than brownish in overall body color and lacks the white rump typical of that species. It also has 24 rather than 22 tail feathers. Except for the presence of ear-tufts, the blue eared pheasant might be confused with the *harmani* race of the white eared pheasant, which has shorter and less highly specialized central tail feathers.

Ecology

Habitats and Population Densities
The blue eared pheasant is associated with the sides of rocky mountains, where there are well-wooded slopes and an abundance of undergrowth. In such areas it ranges up to about 3,050 m (10,000 ft). It reputedly is somewhat less dependent upon water than are the other eared pheasants and has been found well away from any streams, springs, or other sources of fresh water (Beebe, 1918–1922). Blue eared pheasants are associated with the borders of alpine pine, juniper, oak, and birch woods and reportedly never stray out very far onto treeless alpine slopes. However, they may feed out on bushy alpine meadows, returning to tree cover in the evenings to roost. The winter months are spent lower on the mountainsides, and the total altitudinal range of the species is probably between about 2,440 and 3,965 m (8,000 and 13,000 ft). Forests in this species' range apparently attain greater heights than do those farther northeast in the range of the brown eared pheasant. Thus, blue eared pheasants are reportedly able to roost somewhat higher in trees (Beebe, 1918–1922).

Estimates of mean densities in six different study areas range from 12.4 to 57.5 individuals per square kilometer (32.1 to 148.9 per square mile; Li, 1996).

Competitors and Predators
Nothing specific seems to have been written on the major predators of blue eared pheasants, but they are probably essentially the same as those mentioned for the white and brown eared pheasants. Likewise, there are probably no significant competitors in this

species' range, given its very specialized foraging behavior.

General Biology

Food and Foraging Behavior
This species is predominantly herbivorous. Beebe (1918–1922) stated that examination of several crops revealed nothing but the buds and leaves of barberries, stems and roots of young grass, and various kinds of herbs. Cheng (1963) reported that the stomachs of nine individuals consisted of 80 percent vegetable matter (spruce and *Polygonum* seeds, and the leaves of various trees, sedges and herbs) and 20 percent of beetles or their larvae.

Foraging is done in small to large groups. Blue eared pheasants feed on leafbuds and the like during late autumn and winter, and during the warmer parts of the year grub in the soil for roots, bulbs, and possibly other foods. They use trees only for nocturnal roosting (Beebe, 1918–1922).

Movements and Migrations
This species is sedentary.

Daily Activities and Sociality
Like the other eared pheasants, blue eared pheasants are quite social and may be seen in groups of considerable size during the nonbreeding season. Flock sizes of up to 50 or 60 birds have been reported during such periods. However, the birds apparently roost singly, but within sight of one another, on tree branches from 2 to 4 m (6 to 12 ft) aboveground. The middle part of the day is also spent in the shade of the forest trees, with the birds foraging actively during early morning and late afternoon hours (Beebe, 1918–1922).

Social Behavior

Mating System and Territoriality
Like the other eared pheasants, this species is certainly monogamous. Winter flocks seem to be comprised of family groupings. Early in spring these flocks break up into mated pairs. At this time the male begins calling, uttering a call that is loud and hoarse and described variously as sounding something like *ka-ka . . . la!* (Cheng, 1963), *krip-krraah-krrraah!* (Felix, 1964), or somewhat like the note of a peacock

(Beebe, 1918–1922). This call, apparently a challenge call or a mate-attracting call, is usually given soon after sunrise, but sometimes before daybreak, and rarely at midday or at other times during the day. Evidently this call is rarely heard, at least as compared with the calls of the other eared pheasants. The blue eared pheasant is reported to repeat its cry only about five or six times altogether (Beebe, 1918–1922). However, Felix (1964) stated that both the blue and the brown eared pheasants call more often and with stronger voices than does the white.

Nothing is known of territorial sizes or territorial defense in this species, although males reportedly fight with one another during the spring season.

Voice and Display
As noted above, the voice of the male blue eared pheasant is still rather poorly described. Thompson (1976) discussed the voice and especially the displays of the brown and blue eared pheasants in some detail. He stated (pers. comm.) that the courtship postures of the two species were identical and there were also no obvious acoustic differences in the calls, even when comparing sonograms. Felix (1964) likewise indicated that these two species have very similar calls, at least in the case of males.

Postural displays described for the brown eared pheasant also apply to this species, judging from Thompson's comments (figure 30). A detailed comparison of the vocalizations and postural displays of the eared pheasants would be desirable and perhaps of value in assessing evolutionary relationships in the group.

Reproductive Biology

Breeding Season and Nesting
Beebe (1918–1922) stated that the eggs are laid during May or early June. Cheng (1963) gives the breeding period for blue eared pheasant as extending from April to June. The nests are located under the cover of trees or shrubs and are slightly depressed hollows. The clutch size is reported by Cheng (1963) as ranging from 6 to 12 eggs, typically 8, and by Felix (1964) as from 5 to 14 eggs.

Incubation and Brooding
Incubation is performed by the female alone, although the male remains close at hand. Cheng (1963) stated

Figure 30. Display postures of blue eared pheasant, including normal (A) and engorged facial skin (B), and calling posture (C). After photos by the author and John Bayliss.

that under captive conditions a single female can be stimulated to lay as many as 30 eggs during a year. He also noted that the incubation period ranges from 24 to 28 days. Wayre (1969) noted that at the Norfolk Wildlife Park and Pheasant Trust the incubation period has been determined as 28 days. Felix (1964) reported the incubation period as 26–28 days or occasionally as long as 29 days, which would make it the longest incubation period of any of the eared pheasants.

Growth and Development of the Young
Blue eared pheasants show the same rather rapid development of the young that occurs in the other

eared pheasants. Felix (1964) provided measurements on growth of the tail feathers (for blue eared pheasants specifically) and flight feathers (for eared pheasants in general), as well as some weight changes with age that apparently are applicable to all three species.

Evolutionary History and Relationships
The probable relationships of this species have been discussed in the account of the white eared pheasant, which I believe to be a near relative, even though the blue eared pheasant shares long ear-tufts and a generally dark plumage pattern with the brown eared pheasant. Vocalizations and display postures are also

extremely similar in the blue and brown eared pheasants, which comprise an allospecies.

Status and Conservation Outlook

Collar et al. (1994) considered the blue eared pheasant as near-threatened, whereas McGowan and Garson regarded it as safe. This is the only eared pheasant not currently considered endangered or vulnerable, although there is no positive information on which to base a favorable status report. However, there have been several sanctuaries established within the range of the blue eared pheasant. Most of these were developed to protect the habitat of the giant panda, which utilizes similar montane forests, and where much deforestation caused by timbering has occurred (Wang, 1980). For example, in Szechwan forested areas have been reduced by about 30 percent since the 1950s, and in all of China the loss of forested areas in the past three decades amounts to at least 24 percent (Smil, 1983).

BROWN EARED PHEASANT

Crossoptilon mantchuricum Swinhoe 1862 (1863)

Other Vernacular Names

Manchurian eared pheasant; hoki brun, faisan oreillard brun (French); brauner Ohrfasan (German); hoki (Chinese).

Distribution of Species

Historically from southern Chahar in inner Mongolia (west to perhaps eastern Suiyuan because specimens have been taken about 70 km [43 mi] west of Changkiakow, formerly called Kalgan), northern and northwestern Hopeh, and south to southwestern Shansi. The current range is much reduced and fragmented. This species is sedentary and limited to bleak and rocky mountains in shrubs, scrub, coarse grass, or in sites with stands of usually sparse and stunted coniferous or deciduous trees such as birch (Vaurie, 1965). See map 7.

Distribution of Subspecies

None recognized by Delacour (1977).

Measurements

Cheng et al. (1978) reported that eight males had wing lengths of 270–312 mm (10.5–12.2 in) and tail lengths of 518–582 mm (20.2–22.7 in), whereas eight females had wing lengths of 265–290 mm (10.3–11.3 in) and tail lengths of 447–576 mm (17.4–22.5 in). The weights of the males ranged from 1,650 to 2,475 g (3.6 to 5.4 lb) and those of the females from 1,450 to 2,025 g (3.2 to 4.4 lb). Felix (1964) reported male weights of 1,700–2,050 g (3.7–4.5 lb) and female weights of 1,500–1,750 g (3.3–3.8 lb). The eggs average 53 × 39 mm (2.1 × 1.5 in) and have an estimated fresh weight of 44.5 g (1.6 oz). However, David Rimlinger (pers. comm.) estimated the average fresh weight of 58 eggs to be 60 g (2.1 oz).

Identification

In the Field (1,016 mm, 40 in)

This species is easily identified in the field by the combination of long white ear-tufts and a body plumage that is brown, except for the rump and anterior tail areas, which are white. Like blue eared pheasants, brown eared pheasants are forest and forest-edge birds and are more often heard than seen. Vocalizations are apparently much like those described for the white eared pheasant, but the male's challenge call is even more prolonged. The calls of the blue and brown eared pheasants are apparently almost identical.

In the Hand

The presence of long white ear-tufts separates this species from all other pheasants, except the blue eared pheasant. That species is grayish blue on the rump and generally smoke gray over the rest of the plumage, with the exception of the region of the ear-tufts and throat, which are white.

Ecology

Habitats and Population Densities

The brown eared pheasant inhabits subalpine forests of birches, oaks, and pines at elevations of about 1,800–3,500 m (5,905–11,485 ft). During the summer it reaches the highest levels of the treeline and during winter is found somewhat lower, although the seasonal differences are not especially large. The oaks and pines of these forests are not very tall, scarcely reaching 5 m (16 ft), and the birches may be even lower. In Kansu and Shansi Provinces the treeline is at approximately 3,500 m (11,485 ft) and the upper levels of the grass tundra are at about 3,800 m (12,470 ft; Felix, 1964).

Estimates of mean densities range from 0.9 to 7.7 individuals per square kilometer (2.3 to 19.9 per square mile; Li, 1996). Home ranges of birds in the Luyanshan Reserve of Shansi ranged from 2.2 to 128.2 ha (5.4 to 316.6 acres), being largest when the birds initially paired, but diminished to 2.2–6.8 ha (5.4–16.8 acres) as the breeding season approached. In winter brown eared pheasants were found to prefer areas of sunny slopes, good cover, and available food at lower altitudes. During the breeding season they move to higher altitudes where there is open undergrowth, good nesting cover, and available food. After hatching the broods move up mountain ridges and seek dense brush where insects are abundant.

Males show strong territoriality, and monogamy prevails (Zhang, 1995, 1997).

Competitors and Predators
According to Beebe (1918–1922) and Cheng (1963), the natural enemies of the brown eared pheasant consist of foxes, wolves, leopards, and various birds of prey. Of the mammals, the foxes are perhaps most abundant and the leopards the next most common. The brown eared pheasant's defense against raptors is to freeze, but when frightened by mammalian predators it tends to run uphill or head for the nearest cover.

Like the other eared pheasants, this is primarily a digging species, and it probably has no direct competitors.

General Biology

Food and Foraging Behavior
Foods of the brown eared pheasant are believed to consist of tubers, bulbs, acorns, roots, seeds, leaves, stems, shoots, insects, and earthworms. Beebe (1918–1922) described the crop of a bird as being filled with acorns and the gizzard with nearly digested acorns. However, he believed that their usual food is tubers, fine rootlets, and insects and that they obtain their food primarily by digging or grubbing. Beebe observed them foraging in small groups that gathered around a clump of grass and excavated it until it fell down, thus exposing the roots.

Movements and Migrations
As noted above, the seasonal migrations are probably not very great in this species. Beebe (1918–1922) believed their movements to be less than might be expected because brown eared pheasants tend to spend the entire year at median altitudinal levels.

Daily Activities and Sociality
In the course of a day brown eared pheasants probably move out of wooded roosting areas into the grassy meadows and return again each evening. Beebe (1918–1922) found them in flocks of 10–30 birds during the winter. In the course of a single day Beebe observed as many as 33 birds, which were divided into five separate flocks. Even during the winter flocking period he noted that paired birds were

evident, suggesting that permanent pair-bonding is probably present.

Social Behavior

Mating System and Territoriality
All observers of brown eared pheasants in the wild seem to agree that monogamy is the typical mating condition. These are perhaps among the most strongly monogamous of the pheasants. In the spring the flocks break up into pairs, which disperse and probably establish breeding territories. Beebe (1918–1922) observed the male's challenging call being uttered from members of a flock in early spring, when males would utter it from the ground, a boulder, or the branches of a low oak tree. It is apparently not produced during the summer, autumn, or winter.

Voice and Display
Beebe (1918–1922) stated that during calling the male points the bill almost or completely straight upward, elevates the tail moderately, and raises the two central plumes well above the others (figure 31). The call sound like *Trip-c-r-r-r-r-ah!* and begins low and softly, but increases rapidly in volume and continues at times for as long as 60 seconds without apparent interruption. In captive birds the call is less well developed. Beebe noted that the longest duration that he had heard under these conditions was 15 seconds. Thompson (1976), who worked with captive birds, noted that male calling in both brown and blue eared pheasants began in January, rose dramatically in March, and peaked in April, with greatly reduced calling in May, none at all during June and July, and a second calling peak in August that involved both male and females. Thompson noted that during most of the year both sexes call, but that calling in March is almost entirely limited to males. During April courtship and male calling is intense, but calling drops off dramatically after fertilization and egg-laying has begun. The laying cycle in that area (New York) begins in late April or early May and lasts until the latter part of June.

A major feature of courtship in the brown eared pheasant is tidbitting behavior, which is the first phase of male courtship. This is typically followed by male postural display, female crouching and head-weaving behavior, which indicates her readiness to

Figure 31. Copulation sequence of brown eared pheasant, including approach to female (A), followed by mounting (B, C). After sketches by Liu (1986).

mate, and finally mounting and copulation. Tidbitting behavior in eared pheasants is apparently intermediate between two extreme types in which the male "freezes over the food" while calling the hen to him or in which the male repeatedly picks up and drops the food before the female. Lateral postural display consists of enlarging the wattles, drooping the primaries on the side nearer the female until they scrape the ground, erecting the rump and tail-covert feathers, and fanning the tail. The head is pulled in toward the chest, and the body and tail are tipped toward the hen. Copulation may occur after a brief chase or following an invitation posture by the female. After crouching, she weaves her head back and forth in an arc-like manner, with the bill tucked in toward the chest. According to Thompson (1976), this almost invariably stimulates mounting by the male.

Reproductive Biology

Breeding Season and Nesting
Little is known about brown eared pheasants nesting in the wild. However, according to Beebe (1918–1922) nests are invariably placed in the shelter of

pine or birch woods and consist of a simple hollow in the ground or among dead leaves. The clutch size under natural conditions is reportedly 4–14 eggs, or even up to 22 eggs (Lu and Liu, 1983). At least in captivity, the eggs are laid every other day.

In a study at Luyanshan Reserve, 28 nests were found in 1995 and 25 were located in 1996. In 1995 the nest-failure rate was 76 percent and in 1996 it was 73 percent. Most nest failure were caused by human disturbance, but crow and mammal predation also was important, as were nest desertion and nest "parasitism" by eared or common pheasants. Clutch sizes ranged from 4 to 22 eggs, but some of the very large clutches were caused by dump-nesting or parasitic egg-laying. The clutch range of 28 nests found in 1995 was 6–12 eggs, with a mean of 9.4. In one breeding pen a male took over incubation duties on the day prior to hatching, which is the first known example of male incubation by any pheasant species (Zhang, 1995, 1997). In a survey of 106 nests, a mean clutch size was 8.82 eggs, with smaller clutches typical of renesting or first-year females (Z. Zhang et al., 1997).

Incubation and Brooding
The incubation period of the brown eared pheasant is 26–27 days, which is slightly longer than that of the white eared and about the same as that of the blue eared. Incubation is performed by the female alone, but the male remains very close and apparently helps to defend the nest.

Growth and Development of the Young
The average hatching weight of the brown eared pheasant is slightly over 40 g (1.4 oz). Thirteen newly hatched young at the San Diego Zoo ranged in weight from 38.5 to 45 g (1.3 to 1.6 oz), averaging 41.4 g (1.4 oz). By the end of 4 weeks their average weight was 309 g (10.8 oz; David Rimlinger, pers. comm.). Probably later weight changes follow the pattern described for the white eared pheasant.

At least in captivity, male brown eared pheasants attain their full size, spur development, and breeding ability only in their second year, whereas in the blue eared pheasant this apparently occurs in the first year (Thompson, 1976). In the brown eared pheasant the degree of spur development is variable and cannot be used as a certain criterion of sex. Likewise, neither tarsal length nor tarsal diameter are totally reliable criteria, although the facial wattles of males are con-

sistently larger both vertically and horizontally in males than in females. Adult males also have slightly lower-pitched calls and average about 230 g (0.5 lb) heavier than do females (Thompson, 1976).

Evolutionary History and Relationships
This has been discussed above in the section on the white eared pheasant.

Status and Conservation Outlook
The brown eared pheasant is probably more vulnerable than is the white eared pheasant, and it is currently included in the International Council for Bird Preservation list of endangered species. Its status in the wild is not known with certainty (Lu and Liu, 1983). The species may have been extirpated in Hopei Province by 1931, and its situation elsewhere is generally uncertain or unknown (King, 1981). Although a considerable number of brown eared pheasants are currently in captivity, they are nearly all derived from only a few original birds. However, recently there have been importations into England (Wayre, 1975), and some wild-stock birds have been received by the San Diego Zoo. The species is now listed in the first category of protected animals in China, and several reserves have been established for it. Its population now appears to be increasing (Lu and Liu, 1983).

McGowan and Garson (1995) considered the species endangered; Collar et al. (1994) listed it as vulnerable. In Shansi the brown eared pheasant was reported by Li (1996) from 28 localities, as well as three in Hopeh and one in Beijing province. There were perhaps 2,000 birds at Luyanshan Nature Reserve and 1,000 at Pangquangou Nature Reserve as of the early 1990s. However, deforestation, poaching, egg-collecting, and disturbance by mushroom hunters continue to be serious problems. Protection in Shansi, its primary range, has been aided by the naming of this species as the official bird of that province. Like other rare pheasants, the brown eared pheasant is protected nationally by the Wildlife Protection Act of 1988. The total mid-1990s Chinese population may number about 1,000–5,000 birds, most of which are limited to three protected areas of Shansi and one in Hopeh (del Hoyo et al., 1994).

The most recent range estimates for this species (Zhang, 1998) suggest that it is now limited to about 13,600 km^2 (530 mi^2) in three provinces, or less than 1 percent of its original historic range.

Genus *Lophura* Fleming 1822

The gallopheasants are small to medium-sized tropical to montane pheasants, in which the sexes are strongly dimorphic. The males typically have extensive purplish to greenish iridescence dorsally, blackish underparts, and varying amounts of white on the tail, back, and head, which is often in a vermiculated pattern. There are erectile red or blue velvety wattles around the eyes of males, and similar bare orbital areas in females. The wings are rounded, with the tenth primary shortest and the fifth and sixth the longest. The tail is graduated, strongly vaulted, and usually has 16 rectrices (14 in two species and up to 32 in males of one species); the molt is phasianine (centripetal). In two species the tail is unusually short, rounded, and flattened, and in three species the central pair of rectrices is shorter than the third pair. In three species both sexes are crested, whereas in four only the male is crested and three are entirely crestless. Females are brownish to blackish, with varying amounts of spotting or barring. Ten species are recognized here.

Key to Species (and Subspecies of Males) of *Lophura* (in part after Delacour, 1977)

A. Upper parts mostly iridescent bluish, with glossy fringes on feathers.
    B. Lower back coppery red: crested fireback (male).
        C. Lower parts bluish black, with white flank-streaks: Vieillot's crested fireback (*rufa*).
        CC. Lower parts coppery chestnut.
            D. Sides of body and upper breast black and light rufous: Delacour's crested fireback (*macartneyi*).
        DD. Sides of body and breast entirely chestnut rufous.
            E. Smaller (wing under 280 mm [10.9 in]): lesser Bornean crested fireback (*ignita*).

        EE. Larger (wing over 280 mm [10.9 in]): greater Bornean crested fireback (*nobilis*).

BB. Lower back blue.

    C. Not crested.

        D. Tail black; legs grayish green: Salvadori's pheasant (male).[1]

        DD. Tail white; legs red: wattled pheasant (male).

    CC. Crested.

        D. Crest black: imperial pheasant (male).

        DD. Crest white.

            E. Mantle white: Swinhoe's pheasant (male).

            EE. Mantle blue: Edwards' pheasant (male).

                F. All rectrices iridescent blue, the central ones relatively wider, straighter, and more rounded: *edwardsi*.

                FF. Four to 6 central rectrices white, relatively narrower, decurved, and more pointed: *hatinhensis*.

AA. Upper parts not bluish.

    B. Upper parts brown (females).

        C. Uncrested.

            D. Upper parts heavily barred.

                E. Breast plain rufous: Siamese fireback.

                EE. Breast mottled with black and buff: Swinhoe's pheasant.

            DD. Upper parts not barred.

                E. Orbital skin blue: wattled pheasant.

                EE. Orbital skin red.

                    F. Legs gray; yellowish spot present behind the eye: Salvadori's pheasant.

                  FF. Legs red; no yellowish spot present behind the eye.

                        G. Central rectrices dark brown: Edwards' pheasant.

                            H. Underparts generally more brownish: *edwardsi*.

                            HH. Underparts more reddish brown: *hatinhensis*.

                      GG. Central rectrices chestnut brown, with black vermiculations: imperial pheasant.

        CC. Crested.

            D. Orbital skin blue: crested fireback.

            DD. Orbital skin red.

                E. Legs reddish: silver pheasant.

                EE. Legs gray or brownish: kalij pheasant.

  BB. Upper parts gray, black, or black and white.

      C. Uncrested: crestless fireback (both sexes).

          D. Neck and upper back purplish black, vermiculated with gray: Malay crestless fireback (*erythropthalma*).

          DD. Neck and upper back light gray, speckled with black: Bornean crestless fireback (*pyronota*).

---

[1]Racial distinctions of males of this species are still undescribed.

CC. Crested.
    D.   Lower back coppery red: Siamese fireback (male).
    DD. Lower back black or black and white.
        E.   Legs gray or brown; tail usually under 300 mm (11.7 in) and often black; crest usually narrow: kalij pheasant (male).
           F.   Crest white or very pale brown: white-crested kalij pheasant (*hamiltoni*).
           FF. Crest black.
               G.   Plumage entirely black above and below: black kalij pheasant (*moffitti*).
               GG. Plumage variably interspersed with white above or below.
                   H.   Upper plumage entirely black.
                      I. Breast black; rump barred with white: black-breasted kalij pheasant (*lathami*).
                      II. Breast whitish or white.
                          J.   Rump black: black-backed kalij pheasant (*melanota*).
                          JJ. Rump barred with white: Nepal kalij pheasant (*leucomelana*).[2]
                   HH. Upper plumage finely marked with black and white.
                      I. Breast and underparts entirely black: Williams' kalij pheasant (*williamsi*).
                      II. Sides of breast with white shaft-streaks.
                          J.   Central rectrices buffy white on inner web, with black vermiculations: Oates' kalij pheasant (*oatesi*).
                          JJ. Central rectrices often white, completely lacking vermiculations.
                              K.   Black barring of upper parts more extensive, becoming generally slightly darker above: Crawfurd's kalij pheasant (*crawfurdi*).
                              KK. White parts of dorsal feathers closer together, becoming generally lighter above: lineated kalij pheasant (*lineata*).
      EE. Legs red; tail usually over 300 mm (11.7 in) and whitish; crest usually full and black: silver pheasant (male).
           F.   Upper parts with white predominating over black; ground color of central rectrices white.
               G.   Tail 550–730 mm (21.5–28.5 in); hindneck white or nearly so.
                   H.   Black lines on mantle feathers broken and narrow.
                       I. Three to 4 black lines on scapulars and wing-coverts.
                         J.   Lateral rectrices entirely black: Szechwan silver pheasant (*omeiensis*).

[2]The specific epithet is properly spelled as *leucomelanos* according to Sibley and Monroe (1990).

JJ. Lateral rectrices white and black: true silver pheasant (*nycthemera*).

II. Four to 5 black lines on scapulars and wing-coverts: Fokien silver pheasant (*fokiensis*).

HH. Black lines on mantle feathers continuous and wavy: western silver pheasant (*occidentalis*).

GG. Tail 370–620 mm (14.4–24.2 in); hindneck often lightly peppered with black.

H. Hindneck pure white; black barring on tail especially conspicuous.

I. Four to 5 black lines on scapulars and wing-coverts; tail barred but predominantly white: Laos silver pheasant (*beaulieui*).

II. Two black lines on scapulars and wing-coverts; tail barring very wide and the lateral rectrices mostly black: Hainan silver pheasant (*whiteheadi*).

HH. Hindneck peppered with black; tail feather barring not especially conspicuous.

I. Generally lighter above: Rippon's silver pheasant (*ripponi*).

II. Generally darker above: Jones' silver pheasant (*jonesi*).

FF. Upper parts with black predominating over white; central rectrices often becoming buffy.

G. Larger (wing 260–285 mm [10.1–11.1 in]); with a relatively long tail (440–480 mm, 17.2–18.7 in): Ruby Mines silver pheasant (*rufipes*).

GG. Smaller (wing 225–260 mm [8.8–10.1 in]); with a shorter tail (295–415 mm, 11.5–16.2 in).

H. Neck with a wide white border along sides: Annamese silver pheasant (*annamensis*).

HH. Neck thickly barred with black and white; no white border along sides.

I. Tail very short (295–305 mm, 11.5–11.9 in); lateral rectrices nearly black; feathers on sides of neck with three concentric black and white lines: Lewis's silver pheasant (*lewisi*).

II. Tail longer (330–415 mm, 12.9–16.2 in), with less black laterally; neck heavily barred with black and white.

J. Upper parts with slightly narrower black lines; central rectrices buffy: Boloven silver pheasant (*engelbachi*).

JJ. Upper parts with slightly broader black lines; central rectrices pure white: Bel's silver pheasant (*beli*).

SALVADORI'S PHEASANT

Lophura inornata (Salvadori) 1879

Other Vernacular Names
None in general English use; faisan de Salvadori
(French); Salvadori-Fasanhuhn (German).

Distribution of Species
The island of Sumatra, between 610 and 2,440 m
(2,000 and 8,000 ft) in deep mountain forests.
Its distribution is thought to be limited to the
Barisan Range and isolated mountain peaks

from the Ophir districts of Sumatra Barat south
to Mount Dempu in Sumatra Selatan, at elevations of 1,000–2,200 m (3,280–7,220 ft). See
map 8.

Distribution of Subspecies
Lophura inornata inornata (Salvadori): southern
Salvadori's pheasant. The mountains of the southern
half of Sumatra.

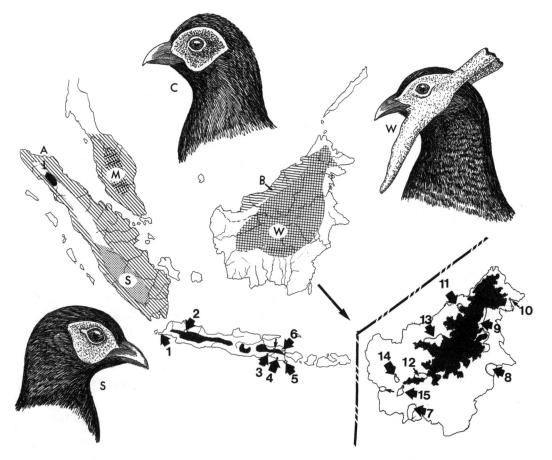

Map 8. Distribution of Atjeh (A) and southern (S) races of Salvadori's pheasant; Bornean (B) and Malay (M) races of crestless
fireback (C); and wattled or Bulwer's pheasant (W). On the Malayan Peninsula the crestless fireback may now be limited to
interior lowland forests (heavy shading), but on Borneo it may extend into Kalimantan an unknown distance (see map 12,
on page 235, for primary forest distribution). The inset maps of Borneo and Java show areas above 500 m (1,640 ft; black),
locations of smaller preserves (small arrows), and national parks or major preserves (large arrows). These preserves are Ujung
Kolon (1), Mount Gede/Pangrango (2), Ranu Darungan (3), Meru Betiri (4), Alas Purwo (5), Baluran (6), Tangun Puting (7),
Kutai (8), Kayan Mentarang (9), Danum Valley (10), Gunung Mulu (11), Bukit Raya/Baka (12), Batutenobang (13), Ronga (14),
and Perai (15).

Lophura inornata hoogerwerfi (Chasen): Atjeh Salvadori's pheasant. Known only from two females collected near Atjeh in northwestern Sumatra. Although still undescribed, color photos of the male suggest no noticeable differences from typical *inornata*. Sometimes considered a separate species (Hoogerwerf's pheasant; van Marle and Voous, 1988; Sibley and Monroe, 1990), but it may not even warrant subspecific recognition (del Hoyo et al., 1994).

Measurements
Beebe (1918–1922) reported that males of *inornata* have wing lengths of 213–227 mm (8.3–8.9 in) and tail lengths of 152–170 mm (5.9–6.6 in), whereas females have wing lengths of 208–228 mm (8.1–8.9 in) and tail lengths of 145–150 mm (5.7–5.9 in). No weights are available. The eggs average 50.8 × 36.2 mm (2.0 × 1.4 in) and the estimated fresh weight is 36.7 g (1.3 oz).

Identification

In the Field (457–584 mm, 18–23 in)
This species is limited to Sumatra. As such, it is unlikely to be confused with any other pheasant found on that island, such as the bronze-tailed pheasant. The almost entirely black plumage of the male is distinctive, whereas the female is chestnut brown, with buffy mottling. Vocalizations have not been described for this species, but probably are similar to those of the other *Lophura* forms, with male crowing and wing-whirring prominent among the acoustic signals.

In the Hand
The dark bluish black male, with no crest and a short tail (to 230 mm [9.0 in]) is distinctive. The similar imperial pheasant has a crest, a longer tail (about 300 mm [11.7 in]), and red legs, whereas the black kalij pheasant also has a crest and reddish legs. Females are similar to those of several other *Lophura* species, but are distinctly mottled with buffy underneath; have greenish gray legs; and have a relatively short, rounded, and blackish brown tail.

Geographic Variation
Geographic variation in males is still undescribed, but the two female specimens so far known from northern Sumatra (*hoogerwerfi*) are less reddish chestnut and more brownish, with black vermiculations more evident than in the nominate form (Delacour, 1977).

Ecology

Habitats and Population Densities
This extremely poorly known pheasant occurs in the montane forests of Sumatra, probably between about 610 and 2,440 m (2,000 and 8,000 ft), at least in the case of the southern subspecies *inornata*. Salvadori's pheasant is said to inhabit flatter ground in the vicinity of mountain peaks, but otherwise its natural history is essentially unknown.

There are no estimates of population density.

Competitors and Predators
There is no information on this.

General Biology

Food and Foraging Behavior
Nothing is known of this in the wild. In captivity Salvadori's pheasants exhibit the same feeding habits as are typical of the firebacks (Howman, 1979).

Movements and Migrations
Nothing of this is known. However, movements are likely to be small or absent in this tropical species.

Daily Activities and Sociality
There is no information on this.

Social Behavior

Mating System and Territoriality
It may be presumed that this species has a polygynous mating system similar to that of its near relatives, but this is only speculation.

Voice and Display
According to Kenneth Fink (pers. comm.), the male's display consists of a rather simple wing-whirring performed while standing in a stiff and erect tiptoe posture (figure 32). The male's yellow eye-ring and similarly colored wattle spot behind the eye become very conspicuous at this time, and the wattle is somewhat enlarged. In spring the males utter a series of clucking calls in early morning, but not before or after the wing-whirring display. Presumably the males perform a tidbitting display and very probably a lateral waltzing display as well, but there are no descriptions of these in the literature.

Figure 32. Display postures of male Salvadori's pheasant, including facial skin engorgement (A) and wing-flapping (B). After photos by Kenneth Fink.

Reproductive Biology

Breeding Season and Nesting
Nothing is known of the breeding season in the wild. In captivity Salvadori's pheasants have recently bred in France (Houpert and Lastere, 1977). They laid two eggs in late June after having been released into an aviary early in May. One egg had been dropped earlier in April, while the birds were still in their winter quarters, apparently having been laid from the elevated perch and broken on the ground below. The female nested in an elevated nesting basket placed close to the pair's perching place, but did not use various artificial nests at ground level. Two apparently unincubated eggs were found on 20 June, one of which hatched on 12 July after an artificial incubation period of 22 days.

Incubation and Brooding
Other than the account just mentioned, there has been only one other published description of breeding, that of Charles Sivelle (Delacour, 1977). He first bred the species in March 1977. The clutch contained two eggs, and Sivelle confirmed an incubation period of 22 days (under incubation conditions of 99.5°F) for this species.

Growth and Development of the Young
Houpert and Lastere (1977) reported that rearing the chick they hatched was not difficult. It readily fed on turkey starter crumbs, with mealworms added, and chopped hard-boiled egg yolk also given occasionally. After 2 months the chick was eating commercial rearing pellets, and at 4 months it was on adult rations of various grains plus rearing pellets. Sivelle noted in an addendum to this paper that the chicks required beak-feeding and initially took small mealworms. Later they began to eat a mixture of game bird crumbs, sunflower meat, milo, and millet, which was supplemented by mealworms.

Evolutionary History and Relationships
This is a highly generalized species of *Lophura;* Delacour (1977) was struck by its "primitive aspect." It seems most likely to be a fairly close relative of the imperial and Edwards' pheasants and the Malay crestless fireback. Delacour considered it to be a transitional form between *edwardsi* and *erythropthalma.* The latter makes more zoogeographic sense, but female plumage patterns are more suggestive of relationships with the Edwards' and imperial pheasants.

Status and Conservation Outlook

There is little information on the status of the Salvadori's pheasant. There are now six protected areas or national parks in Sumatra. Guning Leuser National Park, which contains about 800,000 ha (1.9 million acres; see map 22, page 345), supports the only known population of the northern taxon of this species. Two other preserves of about 1.5 million and 356,000 ha (3.7 million and 879,000 acres) are also located in the uplands of Sumatra, and at least the former (Kerinci-Sebat) is known to support the species. It was observed there in 1986 at about 2,200 m (7,220 ft) on Mount Kerinci above the village of Kerik Tua, at the upper altitudinal limit of its known range. Although a national park, this region has been seriously affected by illegal deforestation. Both taxa of this species are regarded as vulnerable by McGowan and Garson (1995) as well as by Collar et al. (1994).

IMPERIAL PHEASANT

Lophura imperialis (Delacour and Jabouille, 1925)

Other Vernacular Names
None in general English use; faisan impérial (French); Kaiserfasan (German).

Distribution of Species
Mountains of Annam Range, Donghoi Province, southward to northern Quangtri, central Vietnam. There is also one highly improbable record from Yunnan (Li, 1996), and an unlikely report from Laos. See map 9.

Distribution of Subspecies
None recognized by Delacour (1977).

Measurements
The wing lengths of two males were 248 and 252 mm (9.7 and 9.8 in) and their tail lengths were 241 and 300 mm (9.4 and 11.7 in). Four females had wing lengths of 194–234 mm (7.6–9.1 in), with an average of 213 mm (8.3 in) and tail lengths of 176–290 mm (6.9–11.3 in), with an average of 214 mm (8.3 in;

Map 9. Distribution of Edwards' (E), imperial (I), and Swinhoe's (Sw) pheasants and Siamese fireback (S). Locality records of the imperial pheasant (I) and the Vietnamese (V) and nominate (E) taxa of the Edwards' pheasant are shown in lower right inset; see map 21 (page 338) for provincial boundaries. Upper right inset map shows (black) the Swinhoe's pheasant's potential range on Taiwan.

Delacour, 1977; Johnsgard, pers. obs.). Weights are not available. The average egg size is 53 × 40 mm (2.1 × 1.6 in), and the estimated fresh weight is 46.8 g (1.6 oz).

Identification

In the Field (610–762 mm, 24–30 in)

This rare pheasant is limited to northern Vietnam. It might possibly be confused with the Edwards' pheasant, which occurs somewhat farther south. The entirely dark blackish appearance of the male, with its short and dark-colored crest, is distinctive. The species' vocalizations are essentially undescribed, but the male has a conspicuous wing-whirring display. Otherwise the imperial pheasant has been described as rather silent. Females are extremely similar to those of the Edwards' pheasant and probably cannot be safely distinguished in the field.

In the Hand

Male imperial pheasants are unique in their possession of an entirely glossy black plumage; a short, full crest; and a relatively short tail. The very similar Salvadori's pheasant lacks a definite crest and has greenish gray rather than reddish legs. Females are very similar to those of the Edwards' pheasant, but in the imperial pheasant the plumage is more distinctly marked with buffy vermiculations and paler feather edges, which produces a slightly "scaly" appearance, and the posterior head feathers are longer and more crestlike, a condition most obvious on living birds.

Ecology

Habitats and Population Densities

The imperial pheasant was until recently believed to be entirely limited to the rugged limestone mountains west of Dong Hoi in present-day Quang Binh Province (Delacour, 1977), Vietnam, inhabiting relatively impenetrable forest and brush (Delacour and Jabouille, 1925). This area is near the former demarcation line between North and South Vietnam. This area received heavy defoliation applications, so its habitats have probably been seriously affected (King, 1981).

More recent evidence, based on a specimen obtained in 1990 in Cam Xuyen District, Ha Tinh

Province, suggests that imperial pheasants may prefer secondary lowland forest at no more than 100 m (330 ft). The earlier known location for the species was to the south, somewhere in the vicinity of Dong Hoi (Quang Binh Province) and northern Quang Tri Province. The first specimens were obtained there in 1923, in an area of "dense forest and brush of chaotic limestone mountains and rocks" (Delacour, 1977).

There are no estimates of population densities.

Competitors and Predators

This species overlaps locally in the Net River watershed with the closely related Edwards' pheasant (*hatinhensis* taxon), and likely contacts such species as the silver pheasant, red junglefowl, grey peacock pheasant, crested argus, and perhaps the Siamese fireback, some which potentially might be competitors.

The predators of this species have not been identified. However, they probably consist of the usual predatory mammals found in the jungles of Southeast Asia and perhaps some forest-adapted raptors.

General Biology

Food and Foraging Behavior

There is no information on this in the wild. In captivity imperial pheasants are typical of *Lophura* species. They eat a variety of plant and animal materials, but seemingly favor higher protein diets supplemented by grains, fruits, and green materials (Roles, 1981).

Movements and Migrations

Nothing is known of these. However, imperial pheasants probably are highly sedentary, given the minor seasonal temperature variations in this area.

Daily Activities and Sociality

No information is available on wild birds. In captivity imperial pheasants appear to be much like other *Lophura* types in these traits.

Social Behavior

Mating System and Territoriality

There is no information on this in the wild. According to Delacour (1977) imperial pheasants are rather silent, so it is unlikely that large territories are defended.

Voice and Display

Surprisingly little has been written on this. Delacour (1977) stated that the male's courtship is the simple form typical of the kalij group, with wing-whirring and lateral display. No crowing behavior has been mentioned, nor has tidbitting behavior been described. However, it is likely that both do occur.

Reproductive Biology

Breeding Season and Nesting

There is no information from the wild on these topics. In captivity, the initially caught female began laying in April. Nesting occurs on the ground in pockets of dense vegetation. The clutch size in captivity is five to seven eggs, according to Delacour (1977).

Incubation and Brooding

At least under captive breeding conditions, the incubation period is 25 days (Delacour, 1977).

Growth and Development of the Young

There is little specific information on this. However, apparently the young were reared by Delacour fairly easily, being fed on custard and insects. Over the years, the breeding success gradually declined, probably as a result of the high degree of inbreeding. The original male survived from its capture in 1923 until 1940 and was in perfect condition at that time, when World War II brought an end to Cleres, Delacour's chateau in France. During that war the captive stock gradually declined, and by 1959 it had reached a critical point. At that time Carpentier et al. (1975) began to attempt a restoration project, using a single male imperial and a female silver pheasant for breeding. In 1964–1965 a considerable number of hybrids were reared. To these birds were added another pure-bred imperial male and a male that was in part (one-eighth) of Edwards' pheasant ancestry. By 1968 some of the hybrids from these matings produced birds quite close to the imperial pheasant in phenotype.

In addition to these efforts, experiments carried out in Holland using progeny of crosses between the imperial pheasant and the Nepal kalij pheasant have produced birds that also are very close to the imperial pheasant phenotype. Like the imperial pheasant, these birds take almost 2 years to attain full adult plumage and sexual maturity (Roles, 1981). Fertile hybrids have also been reported with the Swinhoe's pheasant, as well as with the Edwards' pheasant (Delacour, 1977). There were still about 24 birds in captive collections in 1982, but most or all of these were of varying degrees of hybrid origin. As of 1997 there were no known pure-bred imperial pheasants in captivity.

Evolutionary History and Relationships

It is apparent from the similar appearance and geographic affinities of the imperial and Edwards' pheasants that they share a fairly recent common ancestry and seem to have barely attained the level of species distinction. Recent studies by Alain Hennache and Ettore Rand (pers. comm.) on mitochondrial DNA of the Vietnamese *Lophura* endemics indicate that only minute differences exist between them, at a level normally associated with subspecific distinction. The imperial pheasant is also obviously closely related to the Salvadori's pheasant and, somewhat more remotely, to the rest of the typical kalij group.

Status and Conservation Outlook

Nothing is known of the current status of the imperial pheasant in Vietnam, but it was certainly greatly affected by habitat changes during the Vietnam War. Maintenance of captive stock is greatly hampered by low fertility and the infusion of foreign genes from related species.

One specimen of the imperial pheasant was reportedly trapped in 1990 by rattan collectors near Ke Go Lake (Ho Ke Go), 20 km (12.4 mi) south of Ha Tinh, Can Xuyen District, southern Ha Tinh Province (Robson et al., 1993; Eames et al., 1994). This is the first evidence of the species' survival since 1923, when Jean Delacour received a live-trapped pair. As noted above, this original pair was supposed to have come from a more southern and remote area of rugged limestone habitat in present-day Quang Binh Province. However, the 1990 record was from secondary lowland forest at 50–100 m (165–330 ft). The imperial pheasant is considered to be endangered by McGowan and Garson (1995) and as critically endangered by Collar et al. (1994). It is not currently protected by any preserves, but there is a proposed preserve at Cat Bin, Ha Tinh Province, where *hatinhensis* has also been reported.

EDWARDS' PHEASANT

Lophura edwardsi (Oustalet) 1896

Other Vernacular Names
Annam kaleege, Vietnamese fireback (*hatinhensis*); faisan d'Edwards (French); Edwards-Fasan (German).

Distribution of Species
This species inhabits damp lowlands and adjoining slopes and ridges of Annamese Vietnam, probably from Ha Tinh south to Da Nang Provinces (see map 21, page 338, for provincial boundaries). It inhabits damp, closed-canopy evergreen forests as well as selectively logged forests with well-developed understory vegetation, from sea level to 300 m (985 ft). See map 9.

Distribution of Subspecies
Lophura edwardsi edwardsi (Oustalet): Edwards' pheasant. Central Vietnam, from Quang Tri Province south to Thua Thien Hue and Da Nang Provinces. Associated with level to gently sloping dense coastal-plain forests having closed canopies and interspersed bamboo patches.

Lophura edwardsi hatinhensis (Vo Quy): Vo Quy's pheasant or Vietnamese pheasant. Known only from Ha Tinh (Nghe Tinh) and Quang Binh Provinces, central Annam, Vietnam. Informally described (on the basis of a single wild-caught male) in 1975, and sometimes considered to represent a distinct species (e.g., Sibley and Monroe, 1990; Collar et al., 1994; McGowan and Garson, 1995). However, adequate evidence as to its species-level separation is still lacking (Vuilleumier et al., 1992; Anderson, 1993; del Hoyo et al., 1994). Associated with level to gently sloping closed-canopy riverine forests of the Annamese lowlands, north of the currently known range of *edwardsi*, but within the known range of *imperialis*.

Measurements
Delacour (1977) reported that males have wing lengths of 220–240 mm (8.6–9.4 in) and tail lengths of 240–260 mm (9.4–10.1 in), whereas females have wing lengths of 210–220 mm (8.2–8.6 in) and tail lengths of 200–220 mm (7.8–8.6 in). A male weighed 1,115 g (2.4 lb) and a female 1,050 g (2.3 lb; Wolfgang Grummt, pers. comm.). The eggs average 45 × 36 mm (1.8 × 1.4 in) and have an estimated fresh weight of 32.2 g (1.1 oz). The male type specimen of *hatinhensis* had a wing measurement of 245 mm (9.6 in), a tail of 270 mm (10.5 in), and weighed 1,100 g (2.4 lb; Vo Quy, 1975).

Identification

In the Field (584–635 mm, 23–25 in)
This species is endemic to central Vietnam. It is likely to be confused only with the even rarer imperial pheasant; both inhabit the Annamese lowlands of central Vietnam. Males of both species are very similar, but the Edwards' pheasant has a white crown and crest, whereas the imperial pheasant is entirely dark bluish black. In both species the legs and facial skin are red. The male's vocalizations are essentially undescribed, but males are said to be rather silent except for a wing-whirring display. Females of the two species cannot be safely distinguished in the field.

In the Hand
This is the only species of *Lophura* in which the males are entirely bluish black except for a white crest (and white central rectrices, in *hatinhensis*). The central pairs of rectrices are blue, wider, and more rounded in nominate *edwardsi*, but are white, narrower, and more pointed in the northern taxon *hatinhensis*. The central rectrices of *hatinhensis* also tend to be more decurved and to extend farther beyond the rest than in *edwardsi*, and the crest may be somewhat longer. Although sometimes called the "Vietnamese fireback," males of *hatinhensis* lack the red upper-part coloration of the fireback group. The Swinhoe's pheasant is rather similar to *hatinhensis* in having white central rectrices, but Swinhoe's males also have white mantles and females have more contrasting upper-part patterning. Males of the sympatric imperial pheasant lack any white feathers. The white rectrices (which evidently vary in number between individuals, sometimes even among siblings) of *hatinhensis* reportedly do not develop until the second year, when breeding is said to initially occur. In contrast, first-year breeding by males of *edwardsi* is typical, although variation in initial breeding by captive-reared birds may occur (Corder, 1996).

Females very closely resemble those of the female imperial pheasant, but have slightly longer tails (at least 200 mm [7.8 in]) and their underparts are not noticeably paler and more grayish than their upper parts. Furthermore, in the Edwards' pheasant the dark brown body plumage is only slightly and finely vermiculated with black, rather than being noticeably vermiculated with black and buff, as is true of the imperial pheasant. The female Salvadori's pheasant is also very similar, but has more buffy spotting underneath, a considerably shorter (to 150 mm [5.9 in]) tail, and greenish gray legs. The female wattled pheasant is also similar, but has blue facial skin.

Separation of females of *edwardsi* and *hatinhensis* is very difficult, but females of the latter form are more reddish in plumage tone. Rozendaal et al. (1991) provided the first description of the female plumage of *hatinhensis* based on a single specimen. He also provided an excellent history of this little-known and puzzling taxon. In at least some captive specimens the third pair of rectrices (from the center) in adult females may be either white or have a long streak of white (Corder, 1996; Davison, 1996). This trait is not typical of females of any *Lophura* species and might simply be an aberrancy resulting from inbreeding because white rectrices may also occur in some captive-bred Swinhoe's pheasants (Weber, 1992). Alternatively, *edwardsi*, *hatinhensis*, and *imperialis* may represent gene pools of great phenotypic variability resulting from recent secondary gene-pool contact and resulting hybridization (Davison, 1996). Research is currently being done to establish the nucleotide sequences of the cytochrome *b* molecule and thus to try determine the degree of genetic differences in these three taxa. However, final results are not yet available (Malone, 1995). Studies of mitochondrial DNA indicate that only minor differences exist among these taxa (Garson, 1998). These studies also suggest that samples from museum skins and wild-caught specimens are different genetically from those of captive stock, probably as a result of the small number of females in the founder population, genetic drift, and gene pollution by other species, especially the Swinhoe's pheasant (Hennache et al., 1997).

New work by Rassmussen (1998) suggests that the imperial pheasant might actually be a hybrid between the Edwards' pheasant and the Annamese race of the silver pheasant (*L. nycthemera annamensis*) rather than one involving the Edwards' pheasant and some other *Lophura*, such as the Siamese fireback.

Ecology

Habitats and Population Densities

Edwards' pheasants are found in the low to moderate altitude and extremely moist forest of the eastern slopes of the mountain of central Vietnam (Assam), occurring from sea level up to about 915 m (3,000 ft). They are associated with very thick undergrowth and liana-covered hillsides and have scarcely been observed by biologists in the wild (Delacour, 1977). Much of this area was extensively sprayed by defoliants during the Vietnam War, and its effects on pheasant and other bird populations remains unknown (King, 1981).

There are no estimates of population densities.

Competitors and Predators

Delacour (1977) stated that this species is sympatric with grey peacock pheasants, green peafowl, red junglefowl, crested argus, and Siamese firebacks. Of these perhaps the congeneric fireback is the most likely competitor, although most of the others mentioned probably also feed on similar foods.

Nothing has been written on predators of Edwards' pheasants. However, they probably include several jungle-adapted felids and mustelids and various large raptors.

General Biology

Food and Foraging Behavior

Although not studied under natural conditions, Edwards' pheasants in captivity seem to be typical of *Lophura*, eating the usual mixture of grains, mash, and green foods (Delacour, 1977).

Movements and Migrations

There is no information on this.

Daily Activities and Sociality

There is no specific information on this. However, there is little reason to believe that it differs from the situation in other forest-dwelling pheasants, with morning and late afternoon foraging and nocturnal roosting in trees. Edwards' pheasants are unlikely to be highly social because that is also atypical of forest-dwelling pheasants.

Figure 33. Display postures of male Edwards' pheasant, including two stages of wing-flapping (A, B), crest-raising (C), and normal resting posture (D). After photos by John Bayliss.

Social Behavior

Mating System and Territoriality

No information is available for wild birds. In captivity Edwards' pheasants seem to be typical *Lophura* species and do not appear to form definite pair bonds.

Territorial behavior is unknown, but the males have been reported to exhibit crowing behavior (repeated *chuck* calls) in the spring (Kenneth Fink, pers. comm.).

Voice and Display

According to Delacour (1977), males of this species simply "whirr their wings, raise their crests, and fluff the feathers of the back in the usual simple way of the kalijs." Photographs of this display by John Bayliss (figure 33) agree with this general description, although more details would be useful. Calling is apparently not associated with wing-whirring.

Reproductive Biology

Breeding Season and Nesting

The breeding season, nest, and clutch size in the wild is unknown. In captivity, Edwards' pheasants are among the earliest to lay. The first female ever brought into captivity (in 1924) laid her first egg under captive conditions in the following year on 23 March. Four more eggs were laid at 2-day intervals, and 10 days later another clutch of five eggs was begun. A third clutch of four eggs followed this, but these proved to be infertile. A second pair laid two eggs in late April and early May. At the same time a male was hybridized with a female Swinhoe's pheasant, and the hybrids proved to be fertile (Delacour, 1977). More recently, hybridization has also occurred with silver pheasants (Lovel, 1977). In general, the clutch size of Edwards' pheasants in captivity seems to range from four to seven eggs.

Incubation and Brooding
The eggs have an unusually short incubation period of only 21–22 days.

Growth and Development of the Young
Reportedly, the young of this species are relatively easy to rear. In 1976, 150 young birds were bred by only 20 breeders. However, infertility is now fairly common and eggs often fail to hatch (Lovel, 1977), presumably as a reflection of the high amount of inbreeding that has occurred since the original importations.

Evolutionary History and Relationships
There can be little doubt that the Edwards' pheasant and imperial pheasant are extremely closely related; one might even argue that they should be considered conspecific. However, besides the minor differences in crest and wing-covert coloration, the imperial pheasant apparently matures only when 2 years old, but the Edwards' pheasant matures its first year. The ecological significance of the delayed maturity in the imperial pheasant is not at all clear, and perhaps these differences are not so hard and fast as they would seem. Both species appear to be derived from an ancestral *Lophura* type of Southeast Asia that has only locally survived in the Vietnam mountains and, in Sumatra, in the closely related form *inornata*.

Status and Conservation Outlook
The northern taxon *hatinhensis* was considered as endangered by McGowan and Garson (1995) and the typical Edwards' form was listed as critically endangered. Both are listed as critically endangered by Collar et al. (1994). The Edwards' taxon has been recently reported at Bach Ma National Park in Ha Tinh Province, where two were captured (Anonymous, 1996). It was also reported from the A Sau

Luoi Valley of Thua Thien Hue Province in 1988 and in the Dakrong District of Quang Tri Province in late 1996 (Eames, 1997). The northern form *hatinhensis* was originally collected at Son Tung, Ky Son Subdistrict, Ky An District, Nghe Tinh (Ha Tinh) Province, and was named for that province. The Net River watershed, located between the common border of Ha Tinh and Quang Binh Provinces and the Net River to the south, was found to support this species during 1994 surveys (Eames et al., 1994). The birds were detected mostly on ridgetops and steep slopes at altitudes of 200–300 m (655–985 ft) in areas of relatively undisturbed closed-canopy forest with palm and saplings in the understory and where selective logging had produced small clearings. It has been recommended that this area be designated as a special-use forest rather than a production forest and that a nature preserve be established. Edwards' pheasant has also been reported from about 10 km (6 mi) north in the Khe Buoi/Ke Go Lake area, both in the Can Xuyen District of southern Ha Tinh Province (Eames et al., 1994). Birds now (1998) present in the Hanoi Zoo reportedly were trapped in the Minh Hoa District of Quang Binh Province. It has been suggested that the nominate Edwards' population may comprise less than 1,000 birds in the wild and the northern form *hatinhensis* may number anywhere from 100 to 10,000 individuals, although much better survey information is needed for both (del Hoyo et al., 1994). A studbook for the Edward's pheasant was established in 1997. Also, a multiauthor review of the current status and future conservation needs of the three Vietnamese *Lophura* endemics is now in preparation, following a special symposium on the group that was held in Hanoi in September 1997.

Although Vietnam was once almost entirely forested, it has now lost over 80 percent of its forest cover as a result of logging, war damage, and land clearing. A national park (Bach Ma, 22,030 ha [54,415 acres]) in Thua Tien-Hue Province, two nature reserves (Pu Mat, 93,400 ha [230,700 acres], in Nghe Anh Province and Vu Quang, 60,000 ha [148,200 acres], in Ha Tinh Province), a watershed protection forest (Ke Go Lake, in Ha Tinh Province), and a cultural and historic site (Phong Nha, in Quang Binh Province) lie within the historic ranges of the Edwards'/Vietnam and imperial pheasants and might afford varying degrees of habitat protection (Eames et al., 1994).

KALIJ PHEASANT

Lophura leucomelana (Latham) 1790

Other Vernacular Names

Kaleej pheasant; faisan leucomèle (French); Schwarz-fasan (German).

Distribution of Species

Himalayas, from the foothills (but usually from about 1,830 m [6,000 ft]) to about 3,660 m (12,000 ft), from about the Indus to northeastern Assam southward to Burma and neighboring western Thailand. This species is sedentary, but moves altitudinally to some extent with the season. It inhabits dense undergrowth of evergreen and deciduous forests near streams and often in thickly overgrown ravines; as well as bamboo, cane, or other thickets; and undergrowth of secondary forest or dense scrub on abandoned plantations (Vaurie, 1965). See map 10.

Distribution of Subspecies (after Ripley, 1961; Vaurie, 1965; Wayre, 1969)

Lophura leucomelana hamiltoni (J. E. Gray): white-crested kalij pheasant. Widespread in the western Himalayas from the Indus and northern Northwest Frontier Province east through the Punjab to western Nepal, from 275 to 3,050 m (900 to 10,000 ft); from tropical moist deciduous and sal forest to the dry temperate zone.

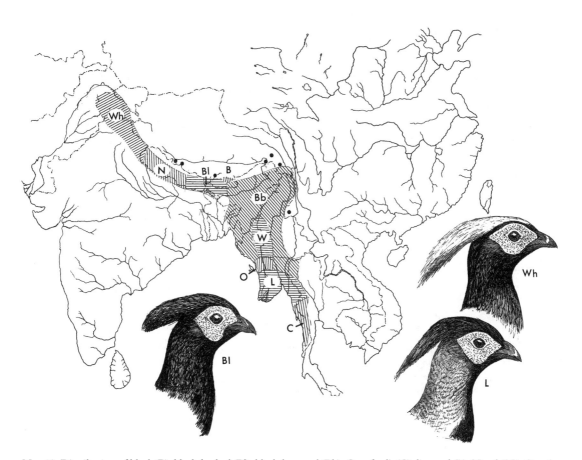

Map 10. Distribution of black (B), black-backed (Bl), black-breasted (Bb), Crawfurd's (C), lineated (L), Nepal (N), Oates' (O), Williams' (W), and white-crested (Wh) races of kalij pheasants. Solid circles indicate peripheral records; a questionable record from eastern Laos is excluded. The breeding range of the black kalij pheasant is still uncertain.

Lophura leucomelana leucomelana (Latham): Nepal kalij pheasant. Limited to Nepal, probably from the Gogra to the Arun Kosi Rivers; in sal, subtropical pine, and moist temperate forest. A few records exist from southern Nepal. This form was also introduced in the Hawaiian Islands in 1962, where it is now expanding in range (Pratt, 1975).

Lophura leucomelana melanota (Hutton): black-backed kalij pheasant. Occurs from Darjeeling, Sikkim, and western Bhutan to the Sankosh River, from the foothills at 105–2,745 m (350–9,000 ft); in tropical semievergreen, subtropical wet, and wet temperate forests.

Lophura leucomelana moffitti (Hachisuka): black kalij pheasant. Range unknown. This form is known only from captive specimens of uncertain origin that were exported from Calcutta occasionally from 1934 to 1949. It is now believed to breed at lower elevations of Bhutan, between the ranges of *melanota* and *lathami* (McGowan and Garson, 1995) or perhaps to the south in India or Bangladesh, as had generally been earlier believed.

Lophura leucomelana lathami (J. E. Gray): black-breasted kalij pheasant. Widespread, in eastern Bhutan, Assam Hills north of the Brahmaputra, Garo, Khasia, Cachar, Naga Hills, Patkoi Hills, Sylhet, Manipur and east to Burma, from 90 to 1,525 m (300 to 5,000 ft); in evergreen, deciduous, and moist temperate forest.

Lophura leucomelana williamsi (Oates): Williams' kalij pheasant. Limited to western Burma, south to

the Chin Hills and central Burma. This form intergrades with *lathami* and *oatesi*.

Lophura leucomelana oatesi (Ogilvie-Grant): Oates' kalij pheasant. Limited to southern Burma in the Arrakan Yomas north to about 20° N. This form intergrades with *lathami* in the west and *williamsi* in the east.

Lophura leucomelana lineata (Vigors): lineated kalij pheasant. Limited to southern Burma in Pegu Yomas, northern Tenasserim, and northwestern Thailand. This form intergrades with *crawfurdi*.

Lophura leucomelana crawfurdi (L. E. Gray): Crawfurd's kalij pheasant. Limited to southern Burma in Tenasserim and western Thailand from approximately 12 to 16° N. This form intergrades with *lineata* in the north.

Measurements

Wing and tail lengths of the subspecies are given in table 24. Ali and Ripley (1978) reported that males of *leucomelana* weigh from 795 to 1,140 g (1.7 to 2.5 lb) and that males of *melanota* weigh from 1,080 to 1,250 g (2.4 to 2.8 lb), whereas *melanota* females weigh from 848 to 1,025 g (1.9 to 2.3 lb) Baker (1928) gave the weights of *hamiltoni* as 1,078 g (2.4 lb) in males and 568–1,022 g (1.3–2.3 lb) in females, those of *leucomelana* as 795–1,135 g (1.8–2.5 lb) in males, and those of *melanota* as 1,078–1,249 g (2.4–2.8 lb) in males and 852–1,022 g (1.9–2.3 lb) in females. The eggs of *melanota* average 48.7 × 37.3 mm (1.9 ×

Table 24
Ranges of wing and tail lengths of kalij pheasants[a]

	Males		Females	
	Wing	Tail	Wing	Tail
hamiltoni	225–250, 8.8–9.8	230–250, 9.0–9.8	203–215, 7.9–8.4	205–215, 8.0–8.4
leucomelana	216–233, 8.4–9.1	250–305, 9.8–11.9	198–211, 7.7–8.2	—
melanota	215–240, 8.4–9.4	238–300, 9.3–11.7	211–222, 8.2–8.7	186–207, 7.3–8.1
lathami	210–240, 8.2–9.4	210–245, 8.2–9.6	205–230, 8.0–9.0	190–225, 7.4–8.8
williamsi	223–247, 8.7–9.6	245–265, 9.6–10.3	206–220, 8.0–8.6	191–202, 7.4–7.9
oatesi	235–295, 9.2–11.5	275–300, 10.7–11.7	205–225, 8.0–8.8	201–234, 7.9–9.1
lineata	220–260, 8.6–10.1	230–345, 9.0–13.5	203–235, 7.9–9.2	220–235, 8.6–9.2
crawfurdi	240–250, 9.4–9.8	270–290, 10.5–11.3	230, 9.0	238, 9.3
Overall	210–295, 8.2–11.5	210–345, 8.2–13.5	198–235, 7.7–9.2	186–238, 7.3–9.3

[a] All measurements are reported in millimeters, followed by the equivalent in inches. A dash denotes no data.

1.5 in) and have an estimated fresh weight of 37.4 g (1.3 oz).

Identification

In the Field (508–737 mm, 20–29 in)
Males of the many races of this species are extremely variable in appearance. However, in areas where possible confusion with the silver pheasant exists kalij pheasants can usually be distinguished by their shorter tails (which are often largely or entirely black), their grayish rather than reddish legs, and a thinner occipital crest. Females cannot be safely distinguished from silver pheasants in the field. The males have harsh crowing calls that are uttered during the breeding season. A drumming sound is produced by wing-whirring, and the species' alarm call is a repeated *whoop-keet-keet*. Kalij pheasants are generally associated with tropical to temperate forests under 1,830 m (6,000 ft), but sometimes occur up to 3,050 m (10,000 ft) in the Himalayas.

In the Hand
Males are likely to be confused only with male silver pheasants. However, kalij males have shorter (under 300 mm [11.7 in]) central tail feathers that only rarely are entirely white and often are entirely glossy black. The darkest races (e.g., *moffitti*) approach the Edwards' pheasant in appearance, but lack white crests and usually have longer (often over 260 mm [10.1 in]) and more pointed tails, as well as less greenish fringes on the upper wing-coverts. Females are similar to those of several other *Lophura* species. However, their more grayish legs separate them from female silver pheasants and their crested condition separates them from most other *Lophura* species.

Geographic Variation
Geographic variation is extremely pronounced among males and is further confused locally by hybridization with the silver pheasant at the eastern edges of the kalij pheasant's range (see figure 35, page 220). In the Himalayas the male variation is clinal in some characters. The brown borders on the back feathers become less conspicuous, and the width of the white tips of the rump and upper tail-covert feathers decreases from west to east, with the white tips disappearing in *melanota* but reappearing farther

east in *lathami*. The length of the crest and the lanceolate feathers of the underparts also decreases from west to east, and the gloss of the upper parts increases and becomes more purplish from west to east (Vaurie, 1965). There is no clear-cut trend in tail-length or wing-length characteristics among the races of this species, with both showing a high range of variability. However, the central tail feathers, which are black in the Himalayan races, become white to buffy white in the more southerly and lowland-adapted forms (*williamsi, oatesi, lineata,* and *crawfurdi*). The feathers of the upper parts and outer tail feathers likewise show increasing amounts of barring or vermiculations, whereas those of the underparts become generally blackish, with the white edging and shaft-streaking diminishing. However, the darkest of all races is *moffitti*, which lacks the white edging on the rump feathers that occurs in the eastwardly adjoining race *lathami*, as well as the pale edging and streaking of the flanks that is found in the races occurring to the north and west. Likewise the southernmost form (*crawfurdi*) has the most reddish legs, approaching those of the silver pheasants, although there is no known contact and opportunities for local hybridization between these forms (Delacour, 1977). Variations in the females are similar to those of the males.

In a study involving male plumage variations among 173 specimens of kalij and silver pheasants, McGowan and Panchen (1994) found that the plumages fell into two major categories. The first were those with darker plumages, usually having white terminal banding on the feathers of the lower rump and back; all of these plumage types were assigned to the kalij pheasant. The second group consisted of the generally lighter plumage phenotypes, as represented by the typical silver pheasant, with black-and-white V-shaped patterning on the body and tail feathers and no white terminal banding dorsally. However, these two extreme phenotypes are linked by intermediate plumage variants, and the geographic divide between the two major types appears to be the Irrawaddy River. McGowan and Panchen suggested that the most primitive, or presumably common ancestral, plumage type occurs at the center of the collective range of these two doubtfully distinct species. On this basis, the presently recognized "kalij" races *oatesi, lineata,* and *crawfurdi* all would fall within the limits of plumage

Figure 34. Display postures of male kalij pheasant, including lateral display (A) and normal (C) postures. Crest-raising (B) and normal crest positions (D) are also shown. After photos by Kenneth Fink.

variations of the silver pheasant, whereas seven of the traditionally recognized silver pheasant subspecies that were studied (*rufipes, occidentalis, ripponi, jonesi, bealieui, nychthemera,* and *fokiensis*) represent a continuous clinal series, without a clear basis for separation into recognizable subspecies.

In a similar study, Hermans (1986) reviewed the plumage variations of the kalij pheasants and emphasized the intergrading aspects of their plumage types. He also pointed out their substantial racial and temporal variations in altitudinal ranges and seasonal migrations, with the northwesternmost forms (*hamiltoni* and *leucomelana*) having the greatest seasonal altitudinal variations (3,000 m, 9,845 ft) and the southeasternmost forms (*lineata* and *crawfurdi*) having the least (700 m, 2,295 ft).

Ecology

Habitats and Population Densities
The nine subspecies of kalij pheasants recognized by Delacour (1977) occur over an extremely wide range of habitats and elevations, from nearly sea level to at least 3,355 m (11,000 ft), and in a variety of tropical to montane forest habitats. Beginning in the western edge of the species' range, the white-crested kalij pheasant occurs from 365 to 3,355 m (1,200 to 11,000 ft), but is most common between 915 and 2,135 m (3,000 and 7,000 ft) and may be found lower in winter and higher in summer. In these medium elevations the dominant trees are pines (mainly *Pinus longifolia*), especially below 1,830 m (6,000 ft), whereas from about 1,525 to 2,745 m (5,000 to

9,000 ft) oak forests (especially *Quercus incana*) predominate. The Nepal kalij pheasant is most common between 1,220 and 1,830 m (4,000 and 6,000 ft), but ranges to 3,050 m (10,000 ft), and inhabits similar montane forest. The black-backed kalij pheasant is most common between 610 and 1,525 m (2,000 and 5,000 ft), but ranges to 2,745 m (9,000 ft), and inhabits subtropical wet and temperate forests receiving up to 320 cm (126 in) of precipitation annually. The black-breasted kalij pheasant is most abundant from the low plains of about 305–915 m (1,000–3,000 ft), but sometimes ranges up to at least 1,830 m (6,000 ft), and is associated with semitropical, wet, evergreen forests and mixed deciduous-evergreen forests receiving from about 229 to 254 cm (90 to 100 in) of precipitation (Bump and Bohl, 1961; Ali and Ripley, 1978).

The more easterly forms of kalij pheasants are also diverse in their relatively tropical habitats. The Williams' kalij pheasant occurs mainly between 305 and 1,830 m (1,000 and 6,000 ft) in bamboo jungles as well as in open forests mixed with bamboo. The Oates' kalij pheasant also inhabits similar brush on bamboo and grass-covered slopes of moderate elevations from about 455 to 915 m (1,500 to 3,000 ft). Crawfurd's and lineated kalij pheasants are generally found in low-elevation forests of tropical to subtropical climate, and occur in rocky ravines, brush-covered slopes, and riverine bamboo thickets, and particularly bamboo jungles between 610 and 915 m (2,000 and 3,000 ft). These two forms sometimes extend into lighter evergreen or deciduous forests at somewhat higher elevations (Baker, 1930; Delacour, 1977).

One of the few density estimates was provided by Fleming (1976), who found 19 kalij pheasants in an area of 0.6 km² (0.25 mi²) in Uttar Pradesh, India.

Competitors and Predators

Bump and Bohl (1961) reported that kalij pheasants do not seem to be very susceptible to predators and they were unable to locate any evidence of predation. Civet cats are known egg predators (Baker, 1935). Competitors include the red junglefowl, males of which have at times been reported to attack kalij pheasants, which seemingly are regularly beaten or even killed by the junglefowl (Baker, 1930).

General Biology

Food and Foraging Behavior

Kalij pheasants are surprisingly omnivorous, eating almost anything from bamboo seeds to small snakes and lizards. However, they have a special fondness for termites, figs, bamboo seeds, forest yams, and the roots of a gingerlike plant (Baker, 1930). Bump and Bohl (1961) also report a wide variety of foods eaten, including seeds, berries, grass, herbs, shrubs, roots, and a diversity of insects, worms, and larvae. Ali and Ripley (1978) mention such specific items as acorns, the ripe fruits of *Pyrus* and *Rosa*, green stems of *Viscum*, pods of *Desmodium*, bulbils of *Dioscorea*, and ripe seeds of *Nyctanthes*, as well as the tops of nettles and ferns, and the fruits of *Polygonum* and *Rubus*.

Foraging is apparently done in rather small groups, perhaps pairs and family units, in the usual scratching and pecking manner of most pheasants. Like junglefowl, kalij pheasants are well adapted for scratching, but they can also dig with their bills for subsurface materials such as roots and tubers.

Gaston (1981b) stated that during the postbreeding period groups of four to six birds are the typical social unit, probably consisting of a pair and their offspring. From October to December larger units of 10–12 individuals are the rule, but from January onward pairs, or males with two or three females, are the most common social groupings.

Movements and Migrations

Kalij pheasants appear to be quite sedentary, although the more northerly forms do undertake some seasonal movements associated with cold weather. They may also make movements of several miles to sources of water during late afternoon hours (Bump and Bohl, 1961).

Daily Activities and Sociality

Shortly after dawn, and again after 4:00 P.M., kalij pheasants forage in overgrown fields or in the vicinity of roads or trails. They are often found in loose groups of two to ten, generally in the vicinity of water, which they also visit regularly. They rest through the heat of the day, normally on the ground. Nighttime roosting is done in fair-sized trees, usually at heights of 6–12 m (20–40 ft) above ground. Typically the same tree is used night after night, unless the birds are disturbed (Bump and Bohl, 1961).

Social Behavior

Mating System and Territoriality

There is considerable disagreement over the mating system of this species (Baker, 1930; Ali and Ripley, 1978). Some observers have seen males in company with two or even three females during the breeding season. However, others have been equally firm in asserting that the birds are monogamous, with the males regularly seen in company with females and their broods. It seems most likely that the male kalij pheasant is facultatively polygynous, that is, he leaves his first mate when she begins incubation, but remains with his latest female to assist in rearing the young or returns to his sole mate when she hatches her brood should he be unsuccessful in fathering additional broods.

Sizes of territories are unknown, but there seems to be little doubt that territorial advertisement is well developed in males. Males have a loud crowing call, which has been described as a loud whistling chuckle or *chirrup*, but perhaps the drumming sound made during the wing-whirring display is equally important. Baker (1930) believed that the sound is made by beating the wings against the sides of the body; he quoted an earlier observation that the sound can be readily imitated by holding a pocket handkerchief by opposite corners and then jerking one's arms apart. The drumming sound has also been compared to the noise made by shaking or flapping a piece of cloth in the wind. Such an imitation will often bring other males on the run, which suggests that the territories may be fairly closely spaced in some habitats.

Voice and Display

In northern India, calling of territorial males occurs from March through May (Gaston, 1981*a*), which probably corresponds to the peak of the laying period. Besides the territorial crowing mentioned above, males and females also have a variety of other vocalizations. When alarmed, both sexes utter a long, squealing whistle, which is often followed by loud and deep clucking notes. Conversational notes among undisturbed birds are also common, including low *kurr-kurr-kurrchi-kurr* sounds that seem to serve as contact signals (Baker, 1930).

Besides the wing-whirring display, kalij pheasants perform a fairly simple lateral courtship by spreading the tail, expanding the facial wattles, waltzing around the female, shaking the tail, and making clucking or booming noises (Delacour, 1977). Tidbitting behavior certainly also is present in the kalij pheasants, but does not seem to have been described in any detail. In the closely related silver pheasant the associated calls are a series of rather rapidly repeated (about five per second), low-frequency notes (Stokes and Williams, 1972).

Lewin and Lewin (1984) have observed the displays of introduced kalij pheasants (probably representing *hamiltoni–leucomelana* intergrade types) on Hawaii. Besides wing-whirring ("wing-fluttering") and lateral ("tail-fanning") displays, they described a "run-jump" display, in which the male approaches a female from several meters away, completing his approach in a series of up to four jumps, and then turning away from her or circling her and sometimes performing wing-whirring. Wing-whirring is mainly performed by mated males, which suggests that it is more hostile than sexual in function; however, this display once was observed in a female. The most commonly observed social group in the Hawaiian study (seen in 49 cases) consisted of a male and female. Only once was a male associated with two females, and males were in attendance on three of six occasions when broods were present.

Reproductive Biology

Breeding Season and Nesting

The breeding seasons of the many subspecies of the kalij pheasant are almost as diverse as their habitats, but invariably include April and May. The white-crested kalij pheasant is said to breed from March to June; the Nepal kalij pheasant from April to June; the black-backed kalij pheasant from March to May; and the black-breasted kalij pheasant from February to October, but mostly in April and May and in July and August (Baker, 1930; Ali and Ripley, 1978). The more tropical forms of Southeast Asia have similar breeding periods. The Williams' kalij pheasant evidently breed at least in April and May, and perhaps from March to June; the Oates' kalij pheasant breeds from March to May; and the lineated kalij pheasant from February to July (Baker, 1930).

The nest itself is a slight hollow, usually in an area of abundant undergrowth and sometimes under an overhanging rock, under a bush, or in a clump of grass. Ample cover and a reasonable proximity to water seem to be the major requirements. The overhead canopy may vary from dense evergreen forest or bamboo jungle to fairly thin wooded cover. In nearly all races the usual number of eggs seems to be from six to nine, with some clutches having as few as five and rarely more than ten (Baker, 1930; Ali and Ripley, 1978). Extremely large clutches of up to 14 or 15 eggs that have been reported would seem to be the result of two females' efforts or other modifications of the normal situation.

Incubation and Brooding

The incubation period of the kalij pheasant may vary somewhat with climate, perhaps taking an average of 20 days in the warmer portions of the range and up to 22 days in the higher and cooler elevations (Baker, 1930). Incubation is performed by the female, with the male apparently taking no role in protecting the nest. However, males have been seen in company with hens leading very tiny chicks, suggesting that as soon as hatching has occurred the male rejoins the family group. Also, a male has even been observed tending a group of small chicks that seemingly lacked a female parent (Baker, 1935).

Growth and Development of the Young

The chicks' flight feathers grow very rapidly, and within a few days they are able to fly almost as well as their parents (Baker, 1930). Renesting is probably rare, but females are known to renest if their first clutch is destroyed prior to hatching (Bump and Bohl, 1961). In captivity, females often lay as many as 25–30 eggs in a season. The birds assume their adult plumage and are able to breed the year following hatching.

Evolutionary History and Relationships

The speciation pattern in all the kalij and silver pheasants is certainly the most complex of any in all the pheasant group, and has been the cause of a vast number of species and subspecies being described, many of which have been based on single specimens. Delacour (1949) was the first person to put these problems into a modern context of subspecies and to try to understand the evolution of the group. Yet, even today there are areas of taxonomic and geographic uncertainties, such as the enigmatic black kalij pheasant and the complex diversity of male plumage variations seen through the ranges of the kalij and silver pheasants.

Status and Conservation Outlook

Although it is possible that some races of this species may be rather rare, the total overall distribution is great and the birds seem to do well in a variety of both original and disturbed habitat types. Kalij pheasants seem to withstand hunting fairly well (Bump and Bohl, 1961) and are highly adaptable and resistant to habitat changes (Yonzon and Lelliott, 1981).

McGowan and Garson (1995) listed the races *lineata* and *crawfurdi* as vulnerable, and *moffitti* as insufficiently known. There are only a scattering of records from China, including Tibet (Li, 1996). Kalij pheasants are local and rare in Pakistan (Roberts, 1991), fairly common in Nepal (Inskipp and Inskipp, 1985), common to very common in India, and rare in Bangladesh (del Hoyo et al., 1994). There is little information from Burma or Thailand, but this southeastern group of subspecies may individually be endangered owing to hunting, snaring, forest burning, and agriculture. Several preserves in Thailand support the species (McGowan and Garson, 1995).

SILVER PHEASANT

Lophura nycthemera (Linnaeus) 1758

Other Vernacular Names
None in general English use; faisan argenté (French); Silberfasan (German); ing-ky, pac-ky (Chinese).

Distribution of Species
This species is found from the mountains of southern China (Yunnan, Fokien, Kwangsi, Kwangtung, and Szechwan) southward through eastern Burma and most of Indochina; it also inhabits the island of Hainan. See map 11.

Distribution of Subspecies
(after Wayre, 1969; Delacour, 1977)

Lophura nycthemera lewisi (Delacour and Jabouille): Lewis's silver pheasant. Mountains of southwestern Cambodia and the border of southeastern Thailand.

Lophura nycthemera annamensis (Ogilvie-Grant): Annamese silver pheasant. Mountain forests of South Vietnam and northeastern Cochin-China.

Lophura nycthemera engelbachi Delacour: Boloven silver pheasant. Limited to the Boloven Plateau, southern Laos.

Lophura nycthemera beli (Oustalet): Bel's silver pheasant. Limited to the higher peaks and ridges of the eastern slopes on the Annamitic Range of Vietnam from Faifoo to near Donghoi.

Lophura nycthemera berliozi (Delacour and Jabouille): Berlioz's silver pheasant. The western slopes and plateaus of the Annamitic Range in central Vietnam. Probably intergrades with *beaulieui*.

Lophura nycthemera rufipes (Oates): Ruby Mines silver pheasant. Burmese highlands of the Ruby Mines District between the Irrawaddy and Salween Rivers

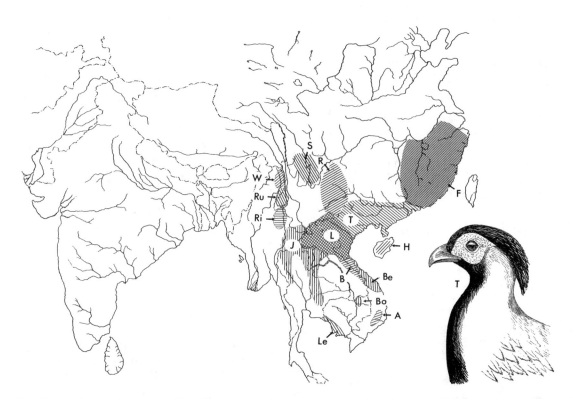

Map 11. Distribution of Annamese (A), Berlioz's (B), Bel's (Be), Boloven (Bo), Fokien (F), Hainan (H), Jones' (J), Laos (L), Lewis' (Le), Rang Jiang (R), Rippon's (Ri), Ruby Mines (Ru), Szechwan (S), true (T), and western (W) races of silver pheasant.

in the northern Shan states; also reported from Yunnan. It interbreeds with *Lophura leucomelana lathami* in the lower valleys, producing unstable hybrids.

Lophura nycthemera ripponi (Sharpe): Rippon's silver pheasant. Southwestern Shan State west of the Salween River (approximately 97–98° E, 20° N).

Lophura nycthemera jonesi (Oates): Jones' silver pheasant. Northern and central Thailand, southwestern Yunnan and southern Shan states east of the Salween River and west of the Mekong.

Lophura nycthemera rongjiangensis Tan and Wu 1981: Rang Jiang silver pheasant. Confined to Szechwan, Guangxi, and southern Kweichow.

Lophura nycthemera omeiensis Cheng, Chang, and Tang 1954: Szechwan silver pheasant. Known only from Szechwan, in the area west and south of the confluence of the Ya Tung and Min Rivers.

Lophura nycthemera occidentalis Delacour: western silver pheasant. Limited to northwestern Yunnan and northeastern Burma east of Myitkyina and Bhama. It interbreeds at lower altitudes with *Lophura l. lathami*.

Lophura nycthemera beaulieui Delacour: Lao silver pheasant. Limited to northern Laos, southeastern Yunnan, western Tonkin, and North Vietnam. It intergrades with *nycthemera* in Tonkin.

Lophura nycthemera fokiensis Delacour: Fokien silver pheasant. From Fokien and Chekiang west to Hunan and Kwangtung.

Lophura nycthemera nycthemera (L.): True silver pheasant. Southeastern China (Kwangtung and Kwangsi) and eastern Tonkin, west to the Red (Yuan) River. It intergrades with *beaulieui* and *fokiensis*.

Lophura nycthemera whiteheadi (Ogilvie-Grant): Hainan silver pheasant. Confined to the mountains of Hainan Island.

Measurements

Wing and tail lengths of the various subspecies are given in table 25. Cheng et al. (1978) reported that five males of *occidentalis* weighed from 1,425 to 1,725 g (3.1 to 3.8 lb) and a female weighed 1,150 g (2.5 lb). Nine males of *beaulieui* weighed 1,500–2,000 g (3.3–4.4 lb), whereas two females weighed 1,160 and

Table 25
Ranges of wing and tail lengths of silver pheasants[a]

	Males		Females	
	Wing	Tail	Wing	Tail
lewisi	240–250, 9.4–9.8	295–305, 11.5–11.9	210–230, 8.2–9.0	230–250, 9.0–9.8
annamensis	225–250, 8.8–9.8	310–355, 12.1–13.8	202–245, 7.9–9.6	215–255, 8.4–9.9
engelbachi	250–270, 9.8–10.5	330–415, 12.9–16.2	230–248, 9.0–9.7	202–272, 7.9–10.6
beli	230–260, 9.0–10.1	340–360, 13.3–14.0	248,[b] 9.7[b]	—
berliozi	255–265, 9.9–10.3	370–450, 14.4–17.6	—	—
rufipes	260–285, 10.1–11.1	400–480, 15.6–18.7	240–257, 9.4–10.0	250–275, 9.8–10.7
ripponi	260–305, 10.1–11.9	433–610, 16.9–23.8	250–270, 9.8–10.5	270–290, 10.5–11.3
jonesi	260–291, 10.1–11.3	440–650, 17.2–25.4	150–270, 5.9–10.5	250–295, 9.8–11.5
omeiensis	276–299, 10.8–11.7	765–800, 29.8–31.2	266, 10.4	353, 13.8
occidentalis	285–296, 11.1–11.5	560–660, 21.8–25.7	246–257, 9.6–10.0	260–283, 10.1–11.0
beaulieui	270–290, 10.5–11.3	430–620, 16.8–24.2	245–270, 9.6–10.5	265–315, 10.3–12.3
fokiensis	261–287, 10.2–11.2	610–730, 23.8–28.5	220–240, 8.6–9.4	250–260, 9.8–10.1
nycthemera	265–297, 10.3–11.6	600–750, 23.4–29.3	240–260, 9.4–10.1	240–320, 9.4–12.5
whiteheadi	245–255, 9.6–9.9	520–530, 20.3–20.7	205–210, 8.0–8.2	225–230, 8.8–9.0
Overall	225–305, 8.8–11.9	295–800, 11.5–31.2	150–270, 5.9–10.5	202–353, 7.9–13.8

[a] Data from Delacour (1977) and Cheng et al. (1978), with one exception (as noted). All measurements are reported in millimeters, followed by the equivalent in inches. A dash denotes no data.

[b] Personal observation.

1,300 g (2.5 and 2.8 lb). Lack (1968) reported the weight of nominate *nycthemera* as 1,150 g (2.5 lb). Baker (1928) reported that males of *rufipes* average 1,362 g (3.0 lb) and females 1,135 g (2.5 lb). The eggs of *nycthemera* average 51 × 39 mm (2.0 × 1.5 in) and have an estimated fresh weight of 42.8 g (1.5 oz).

Identification

In the Field (508–1,270 mm, 20–50 in)
The male silver pheasant is easily recognized in most areas by virtue of its thick, black, decumbent crest that contrasts with a white neck. It has generally black underparts and a very elongated tail that is white centrally and boldly barred with black and white laterally. The male's usual display call is similar to that of a *Phasianus* male, but is more guttural. A loud wing-whirring is also characteristic. Females have somewhat elongated tails, which are olive brown centrally and are heavily marked with brown and black lines outwardly. The flanks are also more heavily marked with bolder patterning than is typical of female kalij pheasants.

In the Hand
The male is easily separated from all species except possibly the kalij pheasant, which sometimes hybridizes with it and produces confusing individuals. However, generally the relatively long and whitish tail (over 300 mm [11.7 in]) and the bold black markings on the lateral tail feathers provide for identification of the silver pheasant. Similarly, the female has a fairly long tail (usually over 225 mm [8.8 in]) that is strongly patterned with dark brown and black lines laterally. Like the males, females are also distinctly crested, which separates them from most *Lophura* species except the kalij pheasant and the crested fireback; this latter species has a vertically rather than posteriorly oriented crest.

Geographic Variation
Geographic variation of the silver pheasant is great, especially in males, and is locally influenced by hybridization with the kalij pheasant. Some of the variation, such as tail length, is clinal. The shortest tail lengths are found in the southernmost forms (*lewisi* and *annamensis*), whereas the longest tails occur in the northernmost race (*omeiensis*). The southernmost

forms also show some traits that approach the plumage conditions found in the kalij pheasant. These include the white shaft-streaks, the white V-shaped markings on the sides and flanks, and the blackish wing- and tail-coverts with relatively narrow white lines. Proceeding northward, the white on the back, upper wing-coverts, and outer tail feathers begins to predominate over the black, making the barring less intense and the bird generally lighter above. However, in *omeiensis* the outer tail feathers are mostly streaked with gray or are even entirely black (on the three outermost pairs of feathers; Delacour, 1977; Cheng, 1979).

Ecology

Habitats and Population Densities
Although closely related to the kalij pheasant, the silver pheasant appears to be adapted to a considerably more grasslike and less forestlike environment. In referring to the race *ripponi*, Baker (1930) described the habitats as consisting of hills covered with a sea of grass, with light deciduous forest, or places where these are mixed with and broken up by ravines and pockets of more dense jungle, often more or less of evergreen type. However, the species prefers broad areas of grassland that are bordered by forests, especially where these areas are rough and broken up by rock outcrops. Its elevational range is from 1,525 to 2,745 m (5,000 to 9,000 ft) and it is most common in Yunnan at about 2,135 m (7,000 ft). In Yunnan the silver pheasant is found in thin oak forests that occur as small patches among higher grasslands and where denser vegetation grows only on stream borders and in large ravines. In such areas the forest patches of stunted oaks dot the grasslands, grasses grow from 0.3 to 0.9 m (1 to 3 ft), and the male's white-and-black plumage pattern blends beautifully with the sun-bleached expanses of grasses (Baker, 1928). In Thailand the local race has been collected as low as 760 m (2,500 ft), but usually occurs above 1,370 m (4,500 ft; Deignan, 1945).

Cheng (1963) states that in China the silver pheasant lives at elevations from 1,500 to 2,000 m (4,920 to 6,560 ft) in forested uplands, bamboo groves, and "straw beds." Beebe (1918–1922) observed the birds in Fokien in mountainous areas where brushy bamboo and sprouting pines were locally present amid a dense vegetational blanket of ferns.

Delacour (1977) reports that the southernmost race *lewisi* inhabits thick evergreen forests of Cambodia above 760 m (2,500 ft), whereas in central Vietnam the more easterly race *annamensis* occurs usually above 1,220 m (4,000 ft) in mountain forests of pines or other evergreens. The Boloven race occurs in very wet forests of southern Laos, between 610 and 1,525 m (2,000 and 5,000 ft). Berlioz's and Bel's races inhabit the mountains of Vietnam, mostly between 610 and 1,525 m (2,000 and 5,000 ft), in fairly dry and relatively damp forests, respectively. The westernmost race, *occidentalis*, inhabits the mountains of Yunnan and Burma between 1,830 and 2,135 m (6,000 and 7,000 ft), where it encounters the black-breasted kalij pheasant and hybridizes with it locally. The true or nominate silver pheasant is found from the foothills of eastern China as high as 1,525–1,830 m (5,000–6,000 ft) and is particularly associated with bamboo woods and evergreen forests. The Hainan race is limited to the damp mountain forests of that island (Delacour, 1977).

Estimates of population densities in China range from 12 to 44 individuals per square kilometer (31.0 to 114.0 per square mile; Gao and Zhang, 1990; Li, 1996).

Competitors and Predators
Nothing specific seems to have been written on these subjects. In some areas the silver pheasant overlaps with and locally hybridizes with the kalij pheasant (figure 35; Delacour, 1948, 1949). It probably also locally overlaps with red junglefowl, common pheasants, and koklass pheasants.

General Biology

Food and Foraging Behavior
Beebe (1918–1922) noted that the specimens of silver pheasants that he examined had eaten insects primarily, as well as a smaller amount of various beetles and occasional flower petals and leaves. Rutgers and Norris (1970) stated that the major foods are berries, fruits, seeds, grain, young shoots and leaves, grass, insects, small reptiles, and worms. In winter silver pheasants subsist on grain, roots, tubers, and bulbs.

Their foraging behavior is essentially like that described for the kalij pheasant, consisting of scratching and occasional digging behavior. However, the sounds of scratching are sometimes loud enough to reveal the birds' locations (Beebe, 1918–1922).

Movements and Migrations
Beebe (1918–1922) was able to detect no special pattern or movements or the location of any roosts, and he believed that the birds wandered about with little definite direction in view. Like the kalij pheasants, the silver pheasants are great runners; they tend to flee uphill and take flight only when they reach the top (Cheng, 1963).

Daily Activities and Sociality
Silver pheasant apparently roost in trees because Cheng (1963) reported that hunters sometimes use torches to aid in hunting perched birds. Foraging is done during the morning and again at dusk, whereas during the middle hours of the day the birds tend to hide.

Silver pheasants evidently move about in small groups while foraging; Baker (1930) quotes a correspondent who observed a flock of seven or eight birds that were flushed from the crest of a grassy ridge, and Beebe observed a group of three (two females and a male) during March.

Social Behavior

Mating System and Territoriality
Nearly all writers agree that this species is polygynous, which is interesting considering the disagreements revolving around the mating system of the closely related kalij pheasant. Perhaps the more definite tendency toward polygyny in this species is related to its more grassland-associated niche as compared to the forest-adapted kalij pheasant. From two to five females per male appear to be typical of this species, with the groups remaining together throughout the year under the domination of the male in typical harem polygyny.

Territories are evidently proclaimed by a combination of crowing and wing-whirring. The crowing is said to resemble that of the common pheasant, but is shorter and deeper in tone (Baker, 1930). Beebe (1918–1922) stated that it is kalijlike, being broken, semiliquid, semiharsh, and guttural.

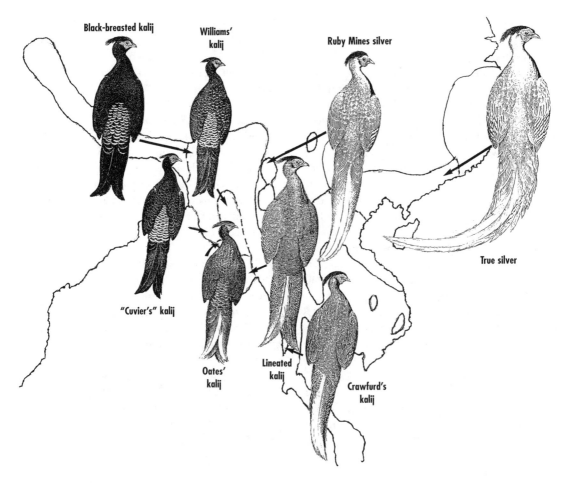

Figure 35. Geographic variation in male plumages of the kalij and silver pheasants superimposed on a range map. Adapted from drawings in Baker (1930).

Voice and Display

Beebe (1918–1922) reported that courtship is typically composed of a definite utterance, combined with wing-whirring. The male approaches the female in an indirect, sidling manner (figure 35), then stops, faces her, stands erect, and utters a two-syllable *ohr-chac* note, with the first syllable accented. Immediately thereafter he whirs his wings and then runs or walks swiftly around the female, spreading his tail and wings and exhibiting them laterally or frontally to her. Wing-whirring may also occur as a sign of suspicion or alarm, but in this case it is typically a short quick beat followed by a long roll. Wing-

whirring behavior has also been observed in females, but its function in females is not well understood. Perhaps it serves as a warning signal or as some type of contact communication. Male wing-whirring is shown in figure 36.

Reproductive Biology

Breeding Season and Nesting

Cheng (1963) reported that in China the breeding season of the silver pheasant begins in April. Although no wild nests have been described, in captivity the birds produce clutches of 4–14 eggs, with an average of

Figure 36. Display postures of the male silver pheasant, including normal (A), facial engorgement with crest-raising (B), and wing-flapping (C). After various sources.

about 24 eggs being produced per female each season. Farther south, the Ruby Mines race of silver pheasant is believed to breed at least in March and April and probably also in May (Baker, 1930). Two females collected in Thailand in mid-May both had bare incubation patches, which indicates a similar period of breeding there (Deignan, 1945). In a Kwangtung sanctuary (Gao and Zhang, 1990), ten nests had an average clutch of 6.9 eggs and an observed maximum of 12 eggs.

Incubation and Brooding

At least in captivity, the incubation period of the silver pheasant lasts 25–26 days. The male takes no part in this, although he may visit his sitting mate once or twice daily during this period (Rutgers and Norris, 1970).

Growth and Development of the Young

For the first 2 weeks after their hatching, the female alone is said to care for the chicks. Later she takes them to join her mate and his harem, although they continue to roost separately with her until they are completely independent. At that time she joins the communal roost with the male and his other females. However, should anything happen to the female, the male assumes parental responsibilities (Rutgers and Norris, 1970). Two years are required for the attainment of full plumage and sexual maturity.

In a study of the growth and development of captive chicks, Gao and Zhang (1990) found that an adult weight plateau was reached by both sexes at about 8 months, although females grew faster initially. Spur length of males increased at a linear rate and could be used for aging purposes for up to 17 months. Males younger than 2 years comprised the most subordinate group of their sex; subordinate males older than 2 years were unable to mate successfully, even when in full breeding plumage.

Evolutionary History and Relationships

There can be little doubt that the kalij pheasant is the nearest relative to the silver pheasant. The two forms probably evolved in southern or tropical and more highly forested habitats (kalij pheasant) versus northern or temperate and more grassy habitats (silver pheasant). The silver pheasant exhibits a higher level of sexual dimorphism than does the kalij pheasant, which seems to be related to a higher propensity for polygyny in conjunction with an edge- and grassland-adapted niche. The strongly contrasting male plumage pattern, with its showy color pattern of black below and white above, is of considerable interest and leads one to wonder about the social importance of white in the male's plumage, especially in terms of its shady woodland habitat.

Status and Conservation Outlook

The relatively broad range of this species and its apparent ability to use woodland edge habitats as well as secondary succession community types such as bamboo thickets probably bodes well for its future. However, McGowan and Garson (1995) listed *annamensis*, *whiteheadi*, and *engelbachi* as endangered and *lewisi* as vulnerable. The two forms *annamensis* (in Vietnam) and *engelbachi* (in Laos) may be most threatened, with populations of only 500–5,000 birds. However, *whiteheadi* (on Hainan Island) and *lewisi* (on the Laotian-Cambodian border) may be only slightly safer, with estimated populations of less than 10,000 birds (del Hoyo et al., 1994; McGowan and Garson, 1995). In mainland China, where eight additional subspecies occur, most have multiple locality records, but *rufipes* is known from only a single locality in Yunnan (Li, 1996).

SWINHOE'S PHEASANT

Lophura swinhoei (Gould) 1862

Other Vernacular Names
Formosan kaleege; faisan de Swinhoe (French);
Formosa-Fasan, Swinhoe-Fasan (German); wa-köe
(Chinese).

Distribution of Species
The island of Taiwan (Formosa) between 305 and
2,135 m (1,000 and 7,000 ft), mostly in primary or
mature secondary hardwood forests (Severinghaus,
1980). See map 9.

Distribution of Subspecies
None recognized by Delacour (1977).

Measurements
Delacour (1977) reported that males have wing
lengths of 250–260 mm (9.8–10.1 in) and tail lengths
of 410–500 mm (16.0–19.5 in), whereas females have
wing lengths of 240–245 mm (9.4–9.6 in) and tail
lengths of 200–220 mm (7.8–8.9 in). Two unsexed
adults averaged 1,100 g (2.4 lb; U. Seals, pers. comm.).
The eggs average 51 × 38 mm (2.0 × 1.5 in) and have
an estimated fresh weight of 40.6 g (1.4 oz).

Identification

In the Field (508–813 mm, 20–32 in)
This species is limited to Taiwan, where the only
other native pheasant is the distinctly different
mikado pheasant. Males of the Swinhoe's pheasant
have white crests, mantles, and central tail feathers,
which provide a color combination unique in the
genus. Females lack crests and have much shorter
tails than do female mikado pheasants. The male
performs a noisy wing-whirring display and utters a
loud, penetrating, plaintive, and rather high-pitched
call, which is apparently associated with mild alarm.
A sharper, rapidly repeated, and high-pitched *deek*
note is uttered in greater alarm.

In the Hand
Male Swinhoe's pheasants are readily identified by
their combination of a white crest, a white mantle,
and pointed white central tail feathers, but otherwise
are almost entirely iridescent purplish black. Females

have red legs, red facial skin, and no apparent crest.
They also have chestnut red outer tail feathers,
whereas the central pair are heavily barred with
brown, black, and buff, as are the wing feathers.
Thus, they are rather similar to the females of the
Siamese fireback, but are less whitish on the underparts and have less rufous on the breast and mantle.

Ecology

Habitats and Population Densities
The habitats of this rare species in Taiwan were best
described by Severinghaus (1980). The species was
mostly observed by him in primary, undisturbed
evergreen tropical forests at elevations of between
1,800 and 2,300 m (5,905 and 7,545 ft). Unlike the
"jungle" environments that were earlier ascribed to
the Swinhoe's pheasant, these forests tend to be shady
and parklike with dappled sunlight penetration and
scattered shrubs and ferns as an understory. Dominant trees consist of several genera of oaks (*Castanopsis, Cyclobalanopsis, Lithocarpus*) and laurels (*Cimmamo-
mum, Actinodaphne, Machilus*). The lowest records of
the species were obtained at 100 m (330 ft) and the
highest at nearly 2,500 m (8,205 ft). The majority of
sightings occurred in primary vegetational habitats
such as mixed forests and coniferous forests, whereas
about one-fifth were made in various secondary habitats such as cassava fields, bamboo plantations, and
natural secondary forests. However, most of these
disturbed habitats showed some characteristics of primary forests, such as mature trees, closed canopies,
lack of grassy areas, or little forest undergrowth. Apparently Swinhoe's pheasants prefer relatively gentle
slopes, but not flat terrain.

Population densities in typical habitats range from
13 to 18 individuals per square kilometer (33.7 to
46.6 per square mile; Li, 1996). One radio-tagged
female occupied a 25-ha (62-acre) home range while
rearing her young (Severinghaus, 1996). During
winter (December) individual males seemed to be
separated by intervals of about 100 m (330 ft), and
similar spacing seemed to occur during March and
April (Severinghaus, 1980). If typical, this is suggestive of a very high density in some habitats.

Competitors and Predators

Except for the mikado pheasant, the Swinhoe's pheasant is primarily sympatric with the Formosan hill partridge. These two species seem to forage together regularly and may actually benefit from the others' presence because they uncover food in different ways. The Swinhoe's pheasant tends to dig with its bill to excavate food, whereas the hill partridge uses kicking movements to dislodge forest litter. The pheasant often digs in areas previously cleared by the partridge, and thus might benefit from its presence (Severinghaus, 1980).

Little is known of the Swinhoe's pheasants' predators. However, according to Severinghaus (1980) the most likely mammalian predator other than man is the ferret-badger (*Melogale moschata*).

General Biology

Food and Foraging Behavior

The Swinhoe's pheasant is evidently quite catholic in its food preferences and a variety of plants have been mentioned as probable or known foods. These include acorns, berries (including *Damnacathus*), flower buds (*Polygonum*), leaves (probably of *Neolitsea* and *Aplenium*), and various other plant parts including those of *Actinodaphne, Cammellia, Cordia, Gardenia*, and the cultivated exotic cassava (*Manihot*). Cassava is the only cultivated exotic on this list of known plant species, and it is presumed but not certain that the birds feed on the starch-filled roots. The seeds of wild taro (*Alocasia*) have also been mentioned as probable foods, as well as unidentified animal materials, earthworms, millipedes, termites, and other insects (Severinghaus, 1980).

Foraging is done in open areas of the forest floor; the birds sometimes use one foot to clear the ground cover, but do not scratch. Digging is the usual method of reaching food, but Swinhoe's pheasants have also been observed jumping to reach items that are just beyond their reach (Severinghaus, 1980).

Movements and Migrations

There is no information on this.

Daily Activities and Sociality

Most foraging activity occurs in early morning and again in late afternoon, with the birds foraging in herbaceous cover of second-growth vegetation or near road edges. Roosting is apparently done singly, although in cold weather as many as eight birds may roost in a single tree. Birds probably go to roost shortly after sunset and have been observed leaving roosts shortly after dawn, with the exact time varying by season. One roost site was about 1–2 m (3–7 ft) aboveground in an area of secondary undergrowth of saplings and small trees within an area of heavier forest. Most winter activities seem to occur before 8:00 A.M. and after 4:00 P.M., at least between mid-December and early April. Later in the spring the periods of morning activity become earlier and the afternoon sightings become later, as might be expected. Little activity occurs between 8:00 A.M. and 4:00 P.M., when perhaps the birds are roosting (Severinghaus, 1980).

Social Behavior

Mating System and Territoriality

Severinghaus (1980) suspected that this species is polygynous under wild conditions, although he was able to obtain no firm data on this and all of the sightings were either single individuals or pairs, rather than harem groupings. He mentioned that Philip Wayre believed the birds to be monogamous, based on experiences with feral birds on Brownsea Island, off the English coast. The relatively dimorphic plumage pattern of the Swinhoe's pheasant might favor the idea of polygyny, although in the similar kalij pheasant there is almost as much plumage dimorphism and that species is commonly reported as monogamous.

Severinghaus (1980) suggested that this species is territorial, but he was able to get no direct information on territorial sizes other than an apparent spacing of males at approximate 100-m (330-ft) intervals along certain census roads. Calling was only rarely heard in a context that might have been related to territorial advertisement. He learned of only one report of fighting among males.

Voice and Display

Severinghaus (1980) described three calls for the Swinhoe's pheasant, including a "murmuring" call, a "sharp" call, and a "plaintive" call. None of these calls fits the criteria that might be expected of a male advertising call. The murmuring call apparently is associated with foraging behavior and the sharp call

Figure 37. Display postures of the male Swinhoe's pheasant, including normal (A), facial engorgement (B), and wing-flapping (C). After various sources.

with intense alarm. The plaintive call was produced only by males, carried at least 90 m (300 ft), and was a high-pitched series of notes sounding like repeated *oot* syllables broken with a pause only at long intervals. This call seemed to be associated with mild alarm.

Stokes and Williams (1972) sonographically illustrated the tidbitting call of this species, which is very similar to that of other *Lophura* types. Wayre (1969) and Delacour (1977) have mentioned the male's typical courtship posture; it is characterized by extreme enlargement of the facial wattles, so that the upper lobe reaches well above the crest of the head (figure 37). The male then performs a series of stiff-legged hops around the female while bobbing his head up and down, which is interspersed with much wing-whirring (Kenneth Fink, pers. comm.).

Several male displays are known in this species (Severinghaus, 1996). These include lateral display with wing-lowering and accompanying *tse*, *tse* notes and wing-whirring followed by *check*, *check* notes. The male also may walk beside a female and circle her while vibrating his wings and spreading his vertically tilted tail or while vibrating or spreading his wings, fanning his tail, and making periodic trilling sounds.

Reproductive Biology

Breeding Season and Nesting

Reports from Taiwan suggest that egg-laying there probably begins in March, or rarely even in February. It extends to May, with a few scattered reports from as late as October. Probably the peak of egg-laying is from March through May, with later records the result of renesting efforts rather than any well-defined second breeding period (Severinghaus, 1980). The clutch size in the wild is from 2 to 12 eggs, but most reports range from 3 to 8 eggs. Reports of larger clutches are largely based on observations in captivity, where clutches of 10 or 12 eggs seem to be fairly common.

Most observers agree that the Swinhoe's pheasant nests on the ground, although about one-quarter indicate that elevated positions are used. Typically the site is described as being next to a fallen tree or in spaces beneath the trunk, between the roots, among rocks, or in clumps of vegetation (Severinghaus, 1980). One elevated tree nest observed by Severinghaus was 5 m (16 ft) aboveground in the depression of a tree stub where the trunk had broken off. This nest was located in a primary hardwood forest and contained five eggs.

Only a few nests of wild birds have been found. However, all were in secure locations well sheltered from rain, often on steep slopes of up to 50–60°. The nests were hidden from above with dense undergrowth, such as ferns or thorny vegetation. Some nests on steep treeless slopes of conifer plantations were entirely hidden by grasses (Severinghaus, 1996).

Incubation and Brooding
The incubation period of Swinhoe's pheasants is apparently of 25 days and is performed by the female alone. There is no evidence of the male defending the nest site.

Growth and Development of the Young
At least in captivity, the chicks of this species are readily raised and require no special attention. Two years are required for the attainment of full plumage in males and for full development of reproductive maturity.

Evolutionary History and Relationships
Zoogeographically speaking, it would seem that the silver pheasant might be this species' nearest relative,

although it is also obviously very close to the Edwards' pheasant in both plumage and ecology. The appearance of white in the male's plumage is of interest and is seemingly nonadaptive as a concealing mechanism in heavy forest cover. However, Severinghaus (1980) suggested that the pattern has a disruptive visual effect that actually might be adaptive in heavy forest habitats.

Status and Conservation Outlook
The Swinhoe's pheasant was considered as vulnerable by the International Council for Bird Preservation (King, 1981). However, its protection has been aided by the formation of a 3,680-ha (9,090-acre) sanctuary of high-altitude habitat in 1974. Severinghaus (1980) judged that as of 1974 the birds were widely distributed on Taiwan and that their numbers were still probably in the thousands. He believed that the Swinhoe's pheasant's preservation will depend on the presence of sufficient habitat in primary hardwood forests. The captive population size of this species is relatively favorable, and it breeds fairly easily under captive conditions. In 1967 and 1968 a limited number of hand-raised birds were released in an area of forest belonging to National Taiwan University in the central mountains. There was some later indications that the birds had survived well (King, 1981).

McGowan and Garson (1995) considered the Swinhoe's pheasant safe, but Collar et al. (1994) regarded it as near-threatened. Severinghaus (1996) stated that the birds are still frequently seen in the 64,000-ha (158,000-acre) Yushan National Park, but habitat is steadily declining outside of such protected areas. The population in Yushan National Park may have numbered about 6,500 birds in the early 1990s. This represents an overall density about one bird per 10 ha (25 acres). Altogether there may have been as many as 10,000 individuals in Taiwan at that time (del Hoyo et al., 1994).

SIAMESE FIREBACK

Lophura diardi (Bonaparte) 1856

Other Vernacular Names
Diard's fireback; faisan prélat (French); Praelat, Prälatfasan (German); kaipha (Native Siamese).

Distribution of Species
Indochina north to central Vietnam and to Luang-Prabang (Laos) on the Mekong River; eastern Thailand as far west as the River Yom and the Klum-Tan Mountains (Wayre, 1964). This species inhabits dense growth from sea level to 610 m (2,000 ft). See map 9.

Distribution of Subspecies
None recognized by Delacour (1977).

Measurements
Delacour (1977) reported that males have wing lengths of 220–240 mm (8.6–9.4 in) and tail lengths of 330–360 mm (12.9–14.0 in), whereas females have wing lengths of 220–240 mm (8.6–9.4 in) and tail lengths of 220–260 mm (8.6–10.1 in). An adult male weighed 1,420 g (3.1 lb) and an adult female weighed 1,025 g (2.2 lb; East Berlin Zoo, Wolfgang Grummt, pers. comm.). Two females at the San Diego Zoo weighed 680 and 800 g (1.5 and 1.8 lb; D. Rimlinger, pers. comm.). The eggs average 48 × 38 mm (1.9 × 1.5 in) and have an estimated fresh weight of 38.2 g (1.3 oz).

Identification

In the Field (610–813 mm, 24–32 in)
Limited to the forests of Southeast Asia, this species is unique in that males have a combination of scarlet facial skin and a posteriorly oriented crest that expands into a comma shape. The long, iridescent tail also droops gracefully downward, and there is a small area of maroon on the back that is less extensive than in the other firebacks. The male produces a whistling call and continually utters a loud, repeated *pee-yu*, as well as performing wing-whirring displays. Females lack an obvious crest, but have upper wing surfaces and elongated central tail feathers that are black with distinctive broad, broken, buffy white barring.

In the Hand
The male's maroon back, red facial skin, and comma-shaped crest serve to identify this species in the hand.

Females resemble those of some other *Lophura* species (e.g., Swinhoe's and Edwards' pheasants), but none of these is so strongly barred on the wings, wing-coverts, and tail feathers or is so uniformly rufous on the upper mantle and breast.

Ecology

Habitats and Population Densities
The Siamese fireback is associated with the densest kinds of cover, including lowland evergreen forests, bamboo brakes, and areas of old cultivation that have been overgrown with *Eupatorium* and thorns (Deignan, 1945). It occurs from sea level to about 610 m (2,000 ft) and is most commonly encountered along roads that have been cut through the jungle (Delacour, 1977).

There are no estimates of population densities.

Competitors and Predators
The Siamese fireback apparently inhabits much the same habitats as does the red junglefowl, although it is probably limited to more dense and generally wetter forests. Predators have not been identified, but probably include a wide array of cats, mustelids, and raptors.

General Biology

Food and Foraging Behavior
Siamese firebacks evidently forage during morning and late afternoon hours, either singly or in family groups (Delacour, 1977). They are believed to forage on all kinds of fruit, berries, insects, worms, and small land crabs, and are thus relatively omnivorous (Baker, 1928). In captivity Siamese firebacks are reported to favor animal foods, such as insects, but can survive on a normal pheasant diet (Howman, 1979). In the wild they are said to search for insects in the vicinity of cattle or buffalo and to forage by scratching about in the forest (Beebe, 1918–1922).

Movements and Migrations
There is no reason to believe that movements of any substantial size occur in this species because there are

very few seasonal changes in temperature or precipitation patterns where the Siamese fireback lives.

Daily Activities and Sociality
Delacour (1977) reported that these birds are found in groups ranging from single males to large family parties while they are foraging in morning and late afternoon hours.

Social Behavior

Mating System and Territoriality
At least in captivity, the Siamese fireback is reportedly monogamous (Delacour, 1977). Perhaps this is also the case in the wild, although there is no information on this point.

Nothing is known of its territoriality, but the male utters a rather loud whistling call and performs a typical *Lophura* wing-whirring display (figure 38). Both of these are probably important acoustic signals in the species' dense forested habitats.

Voice and Display
Besides its whistling call, which is probably a territorial challenge or at least a mate-attracting call, males are said to utter continually a loud, repeated *pee-yu* call (Delacour, 1977). This is perhaps similar to the food call shown sonographically by Stokes and Williams (1972), which consists of a series of equally spaced notes uttered at the rate of about three per second and is associated with tidbitting behavior.

Probably wing-whirring is a major male sexual display, but it is equally probable that lateral display, which would expose the highly colorful rump patches, is also an important part of courtship.

Reproductive Biology

Breeding Season and Nesting
There are relatively few actual dates of nesting from wild birds. Riley (1938) stated that a group of four eggs was collected on 25 May and another egg was obtained on 22 June. There is also a record of a single set of eight eggs obtained on 19 April and an incomplete set of eggs obtained on 2 May. Baker (1928) reported that the first of these two clutches was from a nest found on the ground in a hollow tree and that both of the adults were trapped on the nest.

That observation would favor the view that the birds are indeed monogamous under natural conditions.

Incubation and Brooding
In captivity, the incubation period is 24–25 days, which is typical for the genus *Lophura*. The clutch size in captivity ranges from five to eight eggs. This is probably the normal clutch for wild birds as well, given the limited information from the wild.

Growth and Development of the Young
Nothing is known of this under natural conditions, but in captivity the chicks are said to be easy to rear. However, they do need shelter from direct sunlight and are quite susceptible to diseases such as paratyphoid. The young males become fully colored before they are 1 year old, but their tails increase in length during their second year, when they breed for the first time. However, females may not breed until they are 3 years old (Delacour, 1977).

Evolutionary History and Relationships
Certainly the Siamese fireback is a close relative of the crested and crestless firebacks, and its current range is not distantly removed from that of the Vieillot's crested fireback. Of the entire group, the Bornean crestless fireback would seem to be the most generalized form, based on its simple tail structure and lack of a crest. The Bornean crestless fireback might be an evolutionary ancestral plumage type to the present-day Siamese fireback. The Siamese fireback is also rather kalijlike in posture and proportions, and in that respect seems somewhat transitional between the kalij pheasants and the typical firebacks.

Status and Conservation Outlook
Little is known of the status of this species. However, deforestation in Thailand is going on at a horrendous rate, reducing the surface area of Thailand from 56 percent forested in 1960 to only 25 percent by 1978 (Boonlerd, 1981). About 15 percent of the remaining forest lands, some 3.2 million hectares (8.0 million acres), are now protected as national parks and wildlife sanctuaries, whereas the other 85 percent are composed of national reserved forests that are managed for timber production. The Siamese fireback is known to inhabit at least one of the nation's wildlife sanctuaries, Panom Dongrek, in Srisaket Province,

Figure 38. Display postures of the male Siamese fireback, including normal (A), wing-flapping with facial engorgement (B), and calling (C). After photos by the author and Lincoln Allen.

and almost certainly inhabits several others (Boonlerd, 1981). Some recent efforts have also been made to raise this species in captivity at a breeding center at Bangpra using birds that were obtained from England.

McGowan and Garson (1995) and Collar et al. (1994) regarded this species as vulnerable. In Vietnam it has been reported from one national park, four nature reserves, and a proposed reserve. In Thailand it inhabits several national parks and wildlife sanctuaries, and it is probably is present in several proposed reserves in Laos (McGowan and Garson, 1995).

CRESTLESS FIREBACK

Lophura erythropthalma (Raffles) 1822

Other Vernacular Names
Rufous-tailed pheasant; faisan à queue rousse (French); Gelbschwanzfasan (German); kuang-bestam, meta-merah (Malayan); singgier (Bornean).

Distribution of Species
Lowland forests in Borneo, Sumatra, and the adjoining Malay Peninsula. See map 8.

Distribution of Subspecies
Lophura erythropthalma erythropthalma (Raffles): Malay crestless fireback. Lowland forests of Sumatra and Malay Peninsula north to Kedah.
Lophura erythropthalma pyronota (G. R. Gray): Bornean crestless fireback. Lowland forests of northern Borneo.

Measurements
Delacour (1977) reported that males of *erythropthalma* have wing lengths of 240–250 mm (9.4–9.8 in) and tail lengths of 150–180 mm (5.9–7.0 in), whereas females have wing lengths of 200–220 mm (7.8–8.6 in) and tail lengths of 140–160 mm (5.5–6.2 in). Two males of *erythropthalma* averaged 1,043 g (2.3 lb) and a female weighed 837 g (1.8 lb; various zoo records). No weights are available for *pyronota*. Eggs (*erythropthalma*) average 47.8 × 35.5 mm (1.9 × 1.4 in) and have an estimated fresh weight of 33.2 g (1.2 oz).

Identification

In the Field (406–508 mm, 16–20 in)
This species is limited to lowland forests of Malaysia, where it occurs in company with the crested fireback. The crestless fireback differs from that species in having red facial skin and gray feet, rather than blue facial skin and red feet. The male also lacks a crest and has a short rufous tail that lacks elongated and drooping central feathers. The male utters a low, croaking, and repeated *tooktaroo* call and has a noisy wing-whirring display. Females are extremely dark colored, being almost uniformly glossy bluish black except for a more brownish head and a grayish throat. They also have red facial skin around the eyes.

In the Hand
Male crestless firebacks are the only pheasants with fairly short (under 200 mm [7.8 in]) and rounded tails that are uniformly rufous. Females are likewise the only pheasants that are essentially glossy black except for their more brownish to grayish head and throat and red facial skin around the eyes. Females tend to resemble males of the Salvadori's pheasant, but have a longer (30–33 mm, 1.1–1.3 in) and darker bill and a shorter (maximum 220 mm [8.6 in]) wing.

Geographic Variation
Geographic variation is relatively slight and is mostly limited to males. Those from the northern parts of the range (*pyronota*) have neck and upper back feathers that are light gray with white shaft-streaks and fine black speckles, rather than being purplish black with silvery gray vermiculations. Additionally the feathers of the breast and sides are more lanceolate and have wide white shaft-streaks, whereas the upper tail-coverts are more bluish and the rectrices are black basally. Females of the two races are usually indistinguishable (Delacour, 1977).

Ecology

Habitats and Population Densities
The crestless fireback is limited to primary and tall secondary forests habitats, probably at elevations no higher than 915 m (3,000 ft; Robinson and Chasen, 1936; Smythies, 1981). Beebe (1918–1922) observed the Malayan subspecies in light jungle habitat also occupied by gibbons, red junglefowl, hornbills, babblers, and sunbirds. In Borneo he found crestless firebacks in dense evergreen forest, where the trees were festooned with lianas and lichens and thorns and briers were abundant, including a thorny rattan palm.

There are few estimates of population densities. However, Davison and Scriven (1987) stated that in Malaysia the species inhabits lowland and hill dipterocarp forests growing on level, gently sloping, and steep substrates from sea level up to 200 m (655 ft) and perhaps higher, but not to the upper limits of hill dipterocarp forests. Densities on logged forests with a dense palm undergrowth were fairly high (6.0 birds per square kilometer [15.5 per square mile]). Densities were lower in forests on level ground with mixed dry and swampy terrain (3.0 birds per square kilometer [7.8 per square mile]) and were still lower in forests over level ground in mixed alluvial terrace, dry, and swampy land (0.6 birds per square kilometer [1.6 per square mile]).

Competitors and Predators

The crestless fireback inhabits the same area as the crested fireback and very possibly competes with it. In Borneo it also occurs with the wattled pheasant, which has similar habitats and probably the same foraging requirements. Davison and Scriven (1987) stated that in Malaysia the crestless fireback and crested fireback seem to have nonoverlapping ranges and exclusion of the crestless species occurred when populations of the crested fireback increased.

There are no discussions of possible predators of this species, but Beebe (1918–1922) observed a zebra civet in an area of this species' Bornean habitat. Quite possibly it and other civets of the region are potential predators, as are mongooses and various felids. Beebe reported crestless firebacks to be relatively tame, often frequenting the vicinity of isolated huts or even hamlets, and easily trapped or shot. Beebe judged them to be quite vulnerable to human exploitation.

General Biology

Food and Foraging Behavior

The only definite information on this subject comes from Beebe (1918–1922), who observed that the crops of the crestless firebacks he examined contained about an equal amount of plant and animal material, with termites frequently consumed, whereas ticks and "grubs" were less often encountered. Small, hard berries were found in the crops of two males.

Feeding is done in a fowl-like manner, with much scratching of the forest floor and producing a good deal of noise. However, crestless firebacks seldom utter a sound while foraging and occasionally stop to listen for possible danger (Beebe, 1918–1922).

Movements and Migrations

There is no definite information of movements of crestless firebacks currently available, but

Wayre (1969) concluded that the birds are probably sedentary.

Daily Activities and Sociality

Very few observations have been made on these birds in the wild, but Beebe (1918–1922) observed a pair of birds foraging together. They remained close together but "apparently had no great affection for each other," because they sometimes pecked at one another and twice threatened one another with their spurs. He believed that the unusual development of spurs in females of this species was indicative of a tendency for females to be active in courtship behavior in a more or less reciprocal manner. The birds seem to feed at all hours of the day and to spend the nights roosting well up in trees. Water holes were visited during early morning hours. Beebe said that the sexes were approximately equal in the coveys or flocks that he observed. He did not specify flock size except for a single flock of 22 birds, which was seemingly the largest group he encountered. Robinson and Chasen (1936) stated that flocks of five or six birds are common and are typically composed of a single male and four or five females, although at times two males may be present.

Social Behavior

Mating System and Territoriality

Beebe (1918–1922) judged that perhaps the crestless fireback was polygynous, but tended to question this possibility because of the presence of sharp spurs in the females. No other observers have added new information on this point, although Wayre (1969) suggested that males and females may form a loose bond for much of the year.

Voice and Display

The vocalizations of this species include a vibrating, throat purr; a loud *kak* of alarm; and a repeated, low-pitched *tak-takrau* (Medway and Wells, 1976). David Rimlinger (pers. comm.) informed me that during lateral display the male lowers the near wing, raises the far wing, and enlarges the facial wattles (figure 39). The male also vertically spreads and tilts the tail toward the object of the display. These postures have been observed toward females as well as toward males of other species.

Figure 39. Postures of the male crestless fireback, including display (A) and normal (B, C) postures. After photos by Mr. Tatematsu (A) and Kenneth Fink (B, C).

Reproductive Biology

Breeding Season and Nesting

According to Medway and Wells (1976), eggs have been found in April and in June on the Malay Peninsula. Coomans de Ruiter (1946) has described the nest and eggs of the Bornean subspecies, which are like those of other firebacks. However, very little is known of the natural nesting of either of these two subspecies. A clutch of five eggs was found in June 1965 in Selangor, Malaysia, in a nest placed between the roots of a large tree. Three young were later hatched from this clutch in an incubator (Jarvis and Medway, 1968). Three years later a female from this clutch laid her first eggs in a large aviary. Two nests were made by her, both of which were scrapes in sand among a cluster of pre-cast concrete culvert sections. Six eggs were laid over a period of 12 days. The female built her nest in the usual galliform manner of sideways-throwing of nesting material; although the male approached the nest, he never assisted in this activity. During the egg-laying period the female would spend the nighttime hours perched with her mate. Incubation behavior did not begin until 2 days after the laying of the last egg.

Incubation and Brooding
In the nest studied by Jarvis and Medway (1968), the female brooded both day and night from 26 February to 28 March, leaving the nest only for short periods to feed, drink, and dust-bathe. The eggs proved to be infertile.

Growth and Development of the Young
Crestless firebacks rarely have been reared in captivity. However, Delacour (1977) stated that the chicks are easy to rear and are not susceptible to diseases. The birds do not mature until their second year, but thereafter females may produce up to three clutches in a single breeding season.

Evolutionary History and Relationships
The crestless fireback would certainly appear to be the most generalized form of fireback, and indeed might vie with the Salvadori's pheasant for being one of the most generalized forms of the genus *Lophura*. Its geographic distribution is also appropriate for a centralized and ancestrally derived type of *Lophura*.

Status and Conservation Outlook
Certainly the future of the crestless fireback will depend upon the preservation of adequate areas of primary or mature secondary forests in both Borneo and Malaysia, where much deforestation is occurring at present (Davison, 1981c).

Recent forest management practices in Indonesia offer some hope for forest-dependent pheasants, although recent forest fires have had devastating effects. Currently 27 percent of Indonesia's forests (which cover 113 million hectares [280 million acres]) are at least theoretically protected from commercial exploitation, and additional parks and refuges now (1998) comprise 16 percent of the entire region's land area. In Kalimantan the crestless fireback reportedly occurs in the Barito Ulu Valley (Wheatley, 1996) and in Sarawak at Guning Mulu National Park. On the Malay Peninsula the species occurs in Taman Nagara National Park, Krau Wildlife Reserve, Endau-Rompin Park, and in Pasoh and Panti Forest Reserves (see map 20, page 311). One preserve in Sumatra (Kerinci-Sablat National Park; see map 22, page 345) is known to support the crestless fireback (McGowan and Garson, 1995). These authors considered the race *pyronota* as endangered, whereas *erythropthalma* was listed as endangered in Sumatra and vulnerable in Malaysia. Collar et al. (1994) categorized the species as vulnerable.

CRESTED FIREBACK

Lophura ignita (Shaw and Nodder) 1797

Other Vernacular Names
Vieillot's fireback; fasian noble (French); Rotrücken-fasan (German); ayam pëgar, ayam suil (Malayan); sempidan (Bornean).

Distribution of Species
Dense lowland forests in Malaysia, Sumatra, Borneo, and Banka Island. See map 12.

Distribution of Subspecies
Lophura ignita ignita (Shaw and Nodder): lesser Bornean crested fireback. Kalimantan and Banka Island.

Lophura ignita nobilis (P. L. Sclater): greater Bornean crested fireback. Sarawak and Sabah, in northern Borneo.

Lophura ignita rufa (Raffles): Vieillot's crested fireback. Malay Peninsula to the Isthmus of Kra; Sumatra, except the southeast, where it intergrades with *macartneyi*.

Lophura ignita macartneyi (Temminck): Delacour's crested fireback. Southern Sumatra and Province of Palembang and parts of the Lampongs in southeastern Sumatra.

Map 12. Distribution of Delacour's (D), greater (G), lesser (L), and Vieillot's (V) races of crested fireback. Distribution of primary forest habitat in these areas as of the late 1970s is also shown by fine shading or stippling (Java).

235

Measurements

Delacour (1977) reported that males of *ignita* have wing lengths of 270–280 mm (10.5–10.9 in) and tail lengths of 245–260 mm (9.6–10.1 in), whereas females have wing lengths of 234–254 mm (9.1–9.9 in) and tail lengths of 156–177 mm (6.1–6.9 in). In *nobilis* the males have wing lengths of 280–293 mm (10.9–11.4 in) and tail lengths of 254–285 mm (9.9–11.1 in), whereas females have wing lengths of 252–270 mm (9.8–10.5 in) and tail lengths of 265–295 mm (10.3–11.5 in). Males of *rufa* have wings of 270–300 mm (10.5–11.7 in) and tails of 265–295 mm (10.3–11.5 in), and those of *macartneyi* have wings of 270–300 mm (10.5–11.7 in). Lack (1968) reported the adult weight of *rufa* as 1,800 g (3.9 lb). Beebe (1918–1922) reported the females of the Bornean and Malayan forms to weigh about 1,600 g (3.5 lb), and males of those two forms to average about 2,040 and 2,265 g (4.5 and 5.0 lb), respectively. Riley (1938) reported that five males of *rufa* weighed from about 1,812 to 2,605 g (4.0 to 5.7 lb), with an average of 2,175 g (4.8 lb). The eggs of *ignita* average 54 × 40 mm (2.1 × 1.6 in) and have an estimated fresh weight of 47.6 g (1.7 oz).

Identification

In the Field (610–711 mm, 24–28 in)

The crested fireback, which is limited to Borneo and the Malay Peninsula, is vertically crested and has distinctive blue orbital skin patches in both sexes. Additionally, males have white to cinnamon central tail feathers, which are vaulted and decurved, and brilliant maroon on the lower back and rump. The major vocalizations include a sharp, squirrel-like *chukun, chukan*, uttered as an alarm call; a long squealing call is sometimes also uttered. Wing-whirring is commonly performed as well. Females are generally cinnamon dorsally and have blue facial skin and a short vertical crest. The tail coloration of females varies from bright chestnut to black in different subspecies.

In the Hand

The distinctly vertically crested condition of males, together with their bluish facial skin, maroon rump, and otherwise purplish body iridescence, provides for easy identification. Females have a similar vertical crest and have blue facial skin, which provides a combination that separates them from all other *Lophura* species.

Geographic Variation

Geographic variation in the crested fireback is well marked and is partially clinal. Within Borneo, there is a slight cline in size, with the southern race (*ignita*) somewhat smaller than the northern one. Bornean (and Banka Island) males are separable from those from farther west in that they have central tail feathers that are cinnamon buff to chestnut, rather than pale buff to white. Also, their sides, flanks, and anterior underparts are coppery chestnut, whereas in the more western forms these areas are dark blue with whitish shaft-streaks. Additionally, the more easterly races have whitish rather than crimson legs and two rather than four lobes on the facial wattles; these wattles have no red spots as are present in *rufa*. Females of *rufa* and *macartneyi* have bright chestnut, rather than essentially black, tails (Delacour, 1977) and most distinctly reddish legs, whereas those of *ignita* are sometimes tinged with brown.

Ecology

Habitats and Population Densities

In Malaysia, the crested fireback was found by Beebe (1918–1922) in tangles of thorny palms and enmeshed vines that cover areas of once-cleared lands. These lands are often are close to small streams in low-lying valleys. Likewise, in Borneo the birds inhabit bamboo and other jungle thickets with abundant vines, near jungle rivulets, and in heavily overgrown vegetation. Thick evergreen forests, often near rivers and at relatively low elevations, seem to represent their most typical habitats in both areas.

Davison (1981*a*) has analyzed the habitat requirements of the crested fireback in Malaysia. Of 45 sightings, the majority were in moist forest areas where invertebrate food supplies were plentiful. The birds were seen in nearly every vegetation type and were most often observed close to rivers (40 percent of the sightings were within 100 m [330 ft] of a river), especially in riparian fringes and within limits of the flood forest.

There are few estimates of population densities. However, in Malaysia estimated densities range from 2.67 to 10.67 birds per square kilometer (6.9 to 27.6 per square mile) at the same site, with variations apparently affected by seasonal wandering (Davison and Scriven, 1987). In a second site the estimate was 8.5 birds per square kilometer (22.0 per square mile).

The crested fireback was observed by Davison (1981*a*) only in lowland dipterocarp forests growing on level alluvial riverine terraces near moderate to large rivers.

Competitors and Predators
Davison (1981*a*) has concluded that the crested and smaller crestless species are probably competitors, and that the latter is excluded from habitats occupied by the former in Malaysia. Although crested firebacks seem to require riverine forests, the crestless species can also exist in nonriverine and more hilly forests. Thus, the crestless fireback tends to be displaced into such areas to varying degrees, depending upon the density of crested firebacks in a particular area.

No specific information is available on predators of the crested fireback.

General Biology

Food and Foraging Behavior
Foraging is typically done solitarily, with the birds obtaining plenty of protein-rich invertebrate life by picking up single large food items from the litter surface (Davison, 1981*a*). Among these foods is the crab *Sesarma*, and there are other abundant animal foods in the moist riverine habitats. According to Davison, around the time that the young hatch there is an abundance of fruit, but as they grow the abundance of both fruit and invertebrates declines.

Beebe (1918–1922) noted that among the crested firebacks he examined the crops were filled mostly with leaves and seeds, with insects only occasionally predominating. Among the insects were ants, small beetles, and grubs. In several cases an acornlike fruit was also present.

Movements and Migrations
These large birds seem to be fairly mobile. Davison (1981*a*) noted considerable variations in population densities in a single study area that in part he attributed to the birds' mobility. In some cases individual birds moved up to 300 m (985 ft) in a week. Davison regarded these observations as reflecting a shifting core area within a larger and more stable home range, with the latter seemingly overlapping among individuals. However, the core areas of males were well spaced and perhaps averaged less than 1 ha (2.5 acres)

in size, whereas the total home ranges seemed to be from about 20 to 25 ha (49.4 to 61.8 acres).

Daily Activities and Sociality
Beebe (1918–1922) found that crested firebacks roosted in medium-sized trees, and would move in early morning toward the nearby river. In the middle of the day they would sometimes return to the roost, but on other days they apparently did not. Davison (1981*a*) saw only single birds between January and June, except for a single January sighting of a female and her previous-year's brood. He believed that pairing and laying occurs in Malaysia from June to August and that flocking of broods occurs in September, with a gradual breakup of these groups as the young grow. Although crested firebacks are usually found only in small flocks of up to about 6 birds, one flock of 16 has also been observed (Davison, 1981*a*).

Social Behavior

Mating System and Territoriality
The crested fireback has no loud calls to announce its territory, and wing-whirring is probably the major territorial or self-assertive display of males. It was heard by Davison (1981*a*) only from May to June, during the time when pairs were seen. Thus, this display may be important in attracting mates.

As noted earlier, during the prebreeding period of May to June males maintain rather small core areas within larger home ranges. These core areas tend to be well spaced and perhaps also serve as mating territories. Although the situation in the wild is not very clear, in captivity crested firebacks appear to be monogamous and are best maintained in pairs (Ollson, 1982).

Voice and Display
Beebe (1918–1922) described the male's combination vocal and wing-whirring challenge as sounding like *wooon-k! (whirrrr)*. Baker (1930) described the male's call as a *chukun, chukun,* followed immediately by a whirring sound. The whirring is best imitated by rapidly twirling a small stick in a cleft in which a piece of stiff cloth has been transversely placed.

The typical lateral display of the crested fireback is characterized by a great engorgement of the male's wattles, a high degree of fluffing of the plumage, and a rather slow and measured waltzing around the female (figure 40). This posture appears to be the same or

Figure 40. Display postures of the male crested fireback, including lateral display by Vieillot's race (A), lateral threat by greater Bornean race (B), normal head posture (C), and facial engorgement (D). After photos by the author.

very nearly the same in the two subspecies. Behavior associated with tidbitting has not yet been described.

Reproductive Biology

Breeding Season and Nesting
Davison (1981*a*) judged that the breeding season in Malaysia occurs between June and August, during a period of high fruit availability. Few nests have been found. However, there is an old record of eggs ob-

tained in April, which were taken from a nest of dead leaves, grass, and bamboo-spates under some low bushes in dense evergreen jungle (Robinson and Chasen, 1936). In Borneo, four newly hatched chicks were collected in late July in the Kimanis Bay area (Smythies, 1981).

Incubation and Brooding
At least in captivity, the incubation period of the crested fireback is 24 days. Beebe (1918–1922) ob-

served that a female laid six eggs at intervals of 1 and 2 days, and then incubated for 24 full days. During this time the male performed "guard duty" several yards away from the nesting site. He apparently never closely approached the nest, but guarded it very fiercely. However, as soon as the chicks hatched the male joined his mate; thereafter they were together constantly.

Growth and Development of the Young
Beebe (1918–1922) reported that the male parent took his full share in feeding the young birds hatched by his mate. The male would call them to him for feeding as often as did the hen. Within a week of hatching the young were able to use their wings, and when alarmed they would fly off in various directions rather than squatting down as they had done earlier. By their first autumn the young birds closely resemble their parents. However, although they attain full plumage their first year, crested firebacks do not become reproductively mature until their second or sometimes even their third year (Howman, 1979; Ollson, 1982).

Evolutionary History and Relationships
The crested fireback is probably derived from an ancestral Bornean type similar to the modern crestless species. This perhaps occurred on Sumatra or Malaysia, with a secondary invasion of Borneo by the crested form and of Sumatra and the Malay Peninsula by the crestless form following the completion of speciation.

Status and Conservation Outlook
The future of the crested fireback in its native range depends on the survival of suitable primary forest habitats (see map 12). On the Malay Peninsula about 42 percent of the total land area is still covered by forest (15 percent lowland, 27 percent montane). On Sumatra the percentage is about 57 percent, on Java about 23 percent, and on Borneo (Sarawak, Sabah, and Kalimantan collectively) the various political subdivisions contain from 25 to 70 percent forested habitats (Davison, 1981c). The Malayan race is protected locally by the large Taman Negara National Park (Davison, 1981c); in Sumatra it also inhabits Gunung Leuser National Park (Sumardja, 1981). Four other preserves are present on Sumatra (see map 22, page 345) and there is one proposed preserve, which collectively support at least six pheasant species. In Borneo there are 12 actual or proposed preserves (see map 8), but little is known of their pheasant faunas. The crested fireback was considered vulnerable by McGowan and Garson (1995) and by Collar et al. (1995).

WATTLED PHEASANT

Lophura bulweri (Sharpe) 1874

Other Vernacular Names
Bulwer's pheasant, white-tailed wattled pheasant; faisan de Bulwer (French); Bulwer-Fasan (German); bau-en (Dutch Bornean).

Distribution of Species
Borneo (Sabah, Sarawak, and Kalimantan) in lowland forests and mountain slopes up to about 1,370 m (4,500 ft). See map 8.

Distribution of Subspecies
None recognized by Delacour (1977).

Measurements
Delacour (1977) reported that males have wing lengths of 255–260 mm (9.9–10.1 in) and tail lengths of 450–460 mm (17.6–17.9 in), whereas females have wing lengths of 225–235 mm (8.8–9.2 in) and tail lengths of 175–190 mm (6.8–7.4 in). Four males had weights of 1,470–1,800 g (3.2–3.9 lb), with an average of 1,615 g (3.5 lb), and two females weighed 916 and 1,004 g (2.0 and 2.2 lb; D. Rimlinger, pers. comm.). The eggs average 50.7 × 40.3 mm (2.0 × 1.6 in) and have an estimated fresh weight of 45.3 g (1.6 oz).

Identification

In the Field (559–813 mm, 22–32 in)
This Bornean endemic is unlikely to be confused with any other Bornean pheasants. The male has a uniquely white, fanlike tail that contrasts with an otherwise blackish body and blue facial skin. The female also has bluish facial skin, like that of the crested fireback; however, unlike that species it has no crest. The female is otherwise almost uniformly dark brown. At the peak of the breeding season the males utter a shrill piercing cry, but otherwise are relatively silent. Males also perform a wing-whirring display.

In the Hand
Adult males are instantly recognizable on the basis of their entirely white tail and blue facial skin. Females are almost entirely chestnut brown, with fine vermiculations, a deep chestnut tail (175–190 mm,

6.8–7.4 in), and blue facial skin. Apart from this blue facial skin, females very closely resemble in shape and color the females of the Edwards' and imperial pheasants.

Ecology

Habitats and Population Densities
The wattled pheasant is a submontane species associated with mature forests, which apparently is quite patchily distributed from Sabah south to the upper Barito, Mahakan, and Kabuas Rivers west to Mount Mulu, Sarawak. The wattled pheasant has been seen as low as 150 m (490 ft), but more generally ranges from 300 to 1,500 m (985 to 4,920 ft). Surprisingly, it has been reported from a lowland swamp forest area (Tanjung Puting National Park) in southernmost Kalimantan beyond its known range limits, as well as in the Barito River Valley, central Kalimantan (Wheatley, 1996). The wattled pheasant has been observed at least as high as 1,370 m (4,500 ft) on Mount Mulu, and is apparently usually found in the vicinity of larger rivers and streams (Beebe, 1918–1922; Smythies, 1981).

There are no estimates of its population density.

Competitors and Predators
At least in some areas the wattled pheasant certainly comes into local contact with the crestless fireback, and perhaps also at lower elevations with the crested fireback. However, at present nothing can be said of any competitive interactions that might arise.

Likewise, the species' predators are unknown, although Beebe (1918–1922) mentioned that many of the birds that are caught in snares set by natives are devoured by civet cats unless they are promptly found and removed. He judged that the immaculate white tail of males must make the birds highly conspicuous in their heavily shaded environment. Thus, they are probably more readily taken by predators.

General Biology

Food and Foraging Behavior
Beebe (1918–1922) concluded from the crops that he examined that wattled pheasants are about equally in-

sectivorous and frugivorous. They eat ants, small crickets, and other orthopterans; termites were second only to ants in actual numbers. Small seeds and a nutlike fruit were also present in the crops. Like others of their genus, wattled pheasants feed by scratching in the forest floor, alternating these feeding movements with short periods of intense alertness for possible danger. Captured birds eagerly consume grubs and grasshoppers (Donald Bruning, pers. comm.).

Movements and Migrations

There is no good information on this. However, Beebe (1918–1922) implied that wattled pheasants are apparently localized, sedentary, and usually inhabit the first ridge of jungle stretching back from the upper reaches of a river and extending back no farther than the second ridge. Reports from Kalimantan suggest that seasonal movements occur in response to food resources, with the birds moving into areas following a mass blooming of trees (Donald Bruning, pers. comm.).

Daily Activities and Sociality

Beebe (1918–1922) saw a reputed roosting site of this species, which was near the top of a small hill in low jungle in an isolated, slender, and smooth-barked tree. He observed most wattled pheasants only in the morning and late afternoon hours, which is the usual pattern of daily activity found in most or all tropical pheasants. Beebe believed that the birds made regular morning and evening trips from the hilly jungle areas to river banks to drink each day. He made no comments on group sizes larger than four; he twice saw groups of both parents attending two young.

Social Behavior

Mating System and Territoriality

The high level of sexual dimorphism in this species might suggest that it is polygynous, but as just noted, Beebe (1918–1922) mentioned twice seeing paired birds with young. This would suggest that monogamy is at least occasionally present in the species.

Nothing of certainty can be said of wattled pheasant territoriality. At the peak of the breeding season the male utters a shrill piercing cry (Heinroth, 1938), which is presumably territorial in function or at least may serve to space out males and perhaps attract females.

Voice and Display

In addition to the loud call of the male, which is uttered from a normal walking posture (figure 41), wattled pheasants also utter alarm *kak* notes and a penetrating and rather metallic *kook!*, *kook!* note that may serve as a contact or covey call (Beebe, 1918–1922).

The marvelous display of this species has been described by various authors (Beebe, 1918–1922; Heinroth, 1938; Muller, 1980) and is one of the most remarkable of all pheasants. Its major features are the raising of the white tail feathers into a narrow, almost circular, disklike shape, with no indication of the spreading that Beebe assumed might sometimes take place. The wattles are engorged (Schneider, 1938) into a double structure resembling the elongated tines of a hammer, extending several inches above and below the head, and largely obscuring the head with the exception of the red eye, which is made even more conspicuous by the enlargement of an oval eye-ring. From the side, the beak is wholly obscured by the blue facial wattle, which also expands forward in a shape that almost mimics that of the bill itself (figure 41).

In this posture the bird may perform wing-whirring (figure 42). However, more commonly he simply struts about silently, except for the scratching sounds made by the shafts of the outer tail feathers, which scrape along the ground as the bird moves forward, or, with a hiss, he swivels about while standing in place. The intensity of the posturing varies with the closeness of the male to the female, as does the degree of engorgement and erection of the wattles.

I have not observed a typical waltzing behavior in the wattled pheasant, but Rimlinger (1985) has observed tidbitting behavior. Copulatory behavior has not yet been described for this species, but Rimlinger (1985) has provided detailed information on both male vocalizations and display postures. He also observed lateral display and wing-whirring by females.

Reproductive Biology

Breeding Season and Nesting

Beebe (1918–1922) judged that the breeding season in Borneo is quite prolonged. Immatures obtained in August suggested an approximate April breeding

Figure 41. Display postures of the male wattled pheasant, including calling (A), facial engorgement (B), and full lateral display (C). After photos by the author.

period, although courtship was also observed as late as July. Nests in the wild are still undescribed, but Beebe (1918–1922) believed that the usual clutch probably only contains about two eggs, based on native reports and his observations of broods of only two young.

Incubation and Brooding

A wild-caught female that was bred during 1974 in Mexico laid three eggs in a clump of bamboo in a well-planted aviary. Later C. Sivelle stimulated a female to lay by placing nest boxes in similar bamboo clumps. Sivelle judged from his experience with

Figure 42. Display postures of the male wattled pheasant, including calling (A) and wing-flapping (B). After photos by Lincoln Allen.

captive wattled pheasants that the typical clutch may contain five eggs rather than only two, as Beebe surmised. According to Donald Bruning (pers. comm.), clutches of captive birds may consist of four or more eggs. Females owned by Vern Denton laid most of their eggs in April and early May, with most of a total of 18 eggs laid during 1976 being produced by a single female (Delacour, 1977). One pair that came to the Bronx Zoo in the fall of 1984 produced 12 chicks the following summer during July and August and 16 more between May and September of 1986. No further laying occurred, a pattern of production that has also been seen in other collections. In the early 1990s a pair bred for 2 years at the San Antonio Zoo, but then stopped breeding (Donald Bruning, pers. comm.).

Growth and Development of the Young
Incubation in captivity requires 24–25 days. The young birds are easily raised on game bird crumbles, chick grains, vitamin supplements, and peanut hearts, supplemented later on with fruits such as bananas, apples, and oranges (Delacour, 1977). By the age of 2 months they were feeding on the usual game bird pellets, with small additional amounts of milo, peanut hearts, and herring meal mixed in as well as mealworms, fresh fruit, and medicated water. Young males begin to show their distinctive colors by about 6 months of age. Adults molt their tail feathers with surprising frequency. Some males apparently molt these feathers up to four times a year, but one fertile male molted his twice a year (Delacour, 1977).

Evolutionary History and Relationships
The wattled pheasant is certainly the most distinctive and seemingly remote form of *Lophura*, with no obvious close relatives. The downy young are very much like those of the crested fireback. Thus, it seems most likely that the firebacks are the nearest, if still rather isolated, relatives of this species.

Status and Conservation Outlook
Deforestation of this species' habitat is proceeding rapidly in Borneo; how much it is affecting the welfare of the bird remains unknown. However, the species does inhabit the Gunung Mulu National Park of Sarawak (Davison, 1981c). At present rates, an area roughly the size of Switzerland is being lumbered every year in Indonesia, to say nothing of forests lost

to the massive and uncontrolled fires that occurred during the droughts of 1982–1983 and 1997. However, an extensive forest replanting program exists in Indonesia, which has over 1 billion square kilometers (386 million square miles) of tropical forests, during which more than 11 billion new trees have been planted. Furthermore, 27 percent of the region's existing forests have been designated as limited-production forests, which allow only selective harvesting of trees. As of the late 1970s, the percentages of land mass still covered by primary forests in the Greater Sunda region were: Malay Peninsula, 42 percent; Sumatra, 57 percent; Java, 6 percent; and Kalimantan, 70 percent (see map 12). Currently (mid-1990s) over 10 percent of Indonesia's land area is protected, with 30 national parks and 150 nature reserves. In Sarawak more than 10 percent of the land area is totally protected. On the Malay Peninsula about 46 percent of the land area presently is forested, including 4.7 million hectares (11.6 million acres) of permanent-reserved forest, 1.9 million hectares (4.7 million acres) of protection forest, and 750,000 hectares (1.9 million acres) designated as national parks, state parks, or wildlife preserves.

There are only a few recent sightings of the wattled pheasant, including ones from Gunong Mulu, Sarawak, and Bukit Raya and Ulo Barito, Kalimantan (del Hoyo et al., 1994). There is also a recent breeding record from Brunei (Mann, 1989), a sighting in Sabah's Maliau Basin (Marsh and Gasis, 1990), and a sighting from Gunung Lotung, Sabah (Anonymous, 1989). During the early 1990s wildlife biologists captured six birds in less than a week in Sabah (Donald Bruning, pers. comm.). Wheatley (1996) also listed the wattled pheasant as present in Tanjung Puting National Park, a lowland forest reserve in Kalimantan that is well south of the species' presently known range but that does include some submontane rainforest. Several proposed reserves lie within the species' presumed range in Kilimantan, including Mount Berau, Sankulirang, Long Bangun, Mount Bentuang Karimun, and Meratus Hulu Barabai. The wattled pheasant has also been reported from Lanjak-Entimau National Park, Sarawak, and the Danum Valley Conservation Area, Sabah (McGowan and Garson, 1995). McGowan and Garson estimated a world population of 1,000–10,000 birds and considered the species vulnerable, as did Collar et al. (1994).

Genus *Catreus* Cabanis 1851

The cheer pheasant is a medium-sized montane pheasant in which sexual dimorphism is slight and in which both sexes have long, narrow occipital crests. A large red orbital skin area is present. The plumage is generally gray to buffy, with black barring and spotting, and the highly graduated tail likewise is strongly barred with buff, black, and brown. The wing is rounded, with the tenth primary shorter than the first and the sixth the longest. The tail has 18 rectrices, with the central pair up to five times the length of the outermost pair. The tail molt is phasianine (centripetal). The tarsus is fairly long and spurred in the male. A single species is recognized.

CHEER PHEASANT

Catreus wallichi (Hardwicke) 1827

Other Vernacular Names

Chir pheasant, Wallich's pheasant; faisan de Wallich (French); Wallich-Fasan (German); kahir, chihir (Nepalese); tshi-er (Kumaon and Garwhal).

Distribution of Species

The Himalayas, from Durung Galli and the Hazara District of Afghanistan to the Simla states, Tehri Garwhal, and Nepal. This species inhabits temperate forest, scrub, and meadows between 1,220 and 3,050 m (4,000 and 10,000 ft). See map 13.

Distribution of Subspecies

None recognized by Delacour (1977).

Measurements

Delacour (1977) reported that males have wing lengths of 235–270 mm (9.2–10.5 in) and tail lengths of 450–580 mm (17.6–22.6 in), whereas females have wing lengths of 225–245 mm (8.8–9.6 in) and tail lengths of 320–470 mm (12.5–18.3 in). Males weigh from 1,475 to 1,700 g (3.2 to 3.7 lb) and females from 1,250 to 1,360 g (2.7 to 3.0 lb; Ali and Ripley, 1978). However, Baker (1928) reported the weight of males as 1,192–1,561 g (2.6–3.4 lb; rarely 1,816 g [4.0 lb]) and of females as 900–1,250 g (2.0–2.8 lb). The eggs average 53.4 × 49.3 mm (2.1 × 1.9 in) and have an estimated fresh weight of 71.6 g (2.5 oz).

Map 13. Distribution of eastern (E) and Humes' (H) races of the bar-tailed pheasant (B), and of cheer (C), Elliot's (El), mikado (M), and Reeves' (R) pheasants. The indicated ranges of the Reeves' and bar-tailed pheasants include their overall historic ranges (hatching) and their respective recent ranges (fine and coarse shading). The inset map at right shows the mikado pheasant's distribution on Taiwan (black); the lower inset shows locality records for the cheer pheasant.

Identification

In the Field (965–1,016 mm, 38–40 in)

The cheer pheasant is associated with hilly, broken grassland country and usually occurs in small flocks. The distinctively elongated and buffy brown tail and the short, pointed crest are distinctive; the bar-tailed pheasant is the only remotely similar species that possibly overlaps the extreme eastern end of the cheer pheasant's range. Cheer pheasants have many vocalizations, including a very distinctive crowing, *cher-a-per, cher-a-per, cher, cher, cheria, cheria*. They also utter cackling sounds of repeated *waaak* notes and a sharp alarm note, *tuk, tuk. . . .* Wing-whirring is lacking in this species.

In the Hand

The long (300–600 mm, 11.7–23.4 in) and strongly barred tail together with a straight and tapering brown crest that is directed posteriorly provides an easy combination of traits for in-hand identification. No other long-tailed pheasant is so uniformly buffy brown in body and tail coloration.

Ecology

Habitats and Population Densities

This species occurs over a rather wide altitudinal range in the western Himalayas. It is particularly associated with steep, grass-covered hillsides having scattered trees, especially where rocky crags are also present. Tall grasses, rather than heavily grazed grasslands, are also preferred. In Himachal Pradesh the cheer pheasant's altitudinal range is from about 1,200 to 3,000 m (3,935 to 9,845 ft), or from the subtropical pine forests to the subalpine meadow zones (Gaston et al., 1981). In one small wildlife sanctuary (Chail) a spring density of about 6 pairs per square kilometer (15.5 per square mile) was estimated in 1979 (Gaston and Singh, 1980). A more recent survey of the same area in 1983 provided a density estimate of about 7 pairs per square kilometer (18.1 per square mile; Garson, 1983). Scrub-clearing and single-season cropping with seasonal grazing may tend to favor the cheer pheasant (Garson et al., 1992). However, studies in Himachal Pradesh indicate that a positive relationship exists between cheer pheasant density and both ground cover and scrub cover, especially where scrub cover

ranged from 0.5 to 1.0 m (1.6 to 3.3 ft) in height. Steep hillsides with scrub, grass, and interspersed oaks, pines, and rhododendrons seem to be preferred habitat in India, Pakistan, and Nepal (Kalsi and Kaul, 1997).

In Pakistan this species is now apparently extirpated (Severinghaus et al., 1979). Judging from early literature, its original habitat there evidently consisted of long grasses, thick bushes, precipitous slopes, and tiered cliffs. Roberts (1991) listed the species as rare and in imminent danger of extirpation in Pakistan, in spite of at least four reintroduction efforts. Evidently hand-reared birds suffer high mortality from various native mammalian predators such as cats, civets, and martens, with 70 percent of 132 known mortality causes attributable to this factor (Hussain, 1990). Through the use of responses to tape-recorded calls the presence of cheer pheasants at Pir Chinase, near Muzzafarabad, was established in 1986, the first recent record for Pakistan (Young et al., 1987).

In west-central Nepal, near the eastern edge of the species' range, cheer pheasants have been observed at elevations of about 2,200–2,440 m (7,220–8,005 ft), in open scrubby forest and grassy cliffs. Based on extrapolation from an area of only 0.5 km^2 (0.2 mi^2), a spring estimate of the population density (using male calls as an index) provided a figure of about 8 pairs per square kilometer (20.7 per square mile; Lelliott, 1981*a*).

Population density estimates of this species are facilitated by the prolonged calling period (September to May in Himachal Pradesh) and the fact that evening calling seems to be as prevalent as morning calling (Gaston, 1981*b*). However, perhaps only at the peak of the breeding period are cheer pheasants all distributed as pairs rather than as flocks. Thus, actual numbers may be difficult to determine in this manner. Young et al. (1987) found that broadcasting pre-recorded cheer pheasant calls was most likely to elicit responses before sunrise, especially during late May and early June, when females are incubating and pair members are roosting separately. The mean calling duration during April was about 35 minutes, as compared with about 26 minutes in March and even shorter periods during May and June. These long calling periods during March and April presumably reflect territorial signaling rather than intrapair communication.

Competitors and Predators

Probably few species of pheasants are direct competitors of the cheer pheasant, because it occupies a rather distinctive vegetational stratum. Also, its strong digging feeding behavior is more like that of an eared pheasant or a monal than that of most other pheasants. The cheer pheasant broadly overlaps in altitudinal and ecological range with the Himalayan monal and the koklass pheasant, but both of these species tend to occupy heavier cover than does the cheer pheasant.

Severinghaus et al. (1979) list a variety of bird and mammal predators that occur within the historic range of the cheer pheasant, but none was noted specifically as taking this species. It would seem likely that foxes, jackals, and some of the larger raptors such as Bonelli's hawk-eagle (*Hieraaetus fasciatus*) and booted hawk-eagle (*H. pennatus*) might represent the most serious threats to adult birds.

General Biology

Food and Foraging Behavior

Beebe (1918–1922) noted that in two of "a few" birds he examined, he found an abundance of small, finely ground leaves. However, in general he subscribed to the belief that most of their food comes from digging with their bills, during which they obtain grubs, terrestrial tubers, and the like. Beebe did find the larvae of cockroaches as well as several wireworms in one crop. He also observed birds chasing winged insects. Ali and Ripley (1978) suggested that the cheer pheasant's major foods are roots, tubers, seeds, berries, and various insects and grubs, with grain eaten when it is available.

Foraging is typically done in pairs or sometimes in family groups. Like the Himalayan monal, a single bird or pair may dig a foot or more below the surface until they are almost hidden from view, looking up every few seconds for possible danger (Beebe, 1918–1922).

Movements and Migrations

Judging from its broad vertical range, there are almost certainly major changes in altitudinal distribution in this species. However, the data of Gaston et al. (1981) are apparently insufficient to document this seasonal shift. Baker (1930) stated that in cold weather cheer pheasants may be seen as low as 1,220 m (4,000 ft) and in summer at 3,050 m (10,000 ft) or higher, but on average are to be found between 1,830 and 2,745 m

(6,000 and 9,000 ft). Apparently the birds move around a good deal on their particular hills, but never completely abandon them. Year after year individuals are likely to be found in much the same places.

Daily Activities and Sociality

Cheer pheasants feed in mornings and evenings; unless the day is very cloudy, they remain under cover during the middle of the day. During the night they have been reported to roost on the ground by some observers, but probably more generally they tend to roost in stunted trees, high bushes, or on the summits of high rocks that typically abound in their favored habitats (Baker, 1930). Ali and Ripley (1978) reported that cheer pheasants typically roost in patches of oak forests associated with gullies. As they approach these areas in the evening, they are surprisingly noisy, which would seemingly render them vulnerable to poachers or predators.

Like the monogamous eared pheasants, cheer pheasants tend to be fairly gregarious. Where population densities permit, they are likely to be found in flocks of 5–15 birds, except during the actual breeding period. Much male calling goes on within these flocks, so it is apparent that calling cannot serve as a territorial signal under such cases and must have other unknown social functions.

Social Behavior

Mating System and Territoriality

All authorities agree that the cheer pheasant is entirely monogamous, although the length of the pair bond under natural conditions is unknown. Apparently both sexes often crow both at daybreak and again at dusk, and the call is loud enough to be heard for at least a mile (Baker, 1930).

Territory sizes are still unstudied, but Lelliott (1981a) reported hearing four different individuals calling in mid-May, when most or all birds would be breeding, on an area of only 50 ha (124 acres). Thus, their territories could average no larger than about 12 ha (30 acres), assuming that these calls were from four breeding males.

Voice and Display

The crowing call, which is uttered by both sexes, is loud, rather complex, and has been variously rendered as *chir-a-pir, chir-a-pir chir chir, chirwa, chirwa.* The call is uttered by the male with his head pointing directly upward, in the manner of an eared pheasant. It begins with a series of harsh grating notes in rapid succession, which rise to a crescendo of very high-pitched disyllabic whistles (Wayre, 1969). According to Wayre, captive males often call in duet with eared pheasants, and the calls are sometimes difficult to distinguish. He also stated that the males' visual display is lateral, which is somewhat similar to the posture assumed by true pheasants (*Phasianus*), with the elongated tail being spread widely and tilted toward the female in a manner that also resembles that of *Syrmaticus*. During lateral display the male's head is extended forward and his nearer wing is drooped, much like the lateral display of the common pheasant (Roberts, 1991). Apparently the long occipital crest is never raised in the manner typical of the koklass, and the facial wattles are only moderately enlarged.

Tidbitting calls and behavior, copulatory behavior, and a possible frontal display have not been described for the cheer pheasant. The species is also said to lack wing-whirring displays (Delacour, 1977). However, considering the species' monogamous mating system, these are unlikely to be elaborate or highly conspicuous. Possibly the species' well-developed vocalizations substitute for extensive postural displays, as in the similarly monogamous eared pheasants. Neither group exhibits iridescent or highly colorful plumages. The absence of iridescent coloration in the cheer pheasant is perhaps related both to the reduction of sexual selection pressures associated with monogamy and to the relatively open and grassland-dominated substrate. The cheer pheasant is admirably colored for maximum visual concealment in such habitats.

Reproductive Biology

Breeding Season and Nesting

The breeding season apparently extends from late April to early June. Nesting in lower altitudes often begins near the end of April, with some birds breeding as late as early June at the highest elevations. Clutch sizes are relatively large; 9 to 10 is apparently the usual size, but as many as 13 or 14 have sometimes been reported from nests in the wild. In captivity, cheer pheasants lay clutches of 9–12 eggs, with 15–25 eggs typical in a season (Delacour, 1977).

The nests are usually located at the foot of a boulder on steep hillsides covered with open oak or pine forests. They are often well hidden in grasses, bushes, or bracken in very broken ground. Some nests have been found at the foot of nearly vertical cliffs and in relatively inaccessible sites (Baker, 1930).

Incubation and Brooding

The incubation period is 26 days. All the incubation is performed by the female, although the male remains close at hand. After hatching, the male joins the family and takes an equally strong role in protecting it from any disturbance as does the female (figure 43).

Growth and Development of the Young

Baker (1930) cites a Mr. A. Winbush who encountered a family of newly hatched cheer pheasants. As the young scattered in all directions the two adults rushed toward him with their tails spread, their wings arched, and their neck feathers ruffled. The male approached within 2.5 m (8.0 ft) and continued to threaten him until all the chicks were hidden in the grass. At that point both the adults began to walk away, calling to the chicks all the time. Probably such family groups remain intact through the winter and until the start of the next breeding season. These groups may be the basis for the usual covey size of 5 to 10 or 15 birds. Sexual maturity is attained the year following hatching.

Evolutionary History and Relationships

Delacour (1977) believed that this species is fairly isolated, with varying degrees of similarity to *Syrmaticus*, *Phasianus*, *Lophura*, and *Crossoptilon*. The downy plumage pattern is quite distinctive and somewhat partridgelike. The monomorphic adult plumage is also relatively unusual. In calls and posturing the cheer pheasant has some distinct similarities to the eared pheasants, but otherwise seems much closer to *Syrmaticus* in most respects. I have no strong opinions about the relationships of *Catreus*, and, like Delacour (1977), I believe it to be fairly isolated from other genera.

Status and Conservation Outlook

This species was considered vulnerable by McGowan and Garson (1995) as well as by Collar et al. (1994). The cheer pheasant has been virtually extirpated from Pakistan (Roberts, 1991). In India it occurs in the Great Himalayan National Park, Himachal Pradesh, and in Nepal at Lake Rara National Park, as well as at seven other protected areas of these countries (del Hoyo et al., 1994). As noted earlier,

Figure 43. Postures of the cheer pheasant, including males in display (A) and normal (B) postures, a female (C), and a male tending chicks (D). After photos by M. Ridley (D) and the author.

a seemingly healthy and fairly dense population inhabits the Chail Wildlife Sanctuary, Himachal Pradesh. At this location the population probably numbers more than 1,000 pairs, according to Gaston et al. (1981). Generally these populations are rather irregularly and locally distributed, with fairly good numbers also still present in the Budhil Nalla area, the Kajjiar-Chamba area, and the upper Beas River Valley. The status of the cheer pheasant in Uttar Pradesh is much less understood, but it seems likely that several thousand birds must be present in total. In Nepal it is not believed to occur within any protected areas (Lelliott, 1981a), but it seems to be well distributed in western Nepal (Gaston, 1981a; Roberts, 1981). The species seems fairly resistant to disturbance and sometimes occurs fairly close to villages, but these local populations are typically small and difficult to protect (Gaston et al., 1981). Heavy

grazing and burning of the cheer pheasant's favored grassland habitats are one of the factors contributing to the species' decline. Also, the sedentary nature of the birds makes them fairly susceptible to local overhunting.

Some hand-raised birds have been released in Himachal Pradesh on a forest reserve near Simla, but the results of these efforts are still not very clear. Even in protected areas in India poaching remains a serious problem (King, 1981). In Pakistan a few birds were released in 1978, and 30 more were released in 1979 in the vicinity of Dhok Jiwan (Mirza, 1981a). By 1990 four introductions had been attempted in Pakistan, using fully grown but hand-raised chicks. However, a high level of predation by foxes, civets, cats, and martens has occurred, a situation that is common in reintroduction efforts and that makes such attempts of questionable value.

Genus *Syrmaticus* Wagler 1832

The long-tailed pheasants are small to medium-sized montane pheasants in which sexual dimorphism is well developed and in which the tail is greatly elongated and strongly barred. Both sexes lack crests or ear-tufts. The orbital wattles are red, small to moderate in size, and smooth. Males lack ruffs and their rump feathers are rounded to squarish and often conspicuously patterned, but not disintegrated. The tail is flat, highly graduated, and spread laterally in display. There are 16–20 rectrices, with the central pair sharply pointed and ornamentally barred, but without lateral fringes. The tail molt is phasianine (centripetal). The wing is slightly rounded, with the tenth primary longer than the first and the seventh the longest. The tarsus is fairly long and is spurred in males. Females are mostly strongly marked with brown, black, and white, with white often extensive on the sides and flanks and as shaft-streaks dorsally. Five species are recognized.

**Key to Species (and Subspecies of Males) of *Syrmaticus*
(in part after Delacour, 1977)**

A. Dominant color of plumage dark brown (females).
  B. Throat and foreneck black: Elliot's pheasant.
  BB. Throat and foreneck not black.
      C. Throat and foreneck buffy yellow: Reeves' pheasant.
      CC. Throat and foreneck pale brown.
          D. Upper back with arrow-shaped white shaft-streaks.
              E. General color rufous brown: bar-tailed pheasant.
              EE. General color olive brown: mikado pheasant.
          DD. Upper back lacking white shaft-streaks: copper pheasant.
              AA. Dominant color of plumage not dark brown (males).
  B. Plumage mostly dark blue: mikado pheasant.
  BB. Plumage mostly coppery red or yellow.

    C.   Back buffy yellow, with black feather-edging: Reeves' pheasant.

CC. Upper back coppery red.

        D.   Lower back black and white.

              E.   Sides of neck bluish black: bar-tailed pheasant.[1]

              EE. Sides of neck whitish gray: Elliot's pheasant.

      DD. Lower back coppery red and white: copper pheasant.[2]

            E.   Margins of rump and upper tail-coverts white.

                 F.   Margins of scapulars purplish black; tail predominantly dark brownish red; lower rump almost entirely white: Ijima's copper pheasant (*ijimae*).

                FF. Margins of scapulars white; tail predominantly cinnamon and buffy; brownish bases of white lower rump feathers visible: scintillating copper pheasant (*scintillans*).

            EE. Margins of scapulars and tips of lower back and rump feathers golden; tail very dark brownish red: Soemmering's copper pheasant (*soemmeringi*).

---

[1]See text for racial distinctions.

[2]Two dubious races (*intermedius* and *subrufus*) are excluded from this key.

REEVES' PHEASANT

Syrmaticus reevesi (J. E. Gray) 1829

Other Vernacular Names

Bar-tailed pheasant; faisan vénéré (French); König-fasan (German); chi-ky (Chinese).

Distribution of Species

China, from Kweichow and eastern Szechwan east to Anhwei. Now extirpated from northern China, including inner Mongolia. Feral, captive-bred populations are present in France, the Czech Republic, and a few other European locations, but these may not be self-reproducing. This species is associated with open woodlands of evergreen or deciduous trees such as pines, cypresses, thujas, and oaks and nests in tall grass or bushes such as azaleas (Vaurie, 1965). See map 13, page 246.

Distribution of Subspecies

None recognized by Delacour (1977).

Measurements

Delacour (1977) reported that males have wing lengths of 275–300 mm (10.7–11.7 in) and tail lengths of 1,000–1,600 mm (39.0–62.4 in), whereas females have wing lengths of 235–250 mm (9.2–9.8 in) and tail lengths of 360–450 mm (14.0–17.6 in). Cheng et al. (1978) noted that five males had wing lengths of 262–272 mm (10.2–10.6 in) and tail lengths of 1,020–1,368 mm (39.8–53.4 in), whereas a single female had a wing length of 218 mm (8.5 in) and a tail length of 330 mm (12.9 in). Thirty females averaged 949 g (2.1 lb) and 24 males averaged 1,529 g (3.3 lb; Knoder, 1963). The eggs average 46 × 37 mm (1.8 × 1.4 in) and their estimated fresh weight is 34.8 g (1.2 oz).

Identification

In the Field (711–2,032 mm, 28–80 in)

This species is associated with wooded and hilly areas of central China. Males are easily recognized by their enormously long (to 1,524 mm [60 in]), barred tail and their strongly patterned black-and-white head. Females somewhat resemble those of the common pheasant, but have a more bicolored head pattern, with buffy stripes above the eyes and a buffy throat

and neck that contrasts with a darker crown and ear-patch. Vocalizations include a twittering chuckle, and males repeatedly utter short, piping notes.

In the Hand

The very long, barred tail and the black-and-white head pattern of males make confusion with other species unlikely. Only the Elliot's pheasant has a remotely similar appearance, and in that species the throat and foreneck are blackish rather than white. Females resemble those of several other *Syrmaticus* species, but are unique in having a buffy throat and foreneck and an equally conspicuous pale stripe above the eyes. Their tails are also longer (360–450 mm, 14.0–17.6 in) than those of other *Syrmaticus* species. Young males are female-like, but with more whitish throat markings.

Ecology

Habitats and Population Densities

According to Cheng (1963), this species occupies an altitudinal range in China from 300 to 1,800 m (985 to 5,905 ft). It typically inhabits valleys having steep-sided canyons and where clustered cypresses, pines, or other coniferous trees are to be found.

In a release site in Ohio (a 14-ha [35-acre] island in Tappen Lake, Harrison County), adult Reeves' pheasants used only wooded portions of the island, rather than the more grassy and shrubby areas, prior to late July, when dense ground cover and ripe black-berries had covered the more open areas. The nests were all located in the woody rather than the grassy and shrubby areas (Knoder and Bailie, 1956).

Estimates of densities under natural conditions range from 7.6 to 9.2 individuals per square kilometer (19.7 to 23.8 per square mile), locally reaching 15 per square kilometer (38.9 per square mile) in core areas of preserves (Wu et al., 1993; Li, 1996).

Competitors and Predators

There is no definite information of this in China. However, predation on pen-raised birds in the United States was seemingly high (Seibert and Donohoe, 1965; Korschgen and Chambers, 1970), with avian predators such as great horned owls (*Bubo virginianus*) evidently an important mortality source.

General Biology

Food and Foraging Behavior

Relatively little is known of the Reeves' pheasant's foods in China, but they are known to include legumes, acorns, wild persimmons, radishes, vegetables, and liliaceous tubers (Cheng, 1963). There are several studies of released birds in the United States that indicate a wide diversity of foods consumed. Seibert and Donohoe (1965) reported that Reeves' pheasants mainly consumed wheat, wild cherries, raspberries, brome grass, dogwoods, wild yams (*Dioscorea*), crabgrass, and greenbriers (*Smilax*) in Ohio. Stephens (1966) noted that in Kentucky the birds ate acorns, hawthorn, persimmon (*Diospyros*), panic grass, black gum (*Nyssa*), lespedeza, pokeweed (*Phytolacca*), blueberries, blackberries, waste corn, and wheat, as well as such insects as grasshoppers, beetles, and caterpillars. Korschgen and Chambers (1970) found that in Missouri the birds concentrated especially on sorghum, soybeans, and maize (with these sources comprising over half of the annual diet analysis) plus small amounts of acorns, wheat, grasses and sedges, green forbs, blackberries, cherries, and a very large number of minor plant foods, which totaled well over 100 different types. Animal materials, primarily grasshoppers, comprised only about 5 percent of the estimated annual diet. These food selections seemed to be governed largely by a limited variety of choice native foods.

Movements and Migrations

Reeves' pheasants appear to be highly sedentary in their native China and at least in the southern parts of their range seem to show no tendencies toward a seasonal altitudinal migration. In the more northern areas their movements are probably more irregular and influenced by snowfall patterns (Beebe, 1918–1922).

Daily Activities and Sociality

At least in North America, Reeves' pheasants have been found to roost preferentially in second-growth, pole-sized trees associated with a general absence of ground cover. During winter, the birds usually roost in upper ravines and on south- and east-facing slopes.

The roost sites are from 1 to 6 m (3 to 20 ft) or more in height and on horizontal limbs offering easy escape routes. Several birds may roost in the same tree or in nearby trees, but they do not roost side by side (Korschgen and Chambers, 1970).

Foraging is apparently done in the usual morning and afternoon pattern typical of most pheasants, and Reeves' pheasants also visit sources of grit, such as road sides. Although gregarious during the winter months, in spring these flocks disperse and form groups of two or three birds, usually either a pair or a male and two females (Cheng, 1963).

Social Behavior

Mating System and Territoriality

Although some authors have suggested that this is a monogamous species (e.g., Korschgen and Chambers, 1970), given the considerable size dimorphism of the sexes, substantial sexual plumage differences, and apparent lack of interest in the male for tending the young, there seem to be reasons for considering it at least facultatively polygynous. Knoder and Bailie (1956) reported that on an island where only 6 males were released, at least 11 females nested. The fertility rate of the eggs in 9 of the nests found was 68 percent, which suggests a rather high level of successful mating. However, it is not as high a fertility rate as is typical for pheasants with similar sex ratios, such as wild ring-necked pheasants.

Little is known of territoriality in the Reeves' pheasant. Knoder (1955) did not observe territorial behavior in a large pen initially containing 20 males. However, when these were reduced to eight in early April the pen became approximately divided in half, with three males defending each of the two halves as apparent groups and the two remaining subordinate males being submissive to all the others. Fighting was sometimes observed along the boundary separating the three-male groups, but not among the grouped males.

Voice and Display

Male advertisement of territory or self-assertiveness is attained by a combination of wing-whirring and calling. Typically a bout of wing-whirring is followed by a series of high-pitched chirping calls that may be audible for up to about 185 m (600 ft; Knoder, 1955).

Figure 44. Display postures of the male Reeve's pheasant, including alert "guarding" (A), pre-attack (B), aggressive threat (C), and wing-whirring (D). After sketches by Moynihan (1995).

Postural display before a female is rather distinctive in the Reeves' pheasant. The male will typically approach the female slowly and in concentric circles while erecting the contour feathers and walking in a rather stiff-legged manner. After approaching to within about 2 m (6 ft), a lateral display is performed, with the tail erected and spread. This display may be maintained until the female moves, at which time the male changes to a more directly frontal display by performing a series of stiff-legged bounds toward the female and usually ending up about 0.3 m (1 ft) from her (Knoder, 1955). During the hopping movements the folded far wing is tilted upward somewhat, thereby increasing the visible surface of the male to the female's view (Kruijt, 1962*b*).

Moynihan (1995) has described and illustrated some mainly hostile postures of male Reeves' pheasants, including an alert alarm or "guarding" posture that is assumed when potential threats such as humans are visible. Males also exhibit distinctive threat and pre-attack postures and perform wing-whirring (figure 44). In contrast to the common pheasant, crowing by males often follows rather than precedes wing-whirring. Moynihan believed that females tend to seek out males for mating, and that unlike many polygynous pheasant species male Reeves' pheasants do not form harems, but tend to remain in sexually segregated groups throughout most of the year. Females may thus form same-sex flocks of up to about 30 birds, whereas the all-male groups observed by Moynihan numbered up to about 6. It is possible that an extended period of captive breeding has influenced the social behavior of this population, which has been little studied in China.

Reproductive Biology

Breeding Season and Nesting
Almost nothing is known of the breeding season of the Reeves' pheasant in China. One nest of eight eggs was found in an azalea thicket on a hillside that also was covered by grassy areas, dwarf bamboos, and a sparse growth of pines (Beebe, 1918–1922). In Ohio, the egg-laying period of nine free-ranging females was from 7 April to 7 May and the peak period of hatching was from 21 to 31 May. There was no indication of renesting in this small sample (Knoder and Bailie, 1956). Of 16 nests found on a 14-ha (35-acre) island, 9 were in second-growth saplings and brush; 4 were in sweet clover, rape, and timothy cover; and 3 were in broom-sedge (*Andropogon*; Knoder, 1955).

Incubation and Brooding
Incubation in the Reeves' pheasant is performed by the female alone, and there is no evidence that the male remains close at hand or helps to protect the nest. Based on ten nests produced by free-ranging females, clutch sizes averaged 9.5 eggs and ranged from 6 to 14 (Knoder and Bailie, 1956). Of 16 females that laid their clutches during the previous year in a production-release pen, the clutch size averaged only 4.4 eggs. However, a total of 118 dropped eggs were also found, which suggests a production of about 11 eggs per known laying female (Knoder, 1955). No dump nests or dropped eggs were found among the free-ranging females. Incubation requires 24–25 days, based on observations in captivity. Studies of the Reeves' pheasant in the Tuoda Forest, Kweichow, have shown that nesting there occurs from mid-May to mid-July. Eight nests had 6–9 eggs, averaging 7.3. The hatching success of four clutches was 93 percent (Wu et al., 1993).

Growth and Development of the Young
When a hen with a newly hatched brood is approached, she will typically utter a low call and the chicks will "freeze." As soon as the young are able to fly they will often flush as the female alerts them and fly into heavy cover or alight and run swiftly. Juveniles will often fly 45–90 m (150–300 ft) before landing and when only half-grown are able to fly at least 245 m (800 ft; Knoder, 1955; Knoder and Bailie, 1956).

Based on zoo-raised individuals (David Rimlinger, pers. comm.), newly hatched Reeves' pheasants weigh an average of about 20 g (0.7 oz); eight 1-day-old chicks averaged 21.8 g (0.8 oz). By the time they are 35 days of age they average almost 200 g (7.0 oz). They attain sexual maturity in their first year, but probably remain with their mother for much of the first winter, when adult males may also join these small flocks.

Evolutionary History and Relationships
The genus *Syrmaticus* is certainly fairly close to *Phasianus*, and one might easily argue that the two taxa should be conjoined. However, Delacour (1977) noted that all five species of *Syrmaticus* form a natural group quite distinct from *Phasianus* and that the downy young are quite different in the two genera. Nonetheless, I think it is clear that *Phasianus* and *Syrmaticus* are not very far removed from one another in their evolutionary history.

Interestingly, the Reeves' pheasant has the smallest facial skin area of any of the long-tailed pheasants, but also has the greatest development of tail feathers and the largest amount of white on the head, both of which would seem to be important male signals. This species hybridizes readily with the copper pheasant, but the hybrids have reduced fertility (Delacour, 1977). Hybridization with the common pheasant usually results in sterile females and males with reduced fertility (Knoder, 1963).

Status and Conservation Outlook
The Reeves' pheasant is extensively hunted and trapped in China, not only for food but also for the male's ornamental tail feathers. The species' numerical status in the wild is not currently known, but it seems to be adapted to a variety of open-country or second-growth habitats rather than requiring pri-

mary forest cover. Recent fieldwork by Professor Wei-Shu Hsu indicates that the species is distinctly rare and should be added to the list of endangered species (Eugene Knoder, pers. comm.). McGowan and Garson (1995) considered the species as endangered, whereas Collar et al. (1994) listed it as vulnerable. The feral population in England has essentially disappeared, but the French populations are surviving well, owing largely to the annual release of tens of thousands of birds for hunting purposes. The Reeves' pheasant has also been released with some success in the Czech Republic (Pokorny and Pikula, 1986) and in Hawaii.

By comparison, the situation in China is grim. Li (1996) listed 11 provinces where locality records exist, but the species has now been extirpated from much of its former range. Major remaining populations are centered in Kweichow Province, where 20 sites are known; the population in 1990 was about 500 birds. In Hunan, northwestern Hupeh, and western Anhwei provinces about 1,000 birds were present in 1990. In Kansu, Shensi, and northern Szechwan provinces another 1,000 were estimated. In northeastern Kweichow, southeastern Szechwan, and western Hunan provinces perhaps 500 birds were present. A small population also was present in Kiangsu Province. The most recent estimates from China would suggest that no more than 1,500 birds survive in the wild. A combination of habitat destruction and illegal hunting (mainly for the male's long, ornamental tail feathers) has placed the species at great risk of extinction in the wild (Xu et al., 1990; King, 1992).

BAR-TAILED PHEASANT

Syrmaticus humiae (Hume) 1881

Other Vernacular Names
Hume's pheasant; faisan de Hume (French); Hume-Fasan (German); wit (Burman); loe-nin-koi (Manipuran).

Distribution of Species
India to Thailand, including Assam, northern Burma, and southwestern Yunnan, in wet and moist temperate forests between 1,220 and 3,050 m (4,000 and 10,000 ft).

Distribution of Subspecies
(after Wayre, 1969; Delacour, 1977)
Syrmaticus humiae humiae (Hume): Hume's bar-tailed pheasant. Mountains of Manipur; Naga, Patkoi, and Lushai Hills and northern Burma, west of the Irrawaddy River and south to Mount Victoria.

Syrmaticus humiae burmanicus (Oates): eastern bar-tailed pheasant. Mountains of southwestern Yunnan, western Kwangsi, northern Burma (Kachin and Shan States), and northern Thailand. See map 13, page 246.

Measurements
Delacour (1977) reported that males of *humiae* have wing lengths of 206–225 mm (8.0–8.8 in) and tail lengths of 400–535 mm (15.6–20.9 in), whereas females have wing lengths of 198–210 mm (7.7–8.2 in) and a tail length of 200 mm (7.8 in). Cheng et al. (1978) reported that three males (presumably of *burmanicus*) had wing lengths of 207–213 mm (8.1–8.3 in) and tail lengths of 486–505 mm (19.0–19.7 in), whereas two females had wing lengths of 200 and 210 mm (7.8 and 8.2 in) and one female had a tail length of 195 mm (7.6 in). One male weighed 975 g (2.1 lb) and a female weighed 650 g (1.4 lb). Lack (1968) reported the average adult weight of *humiae* as 1,000 g (2.2 lb), whereas Baker (1928) indicated that males of *humiae* weigh 1,070 g (2.3 lb). The eggs of *humiae* average 48.7 × 35 mm (1.9 × 1.4 in) and have an estimated fresh weight of 32.9 g (1.2 oz). The eggs of *burmanicus* average 46 × 34 mm (1.8 × 1.3 in); the mean mass of 33 *burmanicus* eggs was 27.2 g (1.0 oz; Liu, 1991).

Identification

In the Field (610–914 mm, 24–36 in)
This species is associated with open areas among somewhat forested habitats, especially open oak and pine forests on rocky substrates and with light undergrowth. The dark neck color separates males from Elliot's pheasant, the only other species that might be confused with the bar-tailed pheasant except perhaps the familiar common pheasant, which lacks white scapular stripes. Females have shorter and white-tipped tails, which are also distinctly barred with black. Males apparently lack loud crowing, but do perform wing-whirring displays and utter loud *chuck* notes or contact calls, as well as repeated *buk* calls when alarmed.

In the Hand
Males can be recognized by the combination of a long tail (at least 400 mm [15.6 in]), which is strongly barred with gray, brown, and black, conspicuous white stripes on the scapulars and wing-coverts, and a uniformly dark-colored neck. Only the Elliot's pheasant, which has a whitish neck and whiter underparts, is likely to be confused with the bar-tailed pheasant. Females of the two species are even more similar. However, the female bar-tailed pheasant has fulvous brown rather than black on the front of the neck (sometimes spotted with blackish) and the underparts are barred with fulvous rather than being pure white.

Geographic Variation
Geographic variation among males is slight and possibly clinal. Males of the eastern (*burmanicus*) race are more purplish on the upper parts, the feathers of the lower back are deep black rather than bluish, and have wider white fringes that produce a less distinctly barred effect. Females show no obvious geographic variation (Delacour, 1977).

Ecology

Habitats and Population Densities
According to Beebe (1918–1922), the Hume's race of this species is found in somewhat open jungle

where the trees are primarily oaks and similar species and there is undergrowth and open areas of long grass or mixed grasses and bracken. Typically there are also rock outcrops present, which may be rather scattered or so abundant that the grassy areas form pathways among them. Bar-tailed pheasants range in altitude from about 760 to 1,525 m (2,500 to 5,000 ft) and are usually found in the vicinity of streams (Beebe, 1918–1922). Baker (1928) reported that they range from 1,220 to 2,440 m (4,000 to 8,000 ft), rarely as high as 3,200 m (10,500 ft), in thin and open forest or in mixed scrub and bracken on rocky hillsides. Baker (1930) doubted that bar-tailed pheasants would ever occur as low as 760 m (2,500 ft) except perhaps during the coldest weather, and said that they are more associated with open forests and scrub than with lowland evergreen forests.

The Burmese race of this species has been observed between 1,220 and 1,675 m (4,000 and 5,500 ft) in Thailand. It forages in open hill forests of oaks or pines and spends the rest of its time in dense evergreen forests (Deignan, 1945). In Burma bar-tailed pheasants are said to occur between 1,220 and 2,745 m (4,000 and 9,000 ft) in open rather than in heavy forests as well as on grass-covered hillsides (Baker, 1930).

Estimated densities in China range from 10 to 33 individuals per square kilometer (25.9 to 85.5 per square mile; Li, 1996).

Competitors and Predators

There is no published information on these subjects.

General Biology

Food and Foraging Behavior

Rather little has been written on this. Baker (1930) quoted J. P. Cook as shooting a male out of raspberry bushes and flushing a group of eight or ten birds from a clump of dwarf dates, on which they had apparently been foraging. A Captain Drummond informed Baker that the birds he shot had been foraging mainly on small chestnuts, some kind of red berry, and occasionally on small snails.

Bar-tailed pheasants foraging at dawn and dusk, and the birds spend the rest of the day in heavy wooded cover (Deignan, 1945). They also roost at night in taller forest cover (Davison, 1979*a*).

Movements and Migrations

Nothing is known specifically of this, but there are probably some altitudinal movements associated with seasonal weather changes (Baker, 1930). Davison (1979*a*) estimated a total altitudinal range of 760–3,050 m (2,500–10,000 ft). He said that the birds move to the hilltops in warm, wet weather and nest at least as high as 2,135 m (7,000 ft).

Daily Activities and Sociality

Most observations suggest that bar-tailed pheasants occur in small groups of three to five birds outside the breeding season; these groups frequently consist of a single adult male and varying numbers of females or immatures. Like the Elliot's pheasant, they seem to be relatively nongregarious birds and generally are extremely difficult to observe in their natural habitats.

Social Behavior

Mating System and Territoriality

It seems quite likely that the bar-tailed pheasant is facultatively polygynous under wild conditions, as is the case in captivity. Nothing is known of the species' territorial behavior; a crowing call has been mentioned only by Beebe (1918–1922), but was not noted by Davison (1979*a*). Davison also believed that the bar-tailed pheasant is polygynous and lives in groups of a single male and several females.

Voice and Display

Davison (1979*a*) has provided the only available detailed description of voice and display in this species. Both sexes utter a loud *chuck*, which seems to serve as a contact call, and a repeated *buk* used as a contact call among members of a group, which in louder form seems to serve as an alarm or threat signal. There is also a screeching alarm note and a loud hissing call uttered by males during high intensity lateral display. A tidbitting clucking call is also present.

Males advertise their presence with a wing-whirring display, the sound of which carries more than 30 m (100 ft; figure 45). There is no loud calling associated with this display, but it is most often performed in spring and especially during the first hour after leaving the roost. Calling is usually performed from somewhat raised sites, which may be used repeatedly. The usual threat display is a forward-oriented posture, with the head held low, the tail

Figure 45. Display postures of the male bar-tailed pheasant, including lateral threat (A), facial engorgement (B), frontal threat (C), and intense frontal display (D). Adapted from Davison (1979*a*).

spread, and the body feathers ruffled. A lateral threat, which is very much like that performed by males toward females in courtship, is used between males or by females trying to avoid a courting male. The wing nearer the other individual is lowered, the farther wing is slightly raised, the tail is spread and tilted toward the other bird, and the head is held high with, in males, the wattles distended.

Courtship display among captive bar-tailed pheasants is most frequent in March and April, during the egg-laying period. During tidbitting, the male repeatedly picks up and drops a small object while clucking, which typically brings the female on the run. Two male displays are then performed, the lateral and frontal. Lateral display resembles that just described for threat, but the head is held lower. The male then runs in arcs before the female, always turning away at the end of each run rather than turning toward the hen, as is usually the case in encounters

between males. On one occasion the male was observed to veer toward the female and perform a frontal display that is very similar to the forward intense threat display of males. Similar frontal displays reportedly also occur in the mikado and Elliot's pheasants (Delacour, 1977), but apparently are lacking in typical pheasants and ruffed pheasants.

Prior to copulation the male approaches the crouching hen in an erect posture, with the wattles distended and the nape feathers erected into a peak. No specific postcopulatory displays have been noted (Davison, 1979*a*).

Reproductive Biology

Breeding Season and Nesting
Nests of the bar-tailed pheasant have been found in the wild between March and May. The clutch size of these nests has ranged from 6 to 10, with the average

of five nests being 7.6 eggs (Davison, 1979*a*). In one case the nest was found at the foot of a tree in dwarf-oak vegetation and hidden in a small bush (Baker, 1930). In captivity, females nest on the ground, typically under a shrub or in thick grass if these are available (Davison, 1979*a*).

Incubation and Brooding
Eggs are laid by captive females every second day, typically in the early evening. The normal clutch in this situation is of six or more eggs, although first-year females may produce smaller clutches (Davison, 1979*a*). The incubation period lasts 27–28 days, and the clutch is presumably defended only by the female. There is no indication that the male participates in defense of the brood either.

Growth and Development of the Young
Reportedly bar-tailed pheasant chicks are unusually wild and may be more difficult to rear than those of Elliot's pheasants. At least in one case, they also grew a little more slowly than Elliot's, although the males attained their adult plumage by about 5 months (Delacour, 1977). Sexual maturity is attained in the first year.

Evolutionary History and Relationships
As noted in the account of the Elliot's pheasant below, these two species are very closely related and probably comprise a superspecies.

Status and Conservation Outlook
The bar-tailed pheasant was considered rare as of 1980 (King, 1981), although its actual present status in the wild is uncertain. The Burmese race has evidently been affected little by exploitation in northern Burma and that population was perhaps fairly secure in 1980 (King, 1981). In eastern India the nominate race has long been rare and patchy in distribution, and there have been no sight records for decades (Lamba, 1981). The species' status in Thailand is not currently known, but it is believed to be a rare resident in the extreme northwestern part of that country.

McGowan and Garson (1995) considered both races of the bar-tailed pheasant endangered; Collar et al. (1994) believed the same for the species as a whole. Li (1996) listed 7 sites in Kwangsi and 19 in Yunnan where the race *burmanicus* has been reported. McGowan and Garson (1995) suggested that the Thailand population of this race might number only 250–500 birds, and the nominate race may constitute no more than 1,000 birds. The former has been documented at two Thailand sites, was recently seen at Phu Khieo (Round, 1993), and might also occur in Doi Inthanon National Park. Its presence in Burma is unproven, but *burmanicus* is known in India from two sites in Mizoram and in Murlem and Blue Mountain National Parks (del Hoyo et al., 1994).

ELLIOT'S PHEASANT

Syrmaticus ellioti (Swinhoe) 1872

Other Vernacular Names

Chinese barred-backed pheasant; faisan d'Elliot (French); Elliot-Fasan (German); han-ky (Chinese).

Distribution of Species

Eastern China south of the Yangtze in Kweichow, Kwangsi, Hunan, Kiangsi, Anhwei, Fokien, and Chekiang Provinces. See map 13, page 246.

Distribution of Subspecies

None recognized by Delacour (1977).

Measurements

Delacour (1977) reported that males have wing lengths of 230–240 mm (9.0–9.4 in) and tail lengths of 390–440 mm (15.2–17.2 in), whereas females have wing lengths of 210–225 mm (8.2–8.8 in) and tail lengths of 170–195 mm (6.6–7.6 in). Cheng et al. (1978) reported that three males had wing lengths of 246–257 mm (9.6–10.0 in) and tail lengths of 420–470 mm (16.4–18.3 in), whereas two females had wings of 197–200 mm (7.7–7.8 in) and tails of 176–180 mm (6.9–7.0 in). Seventeen males had an average weight of 1,156 g (2.5 lb), with a range of 1,044–1,317 g (2.3–2.9 lb), and 35 females averaged 878 g (1.9 lb), with a range of 726–1,090 g (1.6–2.4 lb; Gene Knoder, pers. comm.). The eggs average 46 × 34 mm (1.8 × 1.3 in) and have an estimated fresh weight of 30.2 g (1.1 oz).

Identification

In the Field (508–610 mm, 20–24 in)

This Chinese species is associated with thick jungle and ravine vegetation, and is very difficult to observe. Elliot's pheasants have the same shape as the common pheasant, but the long tail is strongly barred with white and brown in males and the sides of the neck are white to grayish white. Vocalizations have not been well described, but during the breeding season the males perform wing-whirring and follow this with a loud, repeated *geke* call. Females have relatively short and blunt-tipped tails, which are tipped with white. They also are distinctly ashy gray on the sides of the neck and in the scapular area.

In the Hand

Males are readily identified by their grayish white necks and their boldly barred tail pattern, with alternating brown, black, and grayish white barring. The lower flanks and underparts are also uniquely white. Females are the only long-tailed (170–195 mm, 6.6–7.6 in) pheasants to exhibit a pure white abdomen and grayish color on the sides and back of the neck. The front of the throat is black.

Ecology

Habitats and Population Densities

Natural habitats of Elliot's pheasants are shrubby areas at moderate elevations in mountains or valleys and dense bamboo thickets and undergrowth of sparse coniferous forests (Cheng et al., 1978). Habitats used in the Leigong Nature Reserve include a variety of evergreen broadleaf, mixed evergreen and deciduous broadleaf forest, deciduous broadleaf forest, and mixed coniferous and broadleaf forest types. These habitats mostly occur between 1,000 and 1,900 m (3,280 and 6,235 ft) elevation and usually have slopes of less than 50° (Ding et al., 1996). Estimated densities in good habitats range from 3.6 to 6.9 individuals per square kilometer (9.3 to 17.9 per square mile; Li, 1996).

Competitors and Predators

Beebe (1918–1922) reported a "half-hearted battle" between an Elliot's and male ring-necked pheasant, but otherwise could learn nothing of this species' associations with any other bird or mammal.

General Biology

Food and Foraging Behavior

In the wild, Elliot's pheasants have been reported feeding on seed pods, seeds, berries, and various kinds of leaves, with a few remains of ants being the only trace of animal foods found in the specimens that Beebe (1918–1922) collected. In captivity the birds seem to do well on the normal pheasant diets (Howman, 1979).

Movements and Migrations

Nothing is known of this in the wild, although Elliot's pheasants are said to be relatively poor fliers and tend to escape on foot. They are said to remain on particular slopes throughout the entire year (Beebe, 1918–1922).

Daily Activities and Sociality

Elliot's pheasants seem to be much like typical *Phasianus* in their normal daily activities. Foraging is done during the daytime hours, probably in the usual morning and late-afternoon manner, and apparently in pairs or at most in small family parties.

In captivity Elliot's pheasants exhibit no indications of social interplay, even when kept in large flocks. Instead they seem to operate as solitary and individual birds, with very little vocal communication (Knoder, 1983).

Social Behavior

Mating System and Territoriality

Nothing is known of this in the wild, but in captivity the most effective sex ratios for high fertility seems to be one or two females per male. With sex ratios of three to five females, fertility was found to drop as much as 24 percent (Knoder, 1983). Presumably Elliot's pheasants are also polygynous in the wild.

There is no clear information on territorial sizes, but in small pens male Elliot's pheasants regularly fight, with one bird eventually killing all the others. Even in pens as large as about 46 × 61 m (150 × 200 ft) the dominant male was found to require a great deal of room in relation to the pen size (Knoder, 1983).

Voice and Display

The vocalizations of this species seem to be relatively uncomplicated and consist mostly of low clucks and chuckles. Both sexes utter a shrill but not loud squeal. This call is used by males during the breeding season as an apparent threat to males in nearby pens. The male's wing-whirring display is a rapid but relatively silent display that is nearly identical to that of the Reeves' pheasant, the firebacks, and the kalij pheasant (Knoder, 1983). This display has been illustrated by Steinbacher (1941; see figure 46). This wing-whirring is commonly followed by a repeated *geke*

call (Cheng et al., 1978). The usual male display to females is lateral, with the neck extended, the body feathers ruffled, and the tail spread and tilted toward the female (Wayre, 1969). In this species, as well as in such relatives as the mikado and the bar-tailed pheasants, the under tail-coverts of one or both sexes may function as a warning or following signal (Davison, 1976). Tidbitting behavior and calling is has been only poorly described. However, Beebe (1918–1922) reported that a male will often call to the hen in a low voice when picking up grain and then spread his tail and flatten his plumage while swelling his wattles as she approaches.

Reproductive Biology

Breeding Season and Nesting

Nothing is known of this species' breeding season in the wild. However, in Ohio Elliot's pheasants were observed to begin laying between 10 and 20 March and reach a peak about the middle of April (Knoder, 1983). By the end of May nearly all egg-laying had terminated, although a few eggs were laid as late as 10 June. On average, eggs were laid every second day. Over the course of a single season females were found to lay an average of 14.3 eggs, but with a maximum of 61 eggs laid by two females in a single pen (Knoder, 1983). Typical clutch sizes in captivity are from six to eight eggs.

Incubation and Brooding

Presumably only the female is involved with incubation and tending the young, although this is not known for free-living birds. Under artificial conditions the incubation period requires 26 days. Knoder (1983) reported a hatchability of 68 percent for more than 2,500 fertile eggs, and a fertility rate of about 52 percent for nearly 5,000 eggs.

Growth and Development of the Young

Elliot's pheasant chicks are relatively easy to raise, and they attain adult plumage at about 16–18 weeks of age (Knoder, 1983). At the San Diego Zoo, 15 newly hatched chicks weighed an average of 21.5 g (0.8 oz), and when 1 month old averaged 127 g (4.4 oz). Four juvenile males averaged 688 g (24.1 oz) at 90 days, and two juvenile females averaged 525 g (18.4 oz) at the same age (David Rimlinger, pers. comm.).

Figure 46. Display postures of male Elliot's pheasant, including normal standing posture (A), facial engorgement (B), lateral display to female (C), and wing-whirring (D). After various sources, including Schenkel (1956–1958).

Evolutionary History and Relationships

The Elliot's pheasant is clearly a very close relative of the bar-tailed pheasant and shows less obvious similarities to the copper pheasant. However, a fairly recent separation of bar-tailed and Elliot's pheasant ancestral stock can be readily visualized in the eastern Himalayas, perhaps during early Pleistocene times, as temperate or montane forest areas were being disrupted by climatic changes.

Status and Conservation Outlook

The Elliot's pheasant is currently considered endangered (King, 1981) and is believed to be becoming increasingly rare in the wild as a result of habitat de-struction. There is no good recent information on its status in the wild, but apparently wild-trapped birds are still fairly common in Chinese zoos. Several nature sanctuaries have been established within the historic range of the Elliot's pheasant, although its actual status in these remains unknown (Knoder, 1983). Efforts are currently under way to census and build up captive populations of this species. It would seem that this species can be rather readily bred in large numbers under captive conditions, with low fertility one of the few problems associated with this approach, according to Knoder. McGowan and Garson (1995) listed the Elliot's pheasant as endangered; Collar et al. (1994) considered it vulnerable.

MIKADO PHEASANT

Syrmaticus mikado (Ogilvie-Grant) 1906

Other Vernacular Names

None in general English use; faisan mikado (French); Mikadofasan (German).

Distribution of Species

Limited to the central mountains of Taiwan (Formosa), between 1,830 and 3,050 m (6,000 and 10,000 ft; Delacour, 1977). See map 13, page 246.

Distribution of Subspecies

None recognized by Delacour (1977).

Measurements

Delacour (1977) reported that males have wing lengths of 210–230 mm (8.2–9.0 in) and tail lengths of 490–530 mm (19.1–20.7 in), whereas females have wing lengths of 187–215 mm (7.3–8.4 in) and tail lengths of 172–225 mm (6.7–8.8 in). Two males averaged 1,300 g (2.8 lb) and two females averaged 1,015 g (2.2 lb; various zoo records). The eggs average 55 × 39 mm (2.1 × 1.5 in) and have an estimated fresh weight of 46.2 g (1.6 oz). Cara Lin Bridgman (1994, pers. comm.) reported that 21 females ranged in weight from 600 to 1,100 g (1.3 to 2.4 lb; average of 4 adults, 700 g [1.5 lb]) and 26 males ranged from 610 to 1,250 g (1.3 to 2.7 lb; average of 7 adults, 905 g [2.0 lb]).

Identification

In the Field (508–813 mm, 20–32 in)

Limited to the island of Taiwan, where it is very rare, the mikado pheasant is unmistakable owing to the male's very long and barred tail and otherwise generally blackish coloration. Females also have a pointed but shorter tail, which is barred with black, brown, and white, and otherwise are generally olive brown above and mottled with white below. Mikado pheasants are found in small groups and are normally quite silent, although during the breeding season the male utters a shrill and drawn-out whistle, *chiri*. Repeated chuckling sounds have also been reported. The females of Swinhoe's and mikado pheasants might be confused, but are readily separated by the red legs of the former and the gray legs of the latter. The Swin-

hoe's sometimes breeds as high as 2,300 m (7,545 ft) elevation, or well within the altitudinal limits of the mikado, which ranges from 1,800 to 3,200 m (5,905 to 10,500 ft; Cara Lin Bridgman, pers. comm.).

In the Hand

The purplish black plumage of the male, which is interrupted on the tail and rump with narrow white barring, is distinctive. Females resemble those of other *Syrmaticus* species, especially the bar-tailed pheasant, but are darker and more olive brown, with white markings more extensive on the mantle and the black and chestnut barring on the tail feathers more conspicuous.

Ecology

Habitats and Population Densities

The mikado pheasant is found at elevations from 1,600 to 3,300 m (5,250 to 10,830 ft) in thick forests having a dense undergrowth of rhododendrons and bamboo. However, it is adaptable to a variety of primary and secondary forest habitats (Severinghaus, 1977, 1978). In much of its range the cliffs are precipitous and relatively inaccessible (Wayre, 1969); thus, the extent of lumbering deforestation has been limited. Evidently the birds prefer areas with slopes greater than 40° and a dense undergrowth of bamboos, other grasses, and ferns (Severinghaus, 1977).

There are few detailed estimates of mikado pheasant population densities. However, Poltack (1972) stated that loggers in one area estimated that about 200 birds in one forest area of 7.8 km² (3.0 mi²), which suggests a very high density of 25.6 birds per square kilometer (66.7 per square mile). Other density estimates range from 18 to 32 individuals per square kilometer (46.6 to 82.9 per square mile; Li, 1996). If an estimate of 10,000 birds for Yushan National Park (del Hoyo et al., 1994) is realistic, the overall density there would be about 15 birds per square kilometer (38.9 per square mile).

Yearly home ranges for two adult males were 7.4 and 1.9 ha (18.3 and 4.7 acres), one juvenile male's range was 6.5 ha (16.1 acres), and a female's range encompassed that of the two males (Bridgman, 1994).

Her range was largest (13.1 ha, 32.4 acres) in spring, smallest (1.6 ha, 4.0 acres) during winter, and intermediate (8.8 ha, 21.7 acres) during fall (Bridgman et al., 1997). Mikado pheasants apparently undertake no apparent altitudinal movements because home ranges are usually within 1 km (0.6 mi) in diameter. Longevity records are limited, but one banded female has survived at least 6.5 years and two males at least 6 years (Cara Lin Bridgman, pers. comm.).

Competitors and Predators
The mikado pheasant occurs with the Swinhoe's pheasant and the common pheasant in Taiwan, but probably has very little contact with either, especially the latter.

There is little information on possible predators. However, these probably consist of large, forest-adapted raptors. W. Goodfellow, who first discovered the species, judged that martens were probably its major enemy (Beebe, 1918–1922).

General Biology

Food and Foraging Behavior
There is no detailed information on this from wild birds. However, in captivity mikado pheasants seem to be rather typical pheasants, in spite of a suggestion that they require considerable amounts of green food in their diet (Howman, 1979). In the wild, the birds are most often observed in April and July, when the wild fruits and seeds upon which they feed are ripe (Wayre, 1969). These foods were said to include strawberries, asters, and ferns, whereas during the rest of the year the species is said to feed on insects as well as on the shoots and buds of plants. Observations on wild mikado pheasants indicate that the birds eat nearly anything, including at least 30 species of plants in nearly as many families, as well as various insects, including beetles (Cara Lin Bridgman, pers. comm.).

Movements and Migrations
There is no information on this. However, other than possible altitudinal movements with the seasons, movements are probably very limited in the mikado pheasant.

Daily Activities and Sociality
Mikado pheasants occur in areas of locally heavy rainfall. During such rainfall they evidently perch in trees, rather than coming down to the ground to feed. At such times as many as nine birds have been seen perching in a single group (Poltack, 1972). Otherwise, mikado pheasants feed in the morning and evening, as do most other pheasants (Delacour, 1977).

Social Behavior

Mating System and Territoriality
At least in captivity, mikado pheasants are polygynous, with two or three females often being fertilized by a single male. Severinghaus (1977) believed that wild birds might be monogamous. There is no information available on territorial behavior in these birds.

Voice and Display
Mikado pheasants are said to be rather silent, except for a shrill whistle that the male utters during the breeding season and a quiet alarm note (Wayre, 1969). The displays of the male have been described by Sahin (1984), who noted that lateral (waltzing) display is present in both sexes and occurs more or less throughout the year. The male's major courtship display consists of an erect frontal display, with the tail held downward rather than raised as in the bar-tailed pheasant. This display was observed only during the mating period and only in males (figure 47). Wing-whirring is also present and appears to function in territorial behavior. Tidbitting, which involves actual food presentation, is also a part of courtship behavior. Sahin and Thomas (1988) suggested that because wing-whirring occurs throughout the year, it is largely territorial in function rather than being mainly sexual. Wing-whirring reaches a peak among males in spring, but also exhibits a second peak during autumn. It also is performed by females, essentially throughout the year. Similarly, both sexes perform lateral display throughout the year, with this posture serving primarily as a hostile signal rather than as a sexual display (Sahin, 1986).

Reproductive Biology

Breeding Season and Nesting
The eggs of the mikado pheasant are laid from the end of February until May, according to Delacour (1977), who presumably was referring to captive birds in France. However, he also stated that mikado pheasants often lay three clutches of eggs between the end

Figure 47. Display postures of the male mikado pheasant, including lateral display (A, B), frontal display (C), and wing-whirring (D). After sketches and photos in Sahin (1984) and Sahin and Thomas (1988).

of March and the middle of July. Beebe (1918–1922) quoted W. Goodfellow, who judged that in Taiwan the birds nest about the end of April and early May, based on the breeding condition of females that he collected.

Few nests have been found in the wild, but Wayre (1969) noted that a forest department assistant in Taiwan told him that he had seen three or four nests. All of them had been constructed of bamboo stalks situated on the trunk or in the branches of a fallen tree, about 1 m (3–4 ft) aboveground. Wayre later tested this idea with captive birds; he found that two females that were given their choice of ground or elevated nesting boxes chose the latter. It is possible that such elevated sites are adaptive in this area of heavy rainfall, which might tend to flood ground nests.

Recent studies by Yao Cheng-te and Cara Lin Bridgman (pers. comm.) have resulted in five more nests being found. All of them have been ground nests that were either hidden under tree roots or by overhanging rocks. One nest had a tunnel-like entrance. The clutches ranged up to five eggs, but most contained two or three. Earlier reports suggest that the usual clutch contains five to eight or perhaps ten eggs, which seems a more typical range for birds of this genus.

Incubation and Brooding
Incubation in the mikado pheasant requires 28 days, and at least in captivity the normal clutch is of about five eggs. There is as yet no indication that the

male participates in nest defense or caring for the young.

Growth and Development of the Young

When hatched, mikado pheasant chicks are relatively large and, like tragopan chicks, are soon able to fly (Wayre, 1969). This precocity is probably adaptive in the forested habitats where the birds are normally found.

Among 22 chicks hatched at the San Diego Zoo, the average weight at hatching was 30.5 g (1.1 oz), and at 1 month of age the chicks averaged 112 g (3.9 oz). Ten males weighed an average of 343 g (12.0 oz) at 60 days, whereas three females averaged 335 g (11.7 oz) at the same age (David Rimlinger, pers. comm.). The males attain their adult plumage in the first year, and initial breeding also occurs at that time. Occasionally, however, the eggs of first-year females are infertile, and better breeding results occur with 2-year-old birds. In captivity, two or three females can be maintained with a single adult male, according to Delacour (1977).

Evolutionary History and Relationships

Presumably the mikado pheasant has been isolated from the other species of *Syrmaticus* since at least the Pleistocene, because Taiwan is located nearly 200 km (125 mi) from mainland China, or well beyond normal pheasant dispersal abilities. Its nearest congeneric relative, zoogeographically at least, is the Elliot's pheasant. However, in general the mikado pheasant appears to be a highly melanized version of the bar-tailed pheasant. It seems likely that all three evolved from a common ancestral type, with the mikado pheasant locally adapting to a deeply forested habitat in which generally dark plumage would be distinctly advantageous. Hybrids between mikado and Elliot's pheasants exhibit fertility only among males, which suggests a fairly prolonged period of genetic separation between them (Delacour, 1977).

Status and Conservation Outlook

The mikado pheasant is currently considered vulnerable, with an estimate of the population in 1975 as perhaps numbering a few thousand individuals (King, 1981). Severinghaus (1978) spent 2 years studying this species in the early 1970s, and made several conservation proposals in 1974 for the mikado and Swinhoe's pheasants. However, these proposals have not yet been followed, and he believes that increased disturbance to the remaining mikado pheasant habitat is likely to occur. Logging has been stopped in one proposed reserve area, and mikado pheasants are also known to occur in the proposed Yu Shan National Park in central Taiwan.

The species was considered secure by McGowan and Garson (1995) and as near-threatened by Collar et al. (1994). Yushan National Park in central Taiwan (64,000 ha, 158,000 acres) is believed to hold about 10,000 birds, or an overall density of one bird per 6.4 ha (15.8 acres), which makes the species somewhat more common than the Swinhoe's pheasant in the same region. Both species are currently protected in three national parks (Yushan, Taroko, and Sheipa) and in several nature reserves. Some very limited illegal hunting probably still occurs, although both of Taiwan's pheasants are now federally protected (Cara Lin Bridgman, pers. comm.).

COPPER PHEASANT

Syrmaticus soemmerringi (Temminck) 1893

Other Vernacular Names
Soemmerring's pheasant; faisan scintillant, faisan de Soemmerring (French); Kuperfasan (German); yamadori (Japanese).

Distribution of Species
Japan in Honshu, Shikoku, and Kyushu. This species is sedentary and inhabits coniferous forest, especially of *Cryptomeria* and cypress, and adjoining mixed forest with dense undergrowth and grassy hillsides in rough and broken country in mountainous regions, but at elevations below 1,370 m (4,500 ft), and in suitable hills near the sea (Vaurie, 1965). See map 14.

Distribution of Subspecies
Syrmaticus soemmerringi scintillans (Gould): scintillating copper pheasant. Northern and central Honshu, north of 35°10′ N; intergrades with *intermedius* and *subrufus*.

Syrmaticus soemmerringi intermedius (Kuroda): Shikoku copper pheasant. Southwestern Honshu and Shikoku; intergrades with *subrufous* in southwestern Shikoku. This form is doubtfully distinct from *scintillans*.

Syrmaticus soemmerringi subrufus (Kuroda): Pacific copper pheasant. Southeastern Honshu, south of *scintillans*; also in southwestern Honshu and southwestern Shikoku, according to Yamashina (1976).

Syrmaticus soemmerringi soemmerringi (Temminck): Soemmerring's copper pheasant. Northern and central Kyushu; intergrades with *ijimae* in the south.

Syrmaticus soemmerringi ijimae (Dresser): Ijima's copper pheasant. Southeastern Kyushu.

Measurements
Wing and tail lengths of the subspecies are given in table 26. Few weights are available, but one adult of unstated sex weighed 907 g (2.0 lb; Stephen Wylie, pers. comm.). The eggs of *scintillans* average 47.7 × 34.9 mm (1.9 × 1.4 in) and have an estimated fresh weight of 32 g (1.1 oz).

Identification

In the Field (508–1,219 mm, 20–48 in)
Limited to Japan, the copper pheasant could only be confused with the other native pheasant, the green pheasant. The coppery color of the male and its very long, brown tail easily allows for separation. Females have shorter, white-tipped tails with little evident barring except for a black band near the tip. Vocalizations of the copper pheasant are only poorly described, but wing-whirring is performed by males during the breeding season.

In the Hand
The entirely coppery brown color of the male and the narrow black barring on the tail are distinctive. Females rather closely resemble those of the bar-tailed and mikado pheasants, but lack arrow-shaped white shaft-streaks on the upper back and are more cinnamon to rufous-colored in general tone.

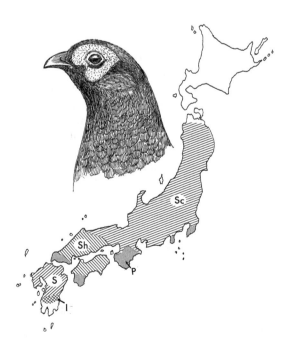

Map 14. Distribution of Ijma's (I), Pacific (P), Soemmerring's (S), Shikoku (Sh), and scintillating (Sc) races of the copper pheasant.

Table 26
Ranges of wing and tail lengths of copper pheasants[a]

	Males		Females	
	Wing	Tail	Wing	Tail
scintillans	205–230, 8.0–9.0	484–752, 18.9–29.3	197–219, 7.7–8.5	164–196, 6.4–7.6
intermedius	207–228, 8.1–8.9	676–845, 26.4–33.0	192–205, 7.5–8.0	155–193, 6.0–7.5
subrufus	205–220, 8.0–8.6	635–815, 24.8–31.8	192–205, 7.5–8.0	164–192, 6.4–7.5
soemmerringi	210–224, 8.2–8.7	655–978, 25.5–38.1	195–220, 7.6–8.6	175–200, 6.8–7.8
ijimae	205–235, 8.0–9.2	640–880, 25.0–34.3	195–216, 7.6–8.4	144–190, 5.6–7.4
Overall	205–235, 8.0–9.2	484–978, 18.9–38.1	192–220, 7.5–8.6	144–200, 5.6–7.8

[a] Data from Delacour (1977), supplemented by personal observations. All measurements are reported in millimeters, followed by the equivalent in inches.

Geographic Variation

Geographic variation is considerable and generally clinal, with southern populations becoming darker and more richly colored, except that the race found in southern Kyushu (*ijimae*) has an almost entirely white lower back and rump, while otherwise being the darkest and most richly colored race. There is a great deal of individual and local variation, and at least two of the subspecies might be considered synonymous when subjected to further study (*intermedius* with *scintillans* and *subrufus* with *soemmerringi*; Wayre, 1964). There are many individuals that exhibit transitional characteristics between the subspecies. Yamashina (1976) has related the origins and distributions of three of the subspecies (*scintillans*, *subrufus*, and *soemmerringi*) to variations in winter temperatures and associated snowfall.

Ecology

Habitats and Population Densities

Yamashina (1976) stated that all the subspecies of the copper pheasant inhabit wooded country and are associated with sloping ground having tall tree cover. In the north, this includes many coniferous forests of cypress and pines, but also sometimes deciduous forests of oaks and beeches. Farther south, the birds typically inhabit broadleaf forests of various types (*Quercus, Castanea, Castanopsis, Machilus*). When in pine woods, copper pheasants especially favor those with well-developed undergrowths of *Lespedeza*, *Cleyera*, various ferns, and several kinds of thorny shrubs. There are no estimates of population densities.

Competitors and Predators

Yamashina (1976) noted that in general the copper pheasant is found in different habitats from the green pheasant, which is associated with grassy plains, farmlands, and light woodlands at low elevations. The two species only uncommonly mix. Beebe (1918–1922) mentioned that foxes and weasels are presumptive enemies of copper pheasants.

General Biology

Food and Foraging Behavior

Copper pheasants feed primarily on acorns and mast of such forest trees as *Castanopsis, Machilus, Cleyera*, and *Castanea*, which are gathered from the forest floor. They sometimes also eat the fruits of *Cleyera, Lespedeza*, and thorny shrubs by flying or hopping up into the trees and plucking them. Animal foods include many kinds of insects, earthworms, and crabs. Young birds exist primarily on animal materials (Ogasawara, 1969; Yamashina, 1976). Beebe (1918–1922) observed that copper pheasants he had collected or examined contained acorns in their crops as well as grubs and the chrysalids of lepidopterans.

Movements and Migrations

There is no specific information on this subject. Beebe (1918–1922) stated that in northern Honshu copper pheasants are forced to descend to lower altitudes during winter, and at that time may outnumber the green pheasant in such regions.

Daily Activities and Sociality

According to Yamashina (1976), copper pheasants are relatively solitary birds, and only seldom form flocks. Beebe (1918–1922) noted seeing a flock of two males and four females, apparently in midwinter. The hen and her young are also said to remain together through the winter. Roosting is done in trees, often on steep slopes, and frequently in pines. One roosting tree observed by Beebe was used by a male and three females or young birds, which roosted no more than 5 m (15 ft) aboveground. Endo (1982) has provided a recent overall summary of the species' general biology.

Social Behavior

Mating System and Territoriality

Although it has not been proven, Yamashina (1976) believed that the copper pheasant is polygynous in the wild. However, Beebe (1918–1922) judged the birds to be monogamous as often as they might be polygynous and said that he had heard reports of males associating with a hen and her brood, which suggests monogamy.

Territory sizes are as yet unestimated. However, Yamashina (1976) reported that the males are highly territorial in spring, with a single male monopolizing a "whole valley." Judging from this account, ridges evidently may form natural boundaries to territories.

Voice and Display

At the start of the breeding season, male copper pheasants begin their wing-whirring display (figure 48). This produces a loud drumming noise that is audible for a considerable distance. Crowing also occurs during the mating season during morning and evening hours (Beebe, 1918–1922). However, Delacour (1977) reported that wing-whirring occurs without the calling. Wattles of the male can be engorged to some degree, but apparently not to the extent typical of *Phasianus*.

Almost nothing has been written of the postural displays of the copper pheasant, which presumably are much like those of the bar-tailed pheasant. This is the only species of *Syrmaticus* that lacks iridescent coloration in males, and it exhibits less sexual dimorphism in plumage coloration than the others.

Reproductive Biology

Breeding Season and Nesting

Delacour (1977) stated that the natural breeding season of the copper pheasant is from the end of March to the beginning of July. Yamashina (1976) reported that the breeding season begins in April or May in the colder areas and about March in more southerly regions. Under captive conditions, the peak of the laying period appears to be from late April to mid-May, with the eggs being laid in late afternoon or evening. In the wild nests are made on the ground, usually beneath a fallen tree, and clutches consist of 7–13 eggs. In captivity females lay from their first year until they are at least 7 years old. However, peak numbers of eggs come from birds between 3 and 6 years old, when from 21 to 31 eggs per year are typical. First-year birds lay only an average of 10 eggs per season, whereas birds from 2 to 7 years old average about 20 eggs.

Incubation and Brooding

Incubation in the copper pheasant requires 24–25 days. The role of the male in defending the nest, if any, is unknown.

Growth and Development of the Young

The artificial rearing of the young is no different from that of the other typical pheasants, according to Yamashina (1976), although he noted that they are highly vulnerable to disease if allowed contact with other poultry. Adult plumage and sexual maturity is attained during the first year.

Evolutionary History and Relationships

The copper pheasant is perhaps the least typical of the *Syrmaticus* species and has perhaps been isolated from the others for the longest period. Some authorities (Yamashina, 1976) consider it to be within the genus *Phasianus*. Although the copper pheasant has been hybridized with the green pheasant in captivity (Hachisuka, 1953) and perhaps also in the wild (Yamashina, 1976; Delacour, 1977), these hybrids reportedly have been infertile. However, hybrids between the copper pheasant and Reeves' pheasant are said to be fertile in the case of males.

Status and Conservation Outlook

The copper pheasant was considered near-threatened by Collar et al. (1994), but as insufficiently known

Figure 48. Display postures of male copper pheasant, including wing-whirring (A), normal posture (B), and facial engorgement (C). After Yamashina (1976) and photos by the author.

by McGowan and Garson (1995). Up to 1 million copper pheasants were once legally shot each year in Japan, and even until 1975 up to 500,000 were shot annually. The birds are now distinctly scarce throughout Japan, but to some extent are supplemented by release of captive-raised stock (Brazil, 1991). Since 1976 it has been illegal to kill female copper pheasants because habitat destruction and greater hunting pressures seem to have been causing population decreases in recent years. As of 1976 there were 2,706 sanctuaries in Japan, covering a total area of 2.6 million hectares (6.4 million acres), which are mainly in the woodlands of mountainous districts. Therefore, many of these areas certainly are protecting this species directly (Yamashina, 1976).

Genus *Phasianus* Linné 1758

The true pheasants are medium-sized, temperate, open-country pheasants in which sexual dimorphism is well developed. The tail is greatly elongated and strongly barred, the rectrices have fringed edges, and erectile ear-tufts are present. The orbital wattles are red, moderate in size, and smooth, with scattered black plumules present. Males have disintegrated rump feathers forming a hairlike texture. The tail is highly graduated, flat to slightly compressed, and has 18 rectrices that are only slightly spread during display. The tail molt is phasianine (centripetal). The wing is rounded, with the tenth primary much longer than the first and about equal to the third. The tarsus is longer than the middle toe and claw and is spurred in males. Males are extensively iridescent over most of their plumage except the abdomen, wing, and tail; the head is strongly metallic colored. Females are rather uniformly brownish and lack white spots or shaft-streaks. Two species are recognized.

Key to Species (and Subspecies of Males) of *Phasianus*[1]

A. Plumage mostly dull brown (females).
    B. Feathers of the mantle spotted with black; underparts distinctly mottled: green pheasant.
    BB. Feathers of the mantle spotted with brown; underparts not mottled: common pheasant.
AA. Plumage mostly iridescent (males).
    B. Underparts entirely greenish: green pheasant.
        C. Smaller (wing 215–225 mm [8.4–8.8 in]) and generally lighter throughout: northern green pheasant (*robustipes*).
        CC. Larger (wing 225–245 mm [8.8–9.6 in]) and generally darker throughout.

[1]Many of these forms represent questionable subspecies, and thus any key is considerably subjective.

D. Back and rump mostly greenish: southern green pheasant (*versicolor*).

DD. Back and rump grayer and bluer: Pacific green pheasant (*tanensis*).

BB. Underparts always with some coppery red: common pheasant.

   C. Lower back, rump, and upper tail-coverts maroon or rufous, sometimes glossed with green; black tail-bars usually narrow.

      D. Without a white neck-ring or only traces present.

         E. Wing-coverts reddish or sandy brown; more purplish throughout (black-necked pheasant group).

            F. Sides and mantle purplish orange; wing-coverts darker brown.

               G. Middle of breast and sides of belly dark purplish green.

                  H. General color more coppery red: southern Caucasian pheasant (*colchicus*).

                  HH. General color paler, more golden orange: northern Caucasian pheasant (*septentrionalis*).

               GG. Middle of breast purplish red bronze: Talich Caucasian pheasant (*talischensis*).

            FF. Sides and mantle golden orange; wing-coverts sandy buff: Persian pheasant (*persicus*).

         EE. Wing-coverts white or nearly so; more reddish throughout (white-winged pheasant group).

            F. Middle of breast and sides of belly dark green.

               G. Flanks narrowly tipped with black; black tail-bars broader: Yarkand pheasant (*shawi*).

               GG. Flanks broadly tipped with black; black tail-bars narrower.

                  H. Chest and breast lighter: Khivan pheasant (*chrysomelas*).

                  HH. Chest and breast darker: Bianchi's pheasant (*bianchii*).

            FF. Middle of breast and sides of belly coppery red.

               G. Chest and upper breast feathers with narrow black edges; center of belly dark brownish: Persian pheasant (*persicus;* of previous group).

               GG. Only the sides of chest and breast edged with black; center of belly light chestnut.

                  H. White collar incomplete and very narrow.

                     I. Flanks spotted with black and purplish blue: Zerafshan pheasant (*zerafschanicus*).

                     II. Flanks spotted with blackish green: Zarudny's pheasant (*zarudnyi*).

                  HH. Collar absent or present only as white spots: Prince of Wales' pheasant (*principalis*).

      DD. With a wide white neck-ring (Kirghiz pheasant group).

         E. Chest glossed with green; narrower white neck-ring: Kirghiz pheasant (*mongolicus*).

         EE. Chest glossed with purple; wider white neck-ring: Syr Daria pheasant (*turcestanicus*).

CC. General color of the lower back, rump, and upper tail-coverts greenish to slate or olive green.
    D. Lower back and rump mostly olive green; tail less heavily barred; no white neck-ring: olive-rumped pheasant (*tarimensis*).
    DD. Lower back and rump grayish to bluish green; tail more heavily barred; usually with white neck-ring (gray-rumped pheasant group).
        E. No white neck-ring or the ring about one-third incomplete.
            F. Dark green of the neck extending to the middle of the chest.
                G. Flanks golden buff; mantle bright sandy red: Zaidan pheasant (*vlangalii*).
                GG. Flanks coppery maroon.
                    H. Mantle bright chestnut; sides more purplish: Stone's pheasant (*elegans*).
                    HH. Mantle and sides more golden: Rothschild's pheasant (*rothschildi*).
            FF. Dark green of the neck bordered by yellow or coppery red of the chest.
                G. Chest and breast feathers broadly margined with black; flanks buffy.
                    H. With a narrow neck-ring and sometimes white eyebrows.
                        I. White eyebrows absent: Sohokoto pheasant (*sohokotensis*).
                        II. Eyebrows present.
                            J. Eyebrows dingy; neck-ring more complete: Shansi pheasant (*kiangsuensis*).
                            JJ. Eyebrows white; neck-ring less complete: Alashan pheasant (*alaschanicus*).
                    HH. No white eyebrows or neck-ring: Kweichow pheasant (*decollatus*).
                GG. Chest and breast feathers narrowly margined with black; flanks darker: Strauch's pheasant (*strauchi*).
        EE. White neck-ring present and complete or nearly so.
            F. Neck-ring narrow and nearly or wholly interrupted in front.
                G. Chest and breast feathers margined with blackish green; general color very pale.
                    H. Scapulars margined with sandy brown.
                        I. Mantle pale golden, with white feather centers: Satchu pheasant (*satscheuensis*).
                        II. Mantle bright golden; no white feather centers: Gobi pheasant (*edzinensis*).
                    HH. Scapulars margined with red maroon: Taiwan pheasant (*formosanus*).
                GG. Chest and breast feathers not margined or only barely margined with blackish green.
                  H. Mantle light golden; scapulars chestnut: Chinese pheasant (*torquatus*).

        HH. Mantle darker golden; scapulars maroon: Tonkinese pheasant (*takatukasae*).

  FF. Neck-ring complete and broad, even in front.

      G. Black patch under eye with a white spot: Manchurian pheasant (*pallasi*).

      GG. Black patch under eye, usually without a white spot.

          H.   General color very pale: Kobdo pheasant (*hagenbecki*).

          HH. General color very dark: Korean pheasant (*karpowi*).

COMMON PHEASANT

Phasianus colchicus Linné 1758

Other Vernacular Names
Black-necked pheasant, Mongolian pheasant, ring-necked pheasant; faisan de Colchide (French); Edelfasan, Jagdfasan (German).

Distribution of Species
Delta of the Volga River, northern Caucasus, and Transcaucasia eastward through northern Iran to Afghan Turkestan, and from Transcaspia and Russian Turkestan eastward through Chinese Turkestan and Mongolia to the Amur Valley and Ussuriland (but locally distributed only in most of these regions), southward through Manchuria, Korea, and China including Taiwan to northern Burma and Tonkin. This species is sedentary, but some migratory movements have been reported in *pallasi*. Introduced in Europe, North America, Hawaii, Japan, Australia, and New Zealand. Inhabits open country in park or farmland, scrubby wastes, along the edges of woods, grassy steppes, hills, and lower mountain slopes with relatively sparse oaks, chestnuts, or pines; also oases of deserts and along rivers and lakes in reed beds or riverside thickets of tamarisks, poplars, wild olives, or of other trees and shrubs; also swamps, edges of rice fields, and in open forest in the mountains and their valleys in western China, where it ascends to 3,965 m (13,000 ft; Vaurie, 1965). See maps 15 and 16.

Distribution of Subspecies (excluding introductions; after Vaurie, 1965; Wayre, 1969)

Black-necked Pheasants (the nominate colchicus group)
Phasianus colchicus colchicus L.: southern Caucasian pheasant. Transcaucasia from Abkhazya east to Georgia (north to Kakhetia and Korelia) to the region of Telavi, northeastern Azerbaijan in the regions between Zakataly and Ismailly, and between Kuba and Khachmas and southern Armenia, possibly also in northwestern Iran. The birds found in Turkey and extreme southeastern Europe are identical with those of Transcaucasia and may be indigenous or may have been introduced. This form has been introduced widely in Europe and elsewhere; birds from Turkey, Bulgaria, and Thrace may be either introduced or native.

Phasianus colchicus septentrionalis Lorenz: northern Caucasian pheasant. Northern Caucasus from Dagestan north to the delta of the Volga River; western limits unknown.

Phasianus colchicus talischensis Lorenz: Talisch Caucasian pheasant. Southeastern Transcaucasia (in Talych north to the lower Kura River), eastward along the southern Caspian districts of Iran to the Babol Sar and Sari in eastern Mazanderan, where it intergrades with *persicus*. Intergrades with nominate *colchicus* in the west.

Phasianus colchicus persicus Severtzov: Persian pheasant. Southwestern Transcaspian region in the valleys of the Atrek and Gurgen Rivers and their tributaries, southeastern coast of the Caspian Sea, west to Ashurada.

White-winged Pheasants (the principalis–chrysomelis group)
Phasianus colchicus principalis P. L. Sclater: Prince of Wales' pheasant. Southern Russian Turkestan and northern Afghanistan, chiefly in the Tejend Valley, west to 80 km (50 mi) east of Askhabad, north to the Kara Kum Desert, south to the Murghab Valley and the oases of Merv, Yelotan, and Panj-deh.

Phasianus colchicus zarudnyi Buturlin: Zarudny's pheasant. Valleys of the middle Amu Daria from about Kerki down to Darganata.

Phasianus colchicus bianchii Buturlin: Bianchi's pheasant. Upper Amu Daria Valley between the Hissar, Alai, Pamir, and Hindukush Mountains in the Bukhara Province.

Phasianus colchicus chrysomelas Severtzov: Khivan pheasant. Delta of the Amu Daria, along the southern shore of the Aral Sea, west to the Ust-wits Plateau and south to the Kara Kum Desert.

Phasianus colchicus zerafschanicus Tarnovski: Zerafshan pheasant. Zerafshan Valley in Bukhara, north to the Zerafshan–Syr Daria Divide, south to the Hissar Mountains.

Phasianus colchicus shawi Elliot: Yarkand pheasant. Sinkiang in the western part of the Tarim Basin from the Khotan Darya west to Yarkand and Kashgar and east from here to Maralbashi, the lower River Aqsu

Map 15. Distribution of Alashan (Al), Bianchi's (B), Chinese (C), Gobi (G), Khivan (Kh), Kirghiz (Ki), Kobdo (Ko), Korean (K), Kweichow (Kw), Manchurian (M), northern Caucasian (N), Persian (Pe), Prince of Wales (PW), Roths-child's (R), Satchu (Sa), Shansi (Sh), Sohokhoto (Soh), southern Caucasian (So), Stone's (St), Strauch's (S), Sungpan (Su), Syr Daria (Sy), Taiwan (T), Talish Caucasian (Tal), Tarim (Ta), Tonkin (To), Yarkan (Y), Zaidan (Zi), Zarudny's (Z), and Zerafshan (Ze) races of the common pheasant. The introduced populations in Europe and Japan (cross-hatched) are of varied racial origins.

and the upper River Tarim. Considered part of Tarim group by Vaurie (1965); intergrades locally with *tarimensis*.

Kirghiz Pheasants (the mongolicus group)

Phasianus colchicus turcestanicus Lorenz: Syr Daria pheasant. Valley of the Syr Daria from the delta and nearby coastal islands of the east coast of the Aral Sea, eastward to western part of Ferghana Valley and on slopes of neighboring mountains to the north and east. Includes *bergii* Zarudny.

Phasianus colchicus mongolicus J. F. Brandt: Kirghiz pheasant. A very large range in northeastern Russian Turkestan, north to 48° N, depressions of Lakes Issyk-kul, Balkash, Ala-kul, Zaisan, eastward to tribu-taries of the River Illi and Dzungaria, from Ebi Nor to Guchen in Chinese Turkestan. Also occurs in ex-treme northern Sikiang.

Tarim Pheasant

Phasianus colchicus tarimensis Pleske: Tarim pheas-ant. Sinkiang in the eastern and southern parts of the Tarim Basin along the River Tarim (east of the range of *shawi*) and regions of Qara Shahr and Bagrach Kol, east to Lop Nor and westward along the course of the Cherchen Darya to the Niya Darya. Intergrades

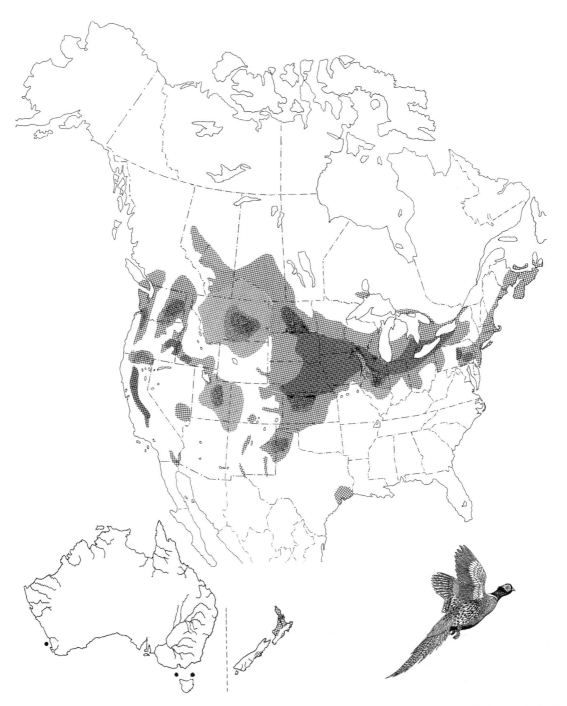

Map 16. Introduced distribution of common pheasant in North America (denser distributions indicated by heavier shading), Australia, and New Zealand. See text for information on Hawaiian and other minor introduced populations.

to the west with *shawi*, which is sometimes considered part of this group.

Gray-rumped Pheasants (the torquatus group)

Phasianus colchicus hagenbecki Rothschild: Kobdo ring-necked pheasant. Isolated in western outer Mongolia from the northern foothills of the Altai north to Lake Khara Usu and the basin of the Khobdo River to Lake Achitu Nor.

Phasianus colchicus pallasi Rothschild: Manchurian ring-necked pheasant. Southeastern Siberia from the upper Amur and Ussuriland south of 44° N to northern Chihli, central Manchuria, and northern Korea above 40° N, intergrading with *karpowi*.

Phasianus colchicus karpowi Buturlin: Korean ring-necked pheasant. Southern Manchuria, south to northern Hopeh and most of Korea, also islands of Tsushima and Quelpart. Intergrades with *pallasi* and *kiangsuensis;* the population is unstable.

Phasianus colchicus kiangsuensis Buturlin: Shansi pheasant. China, north of the Hwang-ho; western and southern Chihli, northern Shansi and Shensi, and adjacent parts of southeastern Mongolia.

Phasianus colchicus alaschanicus Alphéraky and Bianchi: Alashan pheasant. Isolated in oases near the western foothills of Ala-shan Mountains (inner Mongolia).

Phasianus colchicus edzinensis Sushkin: Gobi ring-necked pheasant. Oases of the central Gobi Desert in the valley of the Edzin-gol and near Sokho-nor, Mongolia.

Phasianus colchicus satscheuensis Pleske: Satchu ring-necked pheasant. Isolated in the extreme west of Kansu, north of Nan Shan (Tunhwang) in the basin of the lower Tunhwang-ho and that of lower Shuleh-ho, east to at least Ansi. Also occurs in northern Tsinghai.

Phasianus colchicus vlangalii Przevalski: Zaidan pheasant. Northern Tsinghai, where it is restricted to the marshes of the Zaidan depression.

Phasianus colchicus strauchi Przevalski: Strauch's pheasant. Southern Shensi (Tsinling Range), westward across southern Kansu to Tsinghai, north to central Kansu to at least Wuwei (formerly Liangchow). It also occurs in northern Szechwan.

Phasianus colchicus sohokhotensis Buturlin: Sohokhoto pheasant. Isolated in the Sohokhoto oasis (Mingin), Kansu.

Phasianus colchicus suehschanensis Bianchi: Sungpan pheasant. Northwestern Szechwan from the Min

Shan Range and region of Sungpan, south to Kwanhsien, south of which it intergrades with *elegans*. It also occurs in eastern Tsinghai.

Phasianus colchicus elegans Elliot: Stone's pheasant. Western Szechwan west to 95° E in Tibet and south to the Likiang Range in northwestern Yunnan and northeastern Burma.

Phasianus colchicus rothschildi La Touche: Rothschild's pheasant. Mountains of eastern Yunnan and adjacent Tonkin.

Phasianus colchicus decollatus Swinhoe: Kweichow pheasant. Central China from western Hupeh west to the Red Basin of Szechwan, south to northeastern Yunnan and Kweichow. It intergrades in the east with *torquatus*.

Phasianus colchicus takatsukasae Delacour: Tonkinese ring-necked pheasant. Southeastern Tonkin and adjoining part of China (Kwangsi Province).

Phasianus colchicus torquatus Gmelin: Chinese ring-necked pheasant. Eastern China from Shantung, south to the borders of northern Tonkin, where it intergrades with *rothschildi* and *takatsukasae*.

Phasianus colchicus formosanus Elliot: Taiwan ring-necked pheasant. Island of Taiwan (Formosa) at low and moderate altitudes.

Introduced Populations

Introduced populations representing numerous subspecies (and *P. versicolor* in some cases) exist in Europe from the British Isles, southern Norway, and southern Sweden to Spain, France, Corsica, Italy, Germany, Hungary, Yugoslavia, Greece, and Bulgaria. In Canada, the species occurs in extreme southern British Columbia, southern parts of Alberta, Saskatchewan, Manitoba, and Quebec, plus New Brunswick, Nova Scotia, and locally on Newfoundland. Established populations in the United States occur from Washington's Canadian border south to California and into northeastern Baja California and in the east from Maine through New York, Pennsylvania, New Jersey, Maryland, Indiana, Missouri, Oklahoma, and south to northern Texas, New Mexico, and southeastern Arizona. On the Hawaiian Islands common pheasants have been established on Kauai, Molokai, Lanai, Maui, and Hawaii. In New Zealand they occur over much of the North Island and locally on the South Island. In Japan they are established on Hokkaido and possibly elsewhere. They are local in central Chile (Bay of Coquimbo), on St. Helena, and on Australia's King and Rotnest Islands. They are possibly also established on Flinders Island, as well as on Eleuthera, Bahama Islands. Common pheasants have also been introduced but later extirpated in several areas and unsuccessfully introduced in many other areas (Long, 1981).

Measurements

Wing and tail lengths of many subspecies are given in table 27, as well as the rather limited weight data from the native range in Asia.

Trautman (1982) has tabulated a great deal of weight data on the North American population found in South Dakota (birds of varying racial origins). He reported a yearly average adult male weight of 1,263 g (2.8 lb) and a yearly average adult female weight of 916.5 g (2.0 lb), with sample sizes of 13,124 and 2,071, respectively. Additional weight data on nominate *colchicus* is summarized in Cramp and Simmons (1980). Eggs average about 45 × 36 mm (1.8 × 1.4 in) and the estimated fresh weight is 32.2 g (1.1 oz).

Identification

In the Field (533–889 mm, 21–35 in)

This most widespread species of pheasant is quite variable in plumage, but always has a purplish head and neck, with or without a white neck-ring, a maroon breast tending toward orange on the flanks, and an elongated and barred tail. In areas where confusion with the green pheasant is possible, the brownish rather than greenish tones of the flanks, breast, and mantle are the best distinguishing characteristics. Melanistic mutants especially resemble green pheasants, but are somewhat larger and have longer (over 305 mm [12 in]) tails. Females are generally yellowish buff in color, with a chestnut wash on the upper parts, especially the neck. The tail is long and banded with black and chestnut barring.

Calls of the male are highly varied. However, the male's crowing call is a loud, hoarse, and sudden *Ko-or OK*, *korrk-kok*, or *kok-ok-ok*, with the last syllable very staccato. Wing-whirring also is performed by males during spring.

In the Hand

The common pheasant is likely to be confused only with the green pheasant. Males of the latter species have a shorter tail (to 425 mm [16.6 in]) and are

much more greenish throughout, especially on the upper parts. Females are less strongly marked with black dorsally; the underparts are less distinctly mottled and blocked with darker coloration and normally are pale buff in this area, with only a few dark spots.

Geographic Variation

At least in males, geographic variation in the common pheasant is among the greatest of any species in the pheasant group. Variation may be from clinal to quite abrupt, the latter most typical of isolated populations. In general, from north to south the clines among males exhibit an increase in color saturation, a reduction of the white collar, and a decrease in size. Females are essentially the same throughout, although they tend to be darker and more heavily marked in the easternmost areas (the *torquatus* group; Vaurie, 1965).

Studies by Vaurie (1965) and Delacour (1977) indicate that five groups of subspecies can be recognized, and their taxonomy is followed here. The first, or nominate *colchicus* group, consists of the westernmost forms or "black-necked" pheasants. Males of this group are the most purplish, have brown or buffy upper wing-coverts, and lack white collars or have only faint traces of one. The second group, the "white-winged" pheasants or *principalis–chrysomelas* assemblage, replaces *colchicus* farther east. Males of this group are more reddish and less purplish dorsally and have white or whitish wing-coverts. Some of its included subspecies have incomplete and irregular white collars. Members of the "Kirghiz" or *mongolicus* group, which is found in Turkestan (not Mongolia), also have white on the wings but are more coppery with strong bronze green iridescence and usually have distinct but incomplete white collars. The Tarim Basin group consists of one (Delacour, 1977) or two (Vaurie, 1965) subspecies that are transitional between these first three groups and the more eastern forms. The Tarim pheasant (*tarimensis*) lacks a white neck-ring, has an olive rump, and has light sandy brown wing-coverts. The Yarkand pheasant (*shawi*), which is considered by Delacour (1977) to be part of the preceding group, also lacks a neck-ring, but is generally more white-winged and is distinctly golden yellow on the upper parts. The final group, the gray-rumped pheasants or *torquatus* complex, consists of a large number of subspecies that all have the lower

Table 27

Wing and tail lengths and weights of common pheasants from the native range in Asia

	Males			Females			
	Wing (mm, in)	Tail (mm, in)	Weight (g, oz)	Wing (mm, in)	Tail (mm, in)	Weight (g, oz)	Source(s)[a]
colchicus	238–258, 9.3–10.1	425–536, 16.6–20.9	ave. 1,150, 2.5	210–220, 8.2–8.6	290–310, 11.3–12.1	ave. 850, 1.9	C; D
septentrionalis	250–276, 9.8–10.8	395–406,[b] 15.4–15.8[b]	—	205–228, 8.0–8.9	225–280,[b] 8.8–10.9[b]	—	C
principalis	235–253, 9.2–9.9	390–510, 15.2–19.9	—	208–225, 8.1–8.8	—	—	B; C
zarudnyi	227–244, 8.9–9.5	—	—	213–224, 8.3–8.7	—	—	C
bianchii	240–260, 9.4–10.1	—	956–1,300, 2.1–2.8	219, 8.5	—	710–850, 1.6–1.9	C
chrysomelas	235–250, 9.2–9.8	483,[b] 18.8[b]	—	228, 8.9	315,[b] 12.3[b]	—	C
shawi	233–250, 9.2–9.8	370–515, 14.4–20.1	—	211,[b] 8.2[b]	284,[b] 11.1[b]	—	B
turcestamicus	240–263, 9.4–10.3	—	1,170–1,477, 2.6–3.2	220–240, 8.6–9.4	266–285,[b] 10.4–11.1[b]	740–1,018, 1.6–2.2	C
mongolicus	248–267, 9.7–10.4	510–580, 19.9–22.6	ave. 1,100, 2.4	215,[b] 8.4[b]	312,[b] 12.2[b]	ave. 800, 1.8	C; D
strauchi	225–238, 8.8–9.3	378–595, 14.7–23.2	ave. 1,062, 2.3	196–215, 7.6–8.4	197–275, 7.7–10.7	ave. 835, 1.8	A
tarimensis	240, 9.4	465, 18.1	1,031, 2.3	209–222,[b] 8.2–8.7[b]	241–260,[b] 9.4–10.1[b]	—	E
pallasi	235–245, 9.2–9.6	435–485, 17.0–18.9	1,264–1,650, 2.8–3.6	210–211, 8.2	225–285, 8.8–11.1	880–900, 1.9–2.0	E
karpowi	217–235, 8.5–9.2	424–570, 16.5–22.2	1,000–1,312, 2.2–2.9	194–208, 7.6–8.1	220–290, 8.6–11.3	545–875, 1.2–1.9	E
kiangsuensis	220–235, 8.6–9.2	438–675, 17.1–26.3	1,000–1,100, 2.2–2.4	199–215, 7.8–8.4	238–279, 9.3–10.9	700–1,000, 1.5–2.2	E
suehschanensis	217–232, 8.5–9.0	375–460, 14.6–17.9	770–1,000, 1.7–2.2	—	—	—	E
elegans	205–230, 8.0–9.0	410–465, 16.0–18.1	820–1,250, 1.8–2.7	184–200, 7.2–7.8	235–240, 9.2–9.4	750–800, 1.6–1.8	E
decollatus	230–242, 9.0–9.4	490–576, 19.1–22.5	1,135–1,990, 2.5–4.4	206, 8.0	247, 9.6	625, 1.4	E
torquatus	240–254, 9.4–9.9	425–560, 16.6–21.8	—	208,[b] 8.1[b]	266,[b] 10.4[b]	—	C
Overall	205–276, 8.0–10.8	370–675, 14.4–26.3	770–1,990, 1.7–4.4	184–240, 7.2–9.4	197–315, 7.7–12.3	545–1,018, 1.2–2.2	C

[a] Sources (except where annotated as a personal observation): A = Cheng (1963); B = Vaurie (1965); C = Dementiev and Gladkov (1967); D = Delacour (1977); E = Cheng et al. (1978).

[b] Personal observation.

back and rump light gray (tinged and marked with blue, green, or buff), gray wing-coverts, a mantle and flanks that are always lighter and more yellow than the scapulars and breast, and a buffy olive tail that is heavily barred with black and fringed with pinkish purple. Some of the races have white "eyebrows," and some also have a white collar, but these traits are quite variable (Vaurie, 1965; Delacour, 1977).

Ecology

Habitats and Population Densities

Habitats of the many subspecies vary greatly, but in China consist of three general environments, the overgrown edges of rivers, hilly areas close to large cultivated fields having small bamboo groves and low pine thickets, and flat and level lands cultivated with rice, wheat, or rape (*Brassica;* Cheng, 1963). In the former USSR the primary biotype similarly consists of shrubbery and thickets of bulrushes in river valleys, cultivated terrain, and to some extent brush-covered river valleys of mountains, mostly to elevations between 1,500 and 2,600 m (4,920 and 8,530 ft) and rarely to 3,400 m (11,155 ft; Dementiev and Gladkov, 1967). Desert-adapted subspecies sometimes inhabit quite arid areas of alkaline soils, but here the birds are largely limited to riverine areas or other areas of available fresh water. In the western Palearctic, common pheasants occur from lowlands and broad river valleys to foothills and dry uplands, in areas without deep winter snows or severe cold. In mountains they become limited to narrow wooded valleys and gorges and infrequently occur above 700 m (2,295 ft; Cramp and Simmons, 1980). Similar habitats are used in North America, but there common pheasants are largely associated with cultivated lands (e.g., grains, soybeans, and alfalfa) with nearby grassy and weedy cover or with shrubby areas such as hedges, ditches, marshy edges, woodland borders, and brushy groves. Common pheasants become increasingly limited to irrigated areas in the western and southwestern parts of the North American range. The birds do not thrive in areas of heavy snowfall or in areas of either extreme winter cold or intense summer heat. Common pheasants are especially associated with the "corn belt" and associated calcium-rich soils of central North America (Edminster, 1954). In Hawaii common pheasants are found from sea level to 3,355 m (11,000 ft) in areas where the rainfall varies from

about 25 to 762 cm (10 to 300 in) annually, in all types of soil, topographic, and climatic conditions, and in cultivated areas as well as forested, grassland, desert, or other waste areas (Schwartz and Schwartz, 1951).

Population densities are better known in areas of the common pheasant's introduced range than its indigenous range. For example, in some areas of New Zealand average population densities from 2.3 to 13.8 birds per square kilometer (6.0 to 35.7 per square mile) have been reported (Westerskov, 1963). In Hawaii, the population in the mid-1940s averaged about 7.4 birds per square kilometer (19 per square mile) for the area as a whole, but varied from less than 3.9 to 38.6 birds per square kilometer (10 to 100 per square mile) in some localities (Schwartz and Schwartz, 1951). Similarly great variations in density have been reported in North America, even in small areas. For example, on Pelee Island, Ontario (with a land area of 4,050 ha [10,000 acres]), an introduced pheasant population rose from 36 birds in 1927 to about 12.3 per hectare (5.0 per acre) by 1934 (Stokes, 1954). An enormous, but temporary, buildup of population density to 9.6 per hectare (3.9 per acre) occurred on the 160-ha (397-acre) Protection Island off the coast of Washington within 5 years after pheasants were introduced (Einarsen, 1945). Edminster (1954) judged that in North America, first-class pheasant range should support about one adult per 1.2–1.6 ha (3.0–4.0 acres) in spring, whereas poor range may have one adult per 6.6–8.0 ha (15.0–20.0 acres) in spring. Estimated autumn (both adult and young) densities in first-class range were 2.5 birds per hectare (1.0 per acre) and 0.2–0.5 birds per hectare (0.1–0.2 per acre) in poor range. Decade-long studies in south-central Nebraska resulted in estimates of adult (spring) densities of about 7.4–19.8 per square kilometer (19.2–51.3 per square mile) in three different study areas of good pheasant habitats (Baxter and Wolfe, 1973); these densities would appear to be fairly representative of many midwestern areas. Spring estimates of as high as 30.9 females per square kilometer (80 per square mile) were reported in Iowa during the very high populations of the early 1940s, but more typical densities for the same area were in the range of 15.4 females per square kilometer (40 per square mile) (Kozicky and Hendrickson, 1951). In some parts of the former USSR 20 or more broods are produced per square kilometer (51.8 per square mile; Dementiev and Gladkov, 1967).

Estimated densities in the indigenous range in China evidently vary greatly and mostly range from 0.6 to 64.0 individuals per square kilometer (1.6 to 165.8 per square mile) in various locations (Li, 1996).

Competitors and Predators

Some of the major predators of ring-necked pheasants in North America were mentioned in chapter 5; these consist of a considerable array of raptors and wild mammals, especially foxes. Wagner et al. (1965) reviewed the available information on predators and their effects in North America. They mentioned 12 mammals, 9 birds, and 2 reptiles as known predators on adults or young and 14 mammals, 3 birds, and 2 snakes as known nest predators. Bohl (1964) reviewed the known and probable predators of common pheasants in Korea, whereas Dementiev and Gladkov (1967) mentioned a considerable number of known or presumptive enemies of pheasants in the former USSR. Cheng (1963) said that in China the species' most serious enemy is the civet cat, although wolves, foxes, eagles, owls, and various other predators also may be of significance. Competitors doubtless vary greatly over the enormous native and introduced range of the common pheasant and cannot be easily summarized.

In England, foxes and corvids (crows and magpies) are the primary nest predators; predators collectively may destroy about one-third of all nests. Hedgehogs and badgers are also significant mammalian predators on eggs, whereas foxes are more likely to kill incubating females. By comparison, in several U.S. studies nest predation losses ranged from 20 to 55 percent (overall mean 41 percent; Hill and Robertson, 1988).

General Biology

Food and Foraging Behavior

A considerable amount of information on pheasant foods was provided in chapter 5, from which it is apparent that very great local and seasonal differences exist in the foods of this species. The common pheasant is relatively omnivorous and opportunistic and tends to consume large and energy-rich foods that are easily available, such as cultivated grains, mast, and fruits. The relative abundance of insects and other animal life in the diet also seems to be highly variable. However, in young birds (up to about 4 months old) it is invariably higher than in adults. Ferrel et al.

(1949) reported that 20 birds up to 3 weeks of age averaged 82.9 percent animal foods, 23 birds from 4 to 6 weeks averaged 47.2 percent, 21 birds from 7 to 9 weeks averaged 55.1 percent, 31 birds from 10 to 12 weeks averaged 12.5 percent, and 34 birds from 13 to 16 weeks averaged only 1.8 percent.

Movements and Migrations

Significant movements in the common pheasant seem to be limited to populations in northern areas that are forced out of breeding areas during winter. Cramp and Simmons (1980) summarized the data for the western Palearctic. In Sweden, Norway, Denmark, and Britain movements greater than a few kilometers during a bird's lifetime are unusual, and almost none move more than 10 km (6 mi). However, exceptional cases of movements as great as 40 km (25 mi) have been noted in Sweden, and in Finland one adult male was found to have moved 210 km (130 mi) in 13 months. In North America, various studies have similarly indicated a rather high level of sedentary behavior (e.g., Gates and Hale, 1974, as summarized in chapter 5). However, in the Amur Basin of the former USSR there have been several massive migrations of pheasants, mainly adult males, from China during periodic severe winters (Barancheev, 1965).

Daily Activities and Sociality

Like the more tropical species of pheasant, the common pheasant also exhibits a high level of diurnal periodicity in its activities. Calling activity by males may begin before sunrise. At that time the birds typically leave their roosting sites to forage, which may occupy 2 or 3 hours. Thereafter, the birds typically move to a source of water, particularly in the dry regions. The birds may remain at such sites for only a short time, after which they are likely to spend the warmest part of the day resting, preening, dustbathing, and sleeping in shady areas. Toward evening the birds forage again and frequently also visit sources of grit. Foraging may continue until it is almost dark, when the birds proceed to their roosts. These roost sites may sometimes be in trees, but much more frequently are in dense brushy areas, reed beds, or other areas providing relatively heavy cover.

Sociality patterns probably vary greatly with population density and levels of disturbance, if not other factors. However, at least in North America, some patterns have emerged. The study of Collias and Taber (1951) may be representative. They found that, during winter, common pheasants formed temporary flocks in which individuals moved about and fed together as a relatively coherent unit with shifting membership. Males and females sometimes fed together, but also often formed unisexual groupings. Roosting groups during winter varied from 2 to 24 birds or more, with larger groups typical of very cold weather. The locations of these roosts varied somewhat, although there were favorite roosting sites. Gradually these groupings changed to harems of hens, with each harem dominated by a single male as the breeding season progressed. Shifting of male groups from the period in which they occurred in pairs, trios, or larger groupings to the period of male dispersal and territorial establishment was associated with active increase in testis size, the start of male display toward females, and the onset of intimidation behavior (fighting or threat display) among the females. Both males and females exhibited dominance hierarchies, which seemed to be related to age and perhaps also to weight. All males dominated all females, and males that began to crow and display early in the season generally dominated the males that began later.

Social Behavior

Mating System and Territoriality

A harem mating system is well known for this species; male success in attracting varied numbers of females seems to be related to male-to-male and male-to-female dominance characteristics. Although most males are recognizably territorial, with territories averaging about 1–5 ha (2.5–12.4 acres), a minority of mostly younger males may remain nonterritorial and serve as more mobile "floaters" (Ridley, 1983). Heavier males tend to defend larger territories and are usually older birds (Hill and Robertson, 1988).

Different observers have noted great variations in the sizes of crowing territories, which range from 1.2–1.6 ha (3.0–4.0 acres) to 10–30 ha (25–75 acres; Edminster, 1954). Kozlowa (1947) avoiding calling these areas territories, but instead referred to them as "cruising routes," because she never saw a male expel another from them. She believed that the route was not over 400–500 m (1,310–1,640 ft) and was regularly traced and retraced by males for both foraging and sexual purposes. Taber (1949) agreed that males

have territories, but thought their boundaries were highly plastic and were affected by population density and such local environment features as relative cover and topography.

Voice and Display

Probably the most important and certainly the most conspicuous vocal signal of male common pheasants is their crowing call. This call is loud, sudden, and harsh and typically consists of two or three syllables sounding like *korrk-kok, KO-or OK, ko-koro,* or various other transcriptions. It may carry up to a mile or so under favorable situations and is usually followed by wing-whirring. As many as 12 other adult calls have been distinguished in this species (Heinz and Gysel, 1970). None of these seems to be so clearly associated with sexual display, with the possible exception of hissing, which occurs during intense lateral display and may also follow copulation.

Postural displays of the common pheasant have been described and illustrated by Kozlowa (1947), Taber (1949), Glutz (1973), and Cramp and Simmons (1980). One of the most important is wing-whirring, which normally occurs in association with crowing. The male typically selects a prominent location, draws his body up, pauses, and, sometimes after an inaudible wing-flap, utters his crowing call and almost immediately performs a brief but vigorous wing-whirring (figure 49). During this display the tail may be slightly cocked or held down against the ground as an apparent brace, more commonly the former. These displays may occur every 10–15 minutes during the peak of the display season, but are most common in early morning and late afternoon. During this and other displays the facial wattles are engorged and the ear-tufts are raised. When two displaying males encounter one another they may face each other or walk parallel, they hold their heads high, swell their wattles, and erect the plumage on the back of the neck while uttering a hoarse *krrrah* note. They may also perform a lateral intimidation display, with wing-lowering on the nearer side as well as tail-tilting and partial tail-spreading, but with the head held high rather than low as in courtship. Alternatively, the males may face each other with heads held low, rumps raised, and tails straight out behind, sometimes pecking at grass and uttering purring threat notes. This may grade into actual fighting, with biting and kicking by both males (figure 50). The subordinate male

Figure 49. Display postures of male common pheasant, including normal (A) and facial engorgement (B), crowing (C), wing-whirring (D), and lateral display to female (E). After various sources, especially Glutz (1973).

or loser of an encounter retreats with his feathers sleeked against the body and his wattles retracted. Females may also perform similar intimidation displays to one another (Glutz, 1973).

When displaying sexually to a female, the male assumes a lateral display posture while strutting around the female in semicircles, holding the head in, drooping the nearer wing, tilting the tail toward the female, fluffing the body feathers, and engorging the wattle.

A hissing sound is often associated with this posturing, and the tail feathers may be vibrated to produce a fluttering sound. Tidbitting is also performed, with an associated vocalization of low notes uttered at the rate of about three per second (Stokes and Williams, 1972) while the bird crouches and holds his folded tail high. At least early in the season, copulation is usually preceded by lateral display or tidbitting, but later the male may simply chase the female and at-

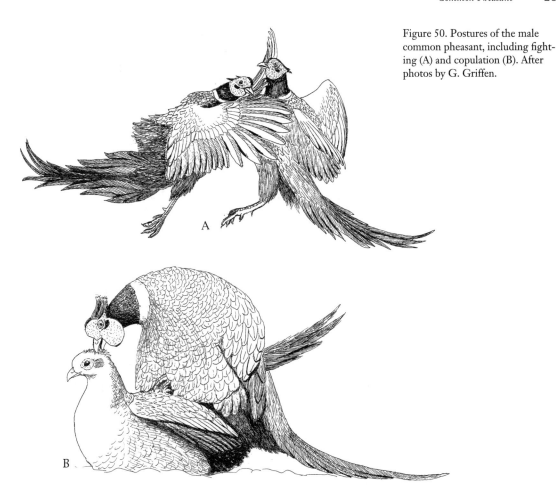

Figure 50. Postures of the male common pheasant, including fighting (A) and copulation (B). After photos by G. Griffen.

tempt to forcibly mount her. Following copulation the male may renew his lateral display, but no other specific postcopulatory displays occur.

Reproductive Biology

Breeding Season and Nesting

The breeding season is highly variable throughout the native and introduced range of the common pheasant. In the majority of the range in North America, the western Palearctic, and Asia common pheasants breed from early or mid-March until early June. A variety of factors probably all contribute to laying phenology. These include age of the female, annual variations in local temperatures, and individual physiological differences among females. Within the general limits set by photoperiodic timing mechanisms, a female's laying times may be set by her physiological condition at the end of the winter and

how rapidly she is able to accumulate the necessary energy reserves associated with egg-laying requirements (Gates and Hale, 1975). Because of persistent renesting efforts, laying and hatching dates may be spread over a period of several months (Wagner et al., 1965), although peak hatching dates are likely to vary from year to year only by a few weeks in most areas.

Common pheasant nests are normally constructed on the ground, in thick grassy, weedy, or shrubby vegetation. Occasionally, however, elevated sites such as straw stacks and old tree nests of other birds or squirrels may be used. Some factors affecting nest-site selection in the common pheasant were discussed in chapter 7. Relative nest concealment, as influenced by the surrounding height and density of the vegetation, seems to be especially important in site selection. There is less evidence that the overall size of the nesting habitat is important and no good evidence that the nest location is significantly related to the distance to the nearest habitat edge (Hoffman, 1973). Nests often appear to be clustered within the presumed limits of a male's crowing territory; perhaps the males adjust their territorial boundaries to include their mates' nest sites (Baskett, 1947; Seubert, 1952). Dumke and Pils (1979) found that females tended to locate their nests less than 1 km (0.6 mi) away from their wintering range and typically at the edges of their prenesting range and the territories of associated males. Hill and Robertson (1988) found that radio-tagged females preferred to nest in woodlands early in the breeding season, but shifted to cereal crop cover later, as these plants grew taller. However, nest densities in differing habitat types were not found to be correlated with relative nesting success.

Eggs are laid at an average rate of 1.4 days per egg for a complete clutch. Clutch sizes are highly variable, but a sample of 236 in Switzerland averaged 11.9 eggs, and 210 English clutches averaged 11.8 eggs (Cramp and Simmons, 1980). A sample of 4,940 clutches from the North American and New Zealand populations of common pheasants had a collective mean of 10.6 eggs, with individual means ranging from 8.6 to 12.6 eggs (Trautman, 1982). A Wisconsin sample of 574 clutches that averaged 11.2 eggs (Gates and Hale, 1975) had statistically significant yearly differences in average clutch size, with a seasonal decline in average size as well. Clutches begun after 15 May in Wisconsin (presumably mostly or all renesting efforts) averaged 10.0 eggs, whereas those begun earlier averaged 12.5 eggs. Up to three renesting efforts have been observed following earlier clutch losses. Seubert (1952) observed that 57 percent of 132 females that had been disrupted from or deserted their first nest established second nests; 7.5 percent of those disrupted from their second nests attempted a third nesting. In a more recent study, Dumke and Pils (1979) found that 69 percent (32 of 47) of the unsuccessful females they studied renested a first time, 41 percent (at least 11 of 27) renested a second time, and 9 percent (1 of 11) renested a third time. All told, these birds averaged 1.8 nests each, and an estimated 75 percent of the females succeeded in producing broods. Four females renested following the loss of broods. Similarly, Penrod et al. (1982) found that four females that had lost their broods at hatching or early in the brooding period renested.

The number of eggs laid in second clutches produced by renesting females is not significantly correlated with the size of the initial clutch. Furthermore, female common pheasants often choose to nest in a different vegetation type for their second nesting efforts. Distances moved between successive nesting attempts may vary from about 40 to 2,300 m (130 to 7,545 ft), averaging in five studies about 370 m (1,215 ft). Home ranges used by renesting females average smaller than those used during initial nestings (10.9 ha [26.9 acres] versus 17.8 ha [44.0 acres]), and these in turn were smaller than prenesting home ranges. Nests are usually placed at the edge of the home range, perhaps to reduce the probability of nest predation by predators that may be attracted to a female's general activities. Only in New Zealand have cases of females rearing two broods in a single breeding season been documented (Hill and Robertson, 1988).

Incubation and Brooding
Incubation in the common pheasant is performed entirely by the female. Under natural conditions it requires from 23 to 28 days, averaging 23.0–23.5 days. Longer periods are associated with varying levels of disturbance to the incubating bird. Chicks are cared for entirely by the female, with no male involvement at all.

Development and Growth of the Young
The young are precocial and largely self-feeding from the outset. They average about 23 g (0.8 oz) at hatching. Males can sometimes be distinguished at hatching by their somewhat larger pale buffy eye-ring and less conspicuous brownish ear-flecks. Positive sex identification can also be made by the presence in 1-day-old or older males of a rudimentary wattle (an unfeathered and nearly unpigmented fold or flap of skin), which is largely hidden by the down but usually evident directly under the eye region and in the loral area above and in front of the eye (Woehler and Gates, 1970). By about 12 days young are able to make their first flights. At 30 days they average about 135 g (4.7 oz), at 2 months about 400 g (14.0 oz), and at 3 months females average 750 g (26.2 oz) and males average 850–950 g (29.8–33.3 oz; Glutz, 1973). By 20 weeks the weight gain in females is essentially finished and their average weight is about 800 g (1.8 lb), while at the same age males are still growing

slightly and average about 1,100 g (2.4 lb; Baxter and Wolfe, 1973).

The brood typically remains with its mother for about 70–80 days before becoming independent. Brood losses between hatching and independence from their mother are usually substantial, often amounting on average to nearly half of the total number of hatched young. Hill and Robertson (1988) found that broods having the largest home ranges suffer the highest mortality rates, and these larger home ranges tend to have lower densities of available insects. The larvae of sawflies and lepidopterans are favored foods for chicks. Both sexes become sexually mature and attain full adult plumage in their first year of life.

Evolutionary History and Relationships

Obviously the nearest relative of the common pheasant is the green pheasant, and the two should be considered no more than allospecies if not indeed simply subspecies. Where both of these forms occur together as a result of introductions, they tend to hybridize and the green pheasant typically suffers (Schwartz and Schwartz, 1951; Kuroda, 1981). The two forms have been considered subspecies by Sibley and Monroe (1990), as well as by the American Ornithologists' Union in recent editions of their *Checklist of North American Birds*. However, the common and green pheasant were considered as full species by Vaurie (1965) and Brazil (1991). Otherwise, the genus *Syrmaticus* clearly seems to be the nearest relative of *Phasianus;* the Reeves' pheasant in particular is extremely suggestive of the *Phasianus* plumage pattern, and the copper pheasant has been considered by some (e.g., Yamashina, 1976) to be congeneric with the common pheasant. I see no major advantage in merging these two genera inasmuch as *Syrmaticus* is already reasonably large. Furthermore there are problems of reduced hybrid fertility involving crosses between *Phasianus* and *Syrmaticus*.

Status and Conservation Outlook

This is one species that conservationists need not worry about. The total world population of pheasants may number in excess of 50 million. Annual harvests in North America alone sometimes approaching 20 million birds, and over 7 million hand-raised birds are released every year in Great Britain for sporting purposes. In some districts of New Zealand there is an annual kill of several thousand birds annually (Westerskov, 1963). A substantial amount of hunting also occurs in Hawaii. The common pheasant population in the former USSR cannot be known with any certainty, but in Tadzhikistan alone the autumn population may be about 1.5 million birds (Dementiev and Gladkov, 1967). Likewise the population and harvest in China are completely unknown. The introduced British population consisted of about 7.5 million birds as of about 1990, making it one of the most common breeding land birds of Britain (Marchant et al., 1990).

GREEN PHEASANT

Phasianus (colchicus) versicolor Vieillot 1825

Other Vernacular Names

Japanese pheasant; faisan versicoloré (French); bunt Fasan, grün Fasan (German).

Distribution of Species

Japan, in park and farmland, brushy sites, and open woods in the plains or lower slopes of the mountains, where it ascends to about 1,065 m (3,500 ft; Vaurie, 1965). Absent from Hokkaido (see map 17). Locally introduced on the Hawaiian Islands (Hawaii, Kauai, Lanai, and perhaps Maui).

Distribution of Subspecies

Phasianus (colchicus) versicolor Vieillot: southern green pheasant. Southwestern Honshu and Kyushu.

Phasianus (colchicus) tanensis Kuroda: Pacific green pheasant. Peninsulas of Izu and Miura in eastern Honshu, the Seven Islands of Izu and the islands of Tanegashima and Yakushima south of Kyushu.

Phasianus (colchicus) robustipes: northern green pheasant. Islands of Sado and Honshu (except re-

gions inhabited by nominate *versicolor* and *tanensis*) and possibly Shikoku.

Measurements

Delacour (1977) reported that males have wing lengths of 196–220 mm (7.6–8.6 in) and tail lengths of 270–425 mm (10.5–16.6 in), whereas females have wing lengths of 195–220 mm (7.6–8.6 in) and tail lengths of 207–275 mm (8.1–10.7 in). The southern race *versicolor* averages slightly larger than *robustipes*, with wing lengths of males of 225–243 mm (8.8–9.5 in) and 215–225 mm (8.4–8.8 in), respectively. Males average about 900–1,100 g (2.0–2.4 lb) and females about 800–900 g (1.8–2.0 lb; Bohl, 1964). Eggs average 43.3 × 34 mm (1.7 × 1.3 in) and have an estimated fresh weight of 27.8 g (1.0 oz).

Identification

In the Field (457–787 mm, 18–31 in)

Limited to Japan, the green pheasant is likely to be confused only with the copper pheasant there, except on Hokkaido, where the common pheasant has been introduced and is expanding in range. Separation from the copper pheasant is easily achieved by the green pheasant's absence of coppery brown head and body coloration in males. Females are more similar, but the female copper pheasant has a white-tipped and otherwise weakly patterned tail. Field separation of female common and green pheasants is probably not possible. Green pheasant male calls are said to be shorter and distinguishable from those of the common pheasant.

In the Hand

The green pheasant is easily confused with the common pheasant, especially the latter's melanistic mutant variety. However, males of the latter form are generally darker and have a longer (over 350 mm [13.7 in]) tail. They also are almost entirely sooty black, with greenish to bluish iridescence on the head, neck, breast, and mantle; heavy black barring on the tail; and olive black wings. Female green pheasants closely resemble those of *colchicus*, but are on average darker and have more distinctively patterned body

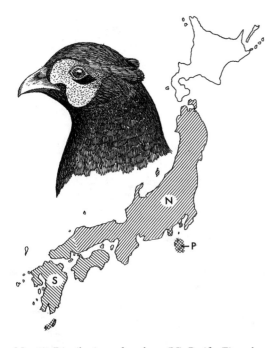

Map 17. Distributions of northern (N), Pacific (P), and southern (S) races of the green pheasant.

coloration. The dark parts of the mantle feathers are black, often with a greenish sheen, and the underparts are more vermiculated and blotched than is true of *colchicus*.

Geographic Variation

Geographic variation in the green pheasant is limited and probably is somewhat clinal, although a good deal of local variation also occurs. In general, birds from more northern areas are lighter and greener than those from farther south, which tend to be darker and bluer. Northern birds may be slightly smaller than southern ones (Delacour, 1977), although this is not a clear-cut trend. In fact, Kuroda (1981) stated that the northern race actually averages slightly larger.

Ecology

Habitats and Population Densities

The northern populations of the green pheasant extend from sea level to about 1,065 m (3,500 ft) elevation, in sparse woods or brush lands near cultivated fields, grassy areas near rivers and cultivated lands, low hilly areas near the coast, and on island mountains having brush and tree cover. The southern population on Kyushu is found in similar habitats, but generally with more luxuriant vegetation and a greater proportion of broadleaf trees. Introduced populations on Hawaii occur at elevations centering between 1,220 and 2,285 m (4,000 and 7,500 ft), with densest populations where there are scattered clumps of koa trees (*Acacia koa*) interspersed with grassy meadows with mixed herbaceous vegetation (Schwartz and Schwartz, 1951).

There do not seem to be any detailed estimates of population densities. However, Kuroda (1981) mapped the local distribution of a flock using a very limited area near the Imperial Palace; his map suggests a density of up to several birds per hectare under highly favorable conditions.

Competitors and Predators

There are only limited ecological contacts between the green pheasant and the copper pheasant in Japan (Yamashina, 1976). These seem to occur primarily during cold weather when the copper pheasant is forced into lower altitudes. Avian predators in Japan include various hawks; additionally crows, ravens, and magpies may be nest predators, as perhaps are weasels

and snakes. However, feral cats and dogs are probably this species' major enemy in Japan (Beebe, 1918–1922; Bohl, 1964).

General Biology

Food and Foraging Behavior

The green pheasant's foods are very similar to those of the common pheasant, with possibly a somewhat higher proportion of wild seeds, fruits, berries, and nuts in the diet and lower proportions of cultivated grains. Fruits include those of such genera as *Rosa*, *Euonymus*, *Viburnum*, *Viscum*, and *Diospyros*, and mast includes acorns, chestnuts, and the seeds of Japanese magnolias. Cultivated grains include millet, wheat, barley, rice, soybeans, and buckwheat; sweet potatoes are also eaten. A considerable array of insects are consumed, many of which are considered harmful. Also, millipedes, crustaceans, and snails have all been reported as food items. Young birds take a high proportion of animal foods (Bohl, 1964).

Foraging is often done in small groups during morning and evening hours near agricultural fields, tea plantations, and open grasslands.

Movements and Migrations

Evidently the only significant movements occurring in the green pheasant are those associated with the higher snowfall areas of northern Honshu, where the birds move to lower elevations and the seashore during severe winter weather. Deep snow may also cause the birds to move into farmyards and feed with domestic fowl. Each day green pheasants make small vertical movements from hillside roost areas into croplands or adjacent areas for foraging; these are followed by a return to the hillsides for resting and dusting. A second movement back to the foraging areas occurs in late afternoon (Bohl, 1964).

Daily Activities and Sociality

Foraging occurs twice a day, with the middle of the day being spent on hillsides, preferably in sandy areas where dusting is also performed. The roosting sites are typically on grassy areas at or near hilltops. However, where this type of cover is lacking green pheasants may roost in brushy vegetation or sometimes in trees (Bohl, 1964).

Large flocks of green pheasants are typical only in winter, when feeding restrictions may force birds into close proximity. Otherwise, only small groups are typically found. During spring the usual grouping is a single male and one or more females. Following the nesting season the male may rejoin his females and their broods or may flock with other males during the autumn and winter seasons (Bohl, 1964). Kuroda (1981) stated that during winter males may remain alone, occur in groups of up to four, or sometimes may be seen as apparently paired birds. Females in winter also often occur in groups of up to six birds; the chicks evidently remain with them only until about September.

Social Behavior

Mating System and Territoriality
Beebe (1918–1922) believed that green pheasants are essentially polygynous, with the breeding season extending from March through May. However, there are marked local variations in the onset of laying, with some beginning as early as January in Kyushu and as late as the latter part of April in northern Honshu (Bohl, 1964). Kuroda (1981) found a small population near Tokyo to be more often monogamous than polygynous.

Green pheasants are said to be territorial, with the males performing crowing calls during the breeding season. According to Kuroda (1981), male calling occurs from mid-March through April and May (in the Tokyo area) and fades out by about the first week of June. He mapped the locations of a considerable number of "suggested" territories, reporting in one case 36 male territories within an area of approximately 16 ha (40 acres). Thus, the males have an average territory of less than half a hectare. This site was in a protected area near the Imperial Palace and probably reflects a very high population density. Territorial behavior in this species was also observed by Maru (1988).

Voice and Display
The crowing call of the male is shorter and somewhat different from that of the common pheasant. A sonogram (Kuroda, 1981) indicates that it is a double-noted call and is rather different in minor acoustic characteristics from that of the common pheasant (cf. sonogram of common pheasant in Cramp and Simmons, 1980, p. 511).

The postural displays of the green pheasant are apparently very much like those of the common pheasant (figure 51). These include a crowing posture followed immediately by wing-whirring and a lateral threat and/or courtship posture that would seem to be identical to that of the common pheasant. Beebe (1918–1922) stated that the courtship display differs in no way in these two species.

Reproductive Biology

Breeding Season and Nesting
The maximum period of egg-laying probably coincides with the peak of crowing behavior in males, that is, April and May in central Japan. The nests are placed on the ground in various cover, but commonly are close to low bushes or at the bases of trees. Nests have been located from sea level to as high as about 915 m (3,000 ft). The usual clutch sizes are from 6 to 12 eggs. Renesting occurs if the first nest is disturbed or destroyed; such nests may be seen from late June until August (Bohl, 1964).

Incubation and Brooding
The female green pheasant incubates the eggs for 23–25 days, without any apparent nest defense on the part of the male.

Growth and Development of the Young
Kuroda (1981) reported that observed brood sizes ranged from 1 to 8 chicks, with a mode of only 2 chicks per brood and an estimated mean of 3.3 for 34 sightings. June was the month in which the largest number of sightings of females with broods were obtained, and by September these family groups had virtually disappeared. This apparent early breakup of family units is surprising and needs confirmation. Like the common pheasant, green pheasants mature in their first year.

Evolutionary History and Relationships
There can be no doubt that this form is extremely closely related to the common pheasant, and that they can be considered no more than allospecies. There is indeed doubt whether they should not be simply considered as subspecies (Goodwin, 1982). Probably green pheasants arrived in Japan via Korea when these land masses were still connected (Kuroda, 1981).

Figure 51. Display postures of male green pheasant, including males in mutual lateral threat display (A), facial engorgement (B), and crowing posture (C). After Kuroda (1981).

Status and Conservation Outlook

The green pheasant is a major game bird in Japan and probably represents the most popular game bird in that country. Annual harvests of more than 500,000 birds a year are now typical, and 100,000 captive-raised birds are released per year. Where the common pheasant has been released in Japan there have been deleterious genetic effects on the native green pheasant population. Therefore, the common pheasant cannot now be legally released there except on Hokkaido and other remote areas not already occupied by the green pheasant (Kuroda, 1981). On Hawaii green pheasants have hybridized with the similarly introduced common pheasant (Goodwin, 1982). There are also a few records of wild hybrids with copper pheasants in Japan. There the two species tend to occupy differing habitats, namely open lowlands in the case of the green pheasant and wooded uplands in the copper pheasant (Brazil, 1991).

Genus *Chrysolophus* J. E. Gray 1834

The ruffed pheasants are small montane pheasants in which sexual dimorphism is strongly developed. The tail is greatly elongated and slightly vaulted. The males have short decumbent crests and ornamental "capes" or ruffs, which can be spread during display. The body feathers of males are mostly either disintegrated and silky, white, yellow, and reds or broad and scaly. The tail is strongly graduated, has 18 rectrices, is strongly barred or mottled, and partially covered by brightly patterned upper tail-coverts. The tail molt is phasianine (centripetal). The wings are relatively short and rounded, with the sixth primary the longest and the ninth and tenth both shorter than the first. The tarsus is relatively long and thin and is spurred in the male. There is a small orbital skin patch, which includes a wattle under the eye that can be expanded during display into a lappet. Females are barred with dark brown and buff, and their orbital area is sparsely feathered. Two species are recognized.

Key to Species of *Chrysolophus* (after Delacour, 1977)

A. Dominant plumage colors yellow and red or white and green (males).
    B. Ruff white, barred with black; abdomen white: Lady Amherst's pheasant.
    BB. Ruff orange, barred with black; abdomen red: golden pheasant.
AA. Dominant plumage color brown (females).
    B. Legs grayish; orbital skin bluish: Lady Amherst's pheasant.
    BB. Legs and orbital skin yellow: golden pheasant.

GOLDEN PHEASANT

Chrysolophus pictus (Linnaeus) 1758

Other Vernacular Names
Painted pheasant; faisan doré (French); Goldfasan (German); kin-ky, ching chi (Chinese).

Distribution of Species
Mountains of central China from southeastern Tsinghai eastward through southern Kansu to southern Shensi north to the Tsinling Range, southward through extreme eastern Tibet, Szechwan, northern Yunnan, Kweichow, western Hupeh, and Hunan to northern Kwangsi. This species inhabits mountain slopes and valleys in habitats well grown with bushes, bamboos, or other dense cover, occasionally in bushes or scrub on tea plantations or terraced fields (Vaurie, 1965). Introduced locally in Great Britain (Galloway, East Anglian Brecks, and South Downs in Hampshire and West Sussex). See map 18.

Distribution of Subspecies
None recognized by Delacour (1977).

Measurements
Cramp and Simmons (1980) reported that males have wing lengths of 190–202 mm (7.4–7.9 in) and tail lengths of 630–765 mm (24.6–29.8 in), whereas females have wing lengths of 179–190 mm (7.0–7.4 in) and tail lengths of 340–372 mm (13.3–14.5 in). The dark variant *"obscurus"* is slightly larger, with male wing lengths of 194–209 mm (7.6–8.2 in) and female wing lengths of 185–193 mm (7.2–7.5 in). Five

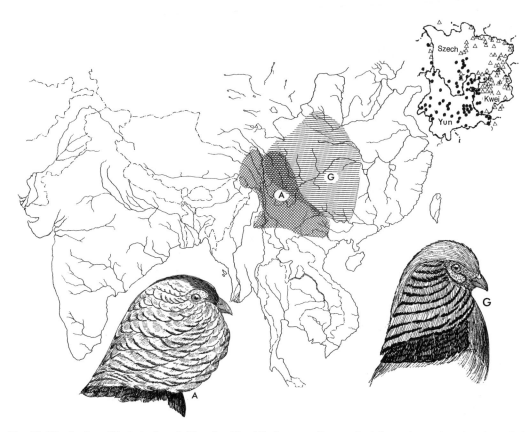

Map 18. Distribution of Lady Amherst's (A) and golden (G) pheasants. See text for information on introduced populations. The inset map shows locality records for the golden pheasant (open triangles) and Lady Amherst's pheasant (solid circles) in the area of contact.

males ranged in weight from 575 to 710 g (1.3 to 1.6 lb), and five females ranged from 550 to 665 g (1.2 to 1.5 lb; Cheng et al., 1978). The eggs average 44 × 34 mm (1.7 × 1.3 in) and have an estimated fresh weight of 28.1 g (1.0 oz).

Identification

In the Field (508–1,168 mm, 20–46 in)
Except for its limited introduced range in Great Britain, this species is found only in the mountains of central China, where it occurs in dense scrub and broken country. Feral birds in Great Britain favor dense woodlands with closed canopies, but forage in clearings. The adult male is unmistakable, owing to its golden nape and the orange red feathers of the underparts and rump. Females are much duller, but are barred in a more distinct pattern than female common pheasants, which is especially evident on the wings and tail. The species' voice is more piercing than that of the common pheasant. The male has a higher-pitched crow, which is either a strident and metallic *chak* or a double *cha-chak*. This metallic sound, which resembles the whetting of a scythe, can be heard for as many as 460 m (1,500 ft) and is repeated every few minutes during the courtship season.

In the Hand
Males can be readily recognized in the hand by their golden cape and rump and brilliant reddish underparts and ornamental tail-coverts. Female golden pheasants are most likely to be confused with those of the Lady Amherst's, but have yellow orbital skin, have yellowish rather than reddish chestnut on the crown, and are less heavily barred with blackish throughout.

Ecology

Habitats and Population Densities
Beebe (1918–1922) found golden pheasants in Hopeh Province on steep, rocky, and barren ridges, with only a dense growth of scrub bamboo present. A similar habitat preference has been described by Cheng (1963). The species evidently avoids heavy forests, open country, and wetland habitats. Their altitudinal range rarely extends above 1,980 m (6,500 ft). In Great Britain introduced birds favor dense plantations of Scots pine and larches, especially those from

15 to 30 years old, and to a lesser extent are found in mixed woodlands. Lowland areas with extensive undisturbed cover are the preferred habitats there (Cramp and Simmons, 1980).

Estimates of densities range from 3.6 to 69.0 individuals per square kilometer (9.3 to 178.7 per square mile), averaging about 29 (75.1; Li, 1996). In a Kweichow nature reserve the densities of birds per square kilometer varied from 7.5 (19.4 per square mile) in shrub and grassland habitats to 45 (116.6 per square mile) in evergreen and deciduous forests (Wu, 1994). In Foping Nature Reserve, Shensi Province, densities varied from 23.8 individuals per square kilometer (61.6 per square mile) during winter to 12.2 territorial males per square kilometer (31.6 per square mile) during the breeding season. Home ranges averaged 37.2 ha (91.9 acres) in winter, 66.6 ha (164.5 acres) in spring, and 102.6 ha (253.4 acres) in summer (Liang, 1997). Females generally have larger average home ranges than males (33.95 ha [83.9 acres] versus 14.29 ha [35.3 acres]), although during winter the males' ranges average larger. Apparently females move about in the males' territories and may remain with some males for several days (Ding et al., 1997).

Competitors and Predators
In China there is apparently no contact between the golden pheasant and the very closely related Lady Amherst's pheasant, which is the only probable serious competitor. The golden pheasant possibly has local contacts with both the Reeves' and common pheasant. Beebe (1918–1922) found no obvious associations with other animals or evidence of any definite predation, but did find some scattered feathers that he judged may have been the result of owl or eagle attacks.

General Biology

Food and Foraging Behavior
In China the species' foods are primarily the leaves and tender shoots of shrubs and low-growing bamboos, the flowers of a species of rhododendron, and some spiders and various insects, especially small beetles (Cheng, 1963). There are no data on the foods of the introduced populations.

Movements and Migrations
In Great Britain, the introduced population is believed to be sedentary (Cramp and Simmons, 1980).

Beebe (1918–1922) judged that there is likely to be relatively little population shifting with the seasons. However, Cheng (1963) noted that in winter golden pheasants move down out of the mountains to lower levels each day to look for food and then go back uphill each evening to roost.

Daily Activities and Sociality

Beebe (1918–1922) found two roosting sites, both of which were on horizontal branches about halfway up pine trees and not more than 3.5 m (12 ft) aboveground. In one case a pair was roosting side by side, while in the other two birds roosted on opposite sides of the tree. The birds are found as single individuals, pairs, or trios throughout the year. Beebe (1918–1922) found no evidence of flocking in this species.

Social Behavior

Mating System and Territoriality

The mating system of the golden pheasant in the wild is not certain. However, in the majority of observed cases the males were apparently associated with only a single female, although occasional polygyny is likely (Beebe, 1918–1922). In captivity polygynous matings are usual. A single male may be easily kept with two or three females (Wayre, 1969) and as many as eight females fertilized per male have been mentioned (Rutgers and Norris, 1970).

During spring, male crowing is the best evidence of territoriality in golden pheasants. Beebe (1918–1922) was able to hear at least four males calling from a single point. Beebe believed that this calling is done from a single point in the immediate vicinity of roosting sites. Calling begins in early morning and evidently carries up to 460 m (1,500 ft), where it might be answered by some other resident male.

Voice and Display

The crowing call of males is a loud, metallic *chak* or *cha-chak*, which lasts about a half a second and is very penetrating. This is uttered by the male only, primarily in spring but rarely also in fall and winter. Other major vocalizations associated with sexual display include a "dance" call by the male, which is a series of varied squeaking, liquid, sharp, and excited-sounding notes given by the male as he is running to display to the female. A "food" call is also presumably given during tidbitting display. Males also utter a very loud hiss during the last phase of lateral display, which may be heard for up to 50 m (165 ft). Golden pheasant also produce several other vocalizations associated with excitement, threat, alarm, and fear (Cramp and Simmons, 1980).

The postural displays of this species are highly developed and have often been described from captive birds (Cheng et al., 1978; Cramp and Simmons, 1980). Beebe (1918–1922) also observed comparable displays in a wild male. Unlike *Syrmaticus* and *Phasianus*, there does not appear to be a wing-whirring display in the golden pheasant. Lateral threat display toward other males seems to be similar to that used in heterosexual encounters; the birds approach one another sideways, with tails spread and bodies held in a position ready to attack. Attacks are made by rapid stabbing with the beak or with kicking of the spurred tarsus. The lateral display toward females is posturally apparently identical to that of the Lady Amherst's pheasant. The male repeatedly runs around the female in wide circles or semicircles and attempts to block her movements; when doing so, he stops and in a sudden rush assumes a static display posture, with his eye held as close to the female's head as possible (figure 52). At the same time, the cape is drawn around the neck on the side nearer the female, which forms a patterned series of concentric circles around the male's eye, and the pupils of the eye are contracted to pinpoints. Simultaneously the wing nearer the female is lowered slightly, the body is tilted toward her to expose the rump and back, and the tail is slanted while the tail-coverts are spread maximally. While this posture is held briefly, the male suddenly utters a sharp hiss. After an extended series of such male displays, the female tends to turn and move away progressively less. The male may follow her or move toward her and attempt to place his foot on her back. At times the female may crouch and invite copulation.

Actual courtship-feeding (tidbitting) by the male has been only rarely observed in the golden pheasant, although Stokes and Williams (1972) reported it for the Lady Amherst's pheasant. While copulating, the male pecks at the female's head, and as soon as cloacal contact is made the female breaks free and flies up to a perch (Cramp and Simmons, 1980). There is relatively little wattle engorgement in this species, which may be related to the fact that the cape display tends to hide nearly all of the male's head below the level of his eyes.

Figure 52. Display postures of male golden pheasant, including lateral display to female (A), normal posture (B), and expanded cape display (C). After various sources.

Reproductive Biology

Breeding Season and Nesting

According to Beebe (1918–1922), breeding in central China begins in April. In Great Britain the eggs are laid in April and May (Cramp and Simmons, 1980). There is almost no information on nesting in the wild, but apparently one clutch was found in a bamboo grove, where the weeds were more than a foot tall and there were many rocky outcrops (Cheng, 1963).

In a study of wild birds at Kuan Kuo Shi Nature Receive in Kweichow Province, 11 nests were found. Eggs were laid on alternate days, and clutch sizes ranged from 5 to 7 eggs (average of 10 nests, 5.8 eggs). Only the females incubated (males usually mated with two females), and incubation lasted 23 days. In two successful nests the hatching rate was 92.8 percent (Wu, 1994).

Incubation and Brooding

In captivity, clutches range from 5 to 12 eggs and the eggs are laid at approximate 24-hour intervals. The male takes no interest in defending the nest or looking after the chicks, but instead tends to distract and drive away the females (Rutgers and Norris, 1970). The incubation period requires 22–23 days, usually the former.

Growth and Development of the Young

The female remains almost totally immobile during incubation, but after hatching she does little other than brood her chicks for the first few days. Later, she calls them to her when she has found a food item and usually offers it beak to beak or sometimes drops it in front of them. The young are apparently able to fly within 12–14 days after hatching. They are cared for by the female until they are about 4 months old or essentially fully grown.

Golden pheasants begin to become sexually active in their first year, when females typically lay. Males of that age are not yet in their final adult plumage or full vigor; thus, probably rather few first-year males are able to mate successfully even though they are sometimes able to fertilize eggs. However, maturity is attained the following spring. Second-year and older males may also court females for a brief period in late summer and cease in September (Cramp and Simmons, 1980).

Evolutionary History and Relationships

Delacour (1977) judged that the genus *Chrysolophus* has affinities both with *Phasianus* and with *Polyplec-* *tron*. *Chrysolophus* resembles the latter in its call notes and the former in its crowing behavior. Certainly the two species of the genus share hissing during intense male display with both *Syrmaticus* and *Phasianus*, but apparently they lack the loud wing-whirring typical of both these genera. Their downy young are more like those of *Phasianus* than *Polyplectron*, and I do not agree that they are distinctly transitional between these two genera. Hybrids of golden pheasants with the Lady Amherst's pheasant are fully fertile, but those with the cheer pheasant are sterile. Hybrids with the silver pheasant have produced sterile females but fertile males, whereas among those with common, green, and Reeves' pheasants only some of the males are fertile (Rutgers and Norris, 1970). This would suggest that both *Lophura* and the *Syrmaticus–Phasianus* group are probably the nearest relatives of *Chrysolophus*.

Status and Conservation Outlook

Little is known of this species' status in the wild, but golden pheasants are regularly snared for their feathers and flesh. The birds are extremely wary, however, and they can apparently survive well under such levels of persecution (Beebe, 1918–1922). McGowan and Garson (1995) considered the species to be secure, but Collar et al. (1994) classified it as near-threatened. Li (1996) described its status as fairly common, but declining. He listed locality records from 13 provinces, primarily Szechwan (42 localities), Shensi (34 localities), Kweichow (30 localities), Hunan (28 localities), and Kwangsi (24 localities). In Great Britain there have been no population estimates since the 1970s, when 500–1,000 pairs were believed present, but the population in Galloway has evidently declined (Marchant et al., 1990).

LADY AMHERST'S PHEASANT

Chrysolophus amherstiae (Leadbeater) 1829

Other Vernacular Names
None in general English use; faisan de Lady Amherst (French); Diamantfasan (German); wokree (Burman); seng-ky, kwa-kwa-chi (Chinese).

Distribution of Species
From about 32° N in eastern Tibet and northern Szechwan south to northeastern Burma and eastward along about 23° N to eastern Kweichow, ascending to 4,575 m (15,000 ft) in the mountains of northwestern Yunnan. This species usually inhabits wooded slopes, bamboos, or other thickets and dense bushes at higher elevations and in colder regions than *C. pictus* (Vaurie, 1965). It has been introduced locally in Great Britain (Bedfordshire and adjoining parts of Buckinghamshire and Hertfortshire). See map 18.

Distribution of Subspecies
None recognized by Delacour (1977).

Measurements
Cramp and Simmons (1980) reported that males have wing lengths of 215–226 mm (8.4–8.8 in) and tail lengths of 830–950 mm (32.4–37.1 in), whereas females have wing lengths of 194–203 mm (7.6–7.9 in) and tail lengths of 286–318 mm (11.2–12.4 in). According to Delacour (1977), tail lengths in males may attain a maximum of 1,150 mm (44.9 in), whereas 375 mm (14.6 in) is the maximum tail length in females. Cheng et al. (1978) reported that five males ranged in weight from 675 to 850 g (1.5 to 1.9 lb) and five females from 624 to 804 g (1.4 to 1.8 lb). Yang (1992) reported that 15 adult males averaged 876.7 g (1.9 lb), as compared with 742.6 g (1.6 lb) in 9 immature males and 602 g (1.3 lb) in 6 females. The eggs average 46 × 35 mm (1.8 × 1.4 in) and have an estimated fresh weight of 31.1 g (1.1 oz).

Identification

In the Field (660–1,702 mm, 26–67 in)
This species is limited to central China south of the range of the golden pheasant. Lady Amherst's pheasant inhabits similar woodlands, scrub, and bamboo habitats, often on rocky ground. The extremely long

and mostly white tail of the male, which has heavy black barring, and the scaly, black-and-white cape pattern are highly distinctive. Vocalizations are very similar to those of the golden pheasant. Likewise, females of the two species probably cannot be safely distinguished in the field, although female Lady Amherst's pheasants have tails that are much more heavily barred with black, buff, and gray.

In the Hand
Male Lady Amherst's pheasants can immediately be recognized by their long (at least 860 mm [33.5 in]) tail, which is strongly barred with black and white, and their scaly black-and-white cape pattern. Females are similar to those of the golden pheasant, but average larger (wing over 180 mm [7.0 in]), are more darkly patterned on the body, and have a tail more heavily barred with black. The orbital skin of female Lady Amherst's pheasants is light blue rather than yellow. Many captive-raised birds show varying degrees of intermediacy in these traits because of hybridization in previous generations.

Ecology

Habitats and Population Densities
In the wild this species is found on high mountain slopes, usually between 2,135 and 3,660 m (7,000 and 12,000 ft), although it occurs locally up to 4,575 m (15,000 ft) in Yunnan. It typically lives in heavy bamboo cover or thorny thickets, but is not usually found in forests (Cheng, 1963). The feral population of Lady Amherst's pheasants in Great Britain is associated with woodlands having dense undergrowth, particularly brambles and rhododendrons (Cramp and Simmons, 1980).

Estimates of densities range greatly from 3 to 140 individuals per square kilometer (7.8 to 362.6 per square mile), averaging about 53 (137.3; Li, 1996). In an area of secondary deciduous forest in Yunnan Province the density of breeding birds per square kilometer was 6.2 (16.1 per square mile), in mixed coniferous and broadleaf secondary forest it was 5.6–5.9 (14.5–15.3) breeders, and in secondary pine forest 4.2 (10.9) breeders. Territories of males during the breeding season were estimated to be extremely small, with two or three females occupying the same area as each male or using nearby areas (Han et al., 1990).

Competitors and Predators
The Lady Amherst's pheasant evidently occurs at about the same level as does the common pheasant and may also have some contacts with the koklass pheasant (Schäfer, 1934). Nothing is known of its predators, although Han et al. (1990) reported blue magpies (*Urocissa erythrorhyncha*) as major egg predators.

General Biology

Food and Foraging Behavior
Beebe (1918–1922) noted that the crops of two males that he examined contained a mass of earwigs of several species, various spiders, small beetles, fern fronds, and 15 bamboo sprouts. It is generally believed that the species favors bamboo sprouts, which is the basis for a local name "*sun-chi*" or "fowl of the buds." When looking for food it searches amid the pebbles and gravel at the base of shady bamboos and may wade in shallow water, turning over pebbles in an apparent search for invertebrates (Cheng, 1963). Cheng et al. (1978) noted that grain, nuts, seeds, and grit were found in nine birds collected in Szechwan and Yunnan.

Movements and Migrations
Lady Amherst's pheasants live at considerably higher altitudes than do golden pheasants. Thus, they have more noticeable seasonal migrations and move to the foothills during severe weather (Cheng, 1963).

Daily Activities and Sociality
The largest groupings of this species are seen in autumn and winter. Groups of one male and one or two females are usual, but as many as 20–30 birds may at times be seen (Cheng, 1963). These groups probably contain several families and include females, young of the year, and some second-year males in full color.

Social Behavior

Mating System and Territoriality
Beebe (1918–1922) observed a male mated to a single female in the wild. However, he also twice observed a male with two females present, which he considered

to be the exception rather than the rule. In captivity Lady Amherst's pheasants are certainly polygynous; in the opinion of Rutgers and Norris (1970), each male should be provided with three or four females.

Territoriality is assumed to exist in this species, inasmuch as Beebe (1918–1922) reported hearing challenge calls in the wild.

Voice and Display
No detailed studies or comparisons with the golden pheasant have been made, but it is believed that the two species have similar vocalizations (Wayre, 1969; Delacour, 1977; Cramp and Simmons, 1980). Likewise, the display repertoires of these two species are virtually the same, if not identical (figure 53). Stokes and Williams (1972) sonographically illustrated the tidbitting call of this species, which consists of three to four calls per second. Each note is very short and has a slight upward inflection, which is very similar to that of *Phasianus*.

Reproductive Biology

Breeding Season and Nesting
Little is known of the breeding season in the wild. However, it is believed to begin early, probably about the beginning of April, and continue through May and into June. There appear to be only two records of clutches taken from the wild; these contained four and seven eggs and were probably incomplete clutches. Both clutches were found in heavy forests, where they had been laid under the protection of a bush (Baker, 1930).

Studies in Gulu, Yunnan Province, suggest that egg-laying there begins in mid-April, with nests typically hidden in thickets or under dead branches (Han et al., 1990). No male involvement in the nesting phase was observed. Six nests were found; the average amount of weight lost by eggs during the 24-day incubation period was 16.5 percent.

Incubation and Brooding
In captivity, clutch sizes of 6–12 eggs are usual, at least for females 2 years old or older. Eggs are probably laid on a daily basis. As with other pheasants, incubation in the Lady Amherst's pheasant begins with the laying of the last eggs. Incubation takes 22–23 days, more often the former. The male plays no role in nest protection.

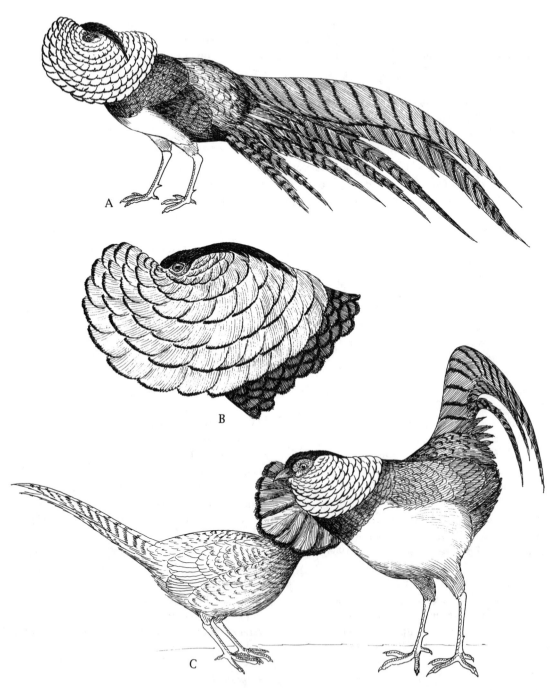

Figure 53. Display postures of male Lady Amherst's pheasant, including lateral display (A), expanded cape display (B), and lateral display to female from opposite side (C). After various sources.

Growth and Development of the Young

As with the golden pheasant, only the female tends the young, which grow at a relatively rapid rate. Seven newly hatched chicks at the San Diego Zoo weighed an average of 24.1 g (0.8 oz) at hatching and 149 g (5.2 oz) at 30 days. At 40 days five of them averaged 207 g (7.2 oz; David Rimlinger, pers. comm.). Males do not acquire their full adult plumage until they are 2 years old, but females often lay eggs their first year. Like golden pheasants, males may also exhibit sexual activity as yearlings, but have limited reproductive success.

Evolutionary History and Relationships

The affinities of *Chrysolophus* with other genera have been discussed in the golden pheasant account. The relationships of the golden and Lady Amherst's pheasants to one another may warrant some additional comments. Hybridization experiments involving these two species have been performed repeatedly. In all cases the conclusion of Phillips (1921), that first-generation hybrids are fully fertile, seems to have been confirmed. However, Phillips also reported that when the F_1 hybrids are bred *inter se*, they exhibited less segregation of male plumage characters than might be expected between two such seemingly closely related species. The segregation was also apparently less than F_2 crosses between the silver pheasant and black-backed kalij that he also bred. Danforth (1950) did some further plumage-inheritance experiments, but concentrated on the male plumage characteristics of *Chrysolophus* × *Phasianus* hybrids. His results were similar to those of Phillips. Baker (1965) found that he could not separate golden and Lady Amherst's pheasants using starch-gel electrophoresis of egg-white proteins. However, he did observe some polymorphism in the albumin component; the golden

pheasant apparently has lost one of three electrophoretically separable types of albumin in the course of speciation. My own interpretation of the available data is that the two forms should be considered a superspecies complex.

It is now apparent that the ranges of the golden and Lady Amherst's pheasants overlap significantly in Szechwan, with wild hybrids sometimes occurring (He et al., 1993). Also, the electrophoretic patterns of their egg-white albumins are virtually identical (Liu and Lu, 1990). Likewise, studies of mitochondrial DNA indicate a very close relationship between these two, with an estimated divergence time of 0.6 million years, as compared with 3.6 million years between *Chrysolophus* and *Phasianus* (Yang, 1992).

Status and Conservation Outlook

Little can be said about the status of the Lady Amherst's pheasant in the wild, but like the golden pheasant it is reportedly very wary and is found in even more remote habitats. Thus, in the absence of other information, it may be assumed that the population is in no obvious danger. Both forms breed extremely freely in captivity, and feral but apparently viable populations have been established in Great Britain. Attempts to introduce the species elsewhere have failed (Long, 1981).

Neither McGowan and Garson (1995) nor Collar et al. (1994) list this species as needing special conservation attention. Li (1996) considered it uncommon in China, with locality records from five provinces, primarily Yunnan. The Lady Amherst's pheasant's status in Burma, which is at the edge of its range, is unknown. The feral population in Great Britain numbered only about 100–200 pairs during the 1970s, and there have been some declines during the 1980s (Marchant et al., 1990).

Genus *Polyplectron* Temminck 1813

The peacock pheasants are small to medium-sized tropical pheasants in which sexual dimorphism is moderately developed. The tail is rounded to graduated, with iridescent bluish green ocelli or banding in males. Bare orbital areas are inconspicuous to only moderately developed in males, wattles are lacking, and crests are short to moderate in length. The tail is variably graduated and has 16–24 rectrices and long, broad upper tail-coverts. The wing is strongly rounded, with the ninth and tenth primaries both shorter than the first and the fourth the longest. In most species the upper wing-coverts, inner secondaries, longer upper tail-coverts, and rectrices are marked with rounded iridescent ocelli or more diffuse iridescence, which is exposed during display. The tail is molted in a pattern beginning with the third from the central pair and proceeding outwardly and inwardly; the second and first pairs are molted between the fourth and fifth and the fifth and sixth pairs, respectively. The tarsus is long and has from one to three short spurs in males. Females are unspurred, considerably smaller than males, duller in color, and, although sometimes having mantle ocelli, usually lack iridescence. Seven species are recognized here.

Key to Species (and Subspecies of Males) of *Polyplectron*

A. Central rectrices with poorly developed ocelli or none; very small bare facial area in adults.
  B. Wing at least 180 mm (7.0 in); 2–5 spurs present (males).
    C. Head and neck gray; mantle with iridescent ocelli: Rothschild's (mountain) peacock pheasant.
    CC. Head and neck brownish black; mantle lacks ocelli: Sumatran bronze-tailed pheasant.
      D. Upper parts less strongly marked: south Sumatran bronze-tailed pheasant (*chalcurum*).

DD. Upper parts more strongly marked: north Sumatran bronze-tailed pheasant (*scutulatum*).

BB. Wing under 180 mm (7.0 in); no spurs present (females).

    C. Head and neck gray: Rothschild's (mountain) peacock pheasant.

    CC. Head and neck brownish gray: Sumatran bronze-tailed pheasant.

AA. Central rectrices with well developed ocelli; large bare facial area in adults.

  B. Mantle and wing-coverts with metallic ocelli; tarsi spurred; head crested (males).

    C. Ventral plumage brown; sides of breast not iridescent; tail without dark subterminal banding or spotting.

      D. Outer tail feathers without ocelli on inner webs; tail under 350 mm (13.7 in): Malay peacock pheasant.

      DD. Outer tail feathers with ocelli on inner webs; tail over 350 mm (13.7 in).

        E. General color dark blackish gray; throat not white; facial skin reddish: Germain's peacock pheasant.

        EE. General color light gray; throat white; facial skin yellowish: grey peacock pheasant.

          F. Ocelli of rectrices surrounded by wide buffy gray bands.

            G. Larger (wing usually over 200 mm [7.8 in]): Ghigi's grey peacock pheasant (*ghigii*).

            GG. Smaller (wing under 200 mm [7.8 in]): Hainan grey peacock pheasant (*katsumatae*).

          FF. Ocelli of tail not surrounded by wide buffy gray bands.

            G. Grayer, with pure white spots and streaks: Lowe's grey peacock pheasant (*bailyi*).

            GG. Browner, with more buffy spots and streaks.

              H. More grayish; the spots more whitish: Himalayan grey peacock pheasant (*bakeri*).

              HH. More brownish; the spots more buffy: Burmese grey peacock pheasant (*bicalcaratum*).

    CC. Ventral plumage black; side of breast iridescent green; tail with dark terminal spotting or subterminal band.

      D. Mantle and wing having numerous ocelli; crest rudimentary: Bornean peacock pheasant.

      DD. Mantle and wing-coverts uniformly iridescent green; distinctly crested: Palawan peacock pheasant.

  BB. Lacking both spurs and crests; mantle iridescence reduced or lacking (females).

    C. Outer tail feathers with definite ocelli on both webs; mantle ocelli often somewhat iridescent.

      D. Posterior margins of dorsal ocelli rounded, with a loose white fringe; body generally grayer: grey peacock pheasant.

      DD. Posterior margins of dorsal ocelli bordered with chevrons or arrow-shaped margins; body generally browner: Germain's peacock pheasant.

    CC. Outer tail feathers with definite ocelli only on outer webs; mantle spotting non-iridescent.

      D. Back feathers and upper tail-coverts lacking ocelli: Palawan peacock pheasant.

DD. Back feathers and upper tail-coverts with blackish ocelli present.
  E. Breast vermiculated gray; ocelli usually with whitish anterior spots: Bornean peacock pheasant.
  EE. Breast vermiculated brown; ocelli usually without whitish anterior spots: Malayan peacock pheasant.

BRONZE-TAILED PHEASANT

Polyplectron chalcurum (Lesson) 1831

Other Vernacular Names
Sumatran peacock pheasant; éperonnier à queue bronzée (French); Bronzeschwanzfasan (German); karo-karo, loekei (Sumatran).

Distribution of Species
The island of Sumatra in mountain forests between 460 and 1,220 m (1,500 and 4,000 ft). See map 19.

Distribution of Subspecies
Polyplectron chalcurum chalcurum Lesson: south Sumatran bronze-tailed pheasant. Mountains of

Sumatra south of the equator, intergrading with *scutulatum*.

Polyplectron chalcurum scutulatum Chasen and Hoogerwerf: north Sumatran bronze-tailed pheasant. Mountains of Sumatra north of the equator, intergrading with *chalcurum*.

Measurements
Delacour (1977) reported that males of *chalcurum* have wing lengths of 162–190 mm (6.3–7.4 in) and tail lengths of 260–380 mm (10.1–14.8 in), whereas females have wing lengths of 150–162 mm (5.9–6.3 in)

Map 19. Distribution of northern Sumatran (N) and southern Sumatran (S) races of bronze-tailed peacock pheasant (Br), and of Bornean (B), Malayan (M), and Palawan (P) peacock pheasants. The indicated Sumatran range of the Malayan peacock pheasant is considered dubious, and the species' known range is now limited to the interior lowland forests of the Malay Peninsula (coarse shading). The enlarged map of Palawan Island shows those areas of steeper slopes above 300 m (985 ft; black) that may still support forest habitats suitable for peacock pheasants as well as recent locality records (arrows). The map of Borneo indicates recent locality records (some unverified) for the Bornean peacock pheasant.

and tail lengths of 180–220 mm (7.0–8.6 in). Two males weighed 425 and 590 g (0.9 and 1.3 lb); four females weighed 238–269 g (0.5–0.6 lb), averaging 251 g (0.5 lb; David Rimlinger, pers. comm.). The eggs average 49 × 36 mm (1.9 × 1.4 in) and have an estimated fresh weight of 35 g (1.2 oz).

Identification

In the Field (356–559 mm, 14–22 in)
The limited Sumatran range of the bronze-tailed pheasant makes it unlikely to be confused with any other species. It is found in heavy cover, often in very mountainous country. Other than some twittering *pitt* calls, almost nothing has been reported of the species' vocalizations. The generally chestnut brown color of the plumage, which lacks definite ocelli, together with the pointed tail should serve for field identification.

In the Hand
The bronze-tailed pheasant is the only peacock pheasant that completely lacks ocelli on the tail feathers, and iridescence is limited to indefinite barring and patches in males. In both sexes the facial area is almost completely feathered, with no obvious bare skin area around the eyes.

Geographic Variation
Geographic variation in the bronze-tailed pheasant is limited and probably is clinal. Birds from northern Sumatra (*scutulatum*) have upper parts that are more strongly marked, with the black barring wider and more distinct.

Ecology

Habitats and Population Densities
This species is found in the montane forests of Sumatra, between about 150 and 1,220 m (500 and 4,000 ft) elevation, where it has not yet been studied in detail.

Competitors and Predators
These are among the smallest of the true pheasants, and are probably dominated by any others that they might encounter in the region. However, nothing is known of bronze-tailed pheasants' competitors or predators.

General Biology

Food and Foraging Behavior
Beebe (1918–1922) examined several crops and found small fruits and insects, but did not provide further specifics. In captivity the bronze-tailed pheasant eats the normal pheasant diet, in addition to peanuts, chopped fruits, seeds, and mealworms (Howman, 1979).

Movements and Migrations
Nothing has been reported on this subject. However, given the equatorial distribution of the species, it is very unlikely that movements of any significance occur.

Daily Activities and Sociality
Nothing has been noted of this in the wild. In captivity bronze-tailed pheasants are highly secretive and spend most of their time under cover or on their perches. They are also extremely quiet birds. Under natural conditions they probably occur in pairs or at most in family groups.

Social Behavior

Mating System and Territoriality
There is no information on this subject in wild birds. Judging from the low degree of sexual dimorphism and the extremely simple male courtship display, only a limited amount of nonmonogamous mating is likely. No crowing or other territorial advertisement behavior has yet been reported for bronze-tailed pheasants.

Voice and Display
Males are said to have a loud and rather harsh call (Rutgers and Norris, 1970), although Delacour (1977) never heard the birds utter anything more than a twittering and repeated *pitt* call. The male's display is laterally oriented, with the tail partially spread in a vertical manner (Delacour, 1977). Roles (1981) is more precise, saying that the male circles the hen with the tail held horizontally, but with the feathers angled into a nearly vertical plane, much like the display of the golden pheasant.

Kenneth Fink and I observed display several times by a single male at the San Diego Zoo. On each occasion the male would approach the female, stop, flap his wings momentarily, then lower his nearer wing directly downward, with the primaries of that wing

Figure 54. Display postures of male bronze-tailed pheasant, including normal posture (A) and wing-flap (B) and lateral freezing (C) sequence. After photos by the author.

scraping the ground in front of its feet, and freeze (figure 54). The tail was not raised or tilted. The only apparent vocalization was a hissing sound, with a click at the end that was probably made with the beak. This hiss is somewhat similar to that made by a male golden pheasant. Unfortunately, it was impossible to see what, if any, response was made by the female. However, the male performed its display in almost precisely the same location each time I saw it, which reminded me of a similar localized activity observed in Rothschild's peacock pheasant.

Recent observations on captive birds by Davison (1985) are more complete than these. He suggests that the major male posturing is lateral, with a lower-ing of the nearer wing and a slight tail-tilting. Other observed male displays were a ruffling of the head and anterior feathers, a crouching run toward the female, head-bobbing (also performed by females), and a "shuffle" that usually occurred between the wing-flap-and-lateral sequence described above. Copulation occurred without associated elaborate display.

Reproductive Biology

Breeding Season and Nesting
There is no information on this in the wild. In captivity, bronze-tailed pheasants lay clutches of two eggs, as do most peacock pheasants. In one case

noted by Delacour (1977) laying occurred in an elevated basket high on an aviary wall; after the eggs were removed, the female laid again less than a month later and a third time during the same season.

Incubation and Brooding
In captivity the incubation period of the bronze-tailed pheasant (under a bantam hen) is 22 days (Delacour, 1977). As far as is known, the male plays no role in reproduction following fertilization.

Growth and Development of the Young
Bronze-tailed pheasant chicks have been reported as fairly easily raised. They require a mixture of the usual pheasant-rearing foods as well as live insects such as ants, ant pupae, and mealworms. The chicks reportedly rarely eat green materials. They attain their full adult plumage in the first year.

Evolutionary History and Relationships
Based on its plumage and display repertoire, there can be little doubt that this is the most generalized of the peacock pheasants and that its nearest relative is *inopinatum*. In the past the genus *Chalcurus* often has been maintained for these two species. Indeed Beebe (1918–1922) listed eight characteristics of plumage and morphology by which *Polyplectron* (*sensu stricto*) differs from *Chalcurus* and in which the former represents a more specialized condition. However, the entire group is fairly easily characterized, and I see no real advantage in separating the two most generalized forms from the remainder.

Status and Conservation Outlook
The bronze-tailed pheasant is probably little bothered by humans for either its flesh or feathers, and the presence of substantial areas of montane forests in Sumatra is believed to provide a degree of security (del Hoyo et al., 1994). McGowan and Garson (1995) considered this species as vulnerable, and Collar et al. (1994) classified it as near-threatened. The bronze-tailed pheasant is still fairly common in some areas, such as the Padang Highlands, and inhabits at least two national parks (Gunung Leuser and Kerinci-Sablat).

ROTHSCHILD'S (MOUNTAIN) PEACOCK PHEASANT

Polyplectron inopinatum (Rothschild) 1903

Other Vernacular Names

Malayan bronze-tailed pheasant, mirror pheasant, mountain pheasant; éperonnier de Rothschild (French); Spiegel-Bronzeschwanzfasan (German).

Distribution of Species

Mountains of the Malay Peninsula from Bukit Fraser and the Semang-ko Pass along the main range to Gunong Uku Kali and Gunong Mengjuang Lebar; Gunong Tahan and Gunong Benom in Pahang. This species inhabits rugged and wooded mountain habitats above 1,065 m (3,500 ft). See map 20.

Distribution of Subspecies

None recognized by Delacour (1977).

Measurements

Delacour (1977) reported that males have wing lengths of 230–255 mm (9.0–9.9 in) and tail lengths of 320–400 mm (12.5–15.6 in), whereas females have wing lengths of 175–190 mm (6.8–7.4 in) and tail lengths of 220–275 mm (8.6–10.7 in). No weights are available. Eggs average 53.2 × 37.0 mm (2.1 × 1.4 in; Vern Denton, pers. comm.) and have an estimated weight of 40.2 g (1.4 oz).

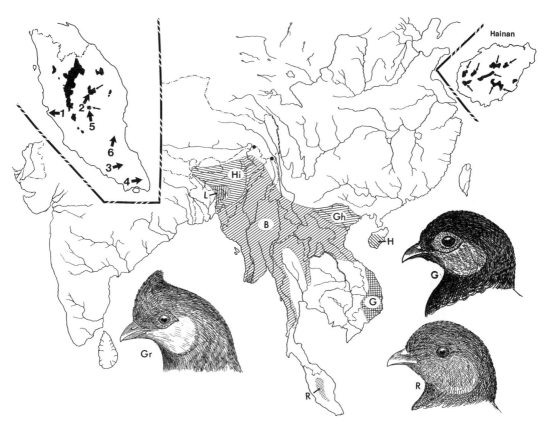

Map 20. Distribution of Burmese (B), Ghigi's (Gh), Hainan (H), Himalayan (Hi) and Lowe's (L) races of grey peacock pheasant (Gr) and of Germain's (G) and Rothschild's (R) peacock pheasants. Solid circles show peripheral records. Upper right inset map shows elevations above 500 m (1,640 ft; black), representing potential distribution of grey peacock pheasant on Hainan Island, with recent locality records shown by arrows. Upper left inset map shows elevations above 900 m (2,955 ft; black) on the Malay Peninsula; the known distribution of the Rothschild's peacock pheasant is limited to these elevations within the Main and Larut Ranges, plus two isolated peaks (arrows). Selected preserves in Malaysia are also shown, including Kuala Selanger (1), Taman Negara (2), Pasoh (3), Panti (4), Krau (5), and Endau Rompin (6).

Identification

In the Field (457–660 mm, 18–26 in)

The Rothschild's peacock pheasant occurs on the Malay Peninsula, where the Malayan peacock pheasant also occurs. However, the latter species is a lowland form unlikely to be found in the high montane habitats of this species. Further, the Rothschild's peacock pheasant lacks a crest and has a face that is fully feathered. The face and head are somewhat more grayish than is the rest of the upper plumage, which tends to be chestnut toned. In both sexes the upper parts are dotted with small, dark ocelli, but those on the tail feathers are limited to the lateral feathers, where they are large and somewhat indefinite. The few reported vocalizations are a low, chuckling, and conversational *chack* call and a quiet, burbling, descending whistle.

In the Hand

Easily identified as a peacock pheasant on the basis of the small ocelli on the mantle, the Rothschild's is the only one that both lacks a crest and extensive bare facial skin, but nonetheless has distinct ocelli on the upper parts.

Ecology

Habitats and Population Densities

Very little is known of the ecology of the Rothschild's peacock pheasant. However, it is said to inhabit the ground story of montane forests above 975 m (3,200 ft) on the Malay Peninsula (Medway and Wells, 1976). Davison and Scriven (1987) found the species only above 900 m (2,955 ft) in tall, lower montane forests, and noted that it has once been reported as high as 1,800 m (5,905 ft) in upper montane forests. All the sites in which this species was encountered were steep, with exposed bedrock in some places, and with some bamboo and stemless climbing palms present. These sites also had thin, mor soils developed over deeper granite-derived materials, but the species has also been reported from sandstone-based soils. The birds were always found on or close to ridge crests.

There are no available estimates of population densities.

Competitors and Predators

No other peacock pheasant occurs at the altitude of the Rothschild's, and it seems unlikely that it would compete with fireback pheasants or with either argus species. Nothing is known with certainty of its predators, although Beebe (1918–1922) judged that civet cats might be an ever-present menace.

General Biology

Food and Foraging Behavior

There is no direct information on this from wild birds except for the crop contents of a single male, which were reported as spiders, white ants, several grubs, and some unidentified creatures. Rothschild's peacock pheasants also eat insects, millipedes, and the fruit of a creeping rattan palm (*Calamus*), according to Beebe (1918–1922). He observed one group of birds scratching in rain-washed gravel, and one of the birds plucking blossoms from a clump of flowers.

Movements and Migrations

There is no information on this. However, given the limited geographic and altitudinal ranges of Rothschild's peacock pheasants, significant movements are highly unlikely.

Daily Activities and Sociality

Beebe (1918–1922) did not observe the Rothschild's peacock pheasant enough to establish their relative sociality or daily activity patterns; indeed a sighting of five birds (three adults and two young) provided his only extended view of the species. Davison and Scriven (1987) stated that the birds are difficult to census because of the absence of any loud calls. Beebe judged that they probably roost in trees free of vines and parasitic plants, and twice he saw single birds apparently sunning themselves on the bare dead limbs of tall trees. Beebe also observed several places where deep holes had been scratched in earth along the sides of the upper slopes of shady ravines, especially on the eastern and northeastern slopes.

Social Behavior

Mating System and Territoriality

Nothing is known for certain of this species' mating system, although its limited sexual dimorphism might suggest monogamy. Ridley (1987) believed that the males of most peacock pheasants desert their mates during or after egg-laying and may remate with a different female, thus exhibiting sequential polygyny. Beebe's (1918–1922) observation of three adults attending two young is interesting; unfortunately he was unable to determine the sexual composition of the adults, although at least one of these was an adult male. Vern Denton (pers. comm.) found no evidence of polygyny in his captive birds when such opportunities were provided. However, recent zoo observations suggest that sequential polygyny may occur (Donald Bruning, pers. comm.).

Although the males may be territorial, they seem to lack any loud advertisement or challenging calls because none has yet been described.

Voice and Display
Beebe (1918–1922) caught a glimpse of display in a wild male, which "stepped into a spot of full sunlight" and momentarily fluffed its plumage, spread and slanted the tail, raised the right wing, and lowered the other. This posture was held only momentarily, and Beebe judged it might be the beginning phase of male display.

In February 1983 I was able to observe briefly a display in a pair of captive birds in the collection of Vern Denton, Livermore, California. The male walked several times in a circular route around a large shrub in the pen; each time as he emerged into the female's full view he stopped and assumed a strong lateral display posture (figure 55), with the tail tilted and spread, the nearer wing lowered, and the farther one raised as he stood motionless for a second or two. No calls were audible from either bird. What struck me as

Figure 55. Display postures of male Rothschild's peacock pheasant, including lateral display to female from farther (A) and nearer (B) sides and crest erection (C). After photos by the author.

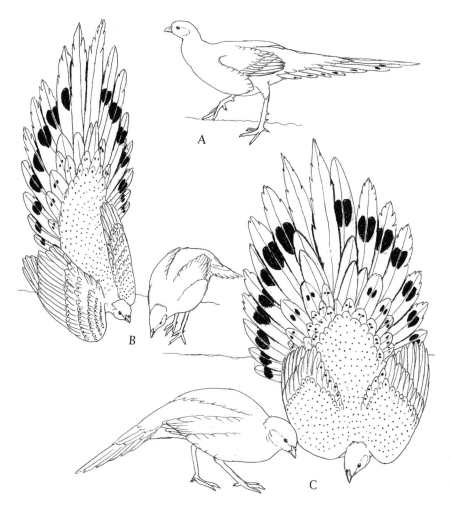

Figure 56. Display postures of the male Rothschild's pheasant, including rapid walking (A) and frontal-lateral display before female (B, C). After Davison (1992) (A) and photos by David Rimlinger (B, C).

especially interesting was the apparent use of his sudden appearance from behind the shrub as a kind of "dramatic entrance." The female seemed to ignore the entire performance. Denton (1978) has described the male displays as "lateral snap postures" rather than full frontal display, as is typical of the grey peacock pheasant. Denton also informed me (pers. comm.) that although the male usually exhibits a "partial" lateral display, when he is actively preparing to breed he has a "full" lateral display. However, an F_1 hybrid with a grey peacock performed the frontal display, whereas a backcross male hybrid to the Rothschild's tended to perform laterally.

Davison (1992) has added useful new information on this species' vocalizations and displays, based on observations of four male and two female captives and their offspring, as well as a hybrid gray × Rothschild's male (of mostly Rothschild's origin). The

male's advertisement call consisted of bursts of one to four notes (usually two); these bursts were uttered at intervals of about 5–6 seconds. The most commonly observed male display consisted of rapid walking with repeated wing-adjusting (figure 56), with the head slightly lowered and held forward. Head-bobbing by the male was also observed frequently, especially when a female walked past or away from the male. Tidbitting behavior by males was also observed, as was a "high-stepping" approach to the female from behind prior to the male's mounting attempts. Wing-flapping was also a common male behavior and often was followed by lateral display, a postural sequence that also occurs in the bronze-tailed peacock pheasant. Ritualized preening was also observed in the male hybrid studied by Davison. During extreme posturing the male hybrid would lower his body so that his breast, bill, and nearer wing would touch or nearly touch the ground. This bird never exhibited an entirely frontal display toward the female, but a fully frontal orientation (figure 56) was observed fairly often among male Rothschild's peacock pheasants at the San Diego Zoo (David Rimlinger, pers. comm.).

Copulatory behavior in the Rothschild's peacock pheasant remains undescribed.

Reproductive Biology

Breeding Season and Nesting
Almost nothing is known of this in the wild, but Beebe (1918–1922) was shown an old nest in a rocky defile at about 975 m (3,200 ft). The nest site was where two large trees had fallen parallel to one another and was close to the hollow formed by the roots tearing out of the ground. Two eggs were present. He judged that late January is the breeding season in Malaysia, because a small chick had been collected in late February.

Denton (1978) stated that a captive female laid her first eggs in March, somewhat later than did his grey peacock pheasants but earlier than the Palawans and about the same time as Germain's. After laying one clutch of two eggs (and their removal) the female Rothschild's waited about 2 weeks before laying another, but the interval was shorter at the start of the breeding season than toward the end. Denton noted that his wild-caught female laid three eggs in 1970, eight in 1971, and several in 1972, when the bird died.

Incubation and Brooding
Apparently two-egg clutches are typical of this species, with the second egg laid 2 days after the first one. The incubation period is 19–21 days, usually the latter.

Growth and Development of the Young
Denton (1978) reported that he was able to raise chicks of this species (or hybrids with grey peacock pheasants) by keeping a bantam chick with them, which would teach them to eat and drink. Egg yolk and chick feed tends to stick to the beak of the bantam chick, which the peacock pheasant chicks then peck at, as they would at their mother's beak. He observed that a group of F_1 hybrids with grey peacock pheasants were almost successful in producing young when only 1 year old; they laid six eggs and hatched two young that died within a week. In birds that more closely approximated pure Rothschild's the period to maturity approached 2 years.

Evolutionary History and Relationships
The Rothschild's peacock pheasant is an extremely interesting species; in nearly all respects it seems transitional between the bronze-tailed pheasant and the more typical *Polyplectron* species. This is true both of the relative development of ocelli on the tail and back and of male posturing. This posturing retains the lateral aspects of the bronze-tailed pheasant display, but begins to exhibit the use of dorsal ocelli in display as well and provides the behavioral precursor for the more fully frontal orientation typical of grey peacock pheasants and their near relatives. This intermediacy in plumage and behavior favors the view that all peacock pheasants should be included in the single genus *Polyplectron*, rather than recognizing the more generalized forms as *Chalcurus*. This view is supported by Denton's (1978) finding that hybrids between Rothschild's and grey peacocks pheasants are fully fertile.

Status and Conservation Outlook
Although there is still very little information, Davison and Scriven (1987) judged that the Rothschild's peacock pheasant is still locally common in its very restricted range on the Malay Peninsula. In addition to occurring throughout the main range, it has recently been recorded in the Larut Range and on Benom and Tahan Mountains to the east and south. It is present in several protected areas, including Taman Negara National Park and Krau Wildlife Reserve, and occurs at Fraser's Hill and Cameron forest areas in the central highlands. The Rothschild's peacock pheasant was classified as vulnerable by McGowan and Garson (1995) and by Collar et al. (1994).

GREY PEACOCK PHEASANT

Polyplectron bicalcaratum (Linnaeus) 1758

Other Vernacular Names

Iris peacock pheasant, chinquis (Chinese) pheasant; éperonnier chinquis (French); Nord-Spiegelpfau (German); tshinquis, chin-tchienkhi (Chinese).

Distribution of Species

Sikkim east through Burma and southeast to Indo-china, plus the island of Hainan. This species inhabits dense evergreen and semievergreen forests to 1,220 m (4,000 ft), rarely higher. See map 20.

Distribution of Subspecies

Polyplectron bicalcaratum bakeri P. R. Lowe: Himalayan grey peacock pheasant. Sikkim, Bhutan, and western Assam south to Manipur and Sylhet, intergrading with nominate *bicalcaratum.*

Polyplectron bicalcaratum bailyi P. R. Lowe: Lowe's grey peacock pheasant. Distribution unknown, but probably western Assam or parts of the eastern Himalayas.

Polyplectron bicalcaratum bicalcaratum (Linnaeus): Burmese grey peacock pheasant. Chittagong, Chin and Kachin Hills, northeastern Assam, Burma east to western Tonkin, south to southern Tenasserim, southwestern Thailand, and central Laos. Also reported from Yunnan and eastern Tibet.

Polyplectron bicalcaratum ghigii Delacour and Jabouille: Ghigi's grey peacock pheasant. Central and North Vietnam north of 16° N and eastern Tonkin, intergrading with *bicalcaratum* in central Laos.

Polyplectron bicalcaratum katsumatae Rothschild: Hainan grey peacock pheasant. The island of Hainan. Considered to be a separate species by Beebe (1918–1922).

Measurements

Delacour (1977) reported that males of *bakeri* have wing lengths of 210–240 mm (8.2–9.4 in) and tail lengths of 350–400 mm (13.7–15.6 in), whereas females have wing lengths of 175–215 mm (6.8–8.4 in) and tail lengths of 230–255 mm (9.0–9.9 in). The race *katsumatae* is smaller, with males having wing lengths of 195–198 mm (7.6–7.7 in) and tail lengths of 285–300 mm (11.1–11.7 in), whereas females have

wing lengths of 160–165 mm (6.2–6.4 in) and tail lengths of 170–190 mm (6.6–7.4 in). Cheng et al. (1978) reported the weights of four male *bicalcaratum* as 660–710 g (1.4–1.6 lb) and of two females as 460 and 500 g (1.0 and 1.1 lb), whereas a single male of *katsumatae* weighed 456 g (1.0 lb). Males of *bakeri* range from about 568 to 910 g (1.2 to 2.0 lb; Ali and Ripley, 1978). The eggs of *bakeri* average 48 × 37.5 mm (1.9 × 1.5 in) and have an estimated fresh weight of 37.3 g (1.3 oz).

Identification

In the Field (559–762 mm, 22–30 in)

This peacock pheasant is fairly widespread in Southeast Asia, occurring in forested habitats up to about 1,830 m (6,000 ft). The birds are generally grayish (males) or brownish (females), with whitish throats and dark ocelli on the back and wings. The male has a bushy crest that is often erected anteriorly in front of the eyes. The tail is generally rounded, with large iridescent ocelli. The species' calls include a loud, whistled *trew-tree'* or *taa-pwi'*. The male also utters a harsh and repeated *putta* call that begins slowly and becomes faster and a sort of "warbling" song, which consists of little chirps that rise and fall and may continue for several minutes. Except on the Malay Peninsula, where it is replaced by the Rothschild's peacock pheasant, this species is not believed to be in contact with any other peacock pheasants.

In the Hand

The grey peacock pheasant is fairly readily identified by the distinctly grayish overall plumage color of males, which is interrupted by numerous iridescent ocelli, and its short, bushy forehead crest and yellowish facial skin. Males of the very similar Germain's peacock pheasant are generally more blackish gray and have reddish facial skin. Females are extremely similar, but female Germain's have more intensely colored facial skin, are more melanistic throughout, and have more distinctly triangular ocelli than do females of the grey peacock pheasant.

Geographic Variation

This is the most widely and variable of the peacock pheasants and the species exhibits variation in both color and size. The smallest and brownest of the races is the easternmost (*katsumatae*), which is isolated on the island of Hainan. The mainland forms also show a clinal trend from gray in the west to brownish in the east (approaching *germaini*, which can be considered the southeasternmost representative of this series), but show little or no size variation. The short crest and ruff of *katsumatae* and its relatively small size also suggest an intermediate link with *germaini*.

Ecology

Habitats and Population Densities

In general, this is a bird of the tropical lowland forests, occurring within the Indian region from the plains and foothills up to about 610 m (2,000 ft) and rarely to as high as 1,830 m (6,000 ft). Grey peacock pheasants are especially common in thick cover along streamside banks and in dense evergreen forests, tangled scrub, and secondary growth or mixed bamboo and thick scrub. A seemingly ideal habitat consists of the heavy undergrowth and small-tree forest developing about 3 or 4 years after a hill rice field has been left uncultivated. Grey peacock pheasants also are associated with ravines in very broken country, where rock outcrops and thick bush cover occur amid virgin forest. At nearly every location that they occur during the breeding season there is water nearby (Baker, 1935).

Estimates of densities in Yunnan range from 1.5 to 2.0 individuals per square kilometer (3.9 to 5.2 per square mile; Li, 1996). The population of the endemic Hainan Island race is also very sparse, with an estimated density of only 1.5 individuals per square kilometer (3.9 per square mile) at Ba Wang Ling (Bawangling) Reserve (Gao, 1991). More recent density estimates for this reserve are 3.7 birds per square kilometer (9.6 per square mile) in 1992 and 1993 (Gao, 1997). Beebe (1918–1922) thought that the birds were locally fairly common; he observed 13 in a walk of about 7 km (4.5 mi), a substantial number in view of the secretive behavior of these pheasants.

Competitors and Predators

There does not appear to be any specific information on these subjects.

General Biology

Food and Foraging Behavior

Beebe (1918–1922) noted that the grey peacock pheasants that he shot near Kachin villages invariably had rice in their crops, as well as small insects. He also noted that they seemed to be especially fond of a small fruit resembling a miniature tomato, which he was evidently unable to identify. According to Ali and Ripley (1978) grey peacock pheasants eat grain, berries, seeds, wild figs, insects, grubs, snails, and other small animals, with termites especially relished and bamboo seeds eaten when they are available. Baker (1930) noted a preference for various figs, wild plums, ber fruit (*Zizyphus*), and bamboo seeds. He considered the birds omnivorous because they often eat invertebrates and sometimes the young shoots of various green crops.

When foraging, the movements of grey peacock pheasants are slow, methodical, and very secretive. They scratch for food in a quiet and restrained manner, move very quietly through the heavy undergrowth around them, and slink under obstructions rather than hopping over them (Baker, 1930).

Movements and Migrations

There is no evidence of significant movements in this species.

Daily Activities and Sociality

Grey peacock pheasants are not social birds, but instead seem to occur in pairs or at most small groups throughout the year. Once the young have left their parents, grey peacock pheasants are said never to occur in flocks (Baker, 1930).

Nothing seems to have been noted of their roosting sites. Beebe (1918–1922) stated that early in the morning and late in the afternoon the male perches on a branch, not its roost, to crow. This occurs nearly every month of the year.

Social Behavior

Mating System and Territoriality

Baker (1930) doubted that this species is polygynous in the wild. Instead he thought that each pair has a well-defined territory from which others are excluded during the breeding season and that the pair bond persisted all year. Baker noted that whenever a bird of one sex is shot or trapped another of the other sex is almost always to be found close at hand. Beebe (1918–1922) also noted that pairs have been recorded in every month and that many seem to be paired all year. However, there is no evidence of the male assisting in the care of the young. Ridley (1987) believed that in most species of *Polyplectron* the pair bond is terminated at about the time of incubation.

Quite certainly the males perform self-advertising displays by crowing, not only during the breeding season but also throughout much of the year (Beebe, 1918–1922).

Voice and Display

The male's crowing call is unusually loud and penetrating; Beebe (1918–1922) described it as *phee-hoo!* It is repeated at intervals varying from every 10 or 15 seconds to every minute or more. He considered this to be a challenge call to other males and an attraction call to females. Baker (1930) called it a loud, chuckling, laughing note. It is usually uttered from an elevated location such as a stump or a low tree branch and sometimes followed by soft wing-flapping sounds. The call has also been described as a musical whistle of two syllables (Wayre, 1969). Another major call is a series of 6–12 croaking calls uttered in rapid succession that sounds more like the noise made by a frog than a bird. This has been variously described as sounding like *wak-wakwak*, *qua-qua-qua*, and *ok-kok-kok*. It is often uttered after some loud noise, such as thunder or a gunshot, and may serve as an alarm note, when it tends to be given as a rapid cackle. Stokes and Williams (1972) illustrated the tidbitting call sonographically; it consists of a very rapid series (about nine per second) of rather high-pitched notes.

The postural displays of the grey peacock pheasant are especially interesting and have been described in detail by various authorities (Pocock, 1911; Beebe, 1918–1922). Pocock noted that although males produce both lateral and frontal displays, the former is rather different from the usual lateral display of pheasants. The nearer wing is scarcely lowered while the farther one is strongly raised, spread to its fullest extent, and pointed forward so that the tip is stretched well in front of the head, which is somewhat retracted. The tail is also fully spread and twisted toward the female to form a nearly vertical fan, thus maximally exposing the ocelli of the tail as well as those of the

upper wing surface to the female's view. The crown feathers are raised and directed forward toward the front of the beak. This display is apparently silent.

The second and more typical display is usually preceded by a melodious chirping whistle. This serves as a food-attraction call to the female, who typically comes running toward the male. The male then faces the hen, raises and spreads his tail and wings, and tilts his body vertically while lowering his head and breast to the ground. The male thus forms a remarkable visual pattern of a somewhat triangular or rounded brown feathered shield, with transverse curving bands of iridescent ocelli (figure 57).

As Pocock (1911) noted, this second posture and visual pattern closely approaches the complex and remarkable frontal display of the argus pheasant. In that species the wing feathers are similarly brought up to meet in the middle above the back, hiding not only the plain-colored back but also even most of the head, which is largely obscured by the primaries of one wing. Therefore, the grey peacock pheasant provides a fine behavioral link between the lateral displays of the Rothschild's peacock pheasant and the more clearly frontal display of the great argus and the true peacocks. Beebe (1918–1922) makes the further point that during the frontal display the iridescent coverts extend beyond the closed primaries to a degree that would not occur during normal wing-spreading and thus further maximize the visual impact of the display. Even the tail is spread to a point that the outermost feathers may be brought to within an inch or two of the ground, forming a complete half-circle of patterned feathers. The female may respond to this display with a similar but less fully developed posture of her own. However, it is more common for her to pay no attention to it. Copulatory behavior has not been specifically described, but probably follows such a display by the male.

Reproductive Biology

Breeding Season and Nesting

In the Indian region, the breeding season of the grey peacock pheasant is from March to June, but mainly occurs in April and May. The nest is usually made in a hollow, often at the foot of a clump of bamboo or in thick bushes among jungle cover. The location is always in very dense jungle, and when placed among bamboos it is always of mixed bamboo and scrub,

Figure 57. Posture of male grey peacock pheasant, including crest erection (A), frontal display (B), lateral display (C), and fronto-lateral display with tidbitting (D). After photos by the author.

rather than in thin and open bamboo forest. A favored site is in tangled secondary growth that is nearly impenetrable close to the ground. Almost invariably the nest will be located no more than a few hundred meters from a stream or pool (Baker, 1930).

According to Beebe (1918–1922), in captivity the nesting season apparently extends from late February to the end of July, but peaks in late March and early April. Captive birds deposit their eggs every 2 days, and a female may produce from 8 to 14 eggs in a single season. Flieg (1973) noted that from three to six clutches were laid per season when clutches were removed, with an average interval of 21 days between clutches. Under captive conditions the clutch size is invariably only one or two eggs, and most but not all observations in the wild support the contention of a two-egg clutch. However, in the race *bakeri* clutches of three or four are not rare (Baker, 1930), and a few clutches of five or six have even been reported. Baker was convinced that these large clutches were not laid by more than one bird, although that would seem difficult to prove.

Incubation and Brooding
In the grey peacock pheasant incubation is performed by the female alone. Baker (1935) believed that the male remains close to the hen while she is sitting, but does not assist in incubation. The incubation period is about 21 days.

Growth and Development of the Young
The care of the young in this and probably the other species of *Polyplectron* is of special interest. The young follow the mother under the cover of her rather long tail, which is held low and somewhat spread. When the female finds a morsel of food, she calls. The chicks dart quickly out to obtain it and as quickly return to the cover of their mother's tail. Furthermore, at least in their first few days, the young are fed directly from the mother's beak rather than being able to pick items up from the ground directly after hatching. Once the chicks have learned to feed themselves, they can catch insects with great success and soon are able to leap into the air and even fly in this pursuit (Beebe, 1918–1922). Sexual maturity is attained during the first year (Flieg, 1973).

Evolutionary History and Relationships
I believe that the grey peacock pheasant and the very closely related *germaini* represent a somewhat transitional form between the more generalized peacock pheasants, the bronze-tailed and Rothschild's, and the more specialized forms in the remainder of the genus. Beyond the superspecies complex of the grey

and Germain's peacock pheasants, the Malayan and Bornean peacock pheasants are probably their nearest relatives. The zoogeographic relationships among this group pose no problems in visualizing the speciation process.

Status and Conservation Outlook
The grey peacock pheasant thrives under conditions of secondary forest succession and thus is likely to remain relatively common indefinitely. It is highly vulnerable to snaring (Beebe, 1918–1922; Baker, 1930), but this would not seem to be a serious threat except very locally. Also, the grey peacock pheasant is apparently not specifically hunted for either its feathers or its flesh. Deforestation is a threat to this and other forest-dwellers in some areas; for example, on Hainan Island there has been a loss of 72 percent of the island's tropical forests since 1949 (Smil, 1983).

McGowan and Garson (1995) considered the Hainan race of this species to be endangered. Li (1996) described its distribution there as being sparse and narrowly confined. Gao (1991) visited nearly all the few remaining areas of potential habitat and found evidence of this race's occurrence in most of them. However, areas of unspoiled montane forests persist only in forest reserves (e.g., Ba Wang Ling, Jian Feng Ling, Bai Shui Ling), and the birds are essentially confined to montane rainforests, montane evergreen broadleaf forests, and ravine rainforests. Hunting on Hainan Island is also a serious conservation problem. A recent population estimate is 2,700 birds in the 740 km^2 (285 mi^2) of forests left on Hainan (Gao, 1997).

GERMAIN'S PEACOCK PHEASANT

Polyplectron germaini Elliot 1866

Other Vernacular Names
Germain's pheasant; éperonnier de Germain (French); Ost-Spiegelpfau (German).

Distribution of Species
South Vietnam north to Quinhon, in damp forests from sea level to 1,220 m (4,000 ft). Also reported once from Thailand (Riley, 1938) and from Cambodia, although Delacour (1977) disputed this. See map 20.

Distribution of Subspecies
None recognized by Delacour (1977). This form is sometimes considered a subspecies of *bicalcaratum* (Beebe, 1918–1922), but is here considered a full species and part of the *bicalcaratum* superspecies complex.

Measurements
Delacour (1977) reported that males have wing lengths of 180–200 mm (7.0–7.8 in) and tail lengths of 250–320 mm (9.8–12.5 in), whereas females have wing lengths of 160–185 mm (6.2–7.2 in) and tail lengths of 220–250 mm (8.6–9.8 in). One male weighed 510 g (1.1 lb) and a female weighed 397 g (0.9 lb; Stephen Wylie, pers. comm.). The eggs average 45 × 35 mm (1.8 × 1.4 in) and have an estimated fresh weight of 30.4 g (1.1 oz).

Identification

In the Field (483–559 mm, 19–22 in)
Limited to Vietnam, this species is the only peacock pheasant found in that area. Both sexes closely resemble the grey peacock pheasant, but they are generally darker and the bare facial skin is distinctly reddish. Males utter a mating call that is a high-pitched *hwo-hwoit*, quickly repeated four to six times. The alarm call is a loud and fast cackle. Germain's peacock pheasants are associated with humid forests up to about 1,220 m (4,000 ft).

In the Hand
The short, bushy crest; rounded tail with numerous ocelli; and the generally brown to brownish gray body coloration separate this species from all other peacock pheasants except the grey. Males of the Germain's are more intensely pigmented and more melanized in general than are males of that species, and their facial skin is red rather than yellow. Females are also very similar, but those of the Germain's have more reddish facial skin, their tail ocelli are better developed, they are less whitish on the throat, and the ocelli on the back and wing-coverts tend to be triangular rather than rounded.

Ecology

Habitats and Population Densities
The habitats of the Germain's peacock pheasant are apparently much like those of the grey peacock pheasant, but are less well described. The birds occur in damp jungles and other forested habitats from sea level up to about 1,220 m (4,000 ft; Beebe, 1918–1922; Delacour, 1977).

There are no estimates of population densities.

Competitors and Predators
No other species of *Polyplectron* occur within the range of this species. However, it may compete to some extent with red junglefowl and *Lophura* pheasants, which probably occupy similar forested habitats. Nothing has been written of possible predators.

General Biology

Food and Foraging Behavior
There is no information on this, but probably this species eats virtually identical foods and forages in a similar manner to the grey peacock pheasant.

Movements and Migrations
There is little reason to believe that any substantial movements occur in this subtropical species.

Daily Activities and Sociality
Like the grey peacock pheasant, this species is relatively nongregarious. The Germain's peacock pheasant apparently occurs as pairs or small family groups in a sedentary and extremely inconspicuous manner.

Social Behavior

Mating System and Territoriality

Most authorities, who apparently made their observations on captive birds, suggest that this species is normally monogamous (Rutgers and Norris, 1970; Roles, 1981). However, there are reports of captive males being mated successfully to two females simultaneously (Roles, 1981).

Nothing is known of territoriality in the wild. Like the grey peacock pheasant the males of this species utter a crowing call during the spring, but it is much harsher and more chuckling than that of the grey. In England, the calling period of the male extends until early June (Keith Howman, pers. comm.).

Voice and Display

As just noted, the crowing call of this species is distinctively loud and harsh and lacks the whistling quality typical of the grey peacock pheasant.

According to Keith Howman (pers. comm.), the Germain's male displays less strongly than does the grey male. He picks up food in the tidbitting display, then tends to retreat as the female approaches, and performs a display that is somewhat intermediate between a lateral and frontal posture, with the tail not tilted so strongly upward as in the grey. Wayre (1969) stated that the male has both a frontal and lateral display and that his call (perhaps the tidbitting call) is a high-pitched musical whistle of three syllables repeated four to six times. Delacour (1977) said that the "love-note" was a repeated and high-pitched *hwo-hwoit*. After a preliminary strutting and stalking around the hen while attracting her attention with a tidbit, he launches into a frontal display (Roles, 1981). Photographs taken by Lincoln Allen (figure 58) suggest a frontal display quite similar to that of the grey, but one that is perhaps more strongly directed laterally and has less wing-spreading and tail-fanning than in the grey. Furthermore, the male Germain's lacks the crest of feathers that in the grey is brought forward above the beak. Copulatory behavior is undescribed.

Reproductive Biology

Breeding Season and Nesting

The breeding season under natural conditions is said to be very extended; Germain's peacock pheasants breed throughout almost the entire year (Delacour,

1977). In captivity (in Florida), the birds begin laying about mid-February, slightly later than the grey peacock pheasants but earlier than Palawan peacock pheasants. Under captive conditions the females lay from three to six one- or two-egg clutches. The intervals between successive clutches averages 21 days, as in the grey and Palawan peacock pheasants (Flieg, 1973). Beebe (1918–1922) reported that four clutches totaling eight eggs are the maximum to be expected from a hen during a single laying season. Nest sites in the wild are undescribed, but are likely to be essentially the same as described for the grey peacock pheasant.

Incubation and Brooding

The incubation period of the Germain's has an average length of 21 days, as in the grey peacock pheasant.

Growth and Development of the Young

Apparently the development of the young follows a very similar pattern to that of the grey peacock pheasant and avicultural techniques for raising them in captivity are essentially the same (Flieg, 1973). The chicks are extremely delicate and almost exclusively insectivorous. If not fed by the mother or foster mother from the bill, they must be hand-fed using forceps or another device until they have learned to pick up food for themselves, which they usually do in a few days (Rutgers and Norris, 1970). Like the grey peacock pheasant, the Germain's attain full adult plumage and apparently also sexual maturity in their first year, although it seems probable that fertility would improve in the second year. Like the other peacock pheasants, the Germain's tends to be long-lived, but is relatively sensitive to cold.

Evolutionary History and Relationships

The Germain's peacock pheasant is obviously a very close relative of the grey peacock pheasant, and indeed Beebe (1918–1922) concluded that they should be considered as only subspecifically distinct. However, Lowe (1925) strongly criticized this position. He said that the specimens that Beebe had based his conclusions on were not typical Germain's and that indeed his plate of *"germaini"* was in fact of an undescribed form of the grey peacock pheasant, which Lowe named *bailyi*. Examination of museum specimens is not convincing, but there seem to be sufficient differences in the voice and displays of live birds

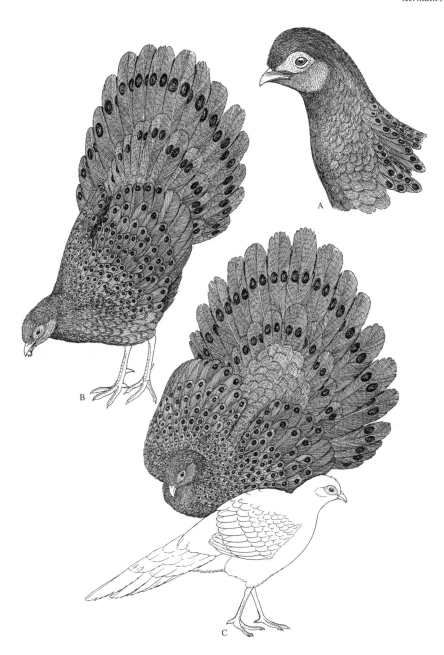

Figure 58. Display postures of male Germain's peacock pheasant, including normal posture (A), tidbitting (B), and full frontal-lateral display (C). After photos by Lincoln Allen.

to warrant calling the Germain's a separate species, although its allopatric range makes any final taxonomic conclusion rather subjective. The two forms have often been hybridized in captivity (Delacour, 1977).

Status and Conservation Outlook
Although the range of this species has undergone great ecological changes due to warfare, it seems likely that the species' ability to utilize dense second-growth following defoliation, deforestation, or logging might have allowed it to survive fairly well.

The Germain's peacock pheasant was classified as vulnerable by McGowan and Garson (1995) and by Collar et al. (1994). In 1990 it was detected regularly during surveys at the large (365-km^2, 141-mi^2) Nam Bai Cat Tien National Park in south-central Vietnam. The species has also been found at the proposed Cat Loc Nature Reserve (McGowan and Garson, 1995).

MALAYAN PEACOCK PHEASANT

Polyplectron malacense (Scopoli) 1786

Other Vernacular Names

Crested peacock pheasant; éperonnier de Hardwicke, éperonnier malais (French); Malaia-Spiegelpfau (German); kuan, kuang (Malayan).

Distribution of Species

Malay Peninsula, north historically to southern peninsular Thailand (now probably extirpated). Perhaps also historically inhabited Sumatra, but this is now considered questionable (van Marle and Voous, 1988). Found below 305 m (1,000 ft) in lowland and foothill forests of interior Malaysia. See map 19.

Distribution of Subspecies

None recognized here. Delacour (1977) considered *schleiermacheri* to be a subspecies of this species, although both Beebe (1918–1922) and Peters (1934) regarded the former as a distinct species, which I follow here.

Measurements

Delacour (1977) reported that males have wing lengths of 200–215 mm (7.8–8.4 in) and tail lengths of 240–250 mm (9.4–9.8 in), whereas females have wing lengths of 180–185 mm (7.0–7.2 in) and tail lengths of 180–190 mm (7.0–7.4 in). One male weighed 680 g (1.5 lb; Riley, 1938). The eggs average 46 × 37 mm (1.8 × 1.4 in) and have a fresh weight of 39–42 g (1.4–1.5 oz; Bruning, 1983).

Identification

In the Field (406–533 mm, 16–21 in)

Limited to the Malay Peninsula, the only other species likely to be encountered in this area is the Rothschild's peacock pheasant, which occurs at higher altitudes and lacks both a crest and orange facial skin. The Malayan peacock pheasant is generally limited to forested habitats under 305 m (1,000 ft). It is usually found in heavy cover, as single birds or pairs. Males have a loud, cackling *kwock-kwock* call that is repeated frequently. Another call is a series of low clucks that descend in pitch and trail off. Females are virtually crestless, have shorter tails than males, and have dark ocelli with buffy borders on their mantles and upper tail-coverts.

In the Hand

Male Malayan peacock pheasants can be identified by the combination of orange facial skin, a long greenish blue crest that is posteriorly oriented, and a generally brownish body studded with many iridescent ocelli on the back and tail. Females have a considerably shorter (under 200 mm [7.8 in]) tail, a pale gray throat, and conspicuous buffy borders to the dorsal ocelli, but otherwise are rather uniformly brown. They approach the female Palawan peacock pheasant in appearance, but the latter lacks the buff-bordered dorsal ocelli and has a more extensive pale gray area on the throat and sides of the face. Females are also very similar to those of the Bornean peacock pheasant, but are less reddish, have a longer tail, and the upper tail-coverts have metallic ocelli, which are sometimes lacking in the Bornean form.

Ecology

Habitats and Population Densities

In Malaysia this species occurs in tall primary and secondary lowland dipterocarp forests, up to at least 180 m (590 ft) elevation. It occurs on level and gently sloping ground and seems to avoid steeply sloping sites. Breeding males are typically associated with dense undergrowth, which is often near water and has bankside vegetation rich in palms and various broadleaf monocotyledons (Davison, 1983*b*). Other authors have reported the species occurring as high as about 305 m (1,000 ft) in Malaysia. Its occurrence in Sumatra is highly doubtful (van Marle and Voous, 1988; McGowan and Garson, 1995). Lowland forest destruction has probably eliminated the Malayan peacock pheasant from Thailand.

McGowan (1991, 1992, 1994) analyzed the behavioral ecology of this species in various study sites on the Malay Peninsula. When he compared microhabitats used by male clusters ("exploded leks") for display sites versus those not used, favored locations were in areas away from rivers and had fewer fallen trees. Furthermore, display scrapes were typically located in vegetationally stable habitats, although it proved impossible to predict their exact location or to anticipate the temporal occurrence of display activities. However, the presence or absence of ground vegetation probably influences such site choices by males. Males prefer to display in small, open areas that are surrounded by dense vegetation (Donald Bruning, pers. comm.).

Within a 256-ha (632-acre) area of primary forest, Davison (1983*b*) found 26 male calling centers or presumed territories. This provides an estimated density of 1 male per 9.8 ha (24.2 acres). Another area had 13 calling centers in a 200-ha (494-acre) study plot, or a mean of 1 male per 15.4 ha (38.0 acres).

Competitors and Predators

There is no specific information on these subjects. Davison and Scriven (1987) found that the Malayan peacock pheasant is altitudinally separated from the Rothschild's on the Malay Peninsula. The Malayan probably has only very limited contacts with the grey peacock pheasant at the extreme northern edge of its range.

General Biology

Food and Foraging Behavior

Beebe (1918–1922) observed wild birds feeding on termites and fly larvae and pupae associated with decaying mollusc shells. Davison (1983*b*) suggested that food supplies affect the male's mating behavior, with more intense activity in years of heavy fruiting of dipterocarp trees. Davison found that *Dracaena* and an unidentified climbing plant with sugary red berries accompanied dipterocarp fruiting; these were included in the diet. Nothing else seems to have been written on their foraging under natural conditions, although according to Beebe (1918–1922), foraging seems to be done by pecking and scratching in the usual pheasant manner.

Movements and Migrations

Apparently at least the males are quite localized on territories for much of the year, whereas females are seemingly more mobile (Davison, 1983*b*). However, there is little reason to believe that any extensive movements occur in the Malayan peacock pheasant.

Daily Activities and Sociality

Davison (1983*b*) made 28 sightings of solitary males, 17 of solitary females, and 6 of solitary birds of undetermined sex, which he interpreted as habitual solitariness in both sexes. There were only four sightings of pairs; two of these were at display areas, whereas the other two were far from any known display area. Most sightings of both sexes were obtained during early morning hours, with the males appearing slightly later than the females, which reflects the earlier attendance by males on their display areas. Beebe (1918–1922) reported seeing Malayan peacock pheasants usually in pairs, with one group of five being the largest he ever observed.

Social Behavior

Mating System and Territoriality

Although most calling has been reported during early morning hours, Davison (1983*b*) noted nocturnal calling during two years when there were synchronous flowering and fruiting periods of dipterocarp trees. In such years a larger number of males were observed calling at his study sites, and there was a larger average number of display areas per calling male. Of 92 such display areas observed, 88 were on old game trails or little-used man-made trails and all were on level or gently sloping ground. The smallest of these was only 400 mm (15.6 in) in diameter, whereas the largest was about 1×2 m (3.3×6.6 ft). These display areas are kept clean by the males, although feathers and droppings accumulate on them. Any single display area might be used for as long as several months, and some are used again in successive years.

Voice and Display

Advertising by males is performed by two types of calls. The first is a short two-noted whistle, with the second note longer and inflected upward. Davison (1983*b*) found that short calls were given about 1 minute apart for up to 30 minutes while males were alone at their display areas or elsewhere in the forest, either on the ground or perched. The long call is a series of loud, grating notes; these initially have intervals of about 5 seconds between notes, but gradually the interval is reduced to about 1.5 seconds. Up to 117 such calls were heard in a series, which lasted up to 3 minutes. Males often replied to one another's long calls by uttering either short calls or long calls of their own. Davison believed that females might sometimes utter long calls, but was unable to prove this. Males also uttered harsh cluck calls, often in long series of up to as many as 230 notes. While at their display sites males also spent some time preening, cleaning the ground surface, and feeding in the immediate vicinity (Davison, 1983b).

Davison saw only two female visits to such calling males and observed display at one of these. When the female approached, the male ceased calling and ruffled his head and neck feathers and performed ritualized preening, courtship feeding, and erect and lateral postures. A culminating frontal display, which has been observed in captive Malayan peacock pheasants, was not seen at this time. Davison (1983b) also observed agonistic display between two males at the edges of their territories. In this case the two males circled one another and called at intervals. Each maintained a lateral posture, with the nearer wing lowered, the farther one raised over the back, and the secondaries and rectrices spread to exhibit the ocelli. Intermittently the two would face one another, with tails spread and heads raised to expose a broad buffy stripe from the throat to the abdomen. The birds circled one another at increasing distances until they were about 40 m (130 ft) apart, when the encounter was abandoned.

More detailed observations of display in captive birds have been provided by Stapel (1976), Bruning (1977), and Davison (1983a). Stapel and Davison have both noted that lateral and frontal displays are performed; although twittering tidbitting calls are uttered, they are relatively quiet. During lateral display hissing and feather-quivering sounds are made. Feather-preening movements ("flagging" in Davison's [1983a] terminology) are frequent. Stapel noted that in one case the male performed a series of preening and dancing movements around the female with a silent lateral display "identical in posture" to that of the Palawan peacock pheasant. During lateral display the crest feathers are extended anteriorly in front of the forehead and the dorsal feathers are strongly raised (figure 59). Davison (1983a) stated that during lateral display toward females the male will tilt his bill downward and variably raise the farther wing above the back, sometimes leaping into the air and hissing as he lands. During tidbitting the male may flap the farther wing or assume a frontal posture with the head lowered until the bill and breast touch the ground. The two wings are variably spread, and the head is twisted to expose only a single eye. Lateral displays predominate over frontal postures, according to Donald Bruning (pers. comm.).

Although copulation has not been seen, observations and films made by J. G. and P. A. Corder have shown that the female begins to perform a distinctive head-lowering and swinging of the head and tail from side to side about 6 or 7 days before laying her single egg; this posture is presumably a mating invitation. The Corders also observed a mutual lateral posturing by the mated pair, which they termed "cross-lateral" display. During this display sequence the two birds circled one another while performing ritualized preening of the head and neck feathers, lifting the farther wing, and ruffling the tail feathers. The male oriented himself so that his spread and tilted tail was positioned directly toward the female's head. Further, the birds would raise their heads while erecting their crests and walk in circles while on tiptoe in small drumming steps. The female uttered soft twittering notes during mutual lateral display, and the male produced single soft whistles. The male's frontal lateral display was typically performed after he placed a tidbit in front of him and as the female approached to receive it. At this time the male's head touched the ground, with his head held parallel to his body and his nearer eye directly facing the female.

Reproductive Biology

Breeding Season and Nesting
The nesting season of the Malayan peacock pheasant in the wild is rather uncertain. Davison (1983b) reported two nests, which he located on 15 March and 2 April. Both contained single eggs; one was on the top of a termite mound about 1.4 m (4.6 ft) above the surrounding ground, whereas the other was on leaf litter at ground level and among upperstory plants. Davison doubted the authenticity of a supposed two-egg clutch reported by Baker (1928) from Thailand, which might actually have come from a grey peacock pheasant. No other nests have been described from the wild.

Bruning (1977, 1983) has reported considerable success in breeding the Malayan peacock pheasant in captivity. He confirmed that a single-egg clutch is

Figure 59. Display postures of male Malayan peacock pheasant, including crest erection and tidbitting (A), lateral display (B), female (for comparison) (C), frontal-lateral display (D), and male rectrix patterning (E). After photos by the author (A), John Bayliss (B), and Delore Jung (D).

normal for this species. Additionally, he found that a female will recycle and lay another egg every 3 or 4 weeks. Under tropical conditions a pair might breed nearly every month of the year.

Incubation and Brooding
Bruning (1977, 1983) noted that the female of one pair of captive birds incubated its single-egg clutch regularly, but that the "parents" would care for and brood the chick, which suggests a greater male role in the nesting and brooding phase than might be considered typical. Bruning reported a 22–23-day incubation period.

Growth and Development of the Young
Bruning (1983) said that newly hatched young weigh from 24 to 28 g (0.8 to 1.0 oz). Because the average fresh egg weight is 40.4 g (1.4 oz), this indicates a hatching weight of about 65 percent of the egg's fresh weight. Chicks lose up to 4 g (0.1 oz) in their first day or so, but gain it back by the third or fourth day. For the next 2 or 3 weeks chicks gain about 2–4 g

(0.07–0.14 oz) per day. By the end of the first month chicks average about 100 g (3.5 oz), or four times their hatching weight; they double that weight during the second month.

One female laid her first egg when only 8 months old. Spurs begin to develop on males when they are 10–12 months of age. However, male chicks as young as 3–8 months old may try to display to adults. Medway and Wells (1976) noted that one hand-reared chick initially displayed at 23 days and that its tarsal spurs began to develop in its ninth month.

Evolutionary History and Relationships
There can be no doubt that the Malayan and Bornean peacock pheasants are extremely closely related; more will be said of that in the discussion of the latter species. Otherwise, this superspecies shows some strong behavioral and morphological similarities with the grey and Germain's peacock pheasants (primarily in most of their plumage characteristics). A few resemblances to the Palawan peacock pheasant can also be perceived, such as in the intensely black underparts and slight tendency toward tail-banding in the Bornean peacock pheasant. The Malayan and Bornean peacock pheasants are unique in the family (if not in the entire Galliformes) in having one-egg clutches. Perhaps this partly reflects their nearly year-long breeding potential. Davison (1983*b*) corre-

lated relative breeding effort in the Malayan peacock pheasant with local food supplies. Thus, it is possible that this species inhabits an environment with fewer suitable foods than would seem typical for pheasants in general; a reduced clutch-size may be one reflection of this.

Status and Conservation Outlook
This species is dependent upon primary forests and well-developed secondary forests in lowland areas, which are fast being removed from the Malay Peninsula (Davison, 1981*c*). Therefore, its long-term conservation outlook must be considered as doubtful. The Malayan peacock pheasant has been reported from five protected areas on the Malay Peninsula, including Taman Nagara National Park; Krau, Sungai Dusun, and Sunkai Wildlife Reserves; and Pasoh Forest Reserve (McGowan and Garson, 1995). However, the species is not known from the additional preserves shown in map 20. As noted earlier, it has been essentially extirpated from Thailand. The Malayan peacock pheasant was classified as vulnerable by McGowan and Garson (1995) and by Collar et al. (1994). An international studbook for the species has been established, which should help provide the most efficient captive breeding possible. As of January 1998, over 350 live specimens were recorded in this studbook.

BORNEAN PEACOCK PHEASANT

Polyplectron schleiermacheri Brüggemann 1877

Other Vernacular Names

None in general English use; éperonnier de Bornéo (French); Borneo-Spiegelpfau (German).

Distribution of Species

Lowland forests of Borneo from Paitan in the extreme north to the southeastern end of the island and southwest an unknown distance in the forests between Banjarmasin and Pontianak. See map 19.

Distribution of Subspecies

None recognized here. This form is considered by Delacour (1977) to be a subspecies of *malacense*, but is regarded by Beebe (1918–1922) as a distinct species. Both its plumage and display characteristics would favor the latter position.

Measurements

Delacour (1977) reported that one male had a wing length of 200 mm (7.8 in) and a tail length of 200 mm (7.8 in), whereas one female had a wing of 165 mm (6.4 in) and a tail of 155 mm (6.0 in). Beebe (1918–1922) reported the male wing length as 200 mm (7.8 in), the male tail length as 190 mm (7.4 in), and the female wing length and tail length as 180 mm (7.0 in) in both cases. No weights or egg measurements are available.

Identification

In the Field (432–508 mm, 17–20 in)

This is the only peacock pheasant present on Borneo, which simplifies identification. Its ecology and behavior are almost undescribed. However, Bornean males are the only peacock pheasants that have nearly black underparts, iridescent green on the sides of the neck and forebreast, and a nape that is greenish to violet with recurved feathers. Females are best recognized by their association with males, but very closely resemble females of the Malayan peacock pheasant.

In the Hand

Males are easily recognized by their distinctive nape, which resembles a Victorian ruff of grizzled gray, black, and metallic green to violet, and their distinc-

tively blackish underparts. Females have slightly shorter tails than those of the Malayan species (about 155 mm [6.0 in] versus 180–190 mm 7.0–7.4 in]), a more reddish overall body coloration, and much darker underparts of almost solid brownish black. Delacour (1977) states that the upper tail-coverts lack blue ocelli in this species, but Beebe (1918–1922) states that these are often well developed and that the gloss in the dorsal ocelli is considerably greater than in *malacense*.

Ecology

Habitats and Population Densities

The Bornean peacock pheasant is limited to the lowland forested areas of Borneo, where it has been only rarely observed by naturalists. Beebe (1918–1922) was able to find three locality records for it, including the type locality of Moera Tewah, southeastern Kalimantan; Paitan, Sabah; and central Sarawak, toward the Kalimantan border. Delacour (1977) reported a fourth locality in the area between Banjarmasin and Pontianak of southwestern Kalimantan.

Nothing is known of its population densities. However, Beebe (1918–1922) judged that Bornean peacock pheasants must be extremely uncommon because the natives he queried were so unacquainted with them.

Competitors and Predators

This species inhabits the same area as the Bornean great argus and a species of civet cat (Beebe, 1918–1922), but the interactions among these are unknown. The wattled pheasant also inhabits the same general area and probably occupies similar habitats, as do crested and crestless fireback pheasants.

General Biology

Food and Foraging Behavior

Nothing is known of this in the wild. In captivity Bornean peacock pheasants are much like the Malayan and others of the genus; they eat a wide variety of foods, but tend toward an insectivorous diet.

Movements and Migrations
There is no information on this subject.

Daily Activities and Sociality
Nothing has been reported on this.

Social Behavior

Mating System and Territoriality
Judging from behavior of captive birds, Bornean peacock pheasants are quite asocial. They must be kept in pairs or alone to avoid birds killing or otherwise harming one another.

Territorial or other self-advertisement calls by males is as yet unreported.

Voice and Display
The little that is known of this derives from a few birds that have been kept by Vern Denton of Livermore, California. He reported (Denton, 1978) that the hand-raised male was highly aggressive and assumed a ruffled-feather posture whenever he sensed an intruder in his domain (figure 60). In this posture the male stalked silently about, presenting a very different appearance from that normally seen in wild-trapped males. Denton did not mention any calling at this time or any crowing at other times. When I observed the bird he remained completely silent, although he retained the aggressive posture constantly, sometimes thrusting the head more forward, but generally resembling a strutting turkey. During intense threat the male would assume a more asymmetric posture, with the spread tail tilted toward me and the farther wing surface raised slightly. A true frontal display was never seen even immediately before an actual attack. During such an attack the bird flew into my face and attempted to strike with his feet. The central iridescent green breast feathers were vertically parted to reveal a contrasting white stripe, which at maximum was almost an inch wide. The crest of the head was also raised to form a nearly triangular shape from the front, and the feathers formed a pointed tip toward the end of the bill, as in the Malayan species. According to Denton (pers. comm.), male display to the female is also lateral and is not maintained for very long. Denton has only observed a frontal display when the male was offering the female a tidbit, when "he had real mating in mind." At that time the head was lowered almost to the ground and the wings were lowered to the sides, but were not fully extended.

Reproductive Biology

Breeding Season and Nesting
Nothing is known of this under natural conditions. However, in captivity Denton (pers. comm.) has found that the female of one pair laid three eggs, all in one-egg clutches one year (1972), whereas an F_1 hybrid female (Bornean × Malayan) laid from one to seven eggs per year over a 5-year period.

Incubation and Brooding
The eggs of the Bornean peacock pheasant are very similar to those of the Malayan. As in the Malayan species, the incubation period is 20–22 days, averaging 21 days (Denton, 1978).

Growth and Development of the Young
Denton (1978) found that young Bornean peacock pheasants were fairly easy to raise. However, they must be kept separate from other species, including other peacock pheasant chicks, because they seem unable to compete with them. The period of sexual maturity and full adult plumage is apparently 2 years. Beebe (1918–1922) believed that first-year females had less fully developed ocelli on the tail-coverts than do adults. Although males develop enough plumage to be sexed by the first summer after hatching, they do not breed until their second year (Beebe, 1918–1922).

Vern Denton (pers. comm.) has raised several hybrids between the Bornean and Malayan peacock pheasant. Apparently the females are as fully fertile and as productive as the full-blooded birds.

Evolutionary History and Relationships
In recent years, this species has generally been considered only as a subspecies of the Malayan peacock pheasant; certainly the two forms are very closely related. However, the highly distinctive male plumage and some minor differences in posturing makes me believe that it is better to consider them as full species or at least as allospecies. The two are as different morphologically as are, for example, the grey and Germain's peacock pheasants. It seems premature to merge the Malayan and Bornean peacock pheasants before more detailed work is done. Certainly there are interesting differences between them.

Figure 60. Display postures by male Bornean peacock pheasant, including lateral display with anterior-posterior head-jerking (A, B), frontal view of crest and breast (C), and maximum lateral display (D). After photos by the author.

Status and Conservation Outlook

Almost nothing can be said with certainty about the status of this elusive species. However, its future surely depends on the fate of the primary lowland and submontane forests of Borneo. The Bornean peacock pheasant is a critically endangered taxon (Collar et al., 1994; McGowan and Garson, 1995), with a population estimated by McGowan and Garson as no more than 1,000 birds. Very little can be said of its current distribution because the species only has been reported from one protected area

(Bukit Baka-Bukit Raya). However, several such locations occur within its presumed range; these include Gunung Palum and Danu Sentorum in Kalimantan, as well as Danum Valley and Tabin in Sabah. O'Brien and Kinnaird (1997) reviewed the distributional status of the Bornean peacock pheasant in Central Kalimantan, using questionnaires and other available information. It appears that the species still occurs but is extremely rare along the Mahakam River of eastern Kalimantan. However, O'Brien and Kinnaird found evidence of its probable continued occurrence at more than 30 locations, mainly at elevations between 100 and 500 m (330 and 1,640 ft), and extending from the Limandau River in the southwest to the Mahakam River in the northeastern portions of south-central and east-central Kali-

mantan. Older records extend from Pontianak near the coast of West Kalimantan to Samarinda on the lower Mahakam River in East Kalimantan, but these lowland sites are now much degraded ecologically.

PALAWAN PEACOCK PHEASANT

Polyplectron emphanum Temminck 1831

Other Vernacular Names

Napoleon's peacock pheasant; éperonnier de Napoléon (French); Palawan-Spiegelpfau (German); sulu maläk, dusan bërtik (Palawan).

Distribution of Species

The island of Palawan in damp, primary forests. See map 19.

Distribution of Subspecies

None recognized by Delacour (1977).

Measurements

Delacour (1977) reported that males have wing lengths of 190–195 mm (7.4–7.6 in) and tail lengths of 240–250 mm (9.4–9.8 in), whereas females have wing lengths of 170–175 mm (6.6–6.8 in) and tail lengths of 165–170 mm (6.4–6.6 in). Two males averaged 436 g (1.0 lb), and two females averaged 322 g (0.7 lb; various museum and zoo records). The eggs average 45 × 36 mm (1.8 × 1.4 in) and have an estimated fresh weight of 32.2 g (1.1 oz).

Identification

In the Field (406–508 mm, 16–20 in)

This species is limited to the islands of Palawan, where it is the only pheasant.

In the Hand

Males are easily identified by their extensive amounts of green iridescence on the neck, mantle, and inner wing-coverts and by their long, erectile crest. The facial skin around the eye of the male is bright red, and the rounded tail has a buffy terminal band. Females have very short crests, and their only iridescence occurs on the ocelli of the outer tail feathers and longer upper tail-coverts. Otherwise the general female coloration is dark brownish black, with a whitish gray throat and with pale gray extending from the throat to an area above the eyes.

Ecology

Habitats and Population Densities

The Palawan peacock pheasant is associated with primary forests of Palawan's coastal plain and, at least in some areas, with the more arid woodlands and scrub of the foothills. Although uncertain, it is possible that the species may be adapted to these secondary forest habitats (King, 1981). Its total altitudinal range is unreported, but the maximum land elevations on this small island are only about 2,000 m (6,650 ft).

Using nearest-neighbor sampling techniques, studies at St. Paul Subterranean River National Park (Caleda et al., 1986; Caleda, 1993) have provided density estimates ranging from 8.5 (in logged areas) to 34 (in forest edge areas) males per square kilometer (22.0 to 88.1 per square mile). Fixed-width transects produced somewhat different results, with primary forests having the highest density of males (25 per square kilometer [64.8 per square mile]) and logged-over areas the lowest (15 per square kilometer [38.9 per square mile]). Areas most heavily used in the park were those having diverse tree and shrub vegetation providing differing cover types for food and shelter throughout the year. McGowan et al. (1989) found evidence of the birds' presence at locations between 400 and 660 m (1,310 and 2,165 ft) elevation. Most encounters were made along primary-forest transects and none occurred in logged areas.

Competitors and Predators

No other pheasants occur on the island of Palawan. It has been suggested that wild cats may be a major predator of this species (Beebe, 1918–1922).

General Biology

Food and Foraging Behavior

Nothing has been written on this subject in wild birds. However, in captivity the birds are typical peacock pheasants, showing a preference for live insects, fruit, seeds, peanuts, and various other foods high in protein or sugars. A survey by Dierenfeld et al. (1998) indicates that peacock pheasants primarily eat insects under natural conditions and thus consume about 35–40 percent protein in their overall diets.

Movements and Migrations

There are probably few if any substantive movements in the Palawan peacock pheasant, given its extremely limited altitudinal range and distribution.

Daily Activities and Sociality

Little has been noted on this subject, but Palawan peacock pheasants seem to occur in pairs even outside of the breeding period (Beebe, 1918–1922).

Social Behavior

Mating System and Territoriality

J. Whitehead (cited in Beebe, 1918–1922) believed these birds to be monogamous in the wild. Referring to captive birds, Jeggo (1975) stated that they are "strictly monogamous."

Territoriality is as yet unproven. However, J. Whitehead (cited in Beebe, 1918–1922) observed that, like the Malayan peacock pheasant and argus pheasant, males have "showing off" arenas that are about 1 m (3 ft) in diameter and often on a hump of earth in some unfrequented part of the forest. This would certainly suggest a territorial attachment, although the manner of territorial advertisement is as yet unstudied.

Voice and Display

Although male crowing behavior may occur in this species, it is as yet undescribed. Postural displays of the Palawan peacock pheasant have been described by Lewis (1939), Jeggo (1975), and Roles (1981). Display typically begins with the male strutting briefly around the female while spreading the feathers of the lower neck and mantle to form a small cape and while holding a food morsel and bobbing the head up and down. The food is then dropped as the female approaches, and the male immediately assumes a full lateral display posture, with the tail fully spread and twisted in a full lateral presentation to the female. The farther wing is also raised vertically and partially opened, and the nearer wing is dropped to the point where the primaries touch the ground (figure 61). The crest is erected and pointed diagonally forward. The head is held in a somewhat retracted position, so that the beak is hidden behind the cape and only the

eye and the white area immediately around it are exposed. This emphasizes the eye in much the same way as in the display of ruffed pheasants and, more interestingly, that of the great argus. The male becomes remarkably flattened and rounded in outline, resembling "a circular plate standing edgewise to the ground and lifted by an invisible stand an inch or two in the air" (Lewis, 1939). At the peak of the display the male may utter a prolonged cry like a soft groan. I have noticed that the lateral display is often preceded by a quick and nearly silent wing-flapping. Also, the display may be followed by a static arched-neck posture that is held for several seconds. As an apparent threat, females sometimes also display laterally in a similar manner to males. Copulation is apparently performed without any specific preliminary posturing; instead the male may simply walk up behind the hen and mount her (Roles, 1981).

Reproductive Biology

Breeding Season and Nesting

In the wild the nesting season is believed to be during December and January, although apparently no nests have ever been described (Beebe, 1918–1922). However, in captivity nesting on the ground and in elevated boxes has been observed. In captivity the birds begin laying (on Jersey Island) at the end of March and may continue until late July or even mid-August (Jeggo, 1975). In Florida they begin in early March (Flieg, 1973). Jeggo (1975) reported that the average production of eggs per female over a 2-year period (involving six layings) was 4.7 eggs per season, whereas Flieg (1973) noted that from three to six eggs per pair could be obtained in a season. The clutch size in captivity is usually two, but sometimes only a single egg is laid, especially in younger females. There is an average interval of 21 days between successive clutches (Flieg, 1973). Under tropical, captive conditions, up to five clutches have been produced per year (Donald Bruning, pers. comm.).

Incubation and Brooding

The incubation period is only 18 (Jeggo, 1975) to 20 (Flieg, 1973) days, probably averaging about 19 days. It is performed by the female alone, but both sexes call to and help feed the chicks (Jeggo, 1975).

Figure 61. Display postures of male Palawan peacock pheasant, including crest-raising and tidbitting (A), tidbitting before female (B), normal head posture (C), and lateral display to female (D). After photos by the author.

Growth and Development of the Young

Like the other peacock pheasants, Palawan chicks must learn to feed from the ground after having first been fed beak to beak by the parent or foster-parent. In the case of a young bird reared by its parents (Jeggo, 1973), for the first few days the hen and chick both stayed well under cover, with the male often in close attendance. Gradually the chick came progressively more into the open when the female called to it to take food, but it still returned to cover as soon as possible. Until the chick was first able to fly to an elevated roost at 13 days, the female brooded it at night on the ground, while the male roosted higher up. When 44 days old the chick was still roosting at night under the female's wing (see figure 13 in chapter 4). By that time it was fully feathered, except for a small area of pin feathers on the center line of its head. By 70 days the development of the juvenile plumage had been completed (Jeggo, 1973).

A sample of eight newly hatched young at the San Diego Zoo weighed 19.6 g (0.7 oz) at hatching and 63 g (2.2 oz) at 29–31 days. At 90 days six young averaged 260 g (9.1 oz; David Rimlinger, pers. comm.). Males become sexually active at 1 year (Flieg, 1973), but may require up to 3 years to attain full plumage. Jeggo (1975) observed that all hand-raised females at least 2 years old laid eggs.

Evolutionary History and Relationships

The Palawan peacock pheasant is geographically the most isolated of the peacock pheasants. The island of Palawan is separated by about 150 km (93 mi) from mainland Borneo, but is connected by numerous intervening small islands and relatively shallow ocean depths that were exposed as dry lowlands as recently as Pleistocene times. I believe that its nearest relative is the Bornean peacock pheasant which, as mentioned earlier, shares some similarities in male plumage patterns, such as a somewhat banded tail and blackish underparts. During the evolutionary process, the Palawan peacock pheasant has apparently lost the ocelli pattern on its wings and back and now has a more uniformly iridescent color. However, the species has retained a spectacular pattern of paired ocelli on

the tail and tail-coverts. Thus, anterior visual attention is now concentrated on the bird's actual eye, which is outlined in white, through the use of a posture that tends to project the eye against an iridescent green backdrop. Evidently this species has also lost all tendencies to perform a frontal display, but instead has evolved a highly specialized lateral posture. The ocelli of the tail and tail-coverts have remained distinctly doubled, rather than tending to merge at the central shaft, as is the case in the Malayan and Bornean peacock pheasants. This latter trend anticipates the development of a single central ocellus of the type characteristic of the true peafowl.

Status and Conservation Outlook

The Palawan peacock pheasant is currently considered an endangered species (Collar et al., 1994; McGowan and Garson, 1995), although its actual population size remains unknown. The entire exploitable forest area of Palawan is restricted to the coastal plain and is under lease. However, as of 1975 only a small proportion of it had been timbered. There were no sanctuaries on the island as of the late 1970s (King, 1981). Currently the species is protected by the St. Paul Subterranean River National Park (54 km^2, 21 mi^2) on the west coast of central Palawan, and it also inhabits a proposed 5,000-ha (12,350-acre) forest reserve (McGowan and Garson, 1995). Nearly all of central Palawan's original lowland forest were logged by the 1980s and the northern forests were then logged, so that little if any primary lowland forest was left by the early 1990s. Surveys in 1987 and 1988 by McGowan et al. (1989) documented the species at several sites in central and northern Palawan, with the population at St. Paul Subterranean River National Park seemingly fairly healthy (Caleda et al., 1986). Although the Palawan peacock pheasant was previously not known to occur in northern Palawan (King, 1981), it is now likely that this less heavily logged region may represent its best hope for survival. A North American studbook is currently being prepared for the species and contains over 500 individuals (Donald Bruning, pers. comm.).

Genus *Rheinartia* Maingounat 1882

The crested argus is a large, tropical pheasant in which sexual dimorphism is well developed. The tail of males is greatly elongated and up to four times the length of the wing. Both sexes have a short occipital crest of stiff, upturned feathers, and most of the body plumage is uniformly brown with blackish barring and buffy freckles or vermiculations. The tail has 12 rectrices; in males it is greatly compressed and strongly graduated, with the central rectrices more than five times the length of the outermost ones. The rectrices and their coverts are marked with spots and streaks of brown, chestnut, white, and gray, but lack definite ocelli. The molt pattern is as described for *Polyplectron*. The wing is strongly rounded, with the first primary longer than any of the four outermost feathers and the fifth and sixth the longest. The tarsus is relatively short and lacks spurs in both sexes. Females differ from males mainly in their shorter and less ornamental tail. A single species is recognized. The proper spelling of this species' generic name is unsettled, but *"Rheinardia"* appears to be correct (Banks, 1993).

CRESTED ARGUS

Rheinartia ocellata (Elliot) 1871

Other Vernacular Names

Rheinard's ocellated pheasant, ocellated pheasant; rheinarte ocellé (French); Perlenpfau (German); tri (Vietnamese).

Distribution of Species

Indochina, mainly in lowland, foothill, and lower montane forests of Vietnam and nearby bordering portions of Laos; also occurs disjunctively in the Malay Peninsula. See map 21.

Distribution of Subspecies

Rheinartia ocellata ocellata (Elliot): Rheinard's crested argus. Central Vietnam from Quinhon in the south (14° N) to Vihn and the Tranninh River (Laos) in the north (19° N).

Rheinartia ocellata nigrescens Rothschild: Malay crested argus. Malay Peninsula, the lower levels of Gunong Benom and Gunong Tahan in Pahan, possibly also the mountains of Trengganu and southern Kelantan.

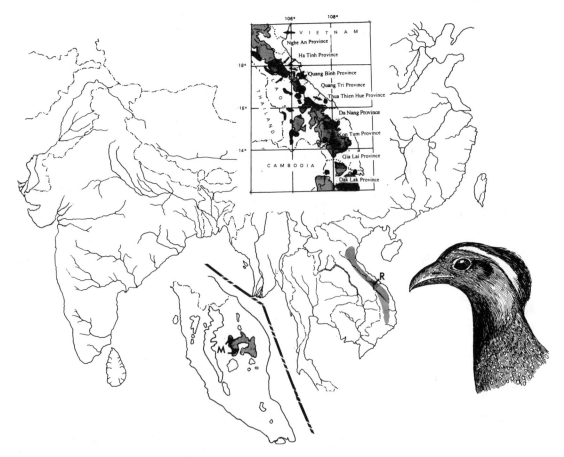

Map 21. Distribution of Rheinard's (R) and Malayan (M) races of the crested argus. Elevations above 500 m (1,640 ft) on the Malay Peninsula are also outlined on the lower inset map, but the known range of the Malayan race is limited to the locations shown by the arrows, mainly in Taman Nagara National Park and nearby peaks (shaded). The upper inset map shows elevations above 500 m (1,640 ft; shaded) in Vietnam and Laos, selected preserves and national parks (black), and some locality records for the crested argus (small arrows). See text for additional recent locality records.

338

Measurements

Delacour (1977) reported that adult males of *ocellata* have wing lengths of 350–400 mm (13.7–15.6 in) and tail lengths of 1,500–1,730 mm (58.5–67.5 in), whereas first- and second-year males have much shorter tails. Females have wing lengths of 320–350 mm (12.5–13.7 in) and tail lengths of 350–430 mm (13.7–16.8 in). Males of *nigrescens* have wing lengths of 370–400 mm (14.4–15.6 in) and tail lengths of 1,500–1,620 mm (58.5–63.2 in), whereas females have wing lengths of 320–340 mm (12.5–13.3 in) and tail lengths of 370–400 mm (14.4–15.6 in). No weights are available. The eggs average about 64.5 × 46 mm (2.5 × 1.8 in) and have an estimated fresh weight of 75.3 g (2.6 oz). Two actual egg weights were 78 and 79 g (2.7 and 2.8 oz), and the newly hatched chicks weighed 53 and 55 g (1.9 and 1.9 oz; Phan, 1996).

Identification

In the Field (762–2,387 mm, 30–94 in)

This extremely rare and elusive species of Malaysia and adjoining Laos and Vietnam is limited to heavy forests. The crested argus is unlikely to be confused with any species except for the great argus, from which males can be distinguished by their short bi-colored crest, their extremely long and broad tail feathers, and their brown rather than reddish legs. Females are also crested, have pale grayish faces that contrast with their darker brown neck and body coloration, and have somewhat barred patterning on the wings and tail. The male's call is a loud *kuau* or *ho-huiho*, which is repeated several times and much like that of the great argus. It is a long, drawn-out call that carries great distances. Soft, sibilant, chuckling notes are uttered by both sexes.

In the Hand

The male's enormously long and broad central tail feathers (130 mm [5.1 in] wide and up to 1,730 mm [67.5 in] long) are unique among birds. Younger males have shorter tails, but these too are relatively broad and are distinctively stippled with tiny buff spots and larger brown spotting. Females have a relatively shorter (350–430 mm, 13.7–16.8 in) tail that is heavily freckled and barred, and a brownish occipital crest that becomes black anteriorly. The somewhat similar great argus female has rather bluish facial skin and no crest.

Geographic Variation

Geographic variation is rather slight in this species; males from the Malay Peninsula are on average darker than those of the nominate form. They also are more regularly spotted, have a longer (85 mm [3.3 in] versus 60 mm [2.3 in]), and whiter crest, underparts that are more spotted with white, and a darker tail. Compared to the nominate form females are slightly brighter, have lighter underparts, and have underparts that are more closely marked with black (Delacour, 1977).

Ecology

Habitats and Population Densities

In Indochina, the ecological range of the crested argus is from sea level to as high as 1,525 m (5,000 ft). According to Delacour (1977), the species is associated with hilly lowland and lower montane forests, on very damp slopes of mountains, especially those slopes below about 900 m (2,955 ft). On the Malay Peninsula this species is known only from Gunong Benom, Gunung Tahan, and Gunung Rabong, as well as one possible locality between these last two areas (Davison, 1980b). All its known locations are on isolated mountain blocks rising above 1,525 m (5,000 ft) in elevation, lying to the east on the main dividing range in the peninsula (Davison, 1977). All observations have been between 790 and 1,080 m (2,590 and 3,545 ft; Davison, 1979c) in these mountains. Studies by Davison (1978a, 1979c) on Gunung Rabong suggest that crested argus were mainly found near the 980-m (3,215-ft) contour level, and their dancing grounds were located in a transitional area between lowland dipterocarp forest and lower montane forest. They seem to occupy a broader range of forests in Vietnam, including primary, seasonal evergreen and sometimes even previously logged forests. These forests are mostly situated under 700 m (2,295 ft) elevation (McGowan and Garson, 1995), but occasionally reach 1,500 m (4,920 ft), as on the Da Lat Plateau (Eames, 1995).

In 1976 the population density estimated by Davison (1978a) for all of Gunong Rabong was about 15 calling males. He said that the best way to express population density was his estimated average distance of 1,100 m (3,610 ft; range 720–1,440 m [2,360–4,725 ft]) between dancing grounds. The total population of this area might be under 50 birds (King, 1981).

Competitors and Predators

Davison (1980*b*) has suggested that competition with the great argus on the Malay Peninsula has forced the crested argus into its highly restricted altitudinal zonation pattern in that area. In Annam there is no contact with any comparably sized pheasant, and perhaps for these reasons the crested argus exhibits a much wider and more altitudinally diverse ecological range. Its predators have not been specifically identified. However, in Malaysia it reportedly inhabits the same areas as tigers and leopards (Beebe, 1918–1922) and undoubtedly encounters many other smaller predators.

General Biology

Food and Foraging Behavior

Beebe (1918–1922) summarized the little information then available, saying that crested argus eat crickets, other insects, and even little frogs, but predominantly insects. However, captive birds will also take grain and other foods of barnyard fowl. Davison (1978*a*) studied 17 droppings collected on the dancing ground of one male. He found that at least 90 percent was nearly digested materials that seemed to be of vegetable origin and fragments of ferns, liverworts, and vegetable fibers were also found. Fruit remains were present in 10 (60 percent) of the droppings, and palm fruit (*Calamus*) remains were found in 6 (40 percent). Invertebrate fragments, perhaps from ants, were observed in 10 (60 percent) of the droppings.

Movements and Migrations

There is no evidence of any significant movements in crested argus. However, Davison (1979*c*) stated that the altitudinal range of these birds outside the breeding season (790–1,080 m, 2,590–3,454 ft) may be somewhat broader than the range of the dancing grounds (820–1,050 m, 2,690–3,445 ft).

Daily Activities and Sociality

During the calling season, males begin about 7:00 A.M. and continue until about noon, depending upon the individual and the day. Males then leave their dancing grounds and only occasionally call in late afternoon at about 5:00 P.M. Nocturnal calling occurs from about 7:00 P.M. onward and is less frequent than during the day. Some nocturnal calling may also be done from roosting sites (Davison, 1978*a*). There is no

evidence of sociality in the crested argus, even out of the breeding season, although females probably remain with their one or two chicks for a prolonged period. Further, when the young became independent they continue to live together for some time (Beebe, 1918–1922).

Mating System and Territoriality

The spacing behavior of calling males indicates territorial behavior; judging from Davison's (1978*a*) sketch map the birds seemed to be maximally dispersed with a rather consistent distance separating them. The male's calls carry considerable distances. They consist of loud and resonant whistles that may be given in response to calls of conspecifics or of great argus and to a lesser extent to a wide variety of other loud noises (Davison, 1978*a*). The dancing grounds of the males are fairly small and consist of cleared areas from less than 1 to 4 m (3 to 13 ft) in length and width. The grounds are level or are sometimes situated on slight rises or humps; they seem to be especially associated with mountain ridge saddles or shoulders. Such sites are very limited, and there is probably considerable competition for suitable dancing grounds. Thus, sites may be used for several years, certainly for at least four seasons (Davison, 1978*a*). Beebe (1918–1922) stated that the male selects a flat and clear area of ground, often at the foot of a great tree on a branch of which the female might perch and watch while he displays. However, this seems unlikely; it is more probable that the female would approach the male on foot.

Voice and Display

Beebe (1918–1922) reported that the major calling season of the Annam race is in March and April, whereas on the Malay Peninsula the calls have been heard both in March (Medway and Wells, 1976; Davison, 1979*c*) and May (Davison, 1978*a*). Davison (1978*a*) reported two types of male calling: short calls and long calls. In his view, the onset of short calling marked the start of the breeding season, and he first heard it on 21 March. This call is trisyllabic and was transcribed as *oo-kia-wau*, with the first syllable slow and humming and the second and third rapid and high-pitched. Males typically call from one to eight times in rapid succession, then pause for about 15 minutes before calling again. The corresponding call of the great argus is disyllabic (lacking the introductory note) and is lower pitched. The longer call consists of a series of 8–17 loud disyllabic notes, with only the first introduced by the humming note. It resembles a similar call of the great argus, but seems more melodious at close range (Davison, 1978*a*). Long calls are apparently given mostly in the morning at various locations away from the display grounds, whereas short calls may be uttered from a few meters away from the dancing ground or from the roost at night. A sibilant, mellow, clucking call was also heard by Davison (1978*a*) from a male standing on its dancing ground.

Davison (1978*a*) observed a limited amount of display on a dancing ground by a male, who threw vegetation fragments about with his beak. The male then stood motionless on his display site with his white crest fully exposed. Over the next 45 minutes the crest was gradually lowered, but the throat and neck feathers remained ruffled. Short calls were uttered at intervals of 7, 11, 18 minutes. One hour after its initial calling the bird fled, having seen Davison. According to Huxley (1941), during calling the head is thrown back and the crest is erect.

Seth-Smith (1932) noted that a captive male displayed in April and May. He raised his crest feathers in such a way that the white area resembled a large powder-puff. The male then ran around the hen, with his broad tail feathers greatly spread. Huxley (1941) also noted the similarity of the raised crest to a powder-puff, which he said is raised in preparatory phase, with the tail spread only somewhat and the wings not dropped. The male then runs past the hen, with the tail feathers spread vertically so that they cover about twice the area of the resting position. The wings are also drooped, the body is slightly inclined toward the female, and the head is held low with the neck outstretched. There is apparently no tendency toward a frontal display in this species (figure 62). Taka-Tsukasa (1929) stated that during the lateral display the bird is shaped like a "straightlined broad tape," and during the run it occasionally stops, lowers its neck a little, and opens its wings. Roles (1981) stated that in intense display the males may leap into the air or pick up stones.

Reproductive Biology

Breeding Season and Nesting
Judging from limited information, the breeding season of the crested argus probably begins in March in both Annam and Malaysia. No nests seem to have been found in the wild by biologists, although Beebe (1918–1922) mentioned that an Annamese native hatched young from two eggs. Although wild birds were brought to France in 1924 by Delacour, it was 7 years later that they first began to breed. Crested argus were bred by Taka-Tsukasa (1929) in Japan in 1928, with the male starting to call in February and the eggs laid from mid-April to the beginning of July. The female nested on the ground, rather than using elevated nest boxes. In 1929 the female laid four two-egg clutches, and the eggs were laid at 2-day intervals. The intervals between the clutches were from 20 to 27 days. These eggs were all laid on the ground, except for a few that were dropped from the perch in 1928 before nesting sites were available. In recent years breeding has been successfully achieved at the Saigon Zoo.

Incubation and Brooding
Incubation has been found to take from 24 to 25 days. Delacour (quoted in Taka-Tsukasa, 1929) also raised this species in France in 1931, with the female nesting in an elevated basket. In this case, three two-egg clutches were produced at about 18-day intervals, and the chicks hatched after 25 days.

Growth and Development of the Young
The primaries of the chicks are already well developed at hatching, and their tail feathers begin to emerge at only 3 days (Taka-Tsukasa 1929). Jabouille (1926, 1930) also raised this species at Hue, Annam, in 1925, with a chick being hatched by its mother and surviving for 40 days. As in *Polyplectron* and *Argusianus*, the chick hid under its mother's tail.

Like the peacock pheasants and the great argus, the young of this species are brooded at night on elevated branches as soon as they are able to fly up to them, with each chick nestling down below its mother's wing and facing in the same direction. Like peacock pheasants, the mother also feeds them beak to beak for the first few days (Delacour, 1977). Judging from photographs by Taka-Tsukasa (1929), the birds lose the last of their downy plumage no more than 60 days after hatching, when they are about the size of bantam hens. Apparently at least 3 years are required for the male to attain nuptial plumage, but the tail continues to increase in length for a few more years (Delacour, 1977).

Figure 62. Display postures of male crested argus, including calling (A), lateral display to female (B), normal head posture (C), crest-raising (D), and maximum wing-dropping and tail-rearing (E). Primarily after photos in Huxley (1941).

Evolutionary History and Relationships

Davison (1980*b*) has postulated the evolutionary history of the crested argus from a common ancestor with the great argus. He judged that since the separation of these two forms the crested argus has lost its frontal display while the great argus has elaborated its frontal posturing. Davison judged that the two species have probably been in prolonged contact on the Malay Peninsula, during which time selection for interspecific differences in displays and male plumage pattern has occurred.

Status and Conservation Outlook

Little is known of this species' current status in Vietnam, but it is believed to be quite common in some areas and more widespread than previously believed (McGowan and Garson, 1995). Eames et al. (1994) reported the crested argus from all of six areas recently surveyed in the Annamese Lowlands of central Vietnam (Kim Quang, Khe Buoi, Cat Bin, Net River watershed, Phuong Nha, and Rao Bong watershed). The species has also been recently recorded south as far as the Da Lat Plateau, Dak Lak Province (Eames, 1995). There are also a few recent records from bordering areas of Laos. Records for the Malaysian race originally included Gunung Tahan, Gunung Rabung, Guning Mandi Angin, and Gunung Gagau in Taman Nagara, but it has more recently also been observed at Gunung Penumpu and Camp Kor in Taman Negara (Mamat and Yasak, 1997).

On the Malay Peninsula the crested argus is known from only a very few localities, and its potential range is not much larger than its known range (Davison, 1977). There is no good evidence of its past or present occurrence in Sumatra (Davison, 1979*d*). Logging in its known range in Malaysia has been done up to about 760 m (2,495 ft) elevation, but this was halted in 1975. Fortunately, logging equipment is unable to operate on the steep slopes encountered at the higher mountain levels used by the crested argus (Davison, 1979*c*).

The race *ocellata* is now regarded as endangered (McGowan and Garson, 1995); Collar et al. (1994) listed the species collectively as vulnerable. The crested argus is considered rare by the International Council for Bird Preservation. A large part of the bird's known range on Gunung Rabong is fortunately in Taman Negara National Park, where the estimated total population may be under 50 birds (King, 1981). Given that situation, it would be appropriate to consider the species as endangered rather than simply rare and to institute whatever conservation measures might be possible.

Genus *Argusianus* Rafinesque 1815

The great argus is a very large, tropical pheasant in which sexual dimorphism is highly developed but iridescence is limited to the ocelli on the wings of the male. The central pair of rectrices of males are extremely elongated, up to four times the length of the outermost rectrices and nearly three times the length of the primaries. The head and neck of both sexes are nearly naked except for the center of the crown and a short occipital crest. The body feathers are mostly spotted and finely barred with brown, buff, and chestnut. The wings are uniquely shaped, with the secondaries longer than the primaries and the primaries gradually declining in length from the first outwardly. A series of iridescent ocelli is present on the inner webs of the secondaries of males. The tail has 12 rectrices, and the molt pattern is as described for *Polyplectron*. The tarsus is relatively long, slender, and lacks spurs in both sexes. Females resemble males, but are duller in plumage, have less highly developed secondaries and rectrices, and have a longer occipital crest. A single species is recognized here, but a second extinct species is recognized by some (e.g., Davison, 1983*c*).

GREAT ARGUS

Argusianus argus (Linnaeus) 1766

Other Vernacular Names
Argus pheasant; argus géant (French); Argusfasan, Arguspfau (German); keee (Dutch Bornean); kuang raya (North Malayan); koeweau (Sumatran).

Distribution of Species
Borneo, Sumatra, and the Malay Peninsula in mature forests between sea level and 1,220 m (4,000 ft). See map 22.

Distribution of Subspecies
Argusianus argus argus (Linnaeus): Malay great argus. The Malay Peninsula, north currently only to about the Thailand border, plus Sumatra.
Argusianus argus grayi (Elliot): Bornean great argus. The island of Borneo.

Argusianus argus bipunctatus (Wood): double-banded great argus. Known only from a single feather of unknown origin. Considered by Delacour (1977) and Beebe (1918–1922) to be a distinct species and by Davison (1983c) to be an extinct form possibly from Tioman Island.

Measurements
Delacour (1977) reported that *argus* males that are at least 7 years old have primary lengths of 450–500 mm (17.6–19.5 in; secondaries 800–1,000 mm [31.2–39.0 in]) and tail lengths of 1,160–1,430 mm (45.2–55.8 in), whereas younger males have progressively shorter wing and tail lengths. Females have primary lengths of 300–350 mm (11.7–13.6 in; secondaries 350–400 mm [13.6–15.6 in]) and tail lengths

Map 22. Distribution of Bornean (B) and Malay (M) races of great argus. The inset map of Sumatra shows elevations above 500 m (1,640 ft) and locations of smaller (small arrows) and major parks or preserves (large arrows), including Mount Leuser (1), Keringi Seblat (2), Southern Barisan Mountains (3), Way Kambas (4), Tigapuluh Mountains (5), Kerumutan (6), Linga Isaq (7), and Berbak (8). Preserves on Borneo and the Malay Peninsula are shown in maps 8 and 20.

of 310–360 mm (12.1–14.1 in). The race *grayi* is slightly smaller; males having primary lengths of 430–470 mm (16.8–18.3 in; secondaries 750–850 mm [29.3–33.2 in]) and tail lengths of 1,050–1,200 mm (41.0–46.8 in), whereas females have primary lengths of 300–340 mm (11.7–13.3 in) and tail lengths of 300–340 mm (11.7–13.3 in). Riley (1938) reported that four males of *argus* weighed from 2,040 to 2,605 g (4.5 to 5.7 lb; average 2,350 g [5.1 lb]), and one female weighed about 1,700 g (3.7 lb). Stephen Wylie (pers. comm.) reported that two males of *argus* weighed 2,043 and 2,725 g (4.5 and 6.0 lb), whereas two females each weighed about 1,590 g (3.5 lb). The eggs of *argus* average about 68 × 44.5 mm (2.7 × 1.7 in) and have an estimated fresh weight of 74.3 g (2.6 oz).

Identification

In the Field (762–2,032 mm, 30–80 in)
Limited to Borneo, Sumatra, and the Malay Peninsula, this species may only be confused with another species on the Malay Peninsula, where the crested argus also occurs. The male great argus is unmistakable owing to his enormously long and relatively narrow central tail feathers and his greatly elongated secondary feathers, which are longer than the primaries. Females somewhat resemble males, but have shorter tails and secondaries and lack the iridescent ocelli. Both sexes have only short crests and reddish feet, which separates them from the crested argus. The male's long call is a loud, musical *kwow-wow* or *kweau* that carries great distances. The other male calls include a short call that is a high-pitched note lasting less than a second and an alarm call. Females also utter repeated *wow* notes. Great argus are associated with heavy tropical forest and are much more likely to be heard than seen.

In the Hand
The incredibly long central tail feathers (over 1,100 mm [42.9 in] in adult males) and the inner secondaries that may reach 1,000 mm (39 in) provide positive identification of males. Females have bluish facial skin, short and brown crests, and pale red legs that are relatively long (tarsus 85–95 mm [3.3–3.7 in]).

Geographic Variation
Known geographic variation in this species is rather slight, with males of the Bornean race slightly smaller

and grayer than those of the nominate form. They are also more reddish orange below, the back is pinky buff rather than yellowish, the spotting of the rectrices and tertials is mostly white rather than white and buff, the upper breast is orange chestnut rather than dark chestnut, and the rest of the underparts are also chestnut rather than brownish, with only fine black and buff vermiculations. Bornean females are lighter brown underneath and have more reddish orange on the neck and back. The single specimen of *bipunctatus* is known only from a primary feather that has a pattern on both webs resembling the inner web of typical *argus*. Unless this is the result of a genetic aberration, it suggests that a population of this species with additional geographic variation may once have existed. Delacour (1977) suggested that perhaps it inhabited Java and has now become extinct, but Davison (1983c) believes that it may have inhabited Tioman.

Ecology

Habitats and Population Densities
Davison and Scriven (1987) found the Malaysian race of this species to be associated with lowland and hill dipterocarp forest sites, but absent in montane forests, coastal gelam (*Melaleuca*) forests, peat swamp forests, and heavily disturbed lowland dipterocarp forests that were highly fragmented and hunted. Toward the northern end of its range the great argus becomes more restricted to hills, which are moister and more nearly evergreen, but in areas where the crested argus was present on midmountain levels this species is absent. Great argus are less common on level or gently sloping country than in hilly areas. In Borneo, the birds also inhabit lowland forest, exclusive of extensively cleared and populated or flat and swampy jungles, and in low montane forests as high as 915 m (3,000 ft; Smythies, 1981).

Densities in Malaysia ranged from about 0.3 males per square kilometer (0.8 per square mile) on low and level sites to 4.5 males per square kilometer (11.7 per square mile) on steep land with many small hills (Davison and Scriven, 1987). On such flat sites as Pasoh and Kuala Lompat the average distances between male display sites was about 450 m (1,475 ft; minimum 260 m [855 ft]), whereas at a hilly site (Ampang) the average was 375 m (1,230 ft; minimum 280 m [920 ft]), with fairly uniform spacing. At Kuala

Lompat 75 percent of the males were found to call no more than 200 m (655 ft) from permanent water and most of their home ranges were on alluvial soils, whereas at the hilly site no part of their habitat was more than 300 m (985 ft) from permanent water (Davison, 1987).

Competitors and Predators
Davison (1980*b*) has suggested that the great argus competes strongly with the crested argus and that the present range of the latter in Malaysia has been affected by this competition. This seems to result in mutual competitive habitat exclusion, although it is not clear which species excludes the other (Davison and Scriven, 1987). In Borneo, Beebe (1918–1922) estimated that the great argus comprised 6 percent of the pheasant population in one rather disturbed area, with the wattled pheasant at 6 percent and the firebacks at 88 percent. However, in a less disturbed area the argus comprised an estimated 65 percent, the wattled pheasant 15 percent, and firebacks the remaining 20 percent. It seems unlikely that any of these smaller species pose a competitive threat to the great argus, although Beebe believed that the spurless argus would have little chance in a fight with the well-armed fireback.

The major enemy of the great argus is certainly humans because the birds are often trapped on their display sites and their feathers are used as ornaments. Beebe (1918–1922) mentioned civet cats and musangs (*Paradoxurus*) as possible predators, but provided no positive evidence. He believed that the loud and prolonged calling of males would tend to place the birds in some jeopardy and said that the great argus probably relies on its acute hearing for warning of danger. The birds can run extremely fast, but are relatively poor fliers.

General Biology

Food and Foraging Behavior
Beebe (1918–1922) could provide little information on the Malayan form of the great argus, other than that they feed on fallen fruit, ants, other insects, slugs, and various shelled molluscs. He found that the Bornean birds seem to eat primarily ants, but also leaves, nuts, and seeds. Surprisingly, Beebe found no evidence of termites in their diet. Davison (1981*e*) observed that Malayan birds consumed a wide variety

of invertebrate and plant materials, based mainly on the analysis of droppings. Fruits of the plant families Palmae, Annonaceae, and Leguminosae predominated the diet, including many climbing species, some from various understory trees, and a few from canopy and emergent species. Termites were found in only 1 of 138 droppings and in one of four gizzards. However, ants were found in many droppings and probably comprise a majority of the invertebrate food, which is consistent with their abundance in the litter fauna.

While foraging, great argus are typically solitary except for females leading young. Foraging birds walk slowly and in a meandering manner, pecking at the leaf litter and sometimes also at the leaves of shrubs. After each peck the bird typically raises its head to look about, presumably for possible danger. Scratching for subsurface materials under the litter was not observed by Davison (1981*e*). Apparently large prey items are favored over smaller ones.

Movements and Migrations
Davison (1981*e*) used radio-telemetry to determine home ranges in two males over periods of several months. He found that the home ranges of these birds were surprisingly small and they had still smaller core areas of intense use. The maximum distance traveled in a single day was 800 m (2,625 ft) for one male and 910 m (2,985 ft) for the other; the core-areas of use (where at least 50 percent of the observations were made) varied monthly from 0.1 to 0.5 ha (0.2 to 1.2 acres). Even outside of the breeding season, which lasts from November to February, the birds had relatively small total home ranges. These collectively averaged about 2.5 ha (6.2 acres) from October to March.

Daily Activities and Sociality
All the evidence would suggest that great argus are relatively solitary (Beebe, 1918–1922; Davison, 1981*e*); they rarely if ever are seen in groups larger than a female and her two young. Davison (1981*c*) observed via telemetry that in November two males foraged for short periods in the morning and evening and otherwise spent their daylight time perched, with the males averaging 81 and 90 percent of the daylight hours thus occupied. Beebe (1918–1922) observed a nighttime tree perch in Borneo and watched a male "climbing" up to it one night, but did not describe its

characteristics. A good deal of nocturnal calling occurs during some months, especially on moonlit nights, when it might go on all night. On cloudy days great argus also sometimes call throughout the day, but the highest calling levels are after sunrise and again between sunset and darkness (Beebe, 1918–1922). Beebe believed that all adult male calling was done from the dancing ground, implying that the birds perch at or very near their own dancing site. However, Davison (1987) noted that some males without dancing areas called from the ground as they wandered about and that some calls are uttered from the roosts.

Social Behavior

Mating System and Territoriality

The great argus is undoubtedly a polygynous species (Davison, 1981d). Probably the only real contacts between adult males and females occur at the dancing ground in conjunction with fertilization.

Territory sizes during the breeding season are apparently fairly large. As indicated by the earlier data on dispersion of male dancing grounds, an average of about 400 m (1,310 ft) separates them. However, it is not known to what extent all of this area may be defended. Davison (1987) found that in one area (Ampang) the display ground of males that died were used in following years by other males, which suggests that the density of calling males might be affected by the availability of suitable hilltop sites. In another area striking topographic features were lacking, and the number of calling males varied with the relative food abundance in the form of fruiting plants.

Voice and Display

The advertisement of male display grounds is done by daily calling. Davison (1982) noted that calling might begin at any time between 6:30 and 10:15 A.M. at two flatland sites, whereas at a favored hilly site it always began before 7:00 A.M. and usually continued until after 11:00 A.M. Males at the latter site called nearly every day. Males in this general region typically call over a long seasonal period, from January or February until sometime between early June and late September. Mating and nesting apparently occur in June (Davison, 1981e).

The male's calls consist of three types of hooting, which are audible for up to about 1 km (0.6 mi), a

yelping alarm call audible for several hundred meters, and various gentle clucking calls (Davison, 1981d). The "long call" is a series of 15–72 hoots, which start as monosyllables but become higher in pitch and disyllabic toward the end. These calls are uttered by adult males throughout the year, but especially at the start and end of the breeding season around February and August. They seem to in part serve as vocal contests between males for dancing ground sites. Long calls also have occasionally been heard from females. The "short call" is a high-pitched, disyllabic note, which usually lasts just under a second. It is uttered every few minutes in bursts of up to 12 calls by males on their dancing grounds. Short calls are also uttered from the roost at night. The number of notes per burst and the number of bursts per hour were found by Davison (1981d) to vary but overlap among six different males; these differences were insufficient to distinguish individuals, although other auditory cues might allow for such separation. Areas over which a single male could be heard varied from about 38 to 145 ha (94 to 358 acres), with the smaller ranges in dense vegetation and level sites and the larger areas from hilltop sites. Males apparently begin to call at about 12 months of age and perfect their calls by about 20 months. Young males may call occasionally and try to make dancing grounds; however, these efforts never result in a large cleaned space, as is typical of adults.

The size of the display site, or dancing ground, is variable. However, it is typically larger than 12 m^2 (129 ft^2) by the end of the display season and rarely may be as large as 72 m^2 (775 ft^2). Each site has one or two regular entrance and exit routes that are used by the male. Davison (1981d) doubted that clearing the sites of debris was related to ground predator detection, but may serve to amplify the sound of foot-stamping in males and may also be visually significant to females. Davison noted that actual cleaning behavior comprised less than 5 percent of the time spent by the male on the site, with inactive periods comprising over 80 percent of the time, calling about 13 percent, preening about 2 percent, and actual postural display a small fraction of 1 percent of the total time.

The postural displays of the great argus have been described by many authors (Beebe, 1918–1922; Seth-Smith, 1925a,b; Bierens de Hann, 1926; Davison, 1982). The most complete is Davison's study, which is the primary basis for the names and descriptions of the displays mentioned below, all of which are associated with male display sites or dancing grounds.

Cleaning of the display site occurs throughout the breeding season. Cleaning may be performed as leaf-throwing with the beak, pecking of overhanging vegetation, bill-scraping around the bases of saplings, and wing-beating with forceful movements that fan away light debris. Of these, leaf-throwing is the most commonly used method. Actual posturing (figure 62 [on page 342]; figures 63, 64) may occur in the absence of any female. However, when females are present, posturing is more prolonged, foot-stamping is more vigorous, and there are more variations in individual display movements.

During head-feather ruffling, the male erects his head, neck, and upper breast feathers to produce a very bushy appearance. This posture seems to represent a state of display readiness and typically precedes foot-stamping (David Rimlinger, pers. comm.). Likewise, general body-shaking typically precedes each renewal of display activity. The first conspicuous posturing during a display sequence is foot-stamping, in which the male walks around the dancing ground with his head held low and neck arched and makes stamping movements that are sometimes audible for up to 25 m (82 ft). These stamping movements are made methodically, at an average rate of nearly three per second (pers. obs.). Stamping continues for a variable period. However, if a female is present, the male may shift to a "tail-high walk" in which he is oriented laterally to the female, with the tail held nearly vertically. The head feathers remain ruffled, and the posture is frequently interspersed with tidbitting movements. These consist of ritualized ground-pecking movements in which the beak sometimes does not actually touch the ground. Alternatively, tidbitting may involve pecking and flicking of leaf litter in the direction of the female, which often attracts her toward the male. No calling is associated with this behavior. Tidbitting may also occur during a "cringing run," in which the male trots in arcs around the female in short and rapid steps. However, on attracting the female, the male is more likely to perform a lateral display. He runs past her with the nearer wing lowered and the farther one raised and then turns and repeats the movement in the opposite direction. In this posture the ocelli of the wing feathers are variably visible. Frequently the male hisses as he performs this display. David Rimlinger (pers. comm.)

has observed a "rush" display not described by other observers, which is performed as the female enters the arena opposite from where the male is standing. He rushes straight toward her while making a hissing sound. This display was observed only occasionally and seemed to be performed instead of or immediately prior to the lateral display. After either a lateral display or intense tidbitting behavior, the male may quickly swerve to face the female and spread both wings fully vertically, which forms an oval radiating fan of feathers with the two longest tail feathers visible behind the wings. The head, which is placed behind one wing, is held so that one eye is visible at the carpal joint. The double wing-fan is essentially funnel shaped, and the ocelli form a radiating series of artificial "eyes" extending from the position of the actual eye. In this extreme posture both legs are synchronously raised and lowered, which alters the pressure of the primaries on one another and the ground and produces a rustling sound, while the tail is simultaneously raised and lowered in a hypnotic rhythm. This rocking and tail-pumping activity occurs at a fixed rate of about three pumps per 5 seconds and is characterized by a slow downward movement of the tail followed by a very fast return to the vertical (pers. obs.). In ten filmed sequences that I examined, the number of such pumps in a single sequence varied from three to nine, and Davison (1982) observed up to 11 such movements. This pumping phase is clearly the climax and it is held by the male for as long as the female remains nearby and in front of him. However, apparently it is not normally followed by attempted copulations (pers. obs.; David Rimlinger, pers. comm.).

Copulation in the great argus has not been described in the literature. However, I observed one instance on 28 February 1983, and David Rimlinger has observed it on three other occasions. In my observation, the female, who had been 6–9 m (20–30 ft) from the foot-stamping male and not paying obvious attention to him, suddenly moved to the rear part of the pair's pen and squatted. The male rushed over to her and mounted immediately. Copulation lasted only a few seconds, with the male opening his wings and grasping her nape while mounted. There were no special displays on the part of either bird afterward. In two earlier observations by David Rimlinger copulation also did not occur during intense male display. Instead, the female walked to the male's arena and squatted down; the male immediately came over

Figure 63. Display postures of male great argus, including foot-stamping (A), tidbitting (B), crest-raising (C), normal head posture (D), and lateral display to female (E). After photos by the author and David Rimlinger.

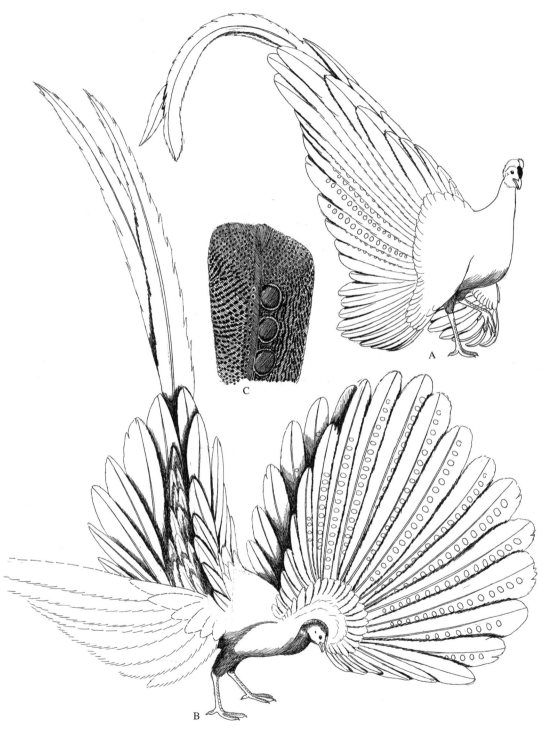

Figure 64. Display postures of male great argus, including rushing (A) and full frontal display (B), showing the tail in fully raised position as well as in lowered position (dotted line). Also shown (C) is detail of a male's ocellus pattern. After photos by the author and David Rimlinger.

and mounted. The sketch in figure 65 shows mounted male balancing with his wings prior to grasping the female's nape.

Davison (1981*d*) believed that females choose among competing males, with or without display areas, based on gross (age-associated) plumage differences or individual display differences rather than on subtle plumage variations or minor vocal differences. According to Davison, female choice provides a good explanation for the evolution of male plumage and behavior patterns.

Reproductive Biology

Breeding Season and Nesting

Very few nests of the great argus have been found in the wild. Beebe (1918–1922) observed a recently hatched nest of the Malayan form at an elevation of 610 m (2,000 ft) on a mountain slope covered by palms and bamboos. He did not provide the exact date of its discovery, which was probably in October or later. Beebe mentioned that chicks have been found in February and August. Davison (1981*c*) found one nest of the Malayan race in June. Medway and Wells (1976) mention clutches obtained on 20 May and 2 July and an egg laid on 27 March by a recently caught female. Less is known of the breeding season of the Bornean form. However, one captive female at the National Zoological Park in Washington, D.C., laid a total of 30 eggs in 15 clutches over a 12-month period. Eggs were laid every month of the year from March 1973 to March 1974, when the male began to molt (Gilbert and Greenwell, 1976). A female of the Malayan race at the San Diego Zoo laid from February to 27 August over a 2-year period. She produced seven eggs in four clutches in 1982 and 17 eggs in nine clutches in 1983. The average interval between eggs was 2 days and the average between clutches was 25 days (David Rimlinger, pers. comm.). Gilbert and Greenwell (1976) noted the usual interval between eggs to be 48 hours, with a maximum of 5 days, whereas the minimum interval between clutches was 2 weeks, with a maximum of 1 month. In 15 clutches, all contained two eggs except for a one-egg clutch and a three-egg clutch.

Incubation and Brooding

The incubation period of the great argus is 24–25 days. One nest incubated by a female had an incuba-

tion period of 24 days 18 hours, whereas artificially incubated eggs required from 24 to 26 days to hatch (David Rimlinger, pers. comm.). There is no indication of any male involvement in nest defense or of the male even maintaining any close proximity to the female's nesting site. At the San Diego Zoo a female laid her eggs in a platform nest 1.5 m (5 ft) aboveground, in preference to ground sites (David Rimlinger, pers. comm.).

Growth and Development of the Young

On hatching, great argus chicks are fed beak to beak by the adult female. Surprisingly, male argus have been observed bill-feeding chicks during their first few weeks (Donald Bruning, pers. comm.). One female at the San Diego Zoo left her nest with both chicks the day after hatching. She continued to feed her young in this way for more than 40 days, when the young were removed from the pen. The chicks brooded with the female on the ground during their first week and then brooded under the female's wing about 0.6 m (2 ft) off the ground on a dead branch. During their second week the chicks were observed to half-climb and half-fly up to this branch, suggesting that true fledging may not occur until sometime later.

Eight chicks weighed an average of 55.4 g (1.9 oz) at hatching and 133 g (4.7 oz) at 1 month. Two chicks weighed an average of 446 g (15.6 oz) at 2 months (David Rimlinger, pers. comm.). Gilbert and Greenwell (1976) commented on the slow growth rates of great argus young as compared with more typical pheasants. They plotted weight changes during the first 33 days after hatching for young of the Bornean subspecies. They noted that males may begin to display at less than 1 month of age and will begin to call at about 12 months. The number of ocelli and the length of the wing and tail feathers continue to increase with each molt for several years after sexual maturity (Davison, 1981*d*).

Evolutionary History and Relationships

The close relationships of the crested argus and great argus have been mentioned in the *Rheinartia* chapter. Both species are certainly derived from ancestral stock in common with the typical peacock pheasant. Brun (1971) discussed the theoretical problems associated with the evolution and ontogeny of the complex male ocellated feather pattern of the great argus.

Figure 65. Display pos-
tures of the male great
argus, including frontal
display (A) and copula-
tion (B). After photos by
Kenneth Fink.

Davison (1981*d*) dealt with the problems of sexual selection as they relate to the evolution of male plumage and behavior. Davison (1983*c*) also discussed the problematic taxonomic disposition of the unique specimen (a single feather) of *bipunctatus*, which he suggested may have been a flightless form that perhaps inhabited the island of Tioman, which has been isolated from the mainland of the Malay Peninsula for about 15,000–20,000 years. However, I have tentatively relegated this form to the subspecific level.

Status and Conservation Outlook
The Malaysian range of this magnificent species is currently being restricted by forest destruction. By 1978 total forest cover had declined from 70 percent to only 42 percent of the land area (Davison, 1981*c*), and lowland forests now cover only about 15 percent of the peninsula. In Borneo (Sarawak, Sabah, and Kalimantan) the lowland forests may still cover at least 40 percent of the land area. Thus, the great argus is probably in less immediate danger there. There are now only a few sanctuaries in the probable Bornean range of this species (Sumardja, 1981), but it does occur in small numbers in the Taman Negara National Park of Malaysia. Since 1980 this species has been reported from 13 protected Thailand sites, and it also has been reported from 17 protected localities in Malaysia (McGowan and Garson, 1995). The species' current situation on Sumatra and Borneo is unknown. McGowan and Garson (1995) categorized the species as vulnerable.

Examples of both races are now being raised in captivity, although very few members of the Bornean race are represented. Under careful avicultural management their numbers are slowly increasing. As perhaps the most remarkable of all the pheasant species in its behavior and plumage, it is especially important that the conservation of the great argus be carefully monitored.

Genus *Pavo* Linné 1758

The peafowl are very large, tropical pheasants in which sexual dimorphism is highly developed, but both sexes exhibit iridescent plumage. Males have a large area of bright orbital skin on the sides of the face. In males, the tail is hidden by extremely elongated and ornamental tail-coverts, which are iridescent and tipped with complex ocelli. The tail itself is flat, graduated, and composed of 18 (females) or 20 (males) rectrices. Molting begins with the second pair from the outermost and proceeds inwardly, with the outer pair being molted just before the innermost pair. The wing is rounded, with the tenth primary shorter than the first and the secondaries shorter than the primaries. In females the upper tail-coverts are less elongated and the plumage is generally less brilliant. The tarsus is relatively long; it is spurred in males and often also in females. Two species are recognized.

Key to Species (and Subspecies of Males) of *Pavo*
(in part after Delacour, 1977)

A. Crest fan shaped, with shafts mostly bare; facial skin white: Indian peafowl.

AA. Crest straight, with barbs from base to tip; facial skin blue and yellow: green peafowl.

    B. Wing-coverts black with a narrow blue border: Burmese green peafowl (*spicifer*).

    BB. Wing-coverts mostly bright iridescent bluish or greenish.

        C. Generally brighter; upper back more "scaly" with golden green; mantle more coppery; lower breast and flanks lighter and brighter: Javanese green peafowl (*muticus*).

        CC. Generally duller; upper breast more coppery; lower breast and flanks duller and darker; mantle more bluish: Indo-Chinese green peafowl (*imperator*).

INDIAN PEAFOWL

Pavo cristatus Linnaeus 1758

Other Vernacular Names

Common peafowl, blue peafowl; paon bleu (French); blauer Pfau (German); mor (Hindi); monara (Ceylonese).

Distribution of Species

Sri Lanka (Ceylon) and India north to Pakistan (Indus River), the Himalayas and the Brahmaputra Valley, east to approximately 95° E. This species inhabits semiopen country from sea level to about 1,525 m (5,000 ft). Introduced and semiferal in many other areas. See map 23.

Distribution of Subspecies

None recognized by Delacour (1977).

Measurements

Delacour (1977) reported that adult (third-year or older) males have wing lengths of 440–500 mm (17.2–19.5 in) and tail-covert lengths of 1,400–1,600 mm (54.6–62.4 in; rectrices of 400–450 mm [15.6–17.6 in]), whereas females have wing lengths of 400–420 mm (15.6–16.4 in) and tail lengths of 325–375 mm (12.7–14.6 in). Males range in weight from about 4,000 to 6,000 g (8.8 to 13.1 lb) and fe-

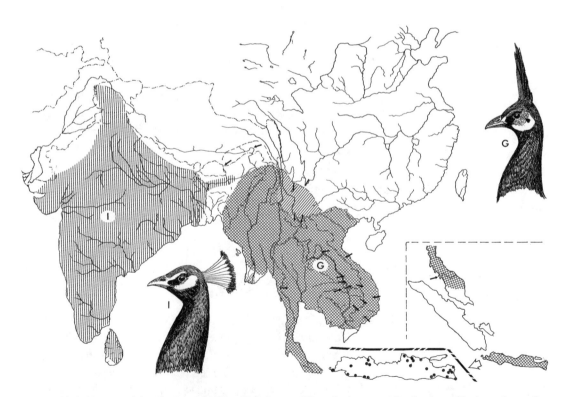

Map 23. Probable original distributions of Indian (I) and green (G) peafowl; current distributions of both species are fragmented, especially the latter's. The hatched area of southwestern China may support a significant but little-studied green peafowl population. Recent records of the green peafowl are indicated by arrows; the inset map of Java shows specific recent locality records (see map 8, on page 198, for locations of preserves).

males from about 2,750 to 4,000 g (6.0 to 8.8 lb; Ali and Ripley, 1978). The eggs average 69.7 × 52.1 mm (2.7 × 2.0 in) and have an estimated fresh weight of 103.5 g (3.6 oz).

Identification

In the Field (914–2,286 mm, 36–90 in)
Peafowl are unmistakable in the field, and any wild peafowl occurring west of Bangladesh will be a member of this species. Both sexes possess whitish cheeks and a tufted, fanlike crest. Females lack the long and iridescent train of males, but their neck and head patterning is very similar. The male's call is a loud, trumpeting *he-on* or *kee-ow*, which carries great distances. Guttural clucking notes are also uttered.

In the Hand
The fanlike crest of both sexes is unique to the Indian peafowl. It occurs in both adult and immature birds, although the crest is smaller in juveniles. The bare facial skin in males and the corresponding area in females is also white.

Ecology

Habitats and Population Densities
In its wild state in northern India, the favored habitats of this species are forests growing along hillside streams, in which the undergrowth consists of ber bushes (*Zizyphus*) and thorny creepers. The bushes grow about 3.0–3.6 m (10–12 ft) apart and spread to form table-shaped tops. These bushes meet one another to form a continuous mass, which allows the birds to move about easily underneath. Higher up in the hill country Indian peafowl are found in open oak forests, where tiny streams run between the hills and each streambank is well covered by bushes, brambles, and reeds. Over much of southern India, where Indian peafowl are protected, they are likely to inhabit any patch of jungle, groveland, or bushes near villages and cultivated areas, and especially thick and high crops such as sugarcane (Baker, 1930). In general, wild birds seem to prefer moist or dry deciduous forests near streams (Ali and Ripley, 1978). Generally Indian peafowl are associated with low plains, usually under 610 m (2,000 ft). However, near the northern edges of the range they have been locally recorded at 1,220 m (4,000 ft) in the Sikkim Hills, at 1,525 m (5,000 ft) in

the Nilghiris area of southern India, and at 1,830 m (6,000 ft) in Simla (Baker, 1930).

There do not seem to be any good estimates of population densities. Johnsingh and Murali (1980) studied an area around the hamlet of Injar, in Tamil Nadu of southern India, which had a total cultivated land area of about 100 ha (247 acres) and an unstated area of fallow land and acacia plantings. They found a population of about 100 adult Indian peafowl, which suggests a density of 1 bird per hectare (0.4 per acre). This group included an estimated 42 breeding females.

Competitors and Predators
Johnsingh and Murali (1980) listed a variety of potential competitors and predators of the Indian peafowl in Injar. Of these the common mongoose (*Herpestes edwardsi*) and jungle cat (*Felis chaus*) were the only mammals listed. Baker (1930) said that the larger cats (leopard and tiger) are generally believed to be serious predators and that he had often found peafowl remains in association with the tracks of one or the other of these species. Stray dogs cause serious losses of chicks in populated areas.

General Biology

Food and Foraging Behavior
Peafowl are believed to be virtually omnivorous (Baker, 1930; Ali and Ripley, 1978); they eat everything from grain and green crops to insects, small reptiles, mammals, and even small snakes. Berries, drupes (such as *Carissa*, *Lantana*, *Zizyphus*), and wild figs (*Ficus*) are apparently favored foods where they are available. Johnsingh and Murali (1980) found Indian peafowl feeding in cultivated fields, on an adjoining acacia plantation, and in fallow lands. They noted the crops of three birds primarily contained plant materials such as leaves, grass seeds, and flower parts. Some *Croton* fruit, *Acacia* seeds, *Cyperus* rhizomes, and rice were also noted, as were various insects such as termites, grasshoppers, ants, and beetles. Indian peafowl usually forage in small groups. These are primarily harem groups during the breeding season and are segregated parties of adult males and females with young outside the breeding season.

Movements and Migrations
There is little information on this subject. However, apparently Indian peafowl use certain traditional

roosting trees, at least where they are protected, and return every night to these. Thus, home ranges are likely to be fairly small and limited to foraging areas radiating from roosts and within easy walking distances from them.

Daily Activities and Sociality

Indian peafowl roost in high, fairly open trees, from which they can see in all directions. When roosting in forests, they select one of the highest trees that is well out in the open (Baker, 1930). The birds generally roost in rather large numbers in such trees (five banyan trees served as roosting site for about 100 birds in the area studied by Johnsingh and Murali). However, during the day they break up into small groups, which for much of the year consist of a male and his harem of three to five females. After leaving the roosting areas the birds move into forest clearings, cultivated fields, or other areas to forage during the early morning hours. The middle of the day is spent in shady sites, often very close to water, where the birds drink and preen at length. Late in the afternoon they forage a second time and return for another drink at dusk before going to roost in the evening.

Social Behavior

Mating System and Territoriality

Most authorities believe that the Indian peafowl is polygynous and that a leklike mating system prevails (Hillgarth, 1984; Rands et al., 1984). Obviously not all males are successful in mating; Sharma (1972) observed a male-to-female sex ratio of 1.7:2.1 in his study area. However, Johnsingh and Murali (1980) noted a sex ratio that seemed to favor females, although they admitted that immature, female-like males probably affected their estimates. Because some are too young and others are too old or otherwise unable to breed, perhaps only half of the females in a given population are actually breeding birds (Sharma, 1972).

Territoriality is curiously developed in this species. Feral males defend breeding territories of 0.5–1.0 ha (1.2–2.5 acres), but they also display in communal leks as small as 0.05 ha (0.1 acres), the latter consisting of open "alcoves" within their larger territories (Hillgarth, 1984).

Voice and Display

The calls of the Indian peafowl are numerous and are still only rather poorly described. Johnsingh and Murali (1980) listed 11 possibly distinct calls, of which three or four are associated with various enemies, one with parent–young relationships, and one with sexual behavior and related aggressive behavior. The male's *kayong-kayong-kayong* call, during which the head and neck are jerked violently, is perhaps associated with dominance display over females. Males also utter a repeated *may-awe* call, especially during the breeding season and particularly in early mornings and evenings. This call is uttered both before and after roosting, so it is probably not a territorial dispersion call.

The displays of this species (figures 66 and 67) can be observed easily in zoo birds. Heinroth (1940) has described copulatory behavior in the species, and the strutting has been described by various observers (Beebe, 1918–1922; Ali and Ripley, 1978). The male's train is erected by the cocking of the rather long but nondecorative rectrices. The lateral tail-coverts extend horizontally and even downward, so that they hide the wings, which partially droop, with the secondaries folded but the primaries extended loosely downward. The whole body is inclined somewhat forward. The head and neck are held erect, in the middle of the radiation pattern of ocelli formed by the tail-coverts. When thus facing a female, the male suddenly performs a quivering shake, which causes the iridescent train to shimmer and the feathers of the wings to rustle audibly. Schenkel (1956–1958) considered this display to be a highly ritualized form of tidbitting behavior, although no food is offered and the bill is not even lowered toward the ground, as in the great argus. Heinroth (1940) considered the display as a kind of general sexual advertisement that is visible at great distances and that would attract any females that might be ready to mate. As a female approaches, the male may actually turn away from her, causing the hen to move around and face him. This may be repeated several times before the female crouches. The male then rushes toward her in a characteristic hoot-dash posture (Ridley et al., 1984), and copulation follows. During mating the train is lowered and mounting occurs in the usual galliform manner. Several recent studies on mate choice among peafowl were summarized in chapter 4.

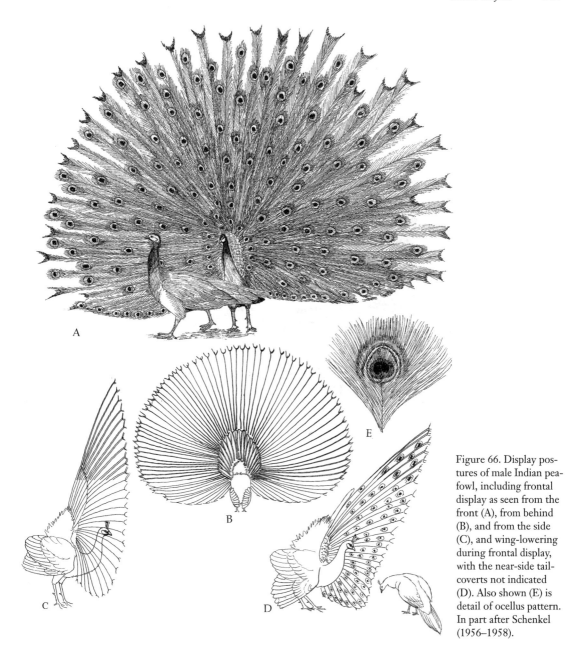

Figure 66. Display postures of male Indian peafowl, including frontal display as seen from the front (A), from behind (B), and from the side (C), and wing-lowering during frontal display, with the near-side tail-coverts not indicated (D). Also shown (E) is detail of ocellus pattern. In part after Schenkel (1956–1958).

Reproductive Biology

Breeding Season and Nesting

Throughout India the nesting season is quite varied, but is apparently always related to the timing of the wet season. In southern India and Sri Lanka nesting reportedly occurs from January to April. Along the foothills of the Himalayas nesting also may occur as early as March or April, but usually begins with the start of the summer monsoon, or about the middle of June. In areas where food is plentiful and there are showers early in the year, Indian peafowl typically breed from January to April. However, in areas where there is a long dry season the birds do not begin to breed until the start of the wet season (Baker, 1930). Sharma (1972) confirmed the start of breeding as

Figure 67. Display postures of male Indian peafowl, including hoot-dash to female (A), calling posture (B), and fighting by males (C). After Petrie et al. (1992) (A), photo by the author (B), and photo by G. Ziesler (C).

coinciding with the onset of rain; in his study area Indian peafowl bred from June to August. Johnsingh and Murali (1980) noted that at Injar (Tamil Nadu) the birds bred from October to December, which was also related to local precipitation patterns.

Nests are typically well concealed and often are located under thorny shrubbery such as *Lantana* or *Zizyphus* (Sharma, 1972). In some low areas, where flooding often occurs, Indian peafowl may nest in elevated sites such as in the crotch of a banyan (*Ficus*) tree (Baker, 1930). Generally the clutch size has been reported to range from three to six eggs, only rarely larger, and with a maximum of eight (Baker, 1930). However, Sharma (1972) reported a range from three to nine eggs in 57 nests and a modal clutch size of six (27 nests).

Incubation and Brooding

Incubation is done by the female alone and lasts for 28–30 days. Sharma (1972) reported a hatching success of 64 percent (206 young from 321 eggs). He said that the female incubates almost 24 hours a day. Hatching usually occurs on the 29th day, with some also on the 30th day.

Growth and Development of the Young

Growth in young peafowl is rather slow, although when the chicks are only 2 weeks old they are able to jump or flutter up to elevated perches for roosting, where they sit on either side of the female and are covered by her wings (Rutgers and Norris, 1970). By 20 days they are able to fly (Sharma, 1972). The chicks are brought to adult roosting areas only when they are nearly 2 months old and fully capable of flight (Johnsingh and Murali, 1980). In the wild, there is evidently a rather high posthatching mortality, with two being the modal number of chicks per brood observed by Johnsingh and Murali (1980). Sharma (1972) also observed a high level of chick mortality and believed that females are incapable of looking after more than five young.

At 4 weeks of age the birds begin to develop crests, and when they are 2 months old the chicks exactly resemble adult females, although they are only half their size (Rutgers and Norris, 1970). The plumage of first-year males is quite variable. By their second year males resemble adult males, but lack ocelli on their train. By their third year males are in full plumage and sexually active, although the length of their train may continue to increase until about the fifth or sixth year (Delacour, 1977). As noted in chapter 4, there is a direct relationship between the length and symmetry of a male's train and his breeding success.

Evolutionary History and Relationships

The relative relationships of *Pavo* to the peacock pheasants, argus pheasants, and Congo peacocks are still somewhat unclear, but it seems likely that *Afropavo* is the nearest living relative of *Pavo*. Boer and van Bocxstaele (1981) confirmed this relationship on the basis of karyotype similarities, but also observed a rather surprising similarity of the chromosomes of both genera to those of guineafowl.

Status and Conservation Outlook

This species is the national bird of India and is under national protection. In some areas the Indian peafowl is extremely common and tame, such as where it is locally protected for religious or sentimental reasons. The species is generally very common in Gujarat and Rajasthan (Ali and Ripley, 1978). There is no reason to consider the species in any danger.

GREEN PEAFOWL

Pavo muticus Linné 1766

Other Vernacular Names
Green-necked peafowl; paon spicifère (French); Ährenträger-Pfau (German); burong merak (Malayan); oodoung (Burman).

Distribution of Species
Bangladesh to Indochina and the island of Java. This species inhabits lowland and moderate-altitude forests and open, parklike areas from the plains to about 915 m (3,000 ft). See map 23.

Distribution of Subspecies
(after Wayre, 1969; Delacour, 1977)
Pavo muticus muticus Linné: Javanese green peafowl. Historically throughout Java and the Malay Peninsula north to the Isthmus of Kra. Now largely limited to Java, where it is endangered.

Pavo muticus imperator Delacour: Indo-Chinese green peafowl. Historically the whole of Indochina, southern Yunnan, and Thailand south to eastern Burma and west to the Salween-Irrawaddy Divide.

Pavo muticus spicifer Shaw and Nodder: Burmese green peafowl. Southeastern border of Assam in the Chittagong and Lushal Hills, western Burma, probably east to the Irrawaddy River.

Measurements
Delacour (1977) reported that adult males of *muticus* have wing lengths of 460–500 mm (17.9–19.5 in) and tail-covert lengths of 1,400–1,600 mm (54.6–62.4 in; rectrices 400–475 mm [15.6–18.5 in]), whereas first- and second-year males have shorter trains. Females have wing lengths of 420–450 mm (16.4–17.6 in) and tail lengths of 400–450 mm (15.6–17.6 in). Males range in weight from about 3,850 to 5,000 g (8.4 to 10.9 lb; Ali and Ripley, 1978). Cheng et al. (1978) reported that three females weighed from 1,060 to 1,160 g (2.3 to 2.5 lb). The eggs average 72.7 × 53.7 mm (2.8 × 2.1 in) and have an estimated fresh weight of 114.9 g (4.0 oz).

Identification

In the Field (1,016–2,438 mm, 40–96 in)
Any wild peafowl occurring from Assam eastward is probably of a green peafowl, although feral individu-als of the Indian peafowl might also occur. The presence of a long, narrow, and columnar crest and the bluish to yellowish facial skin is unique to the green peafowl, which contrasts with the fanlike crest and white facial skin of the Indian peafowl. Otherwise, the two species are rather similar, and their vocalizations are also similar. The male's crowing is not as harsh or penetrating as that of the Indian species, and sounds like *aow-awo*. Likewise, the female's calls are not so loud, but both sexes produce loud, alarm *kwok* notes and guttural clucking sounds.

In the Hand
Any peafowl with blue and yellow facial skin and a long, rather tapering crest is a green peafowl. Otherwise, females of the two *Pavo* species are similar. However, the green peafowl averages slightly larger and its flanks and underparts tend toward blackish, whereas in the Indian peafowl the lower flanks and sides are mostly white or buffy.

Geographic Variation
Geographic variation in this species is limited to plumage coloration and is apparently clinal. The brightest and greenest of the three subspecies is the most southeasterly (Javan) race, whereas the most northwesterly (Burmese) race is distinctly duller and bluer, especially the males. The wing-coverts of the Javanese races are brightly iridescent blue and green, whereas in the Burmese races the coverts are black, with only a narrow bluish border. The secondaries, abdomen, and flanks of the Burmese races are also darker, and the facial skin is less bright in males. Females are also generally duller and have more extensive brownish black in the plumage (Delacour, 1977).

Ecology

Habitats and Population Densities
Baker (1930) described the habitats of the green peafowl as locally quite variable. In various areas these include such diverse habitats as elephant grass, open dry forests, and the densest thorn and bush undergrowth of evergreen forests. In most cases a nearby supply of good and plentiful water seems to

be important. Thus, the birds are often found on the banks of small, clear rivers having an abundance of undergrowth that is not so dense at ground level as to impair their freedom of movement. Green peafowl also are sometimes found in or near rice, mustard, or other cultivated fields. They extend locally up to at least 2,300 m (7,545 ft) and perhaps rarely to 3,000 m (9,845 ft).

Stewart-Cox and Quinnell (1990) estimated a density of 2.3–3.0 individuals per square kilometer (6.0–7.8 per square mile) in a Thailand sanctuary (Hwai Kah Kaeng). Java's Ujung Kolun National Park supports 200–250 birds in its 30,000-ha (74,100-acre) boundaries, suggesting a minimum density of 0.67–0.83 individuals per square kilometer (1.7–2.1 per square mile). The sizes of three male territories in Yunnan ranged from 0.38 to 0.56 km^2 (0.15 to 0.22 mi^2), averaging 0.42 (0.16 mi^2; Yang et al., 1997).

Competitors and Predators

There is no overlap between this species and the Indian peafowl, its most likely serious competitor. Beebe (1918–1922) considered that civet cats are likely a constant menace and noticed finding the half-eaten body of a male with leopard tracks around it. Pythons were also mentioned as possible predators.

General Biology

Food and Foraging Behavior

Beebe (1918–1922) found the crops of these green peafowl to have an abundance of termites, which he considered their major source of animal foods, as well as berries, grass seeds, peppers, flower petals, crickets, grasshoppers, and small moths.

Movements and Migrations

There is little information on this subject, although Beebe (1918–1922) reported that groups of peafowl are extremely sedentary, often occupying the same area of jungle month after month. They feed in various places, but roosting and watering sites are quite regular.

Daily Activities and Sociality

Beebe (1918–1922) observed that green peafowl came regularly each morning to a river, where they would spend about half an hour preening and drinking. After leaving the river in early morning, one

group of birds would go either to a large colony of termite mounds or to a steep valley, where they would disappear and feed for some hours. Toward noon the birds would often rest on a hidden sandbar while usually sleeping or preening, but sometimes drinking and feeding. Late in the afternoon the birds would typically return to the river, where they would sometimes stay until dark.

Roosting is done in jungle trees that are typically tall and dead. The birds reach these sites by flying first into a neighboring tree and then to the top of another, from which they finally fly to their roost. Sometimes four or five will roost in a single tree and rarely as many as seven. In more open jungle, lower trees may be used. These are typically trees that have smooth bark and no branches for some distance aboveground, which is seemingly an antipredatory device.

Social Behavior

Mating System and Territoriality

Green peafowl are a polygynous but still inadequately studied species. Males either defend harems of several females or, more probably, gather to display communally in leks, as does the Indian peafowl. It is likely that the males are territorial during the breeding season, but there is no real information on spacing. It is not known whether the male's loud call serves as a territorial proclamation, although this would seem probable. Beebe (1918–1922) stated that the calls of the green and Indian peafowl are hard to distinguish and that a male green will sometimes call in response to a male Indian peafowl's cry.

Voice and Display

The male's call is a very loud, penetrating *aow-awo* or *waaaa-ak!*, which is repeated several times and uttered with the neck well stretched and the bill somewhat raised. A second call consists of several rapidly repeated notes; it is a more subdued but still penetrating call (Beebe, 1918–1922). This or a similar call sounding like *tak-tak-ker-r-r-oo-oo* was uttered by a bird walking about in search of a companion that had been shot (Ali and Ripley, 1978).

The postural display of this species is almost identical to that of the Indian peafowl. Ali and Ripley (1978) observed copulation to occur when a female quietly approached a displaying male. He uttered a loud scream on seeing her, rushed toward her with drooping wings, and copulated while holding her down with his beak.

Reproductive Biology

Breeding Season and Nesting
In the Indian region, the major breeding season of the green peafowl is from January to April, but breeding from July to September has also been reported (Ali and Ripley, 1978). On the Malay Peninsula (Pahang area) the breeding season seems to be associated with the wet monsoon from November to May, with the birds in full feather during January and February and losing their trains by June or July (Medway and Wells, 1976). In Java, green peafowl in perfect plumage from June to August, at the start of the east monsoon, and they molt in October and November (Beebe, 1918–1922). Hoogerwerf (1949) indicated that nesting on Java occurs from August to October, with the largest number of reported clutches in August.

The nests are constructed on the ground, usually in well-protected locations with good visibility for approaching danger. The clutch size in natural conditions seems to vary from three to six eggs, with up to eight eggs reported under conditions of captivity (Beebe, 1918–1922). Hoogerwerf (1949) said that the clutch usually contains only three or four eggs. Baker (1930) reported a nest in a hollow at the buttressed base of an enormous cotton tree (*Bombax?*) situated under a dense thicket of thorny bushes that was almost impossible to penetrate.

Incubation and Brooding
Incubation is done by the female alone, and it requires 26–28 days. Baker (1928) suggested that in some areas two broods may perhaps be raised in a year, although this seems to be unproven.

Growth and Development of the Young
The rearing conditions and requirements for the chicks of this species are apparently much the same as in the Indian peafowl, although rather little has been written on the subject. In one case it was noted that a 14-day-old chick was able to fly up to the roof of a small house. By 23 days after hatching the crest is already noticeable. Body feathering is apparent at 1 month of age, and when 2.5 months old iridescent colors begin to appear. The young may begin dis-

playing when only 1.5 months old (Ali and Ripley, 1978). Like the argus and peacock pheasants, the chicks tend to hide under their mother's tail while very young and begin to perch on elevated sites at an early age. Sexual maturity is attained by the third year of life.

Evolutionary History and Relationships
The Indian and green peafowl are certainly close relatives, as indicated by the ease of hybridizing them in captivity and the full fertility of the resulting offspring. One hybrid combination (the "Spaulding's peafowl") has been bred for many generations and now appears to be fixed (Delacour, 1977). Certainly the two forms are good species, but they represent an obvious superspecies.

Status and Conservation Outlook
Collar et al. (1994) classified this species as vulnerable. McGowan and Garson (1995) regarded the race *spicifer* as insufficiently known and the other two races as endangered. As of the mid-1990s, the Javan population of *muticus* has been concentrated in Ujung Kulon National Park (see map 8, on page 198, for locations of Javan preserves), with 200–250 birds, and at Baluran Reserve, with about 200 birds. There were also some green peafowl at Alas Purwo National Park (25–50 birds) and some in teak forests and open woodland areas. The total Javan population numbered about 900–1,100 birds in the 1990s (Balen et al., 1995; Balen, 1997). The inset of map 8 shows Javan locality records since 1980. The Indochinese race *imperator* has also been reported from at least seven protected areas in Vietnam, including Nam Bai Cat Tien National Park, where densities exceeding 1 calling male per square kilometer (2.6 per square mile) were reported in 1989–1990 (Le Trong, 1997). This race also has been confirmed from near Bach Ma National Park and in Yok Don Reserve, but may be extirpated from central and North Vietnam. In adjoining Laos *imperator* is very rare and is known from only a few sites in northern and southern Laos, where it receives little protection (Evans, 1997). There is no good information from Cambodia. Birds presumably of the race *imperator* has also been reported recently in the Huai Kha Khang sanctuary of western Thailand, near the Burmese border. In the late 1980s over 200 birds were present there, and in 1993 about 250–270 birds were present (Stewart-

Cox, 1996). As recently as the early 1980s the green peafowl was still locally common in Burma, but there is no recent information available. The species was extirpated from the Malay Peninsula during the 1960s. Surveys in China in recent years have added Yunnan to the areas of major conservation significance for *imperator*. About 800–1,100 birds were probably present there in the mid-1990s, and a few scattered records exist from Tibet (Li, 1996; Wen et al., 1997*a*). It has been reported from 44 counties since first observed in Yunnan in 1917, but has since disappeared from 11 of these. This race is now limited to western, southern, and central Yunnan. Forest fragmentation and the absence of sanctuaries in its major areas of concentration are conservation problems (Wen et al., 1997*b*).

Genus *Afropavo* Chapin 1936

The Congo peacock is a medium-sized tropical pheasant in which sexual differences in weight are slight, but adult plumage dimorphism is considerable, with both sexes exhibiting iridescent plumage and crests. A large area of bluish to reddish skin is present on the sides of the face and upper neck. In males there is a tuft of white bristles in addition to softer black crest feathers. The wing is rounded, with the tenth primary the shortest and the fourth to sixth the longest. The tail has 18 rectrices, is flat and rounded, and is only partially covered by upper tail-coverts. In both sexes, but especially in females, these coverts and most of the mantle feathers are iridescent green. In males the tail and most of the other body feathers are also iridescent, whereas in females they are brownish. The tarsus is relatively long and stout, and is spurred in both sexes, an unusual situation among pheasants. A single species is recognized.

CONGO PEACOCK

Afropavo congensis Chapin 1936

Other Vernacular Names

None in general English use; paon du Congo (French); Kongopfau (German).

Distribution of Species

Rainforests of the east-central Congo Basin, eastward from Boende on the Tshuapa River in the west to the base of the mountain ridge to the west of Lakes Edward and Kivu. In the northern reaches the lower Aruwimi River between Basoko and Banalia, and in the south the vicinity of Lusambo. This species is limited to virgin forests between 365 and 1,525 m (1,200 and 5,000 ft). See map 24.

Distribution of Subspecies

None recognized by Delacour (1977).

Measurements

Verheyen (1965) reported that males have wing lengths of 306–315 mm (11.9–12.3 in) and tail lengths of 206–240 mm (8.0–9.4 in), whereas females have wing lengths of 270–295 mm (10.5–11.5 in) and tail lengths of 169–205 mm (6.6–8.0 in). Two males weighed 1,361 and 1,475 g (3.0 and 3.2 lb), and two females weighed 1,135 and 1,154 g (2.5 and 2.5 lb; David Rimlinger, pers. comm.). The eggs average 59 × 47 mm (2.3 × 1.8 in) and have an estimated fresh weight of 71.9 g (2.5 oz).

Map 24. Overall historic African range (left) and specific locality records (right) of Congo peacock. The probable mid-1990s primary range is indicated by hatching, and two extralimital records are shown as triangles. The historic limits of the Congo Basin equatorial rainforest are also indicated by a broken line.

Identification

In the Field (584–711 mm, 23–28 in)

Limited to the lowland forests of the Congo Basin, this species is unlikely to be confused with any other. Both sexes have iridescent green upper parts, bare pink skin on the foreneck, and erect thin crests somewhat similar in shape to those of the green peafowl. Calls are a complex gobbling, hooting sequence of notes that may be repeated as many as 20 or 30 times.

In the Hand

The peafowl-like shape and appearance of the Congo peacock, but with adults lacking an elongated train of tail-coverts and having no ocelli, provide for easy identification. The adults are smaller than typical peafowl and have a wing length no greater than 330 mm (12.9 in).

Ecology

Habitats and Population Densities

This species is limited to the tropical rainforest area of the Congo Basin, at altitudes from 100 to 1,200 m (330 to 3,935 ft). Cordier (1949) believed that Congo peacocks require semidry forest, avoid lowland areas, and tend to occur where their favorite wild fruits are to be found.

There are no estimates of population densities. However, from all accounts these would appear to be extremely low.

Competitors and Predators

Nothing is known with certainty about these topics in wild birds.

General Biology

Food and Foraging Behavior

Evidently the Congo peacock is a rather omnivorous species, eating a variety of seeds, fruits, and insects. Among the foods that have been found in their crops or gizzards are the drupes of a common broadleaf tree (*Celtis ituriensis*), aquatic insects, and termites. Termite larvae have also been found, which suggests that the birds probably dig into termite mounds in the manner of many tropical pheasants (Verheyen, 1965).

Movements and Migrations

Nothing is known of this subject in the wild.

Daily Activities and Sociality

Little can be said of this with certainty. However, Congo peacock are apparently nonsocial and are more likely to occur in pairs than in harems or other large congregations.

Social Behavior

Mating System and Territoriality

In this species, unlike the other peafowl, the male is believed to form a monogamous pair bond with a female and remain with her through the entire breeding period (Ridley, 1987). Based on captive observations, it is David Rimlinger's (pers. comm.) impression that the bond is stronger than in any other pheasants except the eared pheasants. However, there seems little doubt that at least under such conditions a prolonged and individualized attachment between the sexes is present. The smaller degree of sexual dimorphism in body weight and plumage development than found in typical peafowl would also support this conclusion.

Nothing is known of the degree of territoriality, if any, in the Congo peacock.

Voice and Display

The calls of this species have been variously described. By Chapin's (1938) description the common call is *rro-ho-ho-o-a* followed by a *gowe-gowah*, that many birds may reiterate. The second syllable of the *gowe* portion is more accentuated or higher pitched than the first. From the occasional irregularity of the timing of these notes, it may be possible that two birds may be answering one another. These notes may be repeated 20–30 times, sometimes with short interruptions. The naturalist Charles Cordier (1949) described the first portion of the call as that of the male, *ko-ko-wa*, and the female's response as a higher pitched *hej-hoh*, *hej-hoh*, thus forming a kind of duet. An African native's interpretation of the call was *wai-wai-ekalu-eko-pawola* (Verheyen, 1965). Captive males will often call early in the morning. The head with open bill is brought to the small of the back as the gobbling-hooting call is uttered. Other males may join in and call alternately (Roles, 1981). When alarmed, Congo peacocks crane the neck and utter a clucking note (Jeggo, 1972). The birds also utter

Figure 68. Display postures of male Congo peacock, including frontal display as seen from the side (A) and the front (B), lateral display as seen from the front (C) and the side (D), and threat display as seen from the side (E) and the front (F). In part after sketches by Jeggo (1972).

many low calls while eating, displaying, or simple walking, which perhaps serve as contact calls between paired birds. Some of these calls are very similar to the vocalizations of the Indian peafowl (David Rimlinger, pers. comm.). Some calling occurs at night as well, presumably while the birds are on their roosts, as in typical peafowl.

The displays of the Congo peacock are not yet well understood, but have been described by Jeggo (1972). He stated that the male has two major courtship postures, one frontal and one lateral (figure 68). In the frontal posture the male raises his head, emits a low and drawn-out note, and slightly "inflates" the bare skin of the throat. He then bows while uttering a series of shorter and rapid notes, raises and fans the tail, and drops and tilts the wings slightly outward so that the primaries are visible from the front. The posture somewhat resembles the frontal display of the Indian peafowl, but of course lacking the train of tail-coverts. This posture may be assumed when the male is on a perch or while on the ground as he is facing the female. The female will sometimes respond with a threatlike display, in which she holds her head high with her tail raised and fanned and her wings drooped, and thus strut toward the male.

During lateral display, the male begins with the same call as used during frontal display, but his head is lowered and extended anteriorly in a straight line with the back while the tail is spread and fanned toward the side nearer the female. The wing of the same side is spread and lowered while the other is held in and arched up, which causes the bird to appear tilted toward one side. While in this posture, the bird moves around the female.

Tidbitting behavior is also performed. In this the male holds a morsel in his beak with his head lowered, then calls the female to take it. Similar behavior is performed by both sexes when feeding chicks. This retention of true tidbitting behavior and a lateral display in *Afropavo* is especially interesting in view of their absence in *Pavo*; this suggests that *Pavo* is behaviorally and structurally the more specialized genus.

Both sexes exhibit aggressive behavior in the same manner, by assuming a posture much like that of the strutting display, but with a varying amount of wing-lowering and wing-spreading. In intense aggressive display, as when defending a nest or brood, the female may lower her wings to the point that the carpal joints almost reach the ground (Jeggo, 1972). David Rimlinger (pers. comm.) regarded the courtship and threat display as the same or very similar. He also noted that the male usually spreads his wings at the female or a human only at the most intense moment of posturing. When the wings are spread together with the tail, a full half-circle of iridescent color can be seen when the male is viewed directly from the front.

Reproductive Biology

Breeding Season and Nesting
Nothing definite is known of the breeding season in the wild, and no nests have been described from na-

ture. However, based on captive observations it is probable that the Congo peacock normally nests in trees. It constructs no true nest, but lays its eggs on a flat surface such as a slightly concave platform on a large tree limb, preferably less than 1.5 m (5 ft) aboveground. However, the Congo peacock has also been known to lay its eggs on the substrate surface. The eggs are apparently laid at 3-day intervals. Typically only two or three eggs are laid, although on at least two occasions clutches of four eggs have been reported (Verheyen, 1965). Between 1964 and 1973 a total of 554 eggs in 179 clutches were produced at the Antwerp Zoo, or an average of 3.09 eggs per clutch and an average of 2.2 clutches per pair per year (Lovel, 1976). Up to five or six clutches have been produced by some females in a single year (Van den bergh, 1975).

Incubation and Brooding
Based on observations at Rotterdam Zoo, the incubation period is 27–28 days, whereas at various other zoos the duration has ranged from 25 to 27 days. As she incubates, the female remains completely motionless, lowers her head over her back, and becomes very inconspicuous. The male may spend long periods sitting beside the nest site, but never attempts to incubate. Females leave their incubation for only short periods to feed and drink (Roles, 1981).

Growth and Development of the Young
The young chicks remain in the nest with their mother for up to 2 days, after which she leaves the nest and they flutter down to the ground as well. Both sexes subsequently brood the young chicks and call to them to take bits of food from their beaks. During their first week the chicks eat almost nothing but insects and other animal foods. During the first night that they are out of the nest the chicks may be brooded on the ground by the male while the female roosts on the nest. After that the young are brooded under the wings of the mother while she sits on a low perch. Soon the young are able to reach perches 2 m (6.5 ft) or more in height by half-flying and half-running to reach it. By the sixth day the young can fly up to a height of 5 m (16.4 ft), and after 1 month they brood close to, but not under the wings of, their mother (Verheyen, 1965). Adult plumage is attained in the second year of life. Females become fertile at 1 year, but males only at 18 months. In captivity, a

male can be housed with more than one female, but is apparently interested in only one of them. The other hen tends to be persecuted by the chosen female (Van der bergh, 1975).

Evolutionary History and Relationships
The relationships of the Congo peacock have attracted a good deal of attention and controversy, as summarized by Verheyen (1965). The surprisingly similar appearance of *Afropavo* to a hybrid between a guineafowl and peafowl, as well as some skeletal similarities with guineafowl, have led some people such as Verheyen (1965) to suggest that the Congo peacock may be a connecting form between these two groups. However, the major morphological contribution to its relationship has been provided by Lowe (1939), who did a detailed anatomical study of its bones, muscles, feather tracts, and distinctive plumage pattern. Lowe judged that the plumage pattern of the male was clearly primitive. Also, he detected in its tail-coverts the adumbration pattern of the pavonine ocellations, which can also be seen in the sequence of peacock pheasants beginning with the bronze-tailed, proceeding through the Rothschild's, and on to the typical *Polyplectron* forms. Lowe judged that *Afropavo* may indeed come close to an ancestral peacocklike type that presumably lived during Miocene times in southern Europe. This type may have been driven south during the Pleistocene to a last retreat in the Congo forests, thus having been completely isolated from the other Asian pheasants. Various other recent studies have suggested that *Afropavo* is a relatively isolated form, with only remote affinities to *Pavo* (Gysels and Rabaye, 1962; Hulselmans, 1962). However, the most recent study by Boer and van Bocxstaele (1981) tends to confirm Chapin's (1938) original view, that *Afropavo* and *Pavo* are more closely related to one another than to any other Galliformes. Interestingly, however, an affinity with the guineafowl

group was also suggested by their data. I believe that *Afropavo* is indeed a primitive peafowl, perhaps not very different from the ancestral type that produced the modern *Pavo*, and one that has never acquired (rather than having secondarily lost) the elaborate male ocellation pattern, while at the same time having retained the generalized male pheasant display features (tidbitting and lateral display) that *Pavo* has subsequently lost.

Status and Conservation Outlook
Lovel (1976) summarized the captive status of this species, which is very slowly becoming established in captivity. In the wild, the Congo peacock's current status is essentially unknown, although it is likely that forest clearing and hunting may have reduced its numbers. The Salonga National Park in Zaire, which covers some 3.6 million hectares (8.9 million acres), is in the middle of this species' range and thus provides a degree of protection.

Tribal warfare in the Congo Basin during the 1990s has undoubtedly adversely affected this species' status, but little detailed information exists. Surveys in eastern Zaire from 1989 through the early 1990s confirmed the Congo peacock's existence in about a dozen sites (Hart, 1994). Fieldwork during 1994 and 1995 established that the species occurs between the Lukenie and Sankura Rivers of central Zaire and in the Lomako Forest, the latter location considerably north of the species' previously understood range limits. To the east, the Congo peacock occurs in Maiko National Park in the upper Maiko drainage, where it has been recorded in at least ten locations. It has also been reported from the Kahuzi-Biega National Park Extension, west of Lake Kivu. McGowan and Garson (1995) considered the species' status as insufficiently known, and Collar et al. (1994) classified the Congo peacock as vulnerable.

APPENDIX

Derivations of Scientific and Vernacular Names of Pheasants

Afropavo: Latin, an African peacock.
  congensis: of the Congo River.
Argusianus: Latin, adjective from *Argus,* the original preoccupied name given to the bird.
  argus: in reference to the mythological shepherd with a hundred eyes.
  grayi: after G. R. Gray (1808–1872), one-time curator of birds of the British Museum.
Catreus: Greek, referring to a peacocklike bird.
  wallichii: after Dr. N. W. Wallich (1786–1854), Danish botanist and one-time superintendent of the Calcutta Botanical Garden. The vernacular name cheer is from the Nepalese *kahir,* or *chirir,* an onomatopoeic name.
Chrysolophus: from the Greek *chryseos,* golden, and *lophos,* crest.
  amherstiae: after Sarah, Countess of Amherst (1762–1838).
  pictus: Latin, painted or adorned.
Crossoptilon: from the Greek *krossotos,* fringed, and *ptilon,* down or feathers, referring to the specialized tail feathers.
  auritum: Latin, eared.
  mantchuricum: of Manchuria.
  tibetanum: of Tibet.
Gallus: Latin, meaning a cock. The term gallinaceous is derived from *gallina,* hen.
  lafayetti: after Marquese de Lafayette (1757–1834), French general and statesman.
  sonnerati: after P. Sonnerat (1745–1814), the French naturalist who discovered the species.
  varius: Latin, variegated, in reference to the plumage pattern.

Ithaginis: Greek, legitimate, of a true race.
  cruentus: Latin, stained with blood, in reference to the reddish plumage.
Lophophorus: from the Greek *lophos,* crest, and *phoros,* to bear.
  impeyanus: after Lady Impey, wife of Sir Elilah Impey, first governor of Bengal. The vernacular name is derived from the central Himalayan native name *moonal.*
  lhuysii: after E. Drouyn de Lhuys (1805–1881), president in 1866 of the Societé Imperiale d'Acclimations.
  sclateri: after Dr. P. L. Sclater (1829–1913), English ornithologist.
Lophura: from the Greek, *lophos,* crested, and *ura,* tail.
  bulweri: after H. E. G. Bulwer (1834–1914), governor of Lauban.
  edwardsi: after Prof. A. Milne-Edwards (1835–1900), famous French naturalist and director of the National History Museum in Paris.
  erythropthalma: from the Greek *erythros,* reddish, and *opthalmos,* eye. This species is sometimes placed in the genus *Acomus,* meaning crestless.
  ignita: from the Latin *ignis,* fire, in reference to the back and abdominal plumage.
  hatinhensis: after the Vietnamese province where it was discovered.
  imperialis: Latin, imperial.
  inornata: Latin, nonornamental.
  leucomelana: from the Greek *leukos,* white, and *melas,* black. The vernacular name kaleege is

derived from the native Nepalese name *kalich* or *kalij.* This species is sometimes placed in the genus *Gennaeus,* meaning genuine or good.

nycthemera: from the Greek, *nyctos,* night, and *hemera,* day, in reference to its black-and-white plumage pattern.

swinhoei: after Robert Swinhoe (1836–1877), discoverer of the species and one-time British consul in Formosa.

Pavo: Latin, a peafowl.

cristatus: Latin, crested.

muticus: Latin, docked or curtailed.

Phasianus: from the Greek *phasianos,* a pheasant or bird from Phasis (a river in Colchis).

colchicus: of Colchis, now Mingrelia, an area in western Transcaucasia.

versicolor: Latin, many-colored.

Polyplectron: from the Greek *poly,* many, and *plectrum,* something to strike with.

bicalcaratum: from the Latin *bi,* two, and *calcar,* a spur; two-spurred.

chalcurum: from the Greek *chalkos,* copper, and *oura,* the tail.

emphanum: from the Greek *en,* in, and *phanos,* brightness or light.

germaini: after L. Germain (1827–1927?), who sent the first specimens from China.

inopinatus: Latin, unexpected. The generic name *Chalcurus,* sometimes applied to this species, is explained above.

malaccense: of Malacca.

schleiermacheri: after a Herr Schleiermacher, director of the Hessian Museum in 1877.

Pucrasia: a Latinized onomatopoeic name.

macrolopha: from the Greek *makros,* long, and *lophos,* crested. The vernacular name koklass

is from the native onomatopoeic name *kokla, koklass,* or *phocrass,* from which the generic name *Pucrasia* is also derived.

Rheinartia: after a Captain Rheinard, of the French army in Annam in 1880–1883, who sent the first specimen to Paris.

ocellata: Latin, ocellated.

nigrescens: Latin, dusky.

Syrmaticus: Latin, dragging or trailing behind; from the Latin *syrmia,* a robe with a train, in reference to the long tail.

ellioti: after D. G. Elliot (1835–1915), American ornithologist.

humiae: after Mary Hume (?–1890), wife of Allen Hume, English ornithologist.

mikado: referring to the emperor of Japan.

reevesi: after J. Reeves (1774–1856), English businessman who first brought living specimens to England in 1831.

soemmerringi: after Dr. S. T. v. Soemmering (1755–1830), German professor.

Tragopan: from the Greek *tragos,* a goat, and Pan, the mythical Greek god; literally goat-Pan, in reference to the male's "horns."

blythi: after Edward Blyth (1810–1873), English naturalist and one-time curator of the Calcutta Museum.

caboti: after Dr. Samuel Cabot (1815–1885), American ornithologist.

melanocephalus: from the Greek *melas,* black, and *kephale,* a head.

satyra: satyrlike; a satyr was a Greek and Roman semideity with the horns and hind limbs of a goat.

temminckii: after G. J. Temminck (1775–1858), Dutch ornithologist.

Bibliography

Action, R. W. 1963. Elliot's pheasant program: An evaluation. *Game Res. Ohio* 2:222–234.

Ali, S. 1962. *The birds of Sikkim.* Oxford University Press, Madras.

Ali, S., and S. D. Ripley. 1978. *Handbook of birds of India and Pakistan.* Vol. 2. Oxford University Press, Oxford.

Anderson, E. S. 1993. The Vietnamese pheasant. *WPA News* 41:28–29.

Anderson, W. L. 1964. Survival and reproduction of pheasants and in southern Illinois. *J. Wildl. Manage.* 28:254–264.

Anonymous. 1989. Rediscoveries. *World Birdwatch* 11:4.

Anonymous. 1996. Extinct pheasant rediscovered. *WPA Ann. Rev.* 1995/96:41–42.

Anonymous. 1997. Abstracts of presentations. International Symposium on Galliformes, 8–17 September 1997, Malaka, Malaysia.

Anthony, A. W. 1899. Hybrid grouse. *Auk* 16:180–181.

Arshad, M. I., and M. Zakaria. 1997. Breeding ecology of red junglefowl. Abstract. P. 32 *in* Anonymous, 1997.

Asmundson, V. S., and F. W. Lorenz. 1975. Pheasant-turkey hybrids. *Science* 121:307–308.

Baker, C. M. A. 1965. Molecular genetics of avian proteins. V. The egg white proteins of the golden pheasant *Chrysolophus pictus* L. and the Lady Amherst's pheasant *C. amherstiae* Leadbeater, and their possible evolutionary significance. *Comp. Biochem. Physiol.* 16:93–101.

Baker, E. C. S. 1928. *The fauna of British India, including Ceylon and Burma.* Vol. 5. *Birds.* Taylor and Francis, London.

Baker, E. C. S. 1930. *Game-birds of India, Burma and Ceylon.* Vol. 3. John Bale and Son, London.

Baker, E. C. S. 1935. *The nidification of birds of the Indian Empire.* Vol. IV. Taylor and Francis, London.

Balen, R. van. 1997. Status of conservation of green peafowl on Java. *Tragopan* 7:18–20.

Balen, R. van, D. M. Prawiradlaga, and M. Indrawan. 1995. The distribution of green peafowl *Pavo muticus* in Java. *Biol. Cons.* 71:289–297.

Banks, E. M. 1956. Social organization in red junglefowl hens (*Gallus gallus* subsp.). *Ecology* 37:240–248.

Banks, R. C. 1993. The generic name of the crested argus *Rheinardia ocellata. Forktail* 8:3–6.

Barancheev, L. M. 1965. Migrations of Manchurian pheasants to the Amur region. Pp. 125–132 in *Migrations of birds and mammals.* Nauka, Moscow. (In Russian.)

Baskett, T. S. 1947. Nesting and production of the ring-necked pheasant in north-central Iowa. *Ecol. Monogr.* 17:1–30.

Bates, R. S. P., and E. H. N. Lowther. 1952. *Breeding birds of Kashmir.* Oxford University Press, Oxford.

Baxter, W. L., and C. W. Wolfe. 1973. Life history and ecology of the ring-necked pheasant in Nebraska. Nebraska Game and Parks Commission, Lincoln.

Beebe, C. W. 1914a. Preliminary pheasant studies. *Zoologica* 1(15):261–285.

Beebe, C. W. 1914b. Revision of genus *Gennaeus. Zoologica* 1(17):303–323.

Beebe, C. W. 1918–1922. *A monograph of the pheasants.* Witherby, London. 4 vols. Reprint, as 2 vols., Dover, New York, 1990.

Beebe, C. W. 1936. *Pheasants, their lives and homes.* 2nd ed. Doubleday Doran, New York.

Beer, J. V., and C. Cox. 1981. The production of eggs of the cheer pheasant *Catreus wallichi* for re-introduction programs. Pp. 68–71 *in* Savage, 1981.

Bierens de Hann, J. A. 1926. Die Balz des Argus-fasans. *Biol. Zentralblatt* 46:428–435.

Bisht, M., P. Lakhera, and A. C. Saklani. 1990. Hi-malayan monal pheasant: Current status and habi-tat utilization in Kedarnath Sanctuary, Garhwal Himalaya. Pp. 205–208 *in* Hill et al., 1990.

Biswas, B. 1960–1963. The birds of Nepal. *J. Bombay Nat. Hist. Soc.* 57:278–308, 516–546; 58:100–134, 441–474, 653–677; 59:200–227, 405–429, 807–821; 60:178–200, 388–399.

Biswas, B. 1968. The female of Molesworth's trago-pan, *Tragopan blythi molesworthi* Baker. *J. Bombay Nat. Hist. Soc.* 65:782–784.

Boback, A. W., and D. Müller-Schwarze. 1968. *Das Birkhuhn.* Neue Brehm Bucherei 397. A. Ziemsen, Wittenberg Lutherstadt.

Boer, L. E. M., and R. de van Bocxstaele. 1981. So-matic chromosomes of the Congo peafowl (*Afro-pavo congensis*) and their bearing on the species' affinities. *Condor* 82:204–208.

Boetticher, H. von. 1939. Ueberblick über die Hühn-ervogel und iher Verbreitung. *Fol. Zool. Hydrobiol.* 9:290–299.

Bohl, W. H. 1964. *A study and review of the Japanese green and the Korean ring-necked pheasant.* U.S. Fish and Wildlife Service, Special Scientific Report Wildlife No. 83.

Boonlerd, A. 1981. Thailand: Status of the pheasants. Pp. 39–41 *in* Savage, 1981.

Brazil, M. 1991. *The birds of Japan.* Smithsonian Insti-tution Press, Washington, D.C.

Bridgman, C. L. 1994. Mikado pheasant use of dis-turbed habitats in Yushan National Park, Taiwan, with notes on its natural history. M.Sc. thesis, Eastern Kentucky University, Richmond.

Bridgman, C. L., P. S. Alexander, and L. Chen. 1997. Mikado pheasant home range in secondary growth habitats of Yushan National Park, Taiwan. Ab-stract. P. 68 *in* Anonymous, 1997.

Brisbin, I. L., and A. T. Peterson. 1997. A reappraisal of conservation concerns for the red junglefowl *Gallus gallus:* DNA profiles, historical specimens, and genetic purity. Abstract. P. 79 *in* Anonymous, 1997.

Brun, R. 1971. Problems of ontogenesis and evolu-tion in the feather pattern formation of the argus pheasant (*Argusianus argus*). *Rev. Suisse Zool.* 78:115–134. (In German, English summary.)

Bruning, D. 1977. Breeding the Malay peacock pheasant at the New York Zoological Park. *Avic. Mag.* 83:61–62.

Bruning, D. 1983. Continued breeding success with Malay peacock pheasants, *Polyplectron malacense malacense,* at the New York Zoological Park. *World Pheasant Assoc. J.* 8:62–68.

Brush, A. 1967. Hemoglobins of a ring-necked pheasant × junglefowl hybrid. *Condor* 69:206.

Bump, G., and W. H. Bohl. 1961. *Red junglefowl and kalij pheasants.* U.S. Fish and Wildlife Service, Special Scientific Report Wildlife No. 62.

Bura, A. 1967. Das Juvenilgefieder von *Phasianus colchicus* L., ein Beitrag zur Kenntnis dieser Alterse-tappe des Gefieders. *Rev. Suisse Zool.* 74:301–387. (In German, English summary.)

Burger, G. V. 1966. Observations on aggressive be-havior of male ring-necked pheasants in Wiscon-sin. *J. Wildl. Manage.* 30:57–64.

Burtulin, S. A. 1904. On the geographical distribu-tion of the true pheasants (genus *Phasianus sensu stricto*). *Ibis* Ser. 8 4:377–414.

Burtulin, S. A. 1908. Additional notes on the true pheasants (*Phasianus*). *Ibis* Ser. 9 2:570–592.

Buss, I. O. 1946. *Wisconsin pheasant populations.* Wis-consin Conservation Department Publication 326(A-46).

Caleda, M. 1993. Population ecology of the Palawan peacock-pheasant. *World Pheasant Assoc. J.* 17/18:106.

Caleda, M., R. Lanante, and E. Viloria. 1986. Prelim-inary studies of the Palawan peacock-pheasant. *In* Ridley, 1986 (unpaged).

Carpentier, J., J. J. Yealland, R. Bocxstaele, and W. van den Bergh. 1975. Imperial pheasant hybridiza-tion at the Antwerp Zoo. *Intnl. Zoo Yearb.* 15:100–105.

Catlow, P. 1982. Displays of the Himalayn monal (*Lophophorus impejanus*) in captivity. *World Pheasant Assoc. J.* 7:92–95.

Chapin, J. C. 1938. The Congo peacock. Pp. 101–109 *in* Comptes Rendus du 9ième Congrès Ornithologique International, Rouen.

Chauhan, B. S., and V. Sharma. 1991. Status of west-ern tragopan in Himachal Pradesh, India. *WPA News* 34:25–28.

Cheng, C., J.-L. Li, and F.-L. Lui. 1996. Studies on breeding Chinese monals in captivity. *Acta Zool. Sinica* 42:54–61. (In Chinese, English summary.)

Cheng, T. 1963. *China's economic fauna: Birds.* Science Publishing Society, Peking. (Translated by U.S. Department of Commerce, Washington, D.C.)

Cheng, T. 1979. On subspecific differentiation of the silver pheasant *Lophura nycthemera. World Pheasant Assoc. J.* 4:42–45.

Cheng, T., Y. Tan, T. Lu, G. Bao, and F. Li. 1978. *Fauna Sinica, Series Vertebratica, Aves.* Vol. 4. *Galliformes.* Science Press, Academia Sinica, Peking.

Choudhury, A. 1996. On the trail of Blyth's tragopan. *WPA News* 51:14–16.

Choudhury, A. 1997. New localities for Blyth's tragopan from Nagaland, India. *Tragopan* 6:11–12.

Clark, C. A., Jr. 1964. Life histories and the evolution of megapodes. *Living Bird* 3:149–167.

Collar, N. J., and P. Andrew. 1988. *Birds to watch: The ICBP world checklist of threatened birds.* Pp. 1–303 *in* ICBP Technical Publication No. 8. Smithsonian Institution Press, Washington, D.C.

Collar, N. J., M. J. Crosby, and A. J. Statterfield. 1994. *Birds to watch 2: The world list of threatened birds.* BirdLife Conservation Ser. No. 4. BirdLife International, Cambridge, U.K., and Smithsonian Institution Press, Washington, D.C.

Collias, N. E. 1987. The vocal repertoire of the red junglefowl, a spectrographic classification and code of communication. *Condor* 89:510–524.

Collias, N. E., and E. C. Collias. 1967. A field study of the red junglefowl in north-central India. *Condor* 69:360–386.

Collias, N. E., and E. C. Collias. 1996. Social organization in a red junglefowl *Gallus gallus* population related to evolutionary theory. *Anim. Behav.* 51: 1337–1354.

Collias, N. E., and M. Joos. 1953. The spectrographic analysis of sound signals of the domestic fowl. *Behaviour* 5:175–188.

Collias, N. E., and P. Saichuae. 1967. Ecology of the red junglefowl in Thailand and Malaya with reference to the origin of domestication. *Nat. His. Bull. Siam. Soc.* 22:89–209.

Collias, N. E., and R. D. Taber. 1951. A field study of grouping and social dominance in ring-necked pheasants. *Condor* 52:265–275.

Collias, N. E., D. Hunsaker, and L. Minning. 1966. Locality fixation, mobility and social organization

within an unconfined population of red junglefowl. *Anim. Behav.* 14:550–559.

Collias, N. E., E. C. Collias, and R. I. Jenrich. 1994. Dominant red junglefowl (*Gallus gallus*) hens in an unconfined flock rear the most young over their lifetime. *Auk* 111:863–872.

Coomans de Ruiter, L. 1946. Oologische en Biologische Aanteekeningen over eenige Hoendervogels in de Westerafdeeching van Borneo. *Limosa* 19: 129–140. (In Dutch, English summary.)

Corder, J. 1996. Plumage variation in Vietnamese and Edwards' pheasants. *Tragopan* 4:10–11.

Cordier, C. 1949. Our Belgian Congo expedition comes home. *Anim. Kingdom* 52:99–114, 134–136.

Cottam, C. 1929. The status of the ring-necked pheasant in Utah. *Condor* 13:117–123.

Cracraft, J. 1981. Toward a phylogenetic classification of the Recent birds of the world (class Aves). *Auk* 98:681–714.

Cramp, S., and K. E. L. Simmons, eds. 1980. *The birds of the western Palearctic.* Vol. 2. Oxford University Press, Oxford.

Crowe, T. M. 1978. The evolution of guinea fowl (Galliformes, Phasianidae, Numidinae); taxonomy, phylogeny, speciational biogeography. *Ann. S. Afr. Mus.* 76:43–136.

Crowe, T. M. 1996. Letter to the editor. *WPA News* 46:43.

Crowe, T. M., and P. Bloomer. 1997. Molecular and morphological perspectives on galliform taxonomy: Agreement or conflict? Abstract. P. 64 *in* Anonymous, 1997.

Crozier, G. 1967. Etude de la transferrine des hybrides interspécifiques. *Gallus gallus × Numida meleagris. Ann. Biol. Anim. Biochem. Biophys.* 7:73–78.

Dalke, P. 1937. Food habits of adult pheasants in Michigan based on crop analysis method. *Ecology* 18:199–213.

Danforth, C. H. 1950. Evolution and plumage traits in pheasant hybrids, *Phasianus × Chrysolophus. Evolution* 4:301–315.

Danforth, C. H. 1958. *Gallus sonnerati* and the domestic fowl. *J. Hered.* 49:167–169.

Danforth, C. H., and G. Sandness. 1939. Behaviour of genes in intergeneric crosses: Effects of two dominant genes on colour in pheasant hybrids. *J. Hered.* 30:537–542.

Davison, G. W. H. 1974. Geographic variation in *Lophophorus sclateri*. *Bull. Br. Ornith. Club* 96: 163–164.

Davison, G. W. H. 1976. The function of tail and under tail-covert pattern in pheasants. *Ibis* 118: 123–126.

Davison, G. W. H. 1977. Studies of crested argus. I. History and problems associated with the species in Malaysia. *World Pheasant Assoc. J.* 2:50–56.

Davison, G. W. H. 1978a. Studies of crested argus. II. Gunong Rabong 1976. *World Pheasant Assoc. J.* 3:46–53.

Davison, G. W. H. 1978b. Further notes on *Lophophorus sclateri*. *Bull. Br. Ornith. Club* 98:116–118.

Davison, G. W. H. 1979a. The behaviour of the barred-back pheasant (*Syrmaticus humiae* (Hume)). *J. Bombay Nat. Hist. Soc.* 76:439–446.

Davison, G. W. H. 1979b. The behaviour and ecology of some Malaysian Galliformes, with special reference to *Argusianus argus* (Phasianidae). Ph.D. diss., University of Malaya.

Davison, G. W. H. 1979c. Studies of the crested argus. III. Gunong Rabong, 1977. *World Pheasant Assoc. J.* 4:76–80.

Davison, G. W. H. 1979d. Alleged occurrences of *Rheinartia ocellata* in Sumatra. *Bull. Br. Ornith. Club* 99:80–81.

Davison, G. W. H. 1980a. Galliformes and the Gunung Mulu National Park. *World Pheasant Assoc. J.* 5:31–39.

Davison, G. W. H. 1980b. The evolution of the crested argus. *World Pheasant Assoc. J.* 5:91–97.

Davison, G. W. H. 1980c. The type locality of *Rheinartia ocellata nigrescens* Rothschild. *Bull. Br. Ornith. Club* 100:141–143.

Davison, G. W. H. 1981a. Habitat requirements and food supply of the crested fireback. *World Pheasant Assoc. J.* 6:40–52.

Davison, G. W. H. 1981b. Problems of censusing pheasants in tropical rain forest. Pp. 49–52 *in* Savage, 1981.

Davison, G. W. H. 1981c. Status of the pheasants, Malaysia and Indonesia. Pp. 29–32 *in* Savage, 1981.

Davison, G. W. H. 1981d. Sexual selection and the mating system of *Argusianus argus* (Aves: Phasianidae). *Biol. J. Linn. Soc.* 15:91–104.

Davison, G. W. H. 1981e. Diet and dispersion of the great argus *Argusianus argus*. *Ibis* 123:485–494.

Davison, G. W. H. 1982. Sexual displays of the great argus pheasant *Argusianus argus*. *Z. Tierpsychol.* 58:185–202.

Davison, G. W. H. 1983a. The eyes have it: Ocelli in a rain-forest pheasant. *Anim. Behav.* 31:1037–1042.

Davison, G. W. H. 1983b. Behaviour of Malay peacock pheasant, *Polyplectron malacense*. *J. Zool.* 201:57–66.

Davison, G. W. H. 1983c. Notes on the extinct *Argusianus bipunctatus* (Wood). *Bull. Br. Ornith. Club* 103:86–88.

Davison, G. W. H. 1985. Peacock-pheasant display without ocelli. *Indo-Malayan Zool.* 2:1–7.

Davison, G. W. H. 1987. Ecology and behavior of great argus in suboptimal habitat. Pp. 28–32 *in* Savage and Ridley, 1987.

Davison, G. W. H. 1992. Display of the mountain peacock-pheasant. *World Pheasant Assoc. J.* 15/16: 45–56.

Davison, G. W. H. 1996. Why are *Lophura* pheasants so variable? *WPA Ann. Rev.* 1995/96:34–38.

Davison, G. W. H., and K. Scriven. 1987. Recent pheasant surveys in peninsular Malaysia. Pp. 90–101 *in* Savage and Ridley, 1987.

Deignan, H. G. 1945. The birds of northern Thailand. *Bull. U.S. Natl. Mus.* 186:1–615.

del Hoyo, J., A. Elliott, and J. Sargatal, eds. 1994. *Handbook of the birds of the world*. Vol. 2. *New World vultures to guineafowl*. Lynx Edicions, Barcelona.

Delacour, J. 1945. Note on the eared pheasants (*Crossoptilon*). *Zoologica* 80:30–45.

Delacour, J. 1948. The subspecies of *Lophura nycthemera*. *Am. Mus. Novit.* 1377:1–12.

Delacour, J. 1949. The genus *Lophura*. *Ibis* 91:188–200.

Delacour, J. 1977. *The pheasants of the world*. 2nd ed. World Pheasant Association and Spur Publications, Hindhead, U.K.

Delacour, J. 1978. *Pheasants: Their care and breeding*. T.F.H. Publishing, Neptune, N.J.

Delacour, J., and P. Jabouille. 1925. The birds of Quangtri, central Annam, with notes on others from other parts of French Indo-China. *Ibis* Ser. 12 1:209–260.

Dementiev, G. P., and N. A. Gladkov, eds. 1967. *Birds of the Soviet Union*. Vol. 4. Israel Programme for Scientific Translations, Jerusalem.

Denton, V. 1978. Breeding experiences with Rothschild's and Bornean peacock pheasants. Pp. 318–321 *in* A. C. Risser Jr., L. F. Baptista, S. R. Wyline, and N. B. Gale, eds. Proceedings of the 1st International Birds in Captivity Symposium, Seattle, Washington.

Deraniyagala, P. E. P. 1957. Growth, subspeciation and hybridisation in the Ceylon junglefowl. *Spolia Zeylan* 28:99–106.

Dierenfeld, E., M. Burnett, and C. Sheppard. 1998. Recommendations for diets of captive pheasants. *Michigan Bird and Game Breeders Association Newsletter* 1998(Jan 10):6–11.

Ding, C. 1995. The mating system and breeding ecology of Cabot's tragopan *Tragopan caboti*. *Tragopan* 2:11–12.

Ding, C., and G.-M. Zheng. 1997. The nest-site selection of the yellow-bellied tragopan (*Tragopan caboti*). *Acta Zool. Sinica* 43:27–33. (In Chinese, English summary.)

Ding, C., W. Liang, L. Zhao, and H. Gong. 1997. The ecology of the golden pheasant in the 1996 breeding season. Abstract. P. 70 *in* Anonymous, 1997.

Ding, P., Y. Yang, S. Liang, S.-R. Jiang, and Y. Zuge. 1996. The habitat of Elliot's pheasant in the Leigong Mountain Nature Reserve. *Acta Zool. Sinica* 42:62–68. (In Chinese, English summary.)

Domm, L. V. 1927. New experiments on ovariotomy and the problem of sex inversion in the fowl. *J. Exp. Zool.* 48:385–416.

Dumke, R. T., and C. M. Pils. 1973. Mortality of radio-tagged pheasants on the Waterloo Wildlife Area. Pp. 1–52 *in* Wisconsin Department of Natural Resources, Technical Bulletin No. 72.

Dumke, R. T., and C. M. Pils. 1979. Renesting and dynamics of nest site selection by Wisconsin pheasants. *J. Wildl. Manage.* 43:705–716.

Dunning, J. B. 1993. *CRC handbook of avian body masses*. CRC Press, Boca Raton, Fla.

Durrer, H. 1965. Bau und Bildung der Augfeder des Pfaus (*Pavo cristatus* L.). *Rev. Suisse Zool.* 72:263–411. (In German, English summary.)

Dyck, J. 1992. Reflectance spectra of plumage areas colored by green feather pigments. *Auk* 109:293–301.

Eames, J. C. 1995. Endemic birds and protected area development on the Da Lat Plateau, Vietnam. *Bird Cons. Intnl.* 5:491–523.

Eames, J. C. 1997. Rediscovery of Edwards' pheasant. *Tragopan* 7:3.

Eames, J. C., T. C. Lambert, and C. Nguyen. 1994. A survey of the Annamese lowlands, Vietnam, and its implications for the conservation of Vietnamese and imperial pheasants, *Lophura hatinhensis* and *Lophura imperialis. Bird Cons. Intnl.* 4:343–382.

Edminster, F. C. 1954. *American game birds of field and forest.* Charles Scribner's Sons, New York.

Edwards, W. R., P. J. Mikolay, and E. A. Leitie. 1964. Implications from winter-spring weights of pheasants. *J. Wildl. Manage.* 28:290–297.

Einarsen, A. S. 1945. Some factors affecting ring-necked pheasant density. *Murrelet* 26:2–9, 39–44.

Elliot, D. G. 1870–1872. *A monograph of the Phasianidae, or family of the pheasants.* 2 vols. Published by the author, New York.

Endo, K. 1982. *The Japanese green pheasant.* Kaisei-sha, Tokyo. (In Japanese.)

Errington, P. L., and F. N. Hamerstrom. 1937. The evaluation of nesting losses and juvenile mortality of the ring-necked pheasant. *J. Wildl. Manage.* 1:3–20.

Etchécopar, R. D., and F. Hüe. 1978. *Les oiseaux de Chine, non-passereaux.* N. Boubée, Paris.

Etter, S. L., J. E. Warnock, and G. B. Joselyn. 1970. Modified wing molt criteria for estimating the ages of wild juvenile pheasants. *J. Wildl. Manage.* 34: 620–626.

Evans, T. 1997. Green peafowl *Pavo muticus* in Laos. *Tragopan* 6:9–10.

Felix, F. 1964. *Ohrfasen.* Neue Brehm Bucherei 339. A. Ziemsen, Wittenberg Lutherstadt.

Ferrel, C. M., H. Twining, and N. B. Herkenbaum. 1949. Food habits of the ring-necked pheasant (*Phasianus colchicus*) in the Sacramento Valley, California. *Calif. Fish and Game* 35:51–69.

Fischer, G. L. 1975. The behaviour of chickens. Pp. 454–489 *in* E. S. E. Hafez, ed. *The behaviour of domestic animals.* 3rd ed. Williams and Wilkins, Baltimore, Md.

Fleming, R. L., Jr. 1976. *Birds of Nepal, with reference to Kashmir and Sikkim.* Published by the author, Katmandu, Nepal.

Flieg, G. M. 1973. Breeding the peacock pheasants. *Avic. Mag.* 79:216–218.

Fried, L. A. 1940. The food habits of the ring-necked pheasant in Minnesota. *J. Wildl. Manage.* 4:27–36.

Fumihito, A., T. Miyake, M. Takada, S. Ohno, and N. Kondo. 1995. The genetic link between the Chinese bamboo partridge (*Bambusicola thoracica*) and the chicken and junglefowls of the genus *Gallus. Proc. Natl. Acad. Sci.* 92:11053–11056.

Gao, Y. 1991. Present status of the grey peacock pheasant on Hainan Island. *WPA News* 33:8–10.

Gao, Y. 1997. The ecology of the Hainan hill partridge and grey peacock pheasant on Hainan Island. Abstract. P. 39 *in* Anonymous, 1997.

Gao, Y., and Y. Zhang. 1990. The ecology of the silver pheasant in Dinghashan Biosphere Reserve. Pp. 71–76 *in* Hill et al., 1990.

Garson, P. J. 1983. The cheer pheasant *Catreus wallichii* in Himachal Pradesh. *World Pheasant Assoc. J.* 8:29–39.

Garson, P. J. 1998. Seminar on threatened Vietnamese lowland pheasants. *Tragopan* 8:5.

Garson, P. J., L. Young, and R. Kaul. 1992. Ecology and conservation of the cheer pheasant *Catreus wallichii:* Studies in the wild and the progress of a reintroduction project. *Biol. Cons.* 59:25–35.

Gaston, A. J. 1980. Census techniques for Himalayan pheasants including notes on individual species. *World Pheasant Assoc. J.* 5:40–53.

Gaston, A. J. 1981*a*. The Himalayas: Summary of current knowledge of the status of pheasants. Pp. 33–35 *in* Savage, 1981.

Gaston, A. J. 1981*b*. Field study techniques for censusing pheasants. Pp. 44–48 *in* Savage, 1981.

Gaston, A. J., and J. Singh. 1980. The status of the cheer pheasant *Catreus wallichii* in the Chail Wildlife Sanctuary, Himachal Pradesh. *World Pheasant Assoc. J.* 5:68–73.

Gaston, A. J., M. L. Hunter Jr., and P. J. Garson. 1981. *The wildlife of Himachal Pradesh, western Himalayas.* School of Forest Resources, Technical Notes No. 82, University of Maine, Orono.

Gaston, A. J., A. D. Lelliott, and M. W. Ridley. 1982. Display flight of the male monal pheasant, *Lophophorus impejanus. World Pheasant Assoc. J.* 7:90–91.

Gaston, A. J., K. Islam, and J. A. Crawford. 1983. The current status of the western tragopan. *World Pheasant Assoc. J.* 8:40–49.

Gates, J. M. 1966. Renesting behavior in the ring-necked pheasant. *Wilson Bull.* 78:309–315.

Gates, J. M. 1971. The ecology of Wisconsin pheasant populations. Ph.D. diss., University of Wisconsin, Madison.

Gates, J. M., and J. B. Hale. 1974. Seasonal movements, winter habitat use, and population distribution of an east-central Wisconsin pheasant population. Pp. 1–55 *in* Wisconsin Department of Natural Resources, Technical Bulletin No. 76.

Gates, J. M., and J. B. Hale. 1975. Reproduction of an east-central Wisconsin pheasant population. Pp. 1–70 *in* Wisconsin Department of Natural Resources, Technical Bulletin No. 85.

Gerrits, H. A. 1974. *Pheasants, including their care in the aviary.* Blandford Press, Poole, Dorset, U.K.

Ghose, D., and R. Sumner. 1997. Blyth's tragopan in the Blue Mountains National Park, Mizoram. *Game Bird Breeders' and Conservationists' Gazette* 1997(Aug):26–28.

Ghose, D., and L. Thanga. 1998. Nesting of Blyth's tragopan. *Tragopan* 8:9.

Gilbert, S., and G. A. Greenwell. 1976. An unusually prolific breeding season in the Bornean great argus pheasant. *Intnl. Zoo Yearb.* 16:93–96.

Glenister, A. G. 1951. *The birds of the Malay Peninsula, Singapore, and Penang: An account of the Malayan species with a note on their occurrence in Sumatra, Borneo, and Java, and a list of birds of those islands.* Oxford University Press, Oxford.

Glutz, U. N. von Blotzheim, ed. 1973. *Handbuch der Vogel Mitteleuropas.* Vol. 5. *Galliformes und Gruiformes.* Akademische Verlag, Frankfurt.

Goodwin, D. 1982. On the status of the green pheasant. *Bull. Br. Ornith. Club* 210:35–37.

Göransson, G. 1984. Territorial fidelity in a Swedish pheasant *Phasianus colchicus* population. *Ann. Zool. Fenica* 21:235–238.

Göransson, G., T. von Schantz, I. Froberg, A. Helgee, and H. Wittzell. 1990. Male characteristics, viability, and harem size in the pheasant *Phasianus colchicus. Anim. Behav.* 40:89–104.

Grahame, I. 1971. Breeding the Himalayan blood pheasant. *Avic. Mag.* 77:195–201.

Grahame, I. 1976. The Himalayan blood pheasant: Some further observations. *World Pheasant Assoc. J.* 1:15–22.

Grahn, M. 1992. Intra- and intersexual selection in the pheasant *Phasianus colchicus.* Ph.D. diss., Lund University, Sweden.

Grahn, M. 1993. Mortality in the pheasant *Phasianus colchicus* during the breeding season. *Anim. Behav.* 32:95–101.

Grahn, M., G. Göransson, and T. von Schantz. 1993*a*. Territory acquisition and mating success in pheasants *Phasianus colchicus:* An experiment. *Anim. Behav.* 46:721–730.

Grahn, M., G. Göransson, and T. von Schantz. 1993*b*. Spacing behaviour of male pheasants, *Phasianus* *colchicus*, in relation to dominance and mate acquisition. *Anim. Behav.* 45:93–103.

Gray, A. P. 1958. *Bird hybrids: A check-list with bibliography.* Commonwealth Agricultural Bureaux, Farnham Royal, U.K.

Grummt, W. 1980. Beitrag zur Systematik und Fortpflanzungsbiologie der in Gafanganschaft Gahaltenen Weiben Ohrfasanen, *Crossoptilon crossoptilon* Hodgson. *Milu* 5:103–116. (In German, English summary.)

Guangmei, Z., R. Yin, Z. Zhang, Z. Liu, and H. Zhou. 1989. Courtship display behaviour of Cabot's tragopan. *Acta. Zool. Sinica* 35:332. (In Chinese, English summary.)

Guhl, A. M. 1953. The social behavior of the domestic fowl. Pp. 1–48 *in* Kansas State College of Agriculture, Technical Bulletin No. 73.

Guhl, A. M., and G. J. Fischer. 1969. The behaviour of chickens. Pp. 515–553 *in* E. S. E. Hafez, ed. *The behaviour of domestic animals.* Balliere, Tindall and Cassell, London.

Guiton, P. 1961. The influence of imprinting on the agonistic and courtship responses of the brown leghorn chick. *Anim. Behav.* 9:167–177.

Guy, V. 1982. A new pheasant from Vietnam. Abstract *in* Proceedings of the 2nd International Pheasants in Asia Symposium, Srinigar, Kashmir.

Gysels, H., and M. Rabaye. 1962. Taxonomic relationships of *Afropavo congensis* Chapin 1935 by means of biochemical techniques. *Bull. Soc. R. Zool. Anveris* 26:71–79.

Hachisuka, M. 1953. On hybrids between the green and copper pheasants. *Tori* 13:40–43.

Hall, B. P. 1963. The francolins, a study in speciation. *Bull. Br. Mus. Nat. Hist. (Zool.)* 10:105–204.

Hamerstrom, F. N., Jr. 1936. A study of the nesting habitats of the ring-necked pheasant in northwest Iowa. *Iowa State Col. J. Sci.* 10:173–203.

Han, D. 1990. The ecology of the Joretian koklass pheasant. Pp. 69–70 *in* Hill et al., 1990.

Han, L., L. Yang, and B. Zheng. 1990. Observations on wild breeding ecology of Lady Amherst's pheasant. P. 83 *in* Hill et al., 1990.

Hanebrinck, E. L. 1973. Characteristics and behaviour of a peafowl-guinea hybrid. *Game Bird Breeders' Gazette* 22(2):8–11.

Hanson, W. R. 1970. Pheasant nesting and concealment in hayfields. *Auk* 87:714–719.

Harrison, C. J. O. 1968. A note on the display of the Siamese fire-backed pheasant, *Lophura diardi.* P. 15 *in* 1968 Report of the Pheasant Trust and Norfolk Wildlife Park.

Harrison, C. J. O., and P. Wayre. 1969. The display of the koklass pheasant. P. 20 *in* 1969 Report of Pheasant Trust and Norfolk Wildlife Park.

Hart, J. 1994. Survey and status of the Congo peafowl in eastern Zaire: Progress report (March–June 1994). *WPA Ann. Rev.* 1993/94:44–48.

He, J., J. Li, L. Wei, and J. Zhang. 1993. Notes on some natural hybrids of *Chrysolophus amherstiae* and *C. pictus* from Dayi County, Sichuan. *Zool. Record (China)* 14:239–240. (In Chinese.)

Heinroth, O. 1938. Die Balz des Bulwersfasan. *J. Ornithol.* 86:1–4.

Heinroth, O. 1940. Pfauen- und Truthahnbalz. *Z. Tierpsychol.* 4:330–332.

Heinz, C. H., and L. W. Gysel. 1970. Vocalization behavior of the ring-necked pheasant. *Auk* 87:279–295.

Hennache, A., E. Rand, and V. Lucchini. 1997. Genetic diversity, conservation and phylogenetic relationships of Edwards's pheasant. Abstract. P. 51 *in* Anonymous, 1997.

Henry, C. M. 1955. *A guide to the birds of Ceylon.* Oxford University Press, Oxford.

Hermans, F. 1986. The forgotten kalij pheasant. *WPA News* 11:8–18.

Hewitt, D. 1994. The 1991 census of Galliformes. *WPA News* 43:16–24.

Hickey, J. J. 1955. Some American population research on gallinaceous birds. Pp. 326–395 *in* A. Wolfson, ed. *Recent studies in avian biology.* University of Illinois Press, Urbana.

Hill, D. A., and P. Robertson. 1988. *The pheasant: Ecology, management and conservation.* BSP Professional Books, Oxford, U.K.

Hill, D. A., P. J. Garson, and D. Jenkins, eds. 1990. *Pheasants in Asia, 1989.* World Pheasant Association, Reading, U.K.

Hillgarth, N. 1984. Social organization of wild peafowl in India. *World Pheasant Assoc. J.* 9:47–56.

Hillgarth, N. 1990a. Parasites and female choice in the ring-necked pheasant. *Am. Zool.* 30:227–233.

Hillgarth, N. 1990b. Parasites and sexual selection in pheasants. Ph.D. diss., Oxford University, Oxford.

Hillgarth, N., B. Stewart-Cox, and C. Thouless. 1986. The decline of the green peafowl, *Pavo muticus*. *In* Ridley, 1986 (unpaged).

Hoffman, D. M. 1973. Pheasant nest site selection. Pp. 1–27 *in* Colorado Division of Wildlife, Special Report No. 32.

Holmgren, S. T. 1974. Raising satyr and Temminck's tragopans. *Game Bird Breeders' Gazette* 21(11): 11–14; 21(12):27–30; 22(2):14–17; 22(3):8–11.

Hoogerwerf, A. 1949. Bijdrage tot de Oologie van Java. *Limosa* 22:1–289.

Hoogerwerf, A. 1950. De Avifauna van Tjiboda en Omgreving, inclusief het Natuurmonument Tjibodas-Gn. Gebe (West-Java). *Limosa* 23:1–158.

Houpert, R., and R. Lastere. 1977. Captive breeding of the Salvadori's pheasant. *World Pheasant Assoc. J.* 2:100–103.

Howman, K. C. R. 1979. *Pheasants: Their breeding and management.* K. and R. Books, Edlington, U.K.

Howman, K. C. R. 1993. *Pheasants of the world: Their breeding and management.* Hancock House, Blaine, Wash.

Hudson, G. E. 1955. An apparent hybrid between the ring-necked pheasant and the blue grouse. *Condor* 57:304.

Hüe, F., and R. D. Etchécopar. 1970. *Les oiseaux du proche et du moyen Orient.* N. Boubée, Paris.

Hulselmans, J. L. J. 1962. The comparative myology of the pelvic limb of *Afropavo congensis* Chapin 1936. *Bull. Soc. R. Zool. Anveris* 26:24–70.

Hussain, M. A. 1990. Re-introduction of cheer pheasants in Margalla National Park, Pakistan: Release of cheer poults and their survival (1978–1989). Pp. 228–232 *in* Hill et al., 1990.

Huxley, J. 1941. The display of Rheinart's pheasant. *Proc. Zool. Soc. Lond.* 111:277–278.

Ingram, C. 1955. The order in which the remiges and rectrices are moulted in certain birds. Pp. 270–277 *in* Acta XI Congress of International Ornithologists, Basel.

Inskipp, C. 1989. *Nepal's forest birds: Their status and conservation.* ICBP Monograph No. 4. BirdLife International, Cambridge, U.K.

Inskipp, C., and T. Inskipp. 1985. *A guide to the birds of Nepal.* Smithsonian Institution Press, Washington, D.C.

Islam, K. 1983. Distribution, habitat and status of the western tragopan in Pakistan. Pp. 37–44 *in* Proceedings of the Jean Delacour IFCB Symposium on Breeding Birds in Captivity, 22–27 February, Los Angeles, Calif.

Islam, K. 1987. Distribution and status of the western tragopan in Pakistan. Pp. 37–44 *in* Savage and Ridley, 1987.

Islam, K. 1992. Evolutionary history and speciation of the genus *Tragopan*. Ph.D. diss., Oregon State University, Corvallis.

Islam, K., and J. Crawford. 1988. Habitat use by western tragopans *Tragopan melanocephalus* (Gray) in northeastern Pakistan. *Biol. Cons.* 40:101–115.

Islam, K., and J. Crawford. 1996. A comparison of four vocalizations of the genus *Tragopan*. *Ethology* 102:481–494.

Islam, K., and J. Crawford. 1998. Comparative displays among four species of tragopans and their derivation and function. *Ethol., Ecol. and Evol.* 10:17–32.

Jabouille, P. 1926. La reproduction du rheinarte ocellé. *L'Oiseau* 7:227–229.

Jabouille, P. 1930. Le phénix fabuleux de la Chine et le faisan ocellé d'Annam. *L' Oiseau* 11:220–232.

Jarvis, C., and Lord Medway. 1968. Sideways throwing and associated reproductive behaviour in the crestless fireback pheasant, *Lophura erythropthalma*. Pp. 13–14 *in* 1968 Report of the Ornamental Pheasant Trust and Norfolk Wildlife Park.

Jeggo, D. 1972. Congo peafowl, *Afropavo congensis*. Pp. 43–49 *in* Jersey Wildlife Preservation Trust, 9th Annual Report.

Jeggo, D. 1973. Preliminary notes on the Palawan peacock pheasant (*Polyplectron emphanum*) breeding programme at the Jersey Zoological Park. Pp. 76–81 *in* Jersey Wildlife Preservation Trust, 10th Annual Report.

Jeggo, D. 1975. Breeding the Palawan peacock pheasant at Jersey Zoological Park. *Avic. Mag.* 81:8–12.

Jenkins, D., ed. 1993. *Pheasants in Asia, 1992.* World Pheasant Association, Reading, U.K.

Jewett, S. C. 1932. An unusual gallinaceous hybrid (*Phasianus × Dendragapus*). *Condor* 34:191.

Jia, C., G. Zheng, X. Zhou, and H. Zhang. 1997. Home range and habitat selection of blood pheasant. Abstract. P. 67 *in* Anonymous, 1997.

Johnsgard, P. A. 1973. *Grouse and quails of North America.* University of Nebraska Press, Lincoln.

Johnsgard, P. A. 1982. Etho-ecological aspects of hybridization in the Tetraonidae. *World Pheasant Assoc. J.* 7:42–57.

Johnsgard, P. A. 1983*a*. Hybridization and zoogeographic patterns in pheasants. *World Pheasant Assoc. J.* 8:89–98.

Johnsgard, P. A. 1983*b*. *The grouse of the world.* University of Nebraska Press, Lincoln.

Johnsgard, P. A. 1988. *The quails, partridges and francolins of the world.* Oxford University Press, Oxford.

Johnsingh, A. J. T., and S. Murali. 1980. The ecology and behaviour of the Indian peafowl (*Pavo cristatus* Linn.) of Injar. *J. Bombay Nat. Hist. Soc.* 75(Suppl.): 1069–1079.

Johnson, F. E. B. 1964. Sonnerat's junglefowl (*Gallus sonnerati* Temminck). P. 18 *in* 1964 Ornamental Pheasant Trust Report.

Johnson, R. A. 1963. Habitat preference and behavior of breeding junglefowl in central western Thailand. *Wilson Bull.* 75:270–272.

Juhn, M. 1937. The races of domestic fowl. *Nat. Geog.* 51(4):379–452.

Juhn, M. 1938. Emergence orders and growth rates in the juvenile plumages of the brown leghorn. *J. Exp. Zool.* 77:467–489.

Kabat, C., D. R. Thompson, and F. M. Kozlik. 1950. Pheasant weights and wing molt in relation to reproduction with survival implications. Pp. 1–26 *in* Wisconsin Conservation Department, Technical Bulletin No. 2.

Kalsi, R., and R. Kaul. 1997. Density index and habitat associations of the cheer pheasant *Cateus wallichii* in Himachal Pradesh, India. Abstract. P. 43 *in* Anonymous, 1997.

Khaling, S. 1997. Satyr tragopan in the Singhalila National Park, Darjealing, India. *Tragopan* 6:13–14.

King, D. F. 1992. Reeves's pheasant in close encounter. *WPA News* 37:12–16.

King, D. F., and E. C. Dickinson. 1975. *A field guide to the birds of Southeast Asia.* Houghton Mifflin, Boston, Mass.

King, W. B., ed. 1981. *Endangered birds of the world: The ICBP red data book.* Smithsonian Institution Press and International Council for Bird Preservation, Washington, D.C.

Knoder, C. E. 1955. Reeves' pheasant investigation: Production-release pen study. *Ohio Wildl. Invest.* 6(1):48–51, 116–117.

Knoder, C. E. 1963. Genetic analysis of hybrids between the avian genera *Syrmaticus* and *Phasianus. Game Res. Ohio* 2:215–222.

Knoder, C. E. 1983. Elliot's pheasant conservation. *World Pheasant Assoc. J.* 8:11–28.

Knoder, C. E., and R. M. Bailie. 1956. The ecology and breeding biology of the Reeves' pheasant on Tappan Island, 1955. *Ohio Wildl. Invest.* 7(3):36–45.

Koelz, W. N. 1954. Ornithological studies: New birds from Iran, Afghanistan and India. Privately printed.

Korschgen, L. J. 1964. Foods and nutrition of Missouri and midwestern pheasants. Pp. 159–181 *in* Transactions of the 29th North American Wildlife Conference.

Korschgen, L. J., and G. D. Chambers. 1970. Propagation, stocking and food habits of Reeves' pheasants in Missouri. *J. Wildl. Manage.* 34:274–282.

Kozicky, E. L., and G. O. Hendrickson. 1951. The production of ring-necked pheasants in Winnebago County, Iowa. *Proc. Iowa Acad. Sci.* 58: 491–495.

Kozlowa, E. W. 1947. On the spring life and breeding habits of the pheasant in Tadjikstan. *Ibis* 89: 423–428.

Kruijt, J. P. 1962*a*. On the evolutionary derivation of wing display in Burmese red junglefowl and other gallinaceous birds. *Symp. Zool. Soc. Lond.* 8:25–35.

Kruijt, J. P. 1962*b*. Notes on wing display in the courtship of pheasants. *Avic. Mag.* 69:11–20.

Kruijt, J. P. 1964. Ontogeny of social behaviour in Burmese red junglefowl (*Gallus gallus spadiceus* Bonnaterre). *Behaviour* 12(Suppl.):1–201.

Kruijt, J. P. 1966. The development of ritualized displays in junglefowl. *Phil. Trans. R. Soc. Lond. B* 21: 479–484.

Kuroda, N. 1981. The Japanese green pheasant *Phasianus (colchicus) versicolor* in Japan. *World Pheasant Assoc. J.* 6:60–72.

Lack, D. 1968. *Ecological adaptations for breeding in birds.* Oxford University Press, Oxford.

Lamba, B. S. 1981. Eastern India: Preliminary review. Pp. 36–38 *in* Savage, 1981.

Lamba, B. S. 1987. Pheasant census survey of Kashmir Valley. Pp. 52–54 *in* Savage and Ridley, 1987.

La Touche, J. D. D. 1900. Notes on the birds of north-west Fokien. *Ibis* Ser. 7 6:34–60.

Laughrey, A. G., and R. H. Stinton. 1955. Feeding habits of juvenile ring-necked pheasants on Pelee Island, Ontario. *Can. Field Nat.* 69:59–65.

Lelliott, A. D. 1981*a*. Cheer pheasants in west-central India. *World Pheasant Assoc. J.* 6:89–95.

Lelliott, A. D. 1981*b*. Studies of Himalayan pheasants in Nepal, with reference to their conservation. M.Sc. thesis, University of Durham, England.

Lelliott, A. D., and P. B. Yonzon. 1980. Studies of Himalayan pheasants in Nepal. *World Pheasant Assoc. J.* 5:11–30.

Lelliott, A. D., and P. B. Yonzon. 1981. Pheasant studies in Annapurna Himal. Pp. 53–55 *in* C. Savage, ed. *Pheasants of Asia, 1979*. World Pheasant Association, Exning, Suffolk, U.K.

Leonard, M., and A. G. Horn. 1995. Crowing in relation to status in roosters. *Anim. Behav.* 49:1283–1290.

Le Trong, T. 1997. Green peafowl *Pavo muticus* in Vietnam. *Tragopan* 7:16–17.

Lewin, V., and G. Lewin. 1984. The kalij pheasant, a newly established game bird on the island of Hawaii. *Wilson Bull.* 96:634–646.

Lewis, J. S. 1939. Courting display of Napoleon's peacock pheasant (*Polyplectron napoleonis*). *Avic. Mag.* Ser. 5 4:233–235.

Li, X. 1991. *Crimson-bellied tragopans*. International Academic Publishers, Beijing.

Li, X. 1996. *The gamebirds of China: Their distribution and status*. International Academic Publishers, Beijing.

Liang, W. 1997. Activity and habitat selection of golden pheasant, *Chrysolophus pictus*. *Tragopan* 7:6.

Ligon, J, D., and M. Zwartjes. 1995*a*. Ornate plumage of male red junglefowl does not influence mate choice by females. *Anim. Behav.* 49:117–125.

Ligon, J, D., and M. Zwartjes. 1995*b*. Female red junglefowl choose to mate with multiple males. *Anim. Behav.* 49:127–135.

Ligon, J, D., R. Kimball, and M. Merola-Zwartjes. 1998. Mate choice by female red junglefowl: The issues of multiple ornaments and fluctuating asymmetry. *Anim. Behav.* 55:41–50.

Lill, A., and D. G. M. Wood-Gush. 1965. Potential ethological isolating mechanisms and assortative mating in the domestic fowl. *Behaviour* 25:16–24.

Liu, R. 1986. Mating behaviour of the brown eared pheasant. *In* Ridley, 1986 (unpaged).

Liu, R., and T. Lu. 1990. The taxonomy of ruffed pheasants. Pp. 95–97 *in* Hill et al., 1990.

Liu, X. 1991. *Syrmaticus humiae* (Hume). Pp. 314–327 in *The rare and endangered gamebirds of China*. Fujian Science and Technology Press, Fuzhou.

Long, J. L. 1981. *Introduced birds of the world*. A. H. and A. W. Reed, Sydney.

Lovel, T. W. I. 1976. The present status of the Congo peacock. *World Pheasant Assoc. J.* 1:48–57.

Lovel, T. W. I. 1977. A stud book for the Edwards's pheasant. *World Pheasant Assoc. J.* 2:97–99.

Lowe, P. R. 1925. Some notes on the genus *Polyplectron*. *Ibis* Ser. 12 1:476–484.

Lowe, P. R. 1939. Some preliminary notes on the anatomy and systematic position of *Afropavo congensis* Chapin. Pp. 219–230 *in* Proceedings of the 9th International Ornithological Congress, Rouen.

Lu, T., and R. S. Liu. 1983. Brown eared pheasant: Studies of its ecology and biology. *Acta Zool. Sinica* 29:278–290. (In Chinese, English summary.)

Lu, T., and R. S. Liu. 1986. An ecological study on the Chinese monal pheasant, *Lophophorus lhuisii*. *Acta Zool. Sinica* 32:273–279.

Lu, X. 1997. Study of habitat selection and behaviour of Tibetan eared pheasant. *Tragopan* 7:5–6.

Ludlow, F. 1951. The birds of Kongbo and Pome, southeast Tibet. *Ibis* 93:574–578.

Ludlow, F., and N. B. Kinnear. 1944. The birds of southeastern Tibet. *Ibis* 86:348–389.

Mace, G. M., and R. Lande. 1991. Assessing extinction threats: Toward a reevaluation of *iucn* threatened species categories. *Conserv. Biol.* 5:148–157.

MacMullan, R. A. 1960. *Michigan pheasant populations*. Michigan Department of Conservation, Game Division Report No. 2277.

Mainardi, D. 1963. Immunological distances and phylogenetic relationships in birds. Pp. 103–114 *in* Proceedings of the XIII International Ornithological Congress, Ithaca, N.Y., Vol. 1.

Mallet, J. J. 1973. Notes on the breeding of eared pheasants at the Jersey Zoological Park. Pp. 74–75 *in* Jersey Wildlife Preservation Trust Annual Report No. 10.

Mallinson, J. J. C. 1979. The establishment of viable captive populations of endangered bird species with special reference to the white eared pheasant *Crossoptilon crossoptilon* at the Jersey Zoological Park. *World Pheasant Assoc. J.* 4:81–92.

Mallinson, J. J. C., and K. M. Taynton. 1978. White eared pheasant studbook number two. *Dodo* 15:92–96.

Malone, E. 1995. The endemic Vietnamese *Lophura* pheasants: Conservation and taxonomy. *Tragopan* 3:2–3.

Mamat, I. H., and M. N. Yasak. 1997. The status and current distribution of the crested argus (*Rheinardia ocellata nigrescens*) in peninsular Malaysia. Abstract. P. 31 *in* Anonymous, 1997.

Mann, C. F. 1989. More notable bird observations from Brunei, Borneo. *Forktail* 5:17–22.

Manning, J. T. 1989. Age advertisement and the evolution of the peacock's train. *J. Evol. Biol.* 2:299–313.

Manning, J. T., and M. A. Hartley. 1991. Symmetry and ornamentation are correlated in the peacock's train. *Anim. Behav.* 42:1020–1021.

Marchant, J. A., R. Hudson, S. P. Carter, and P. Whittington. 1990. *Population trends in British breeding birds.* British Trust for Ornithology, Thedford, U.K.

Marien, D. 1951. Notes on some pheasants from southwestern Asia, with remarks on molt. *Am. Mus. Novit.* 1518:1–25.

Marsh, C., and J. Gasis. 1990. Conservation in Malaysia. An expedition to Sabah's lost world: The Maliau Basin. *Malay Nat.* 43(3):15–22.

Maru, T. 1988. The territory of the Japanese green pheasant, *Phasianus versicolor. Strix* 7:149–158. (In Japanese, English summary.)

Mateos, C., and J. Carranza. 1995. Female choice for morphological features of male ring-necked pheasants. *Anim. Behav.* 49:737–748.

Mateos, C., and J. Carranza. 1996. On the intersexual selection for spurs in ring-necked pheasants. *Behav. Ecol.* 7:362–369.

Mateos, C., and J. Carranza. 1997a. Signals in intrasexual competition between ring-necked pheasant males. *Anim. Behav.* 53:471–485.

Mateos, C., and J. Carranza. 1997b. The role of bright plumage in male-male interactions in the ring-necked pheasant. *Anim. Behav.* 54:1205–1214.

Mayr, E., and D. Amadon. 1951. A classification of Recent birds. *Am. Mus. Novit.* 1496:1–42.

McAtee, W. L. 1945. *The ring-necked pheasant and its management in North America.* American Wildlife Institute, Washington, D.C.

McCabe, R. A. 1949. A ten-year study of refuge populations of ring-necked pheasants. Ph.D. diss., University of Wisconsin, Madison.

McGowan, J. K. 1991. Ecology and behaviour of the Malaysian peacock pheasant: A threatened tropical rain forest species. *WPA News* 34:11–14.

McGowan, J. K. 1992. Social organisation in the Malaysian peacock pheasant. Ph.D. diss., Open University, U.K.

McGowan, J. K. 1994. Display dispersion and microhabitat use by the Malaysian peacock pheasant *Polyplectron malacense* in peninsular Malaysia. *J. Trop. Ecol.* 10:229–244.

McGowan, J. K., and P. J. Garson. 1995. *Pheasants: Status survey and conservation action plan, 1995–1999.* IUCN, Gland, Switzerland, and World Pheasant Association, Reading, U.K.

McGowan, J. K., and A. L. Panchen. 1994. Plumage variation and geographical distribution in the kalij and silver pheasants. *Bull. Br. Ornith. Club* 114:113–123.

McGowan, J. K., I. R. Hartley, and R. P. Girdler. 1989. The Palawan peacock pheasant habitat and pressures. *World Pheasant Assoc. J.* 14:80–100.

Medway, Lord, and D. R. Wells. 1976. *The birds of the Malay Peninsula.* Vol. 5. Witherby, London.

Miller, D. B. 1978. Species-typical and individually distinctive acoustic features of crow calls of red junglefowl. *Z. Tierpsychol.* 47:182–193.

Mirza, Z. B. 1976. Pheasant restoration in Pakistan. Pp. 20–24 *in* 1976 Pheasant Trust and Norfolk Wildlife Park Annual Report.

Mirza, Z. B. 1981a. Status of the pheasants in Pakistan. Pp. 27–28 *in* Savage, 1981.

Mirza, Z. B. 1981b. Cheer pheasant release programme in Pakistan. Pp. 72–74 *in* Savage, 1981.

Mirza, Z. B., A. Aleem, and M. Asghar. 1978. Pheasant surveys in Pakistan. *J. Bombay Nat. Hist. Soc.* 74:292–296.

Mitchell, P. C. 1911. On longevity and relative viability in mammals and birds, with a note on the theory of longevity. *Proc. Zool. Soc. Lond.* 1911:425–548.

Morejohn, G. V. 1968a. Study of plumage of the four species of the genus *Gallus. Condor* 70:56–65.

Morejohn, G. V. 1968b. Breakdown of isolation mechanisms in two species of captive junglefowl. *Evolution* 22:576–582.

Morris, D. 1957. Courtship of pheasants. *Zoo Life* 12:8–13.

Moynihan, M. 1995. Social structures and behavior patterns of captive and feral Reeves' pheasants *Syrmaticus reevesi* in France. *Alauda* 63:213–228.

Mueller, C. F., and H. C. Seibert. 1966. Wing and tail molt in Reeves' pheasant. *Ohio J. Sci.* 66:489–495.

Muller, K. A. 1980. Now you see him, or did you? *Zoonooz* 53:12–14.

Nelson, M. M., R. A. Chesness, and S. W. Harris. 1960. Relationship of pheasant nests to hayfield edges. *J. Wildl. Manage.* 24:430.

Norapuck, S. 1982. Survey of rarest pheasants of Thailand. Abstract *in* Proceedings of the 2nd International Pheasants in Asia Symposium, Srinigar, Kashmir.

O'Brien, T. G., and M. F. Kinnaird. 1997. Wildlife Conservation Society surveys for Bornean peacock pheasant. *Tragopan* 6:5–6.

Ogasawara, K. 1969. Winter habitats and food habits of the green and copper pheasants. *Misc. Rep. Yamashina Inst. Ornithol. Zool.* 5(4):351–363. (In Japanese, English summary.)

Ogilive-Grant, W. R. 1893. *Catalogue of the game birds in the collection of the British Museum.* Vol. 22. British Museum of Natural History, London.

Ollson, M. 1982. Firebacks: Their captive management and propagation. *Game Bird Breeders' Gazette* 31(10/11):8–10.

Olsen, D. W. 1977. *A literature review of pheasant habitat requirements and improvement methods.* Utah State Department of Natural Resources, Publication No. 77-7.

Olson, S. L. 1980. The significance of the distribution of the Megapodiidae. *Emu* 80:21–24.

Paludan, K. 1959. Results of pheasant marking in Denmark, 1949–55. *Dan. Rev. Game Biol.* 4:1–23.

Papeschi, A., A. Hoodless, J. P. Carroll, and F. Dessi-Fulghari. 1997. A possible role of male ornaments as predictors of survival and reproductive success in common pheasant. Abstract. P. 59 *in* Anonymous, 1997.

Penrod, B., M. Dixon, and J. Smith. 1982. Renesting by ring-necked pheasants after loss or separation from their brood. *N.Y. Fish and Game J.* 29:209–210.

Peters, J. L. 1934. *Check-list of birds of the world.* Vol. 2. Harvard University Press, Cambridge, Mass.

Petrie, M., and T. Halliday. 1994. Experimental and natural changes in the peacock's (*Pavo cristatus*) train can affect mating success. *Behav. Ecol. and Sociobiol.* 35:213–217.

Petrie, M., T. Halliday, and C. Saunders. 1991. Peahens prefer males with elaborate trains. *Anim. Behav.* 41:323–331.

Petrie, M., T. Halliday, H. Budgey, and C. Pierpoint. 1992. Multiple mating in a lekking bird: Why do peahens mate with more than one male and with the same male more than once? *Behav. Ecol. and Sociobiol.* 31:349–358.

Phan, V.-L. 1996. The first crested argus *Rheinardia ocellata* hatched in the Saigon Zoo and Botanical Gardens. *WPA News* 50:35–36.

Phillips, J. C. 1921. A further report on species crosses in birds. *Genetics* 6:366–383.

Pocock, R. I. 1911. The display of the peacock pheasant. *Avic. Mag.* Ser. 3 2:229–237.

Poda, J. N. 1985. Une hybridation naturelle entre un coq *Gallus gallus* et une pintada *Numida meleagris. Notes Doc Volta* 16:25–30.

Pokorny, F., and J. Pikula. 1986. Biology of *Syrmaticus reevesi. Prir. Pr. Cesk. Akad. Ved.* 20:1–66.

Poltack, D. 1972. In search of the mikado pheasant. Pp. 6–8 *in* 1972 Annual Report of the Pheasant Trust and Norfolk Wildlife Park.

Pratt, T. K. 1975. The kalij pheasant on Hawaii. *Elepaio* 36:66–67.

Randi, E., G. Fusco, R. Lorenzi, and T. M. Crowe. 1991. Phylogenetic relationships and rates of allozyme evolution within the Phasianidae. *Biochem. System. and Ecol.* 19:213–222.

Rands, M. R. W., M. W. Ridley, and A. D. Lelliott. 1984. The social organisation of feral peafowl. *Anim. Behav.* 32:830–835.

Rassmussen, P. 1998. Is the imperial pheasant, *Lophura imperialis*, a hybrid? Work in progress and a call for information. *Tragopan* 9:8–10.

Ridley, M. W. 1983. The mating system of the pheasant *Phasianus colchicus.* Ph.D. diss., Oxford University, Oxford.

Ridley, M. W., ed. 1986. *Pheasants in Asia, 1986.* Proceedings of the 3rd International Symposium, Chaing Mai, Thailand, 1986. World Pheasant Association, Reading, U.K.

Ridley, M. W. 1987. Relevance of social organization to conservation in the pheasant family. Pp. 115–118 *in* Savage and Ridley, 1987.

Ridley, M. W., and D. A. Hill. 1987. Social organisation in the pheasant (*Phasianus colchicus*): Harem formation, mate selection and the role of mate guarding. *J. Zool.* 211:619–630.

Ridley, M. W., A. D. Lelliott, and M. R. W. Rands. 1984. The courtship display of feral peafowl. *World Pheasant Assoc. J.* 9:57–67.

Riley, J. H. 1938. Birds from Siam and the Malay Peninsula in the U.S. National Museum collected by Dr. Hugh M. Smith and William L. Abbott. *Bull. U.S. Natl. Mus.* 172:1–581.

Rimlinger, D. S. 1984. Display behaviour of the Temminck's tragopan. *World Pheasant Assoc. J.* 9:19–32.

Rimlinger, D. S. 1985. Observations on the display behaviour of the Bulwer's pheasant. *World Pheasant Assoc. J.* 10:15–26.

Rimlinger, D. S., and P. Whitman. 1986. Observations on the breeding and behaviour of the Chinese monal in captivity. *In* Ridley, 1986 (unpaged).

Rimlinger, D. S., H. F. Landel, C. Y. Cheng, and G. Guo. 1997. Natural history of a marked population of Chinese monals in Sichuan Province. Abstract. P. 40 *in* Anonymous, 1997.

Ripley, S. D. 1961. *A synopsis of the birds of India and Pakistan, together with those of Nepal, Sikkim, Bhutan and Ceylon.* Bombay Natural History Society, Bombay.

Roberts, J. O. M. 1981. Status of the pheasants of Nepal. Pp. 22–26 *in* Savage, 1981.

Roberts, T. 1991. *The birds of Pakistan.* Vol. 1. Oxford University Press, Oxford.

Roberts, T. J. 1970. A note on the pheasants of west Pakistan. *Pak. J. For.* 20:319–326.

Robertson, W. B., Jr. 1958. Investigation of ringnecked pheasants in Illinois. Pp. 1–138 *in* Illinois Department of Conservation, Bulletin No. 1.

Robinson, H. C., and F. N. Chasen. 1936. *The birds of the Malay Peninsula.* Vol. III. *Sporting birds: Birds of the shore and estuaries.* Witherby, London.

Robson, C. R., J. C. Eames, C. Nguyen, and V. L. Truong. 1993. Further recent records of birds from Viet Nam. *Forktail* 8:25–52.

Roden, G. S. 1899. The courting dance of the moonal pheasant. *J. Bombay Nat. Hist. Soc.* 12:573–574.

Roles, D. G. 1981. *Rare pheasants of the world.* Spur Publications, Liss, Hampshire, U.K.

Round, P. D. 1993. Recent reports: February–June 1993. *Bangkok Bird Club Bulletin* 10(7):10.

Rozendaal, F., C. Nguyen, V. L. Truong, and D.-L. Nagia. 1991. Notes on Vietnamese pheasants, with description of female plumage of *Lophura hatinhensis. Dutch Birding* 13:12–15.

Ruan, X. D., H. Lin, and C. Cheng. 1993. Study on breeding behaviour of the Chinese monal in captivity. *World Pheasant Assoc. J.* 17/18:25–36.

Rushen, J. 1982. The peck orders of chickens: How do they develop and why are they linear? *Anim. Behav.* 30:1129–1137.

Rutgers, A., and K. A. Norris, eds. 1970. *Encyclopedia of aviculture.* Vol. 1. Blandford Press, London.

Sahin, R. 1984. Zur Balz des Mikado-Fasans (*Syrmaticus mikado*) in Gefangenschaft. *J. Ornithol.* 125:15–23.

Sahin, R. 1986. Zur lateral Präsentation des Mikado-Fasans (*Syrmaticus mikado*) in Gefangenschaft. *J. Ornith.* 126:213–215.

Sahin, R., and E. Thomas. 1988. Zum Flügelschwirren der Mikado-Fasans (*Syrmaticus mikado*). *J. Ornith.* 129:325–341.

Sathyakumar, S., S. N. Prasad, G. S. Rawat, and A. T. Johnsingh. 1993. Ecology of kalij and monal pheasants in Kedarnath Wildlife Sanctuary, western Himalaya. *World Pheasant Assoc. J.* 17/18:87–88.

Sato, T., T. Ishii, and Y. Hirai. 1967. Genetic studies on serum protein in Phasianidae. 3. Immunoelectrophoresis comparison of the sera of domestic fowl, guinea fowl, and their hybrids. *Jap. J. Gen.* 42:51–59.

Savage, C., ed. 1981. *Pheasants in Asia, 1979.* Proceedings of the 1st International Pheasant Symposium. World Pheasant Association, Exning, Suffolk, U.K.

Savage, C., and M. W. Ridley, eds. 1987. *Pheasants in Asia, 1982.* Proceedings of the 2nd International Symposium on the Pheasants in Asia. World Pheasant Association, Reading, U.K.

Schäfer, E. 1934. Zur Lebenweise der Fasanen des chinesisch-tibetischen Grenzlandes. *J. Ornithol.* 82:487–509.

Schantz, T. von, M. Grahn, and M. Göransson. 1994. Intersexual selection and reproductive success in the pheasant *Phasianus colchicus. Am. Nat.* 144:510–527.

Schenkel, R. 1956–1958. Zur Deutung der Balzeistungen einiger Phasianiden und Tetraoniden. *Ornithol. Beobacht.* 53:182–201; 55:65–95.

Schick, C. 1952. *A study of pheasants on a 9000-acre prairie farm, Saginaw County, Michigan.* Michigan Department of Conservation, Lansing.

Schneider, A. 1938. Bau und Erection der Hautlappen von *Lobiophasis bulweri. J. Ornithol.* 86:5–8.

Schwartz, C. W., and E. R. Schwartz. 1951. An ecological reconnaissance of the pheasants of Hawaii. *Auk* 68:281–314.

Seibert, H. C., and R. W. Donohoe. 1965. *The history of the Reeves' pheasant program in Ohio.* Ohio Game, Monograph No. 1.

Serebrovsky, A. S. 1929. Observations on interspecific hybrids of the fowl. *J. Gen.* 21:237–240.

Seth-Smith, D. 1925a. The argus pheasant and its display. *Avic. Mag.* Ser. 4 3:175–179.

Seth-Smith, D. 1925b. On the display of the argus pheasant (*Argusianus argus*). *Proc. Zool. Soc. Lond.* 95:323–325.

Seth-Smith, D. 1932. The display of Rheinart's pheasant. *Avic. Mag.* Ser. 4 10:122–123.

Seubert, J. L. 1952. Observations on the renesting behavior of the ring-necked pheasant. Pp. 305–329 *in* Transactions of the 17th North American Wildlife Conference.

Severinghaus, S. R. 1977. A study of the Swinhoe's and mikado pheasant in Taiwan with recommendations for their conservation. Ph.D. diss., Cornell University, Ithaca, N.Y.

Severinghaus, S. R. 1978. Recommendations for the conservation of the Swinhoe's and mikado pheasants in Taiwan. *World Pheasant Assoc. J.* 3:79–89.

Severinghaus, S. R. 1979. Observations on the ecology and behavior of the koklass pheasant in Pakistan. *World Pheasant Assoc. J.* 4:52–69.

Severinghaus, S. R. 1980. Swinhoe's pheasant in Taiwan. *Living Bird* 18:189–209.

Severinghaus, S. R. 1996. Swinhoe's pheasant in Yushan National Park. *WPA News* 60:21–28.

Severinghaus, S. R., Z. B. Mirza, and M. Asghar. 1979. Selection of a release site for the re-introduction of cheer pheasants in Pakistan. *World Pheasant Assoc. J.* 4:100–115.

Shamblin, B. 1997. Vietnam's endangered and threatened pheasants. *Game Bird Breeders' and Conservationists' Gazette* 1997(Oct):10–17.

Sharma, I. 1972. Ecological study of breeding of the peafowl, *Pavo cristata*. *Alauda* 40:378–384. (In French, English summary.)

Sharma, V. 1993. Western tragopan in captivity. *WPA News* 40:24.

Shi, H., G.-M. Zheng, H. Jiang, and Z.-K. Wu. 1996. The study of habitat selection in Temminck's tragopan. *Acta Zool. Sinica* 42:90–94. (In Chinese, English summary.)

Sibley, C. G., and J. E. Ahlquist. 1972. A comparative study of the egg white proteins of non-passerine birds. *Peabody Mus. Nat. Hist. Bull.* 39:1–276.

Sibley, C. G., and J. E. Ahlquist. 1990. *Phylogeny and classification of birds: A study in molecular evolution.* Yale University Press, New Haven, Conn.

Sibley, C. G., and B. L. Monroe Jr. 1990. *Distribution and taxonomy of birds of the world.* Yale University Press, New Haven, Conn.

Sivelle, C. 1979. Tragopans. *Avic. Mag.* 85:199–209.

Skervold, H., and A. F. Mjelstad. 1992. Capercaillie-chicken hybrids. *J. Anim. Breed. Genet.* 109:149–152.

Smil, V. 1983. Deforestation in China. *Ambios* 12(5):226–231.

Smythies, B. E. 1953. *The birds of Burma.* 2nd ed. Oliver and Boyd, Edinburgh.

Smythies, B. E. 1981. *The birds of Borneo.* 3rd ed. Sabah Society and Malayan Nature Society, Kuala Lumpur.

Snow, D. W. 1968. *An atlas of speciation in African non-passerine birds.* British Museum of Natural History, London.

Stapel, C. 1976. Some observations on behavior and display of peacock pheasants. *World Pheasant Assoc. J.* 1:109–112.

Steinbacher, G. 1941. Das Flugelschlagen der Fasane. *Zool. Gart. N. F.* 13:233–236.

Steiner, H. 1945. Ueber letal Fehentwicklung der Zeiten Zackommenschafts-Generation bei tierischen Artbastarden. *Arch. Julius Klaus-Stift. Verebungforsch. Socialanthropol. Rossenhg.* 20:236–251.

Stephens, C. H. 1966. *Reeves' pheasant investigations in Kentucky.* Kentucky Department of Fish and Wildlife Resources, Game Management Technical Series No. 15.

Stewart, G. 1994. Sahahan pheasantry. *WPA News* 43:25.

Stewart-Cox, B. J. 1996. Green peacocks reclaim traditional lands. *WPA News* 51:1–4.

Stewart-Cox, B. J., and R. Quinnell. 1990. Using calls, footprints and sightings to survey green peafowl in western Thailand. Pp. 129–137 *in* Hill et al., 1990.

Stock, A. B., and T. D. Bunch. 1982. The evolutionary implications of chromosome banding pattern homologies in the bird order Galliformes. *Cytogenet. Cell Biol.* 34:136–148.

Stokes, A. W. 1954. *Population studies of the ring-necked pheasant on Pelee Island, Ontario.* Ontario Department of Lands and Forests, Technical Bulletin No. 4.

Stokes, A. W., and H. W. Williams. 1972. Courtship feeding in gallinaceous birds. *Auk* 89:177–180.

Stresemann, E. 1965. Die Mauser der Huhnervogel. *J. Ornithol.* 106:58–64.

Strode, D. H. 1941. The 1940 pheasant nesting study in Wood County, Ohio. *Ohio Wildl. Res. Sta. Rel.* 157:1–34.

Sullivan, M. S. 1992. Social and sexual preferences of red junglefowl. Ph.D. diss., Oxford University, Oxford.

Sumardja, E. A. 1981. First five national parks in Indonesia. *Parks* 6(2):1–4.

Sun, Y.-H. 1995. Discovering the winter food of Cabot's tragopan in Wuyanling, Zhejiang. *WPA News* 47:10–12.

Sun, Y.-H., and Y. Fang. 1997. The ecology and behaviour of the blood pheasant at Lianhuashan. Abstract. P. 86 *in* Anonymous, 1997.

Sutter, E. 1971. Ausbildung und Mauser der Flugelgefieders beim juvenilen Jagdfasan *Phasianus colchicus*. *Ornithol. Beob.* 68:79–222.

Taber, R. D. 1949. Observations on the breeding behavior of the ring-necked pheasant. *Condor* 51: 153–175.

Taka-Tsukasa, N. 1929. The breeding of Rheinart's argus pheasant in Japan. *Avic. Mag.* Ser. 5 7:306–309.

Taka-Tsukasa, N. 1933–1943. *The birds of Nippon.* I. Galli, Tokyo.

Tan, Y., and Z. Wu. 1981. A new subspecies of silver pheasant from Guizhou, China. *Zool. Res.* 2:301–305. (In Chinese.)

Thompson, J. A. M. 1996. New information about the presence of the Congo peafowl. *WPA News* 50:3–8.

Thompson, L. E. 1976. Methods of sexing eared pheasants. *Avic. Mag.* 82:39–50.

Trautman, C. G. 1950. Determining the age of juvenile pheasants. *S.D. Conserv. Dig.* 17(8):8–10.

Trautman, C. G. 1952. Pheasant food habits in South Dakota. Pp. 1–89 *in* South Dakota Game, Fish, and Parks Department, Technical Bulletin No. 1.

Trautman, C. G. 1982. History, ecology and management of the ring-necked pheasant in South Dakota. Pp. 1–118 *in* South Dakota Game, Fish, and Parks Department, Technical Bulletin No. 7.

Van den bergh, W. 1975. Breeding the Congo peacock at the Royal Zoological Society of Antwerp. Pp. 75–86 *in* R. D. Martin, ed. *Breeding endangered species in captivity.* Academic Press, London.

van Marle, J. G., and K. H. Voous. 1988. *The birds of Sumatra.* British Ornithologists' Union Check-list No. 10.

Vaurie, C. 1965. *The birds of the Palearctic fauna: Non-passerines.* Witherby, London.

Vaurie, C. 1972. *Tibet and its birds.* Witherby, London.

Verheyen, R. 1956. Contribution de l'anatomie et à la systématique des Galliformes. *Bull. Inst. Roy. Sci. Nat. Belg.* 32(42):1–24.

Verheyen, R. 1962. Monographie du paon congolais. *Bull. Soc. R. Zool. Averis* 26:3–94.

Verheyen, R. 1965. *Der Kongofau (Afropavo congensis).* Neue Brehm Bucherei 351. A. Ziesmen, Wittenberg Lutherstadt.

Vo Quy. 1975. *Birds of Vietnam.* Vol. 1. Hanoi. (In Vietnamese.)

Vuilleumeier, F., M. Le Croy, and E. Mayr. 1992. New species of birds described from 1981 to 1990. *Bull. Br. Ornith. Club* 112A:267–309.

Wagner, F. H., C. D. Besadny, and C. Kabat. 1965. Population ecology and management of Wisconsin pheasants. *Wis. Conserv. Bull.* 35:1–168.

Wang, H.-P. 1980. Nature conservation in China: The present situation. *Parks* 5(1):1–10.

Wayre, P. 1964. Display of the common koklass (*Pucrasia m. macrolopha* Lesson). P. 13 *in* 1964 Report of the Ornamental Pheasant Trust.

Wayre, P. 1969. *A guide to the pheasants of the world.* Country Life, London.

Wayre, P. 1975. Breeding endangered pheasants in captivity as a means of ensuring their survival. Pp. 87–97 *in* R. D. Martin, ed. *Breeding endangered species in captivity.* Academic Press, London.

Wayre, P. 1978. The monals, genus *Lophophorus*. *Pheasant Trust and Norfolk Wildl. Pk. News* 1978 (Mar):4–5.

Weber, R. 1992. The Swinhoe's pheasant. *WPA News* 37:29–30.

Weigand, J. P., and R. G. Janson. 1976. *Montana's ring-necked pheasant: History, ecology and management.* Montana Department of Fish and Game.

Wei-Shu, H. 1982. The pheasants of China. Abstract *in* Proceedings of the 2nd International Pheasants in Asia Symposium, Srinigar, Kashmir.

Wen, X., X. Yang, and L. Lan. 1997*a*. Status and conservation of green peafowl in China. *Tragopan* 7:6.

Wen, X., X. Yang, and L. Yang. 1997*b*. The status and conservation of green peafowl in China. Abstract. P. 47 *in* Anonymous, 1997.

Westerkov, K. 1957. *Growth and moult of pheasant chicks.* New Zealand Department of Internal Affairs, Wildlife Publication No. 47.

Westerkov, K. 1963. *Evaluation of pheasant liberations in New Zealand based on a 12-year banding study.*

New Zealand Department of Internal Affairs, Wildlife Publication No. 71.

Westerkov, K. 1974. Distribution and life-history of the ring-necked pheasant (*Phasianus colchicus karpowi*) in Korea. *N.Z. Wildl.* 46:7–9, 11–19.

Wetmore, A. 1960. A classification for the birds of the world. *Smithson. Misc. Collect.* 139(11):1–37.

Wheatley, N. 1996. *Where to watch birds in Asia.* Princeton University Press, Princeton, N.J.

Whittington, P., D. Benstead, N. Bean, and D. Showler. 1994. Further searches for the western tragopan: The 1994 survey in Palas Valley. *WPA News* 46:5–9.

Wittzell, H. 1991. Natural and sexual selection in the pheasant *Phasianus colchicus*. Ph.D. diss., Lund University, Sweden.

Woehler, E. E., and J. M. Gates. 1970. An improved method of sexing ring-necked pheasants. *J. Wildl. Manage.* 34:228–231.

Wolters, H. E. 1975–1982. *Die Vogelarten der Erde.* Paul Parey, Hamburg.

Wood-Gush, D. G. M. 1954. The courtship of the brown leghorn cock. *Br. J. Anim. Behav.* 2:95–102.

Wood-Gush, D. G. M. 1955. The behaviour of the domestic chicken: A review of the literature. *Br. J. Anim. Behav.* 3:81–110.

Wood-Gush, D. G. M. 1956. The agonistic and courtship behaviour of the brown leghorn cock. *Br. J. Anim. Behav.* 4:133–142.

Wood-Gush, D. G. M. 1971. *The behaviour of the domestic fowl.* Heinemann, London.

Wu, Y., and J. T. Peng. 1996. Breeding ecology of the white eared pheasant (*Crossoptilon crossoptilon*) in western Sichuan, China. *J. Yamashina Inst. Ornith.* 28:98–102.

Wu, Z. 1994. Breeding biology of the golden pheasant. *WPA News* 45:35.

Wu, Z., Z. Li, Z. Yu, and H. Jang. 1993. Study of Reeves's pheasant in Tuoda Forest, Gizhou, China. *WPA News* 39:7–11.

Xu, W., Z. Wu, and Z. Li. 1990. Current status of the Reeves's or white-crowned long-tailed pheasant in China. Pp. 31–32 *in* Hill et al., 1990.

Yahya, H. S. A. 1993. Habitat preferences of monal pheasants. *World Pheasant Assoc. J.* 17/18:89–90.

Yamashina, Y. 1976. Notes on the Japanese copper pheasant *Phasianus soemmeringii*. *World Pheasant Assoc. J.* 1:23–42.

Yang, L. 1992. *The Chinese phasianids: Lady Amherst's pheasant.* China Forestry Publishing House, Beijing. (In Chinese, English summary.)

Yang, L., X. Wen, and X. Yang. 1994. On the taxonomy of blood pheasant (*Ithaginis*). *Zool. Res. (China)* 15:21–30. (In Chinese, English summary.)

Yang, X., X. Wen, and L. Yang. 1997. A preliminary study of the habitat use and behaviour of green peafowl in spring in China. *Tragopan* 7:6–7.

Yin, T. 1970. Record of the Himalayan monal *Lophophorus impeyanus* (Latham) in Burma. *J. Bombay Nat. Hist. Soc.* 67:328–330.

Yonzon, P. B., and A. D. Lelliott. 1981. Human interference in ecosystems. Pp. 55–62 *in* Savage, 1981.

Young, L., P. J. Garson, and R. Kaul. 1987. Calling behaviour and social organisation of the cheer pheasant: Implications for survey techniques. *World Pheasant Assoc. J.* 12:30–43.

Young, L., G. M. Zheng, and Z. W. Zhang. 1991. Winter movements and habitat use by Cabot's tragopan *Tragopan caboti* in southeastern China. *Ibis* 113:121–126.

Zeliang, D. K. 1981. Blyth's tragopan breeding centre, Khima Nagaland. Pp. 88–91 *in* Savage, 1981.

Zhang, C., X. Zhu, and B. Pang. 1997. *Birds of China.* China Forestry Publishing House, Beijing.

Zhang, Z. 1995. Brown eared pheasant project report. *WPA News* 49:14–16.

Zhang, Z. 1997. Studies on habitat and management of the brown eared pheasant in China. *Aviculturists and Wildl. Conservationists' Gazette* 1997(Feb):34.

Zhang, Z. 1998. The distributional range of broad eared pheasant *Crossoptilon mantchuricum*. *Tragopan* 9:5–10.

Zhang, Z., G. Zheng, X. Yang, and J. Wu. 1997. Clutch size and nesting survival of brown eared pheasant. Abstract. P. 38 *in* Anonymous, 1997.

Zheng, G. 1988. A review of biological studies of Phasianidae in China. *World Pheasant Assoc. J.* 13(1987/88):13–20.

Zheng, G., and Z. Zhang. 1993. The distribution and status of pheasants in China. Pp. 15–19 *in* Jenkins, 1993.

Zheng, G., R. Yin, Z. Zhang, Z. Liu, and H. Zhou. 1989. Courtship display behaviour of Cabot's tragopan, *Tragopan caboti*. *Acta Zool. Sinica* 35:328–332.

Zuk, M., K. Johnson, R. Thornhill, and J. D. Ligon. 1990*a*. Mechanisms of female choice in red jungle-fowl. *Evolution* 44:477–485.

Zuk, M., R. Thornhill, J. D. Ligon, K. Johnson, S. Austad, S. H. Ligon, N. W. Thornhill, and C. Costin. 1990*b*. The role of male ornaments and courtship behavior in female mate choice of red junglefowl. *Am. Nat.* 136:459–473.

Zuk, M., S. L. Pompa, and T. S. Johnsen. 1995. Male courtship displays, ornaments and female mate choice in captive red jungle fowl. *Behaviour* 132:821–836.

Index

This index includes only pheasants—the currently used scientific and English vernacular names of pheasant genera, species, species groups, and subspecies. Because the vernacular (common) names are used more often in the text, those entries are more complete than the corresponding scientific-name entries. Common names ending with the word "pheasant" are indexed under the second-to-last term (e.g., Hume's bar-tailed pheasant is found under "bar-tailed pheasant"). Common names ending in anything other than "pheasant" are indexed under the last term (e.g., Burmese red junglefowl is under "junglefowl"). **Page numbers in boldface** refer to the principal account of the genus or species. *Page numbers in italics* point to any maps or figures that are outside the taxon's principal account. "Pl." indicates the number of a color plate. Neither the front matter nor the appendix has been indexed here.